Agricultural Structures and Mechanization

Agricultural Structures and Mechanization

Editor

José Pérez-Alonso

MDPI • Basel • Beijing • Wuhan • Barcelona • Belgrade • Manchester • Tokyo • Cluj • Tianjin

Editor
José Pérez-Alonso
University of Almería
Spain

Editorial Office
MDPI
St. Alban-Anlage 66
4052 Basel, Switzerland

This is a reprint of articles from the Special Issue published online in the open access journal *Agriculture* (ISSN 2077-0472) (available at: https://www.mdpi.com/journal/agriculture/special_issues/Agricultural_Structures_Mechanization).

For citation purposes, cite each article independently as indicated on the article page online and as indicated below:

LastName, A.A.; LastName, B.B.; LastName, C.C. Article Title. *Journal Name* **Year**, *Volume Number*, Page Range.

ISBN 978-3-0365-5583-6 (Hbk)
ISBN 978-3-0365-5584-3 (PDF)

Contents

About the Editor

José Pérez Alonso

José Pérez Alonso (Agricultural engineer) graduated as an Agricultural Engineer and received his doctorate from the Polytechnic University of Valencia (Spain). He currently works as a full professor, attached to the Department of Engineering at the University of Almería (Spain). He has developed his teaching and research work for 28 years. He participed in numerous research projects, obtaining results published in journals included in the Journal Citation Reports (35 articles, 16 Q1 and 12 Q2 of the JCR Science Edition), patents (4), utility models (1), as well as communications in conferences (internationals and nationals (>70)). His main lines of research are as follows: design of greenhouses, risk prevention in workplace greenhouses and alternative energies applied to greenhouses. He has been the director of the Agricultural Research Center of the University of Almería for three and a half years. Likewise, he was the secretary of the Engineering Department of the University of Almería for 6 years and 8 months.

Preface to "Agricultural Structures and Mechanization"

Due to the importance of being able to feed the entire world population in a sustainable way, it is necessary to apply all the technological innovations that have been developed through research and technological development in the different food production systems. Specifically, it is interesting to collect the technological innovations in the field of agricultural structures and mechanization, and determine their role in the production of high-quality food, so that they are available to society. For this reason, the editorial MDPI created a Special Issue on this subject in the journal Agriculture, which has made it possible to compile 26 scientific articles, which will now be jointly published.

The works that are compiled in this book deal with interdisciplinary studies in which the following fields of research are addressed: agricultural machinery, precision agriculture, harvest technology, post-harvest technology, sensors, automation, ergonomics, energy and the environment.

Finally, we would like to thank all the authors who have decided to make their research known through this work, and thus put their developments and new technologies at the service of all the working areas of society involved in food production.

José Pérez-Alonso
Editor

agriculture

MDPI

Article

A Novel Semiautomatic Flesh Peeling and Seed Removing Mechanism for Dried Longan

Cheng-Han Li [1], Yu-Ting Hsu [1], Chun-Hung Hsieh [2] and Ching-Wei Cheng [3,*]

1 Department of Bio-Industrial Mechatronics Engineering, National Chung Hsing University, Taichung 402, Taiwan; cghan@smail.nchu.edu.com.tw (C.-H.L.); atin4100040063@gmail.com (Y.-T.H.)

2 Department of Information Management, National Taichung University of Science and Technology, Taichung 404, Taiwan; guu@nutc.edu.tw

3 College of Intelligence, National Taichung University of Science and Technology, Taichung 404, Taiwan

* Correspondence: cwcheng@nutc.edu.tw; Tel.: +886-4-2219-5795

Abstract: Longans, which are produced in tropical countries, are rich in nutrients and high in commercial value. To extend their shelf life, fresh longans are roasted to dry them out. For the convenience of cooking applications, the dried longan pulp is peeled and seeded beforehand. Presently, this is done manually; thus, this research aims to automate this process. A cutting tool with a serrated bending angle is used to cut the dried longan pulp, and then the longan is fixed by a lifting mechanism. The pulp is breached with a cutting tool and this breach is pushed up against a 6 mm hole in the silicone soft plate of the lifting mechanism to separate out the seed by squeezing. Commercially available dried longans are used in this experiment. The cutting surface width of the cutter designed in this study is 20 mm, and the serration bending angle is 30°. This cutter operates at a speed of 29 cm/s on dried longan pulp, attaining a success rate of 85%. The lifting mechanism also presented in this study fixes the dried longan and uses a squeezing motion to separate the flesh from the seed with a silicone soft plate. Here, a lifting speed of 28 mm/s achieves the highest rate of success at peeling, and the success rate of pulp-and-seed separation can reach 86.7%.

Keywords: dried longan; lifting mechanism; silicone soft plate

Citation: Li, C.-H.; Hsu, Y.-T.; Hsieh, C.-H.; Cheng, C.-W. A Novel Semiautomatic Flesh Peeling and Seed Removing Mechanism for Dried Longan. *Agriculture* **2021**, *11*, 48. https://doi.org/10.3390/agriculture11010048

Received: 10 December 2020
Accepted: 8 January 2021
Published: 10 January 2021

1. Introduction

Longan (Dimocarpus longan Lour), an evergreen tree belonging to the Sapindus family, is grown as a commercial fruit crop [1,2]. Longan fruit is rich in nutrition and commercial value and is produced in tropical countries, especially Thailand and China [2,3]. Longan pulp can be used in traditional Chinese medicine to promote blood metabolism, relieve insomnia, and prevent amnesia because it contains many biologically active compounds and nutrients [4,5]. Taiwan is currently an important source of longan in Asia. The longan growing season is mostly concentrated in summer. Freshly harvested longan is usually high in temperature and moisture because of the weather, causing the quality of the fruit to deteriorate rapidly [3,6]. The freshness period is short, and the pulp will slowly ooze juice and deteriorate. This phenomenon is called pulp autolysis or pulp decomposition and greatly affects the pulp's quality, freshness, and value [7]. To extend its shelf life and facilitate storage, longan is usually processed. Typically, it is dried out through baking; dried longan can be stored for a longer time and made into canned food, wines, or jams. Separate baking processes are used for shelled and unshelled longan. Traditionally in Thailand, longan is first harvested, then shelled, and then baked [8]. The shelled longan is dried in hot air at 70 °C over 12–15 h. The final moisture content of the longan can reach 22%. The dried longan pulp will be golden in color [8,9]. It takes approximately 48–52 h to dry the shelled longan in hot air at 75 °C, and the final moisture content can be reduced to 22% [9]. In Taiwan, shelled longan is usually roasted directly through one of two methods:

hot air baking or smoking. In each method, the dried longan will have a smoky fragrance and its pulp will appear dark brown after removing the shell.

Presently, the separation of the seeds and pulp of dried longan is still done manually by workers. The longan pulp contains high sugar content; after being dried in hot air, the pulp is caramelized and highly viscous. The shape of the fruit is irregular, complicating the process of seed removing. During manual seeding, the longan is typically heated beforehand. A knife is used to cut a hole in the pulp, and the seeds are removed. The operational process is slow and requires a lot of human resources. Because of changes in the rural population structure (particularly the aging of the population engaged in agriculture), human resources have been greatly reduced, and manual separation of the seed and the pulp may incur food hygiene and safety concerns. If machines can replace artificial longan deseeding operations, they can effectively reduce costs and improve production efficiency, food hygiene, lack of human resources, and other issues.

Longan-related research mostly focuses on its physical and chemical properties. Research on seeding machinery is quite rare, and there is no fully commercialized system available on the market. Xie et al. (2006) developed a seed removing machine for fresh shelled longan using air pressure as a power source to drive a pneumatic cylinder, which then drives a triangular blade to cut the freshly shelled longan together with its pulp [10]. The longan seeds are then pushed out by the pneumatic ejector bar to complete the work of removing them. The size of the cut for the fresh longan seeds will be affected by the thickness of the blade. If the cutting blade is too thick or too thin, it may cause the fresh longan cut to be incomplete, preventing the seeds from being removed. After the fresh longan is cut, the ejector bar's diameter and speed will also affect the seeding success and pulp loss rates. Pen et al. (2012) used simulated finger-drawing arcs to peel off dried longan flesh and a pneumatic cylinder as a power source to drive the triangular cam [11]. The cam controls the opening and closing of the L-shaped bracket, whose two ends are equipped with symmetrical pulp peeling arc blades. The two arc blades will cut a hole in the longan pulp for the seeds to be ejected using a rod ejector. The depth of occlusion between the two arc blades and the dried longan pulp, the opening and closing speed of the L-shaped bracket, and the pushing speed of the longan seeds' ejector bar will all affect the pulp peeling success rate.

This study aims to design a set of semiautomatic dried longan pulp peeling machines using a suitable cutter. The relationship between the shape and the speed of the cutter and the dried longan cutting success rate is also discussed. A lifting mechanism is designed using a silicone soft board to complete the seed removing operation. The dried longan is extruded through this board to peel the pulp and remove the seeds. The best speed and success rate of this process were found. The semiautomatic machine designed in this study can serve as a reference for future research on dried longan processing machinery.

2. Materials and Methods

2.1. Dried Longan Sample

Mature longan can be conical, heart-shaped, spherical, or other shapes. Some varieties have a significantly longer appearance on one side. After the drying process, the fruit's appearance will become wrinkled and rough. The dried longan's shape will be irregular, such that its flesh will be close to the inner core and appear approximately spherical or oval. Commercially available dried longan was used as our experimental material in this study. All experimental samples were stored under refrigeration. Before use, they were placed at room temperature to regain moisture, before being heated to 60.0 °C in an oven to reduce the pulp's viscosity. The dried longan was used to test the flesh peeling and seed removing machine developed in this study. The average diameter of longan samples (pulp containing seeds) was 17.20 mm and that of dried longan seeds was 14.95 mm.

2.2. Experimental Equipment

The cutting tool of the dried longan pulp peeling machine is driven by a motor driver (VNH2SP30) and a direct current (DC) motor (JGB37-550) that has the maximum power of 60 W. The lifting mechanism for the peeling of the pulp and removal of seeds is controlled by a stepping motor driver (AZD-CD, Oriental Motor Co., Ltd., Taiwan) and a stepping motor (AZM66AC-TS7.2, Oriental Motor Co., Ltd.). Once in position, the lifting mechanism is used to clamp, fix, and squeeze the dried longan, and a silicone soft board is used to peel the pulp and remove the seeds. The DC motor is controlled using an Arduino UNO controller. The positions of the stepping motor and lifting mechanism are controlled using a computer programming interface. The power consumption of the semiautomatic flesh peeling and seed removing machine is approximately 480 W. The soft board is 50 × 50 mm in size and is made of C-25 silicone (Asia Silicone Chemical Materials Co., Ltd., Taichung, Taiwan) with the tensile strength of 50 kg/cm^2, elongation of 750%, and hardness of 25 HA.

2.3. Semiautomatic Dried Longan Pulp Peeling Machine

The semiautomatic dried longan pulp peeling and seed removing machine is based on aluminum extrusion. A stepping motor drives the eccentric shaft turntable in Figure 1a. The stepping motor's control can accurately clamp the dried longan. When the eccentric shaft turntable motor rotates upward, the dried longan-fixing jaws will open. At this time, the dried longan is placed on the pedestal. The eccentric shaft turntable is turned downward, and the jaws will clamp the dried longan. After the dried longan is fixed, the DC motor starts to drive the cutting tool's reciprocating link to cut the pulp. Because of the limited size of this machine, the longan pulp has been cut, and a silicone soft plate is installed along the aluminum extrusion groove, as shown in Figure 1a,b. To peel the pulp, the eccentric shaft turntable is turned up to the top, the dried longan in the holder is cut and squeezed with the silicone soft plate, and the longan seeds will pass through the hole. The longan pulp and seeds will then be separated; finally, the longan pulp remains under the silicone soft board, and its seeds will be placed on the top through the hole in the soft board, as shown in Figure 2.

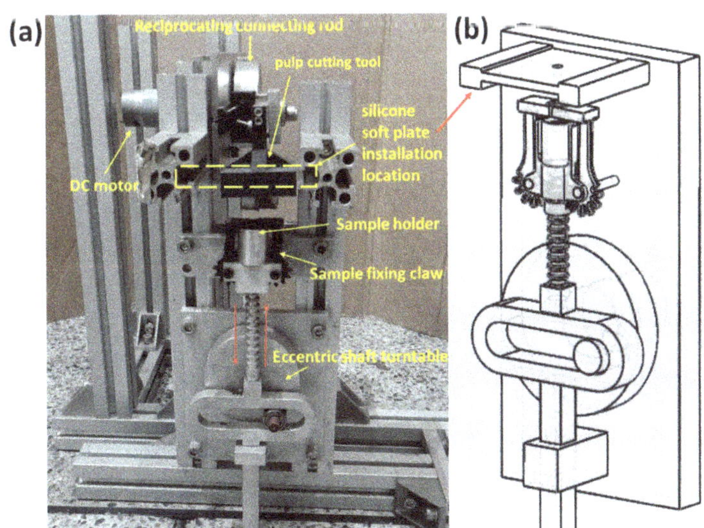

Figure 1. (**a**) Semiautomatic dried longan peeling machine. (**b**) Schematic of the installation position of the silicone soft board for peeling.

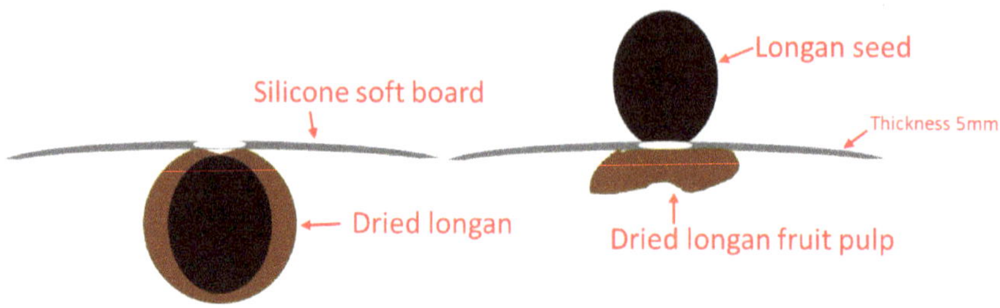

Figure 2. After the dried longan is cut, the hole cut in the pulp and the silicone soft board are squeezed against each other. The seeds and the flesh will separate; the pulp will stay under the silicone soft plate while the longan seeds will be extruded to the other side.

2.4. Design of the Dried Longan Pulp Cutting Tool

For manual longan peeling, the larger the hole in the pulp, the easier it is to remove the seeds. In this study, a longan cutting tool was designed, and the bending angle and width of the knife's cutting surface were changed, as shown in Figure 3. A knife drive mechanism was designed to perform a hole cutting test on the dried longan pulp. The formation and separation of chips in the cutting process are due to the material being subjected to the maximum shear stress; thus, we analyze it using shear-stress theory. When the tool and the cutting workpiece move relative to each other, the material is cut by shear force. This cutting surface is called a shear surface. For larger shear angles, the chip layer is thinner, resulting in smaller friction. The required cutting force is also smaller. By contrast, smaller shear angles make the chip layer thicker, resulting in larger friction. The cutting force is relatively large. Thus, this study designed a cutting tool for dried longan pulp.

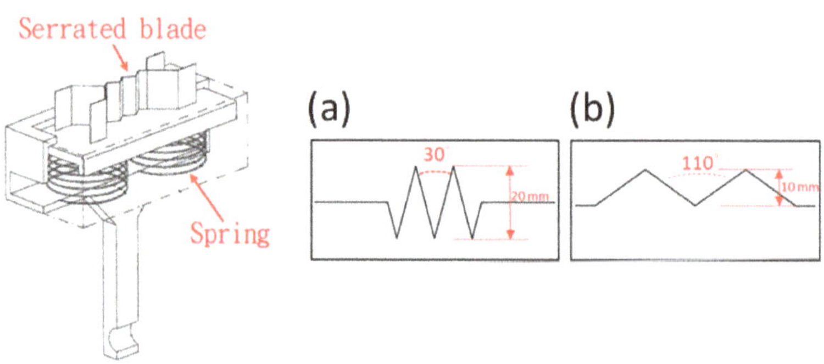

Figure 3. A schematic of a cutting tool for dried longan pulp. (**a**) A cutter with the blade width of 20 mm and 30° serrated bending angle. (**b**) A cutter with the blade width of 10 mm and 110° serrated bending angle.

Our cutting tool is made of a spring and aluminum foil, as shown in Figure 3. To prevent the longan seeds from being squeezed excessively by the cutter during the pressing process, the spring is used as an extrusion buffer and helps to make the knife edge fit more closely with the longan pulp. In this study, two types of knives were made for testing. For the first, the cutting surface width is 20 mm, and the serration bending angle is 30°; for the second, the cutting surface width is 10 mm, and the serration bending angle is 110°. Figure 3 shows the comparison of the two knives. At the same cutting speed, different knives' specifications are tested to find the most suitable for cutting dried longan pulp.

2.5. Design of the Cutting Tool's Driver Mechanism

Cutter actuation adopts the scotch yoke mechanism, and a DC motor is used to convert circular motion into linear reciprocating motion. The linear reciprocating mechanism is designed to match the cutter-driven stroke shown in Figure 4. When the motor drives the disk to rotate, the shaft drives the rod to make a linear reciprocating motion to perform the opening operation. When the disk rotates through 0°–180°, the driving rod moves forward and produces a maximum stroke of 45 mm at 180°; the rotation range is between 181° and 360°, and the rod returns to the origin at 360°. The eccentricity of the eccentric shaft turntable is 34.3 mm, and the maximum stroke of the cutter is 68.6 mm.

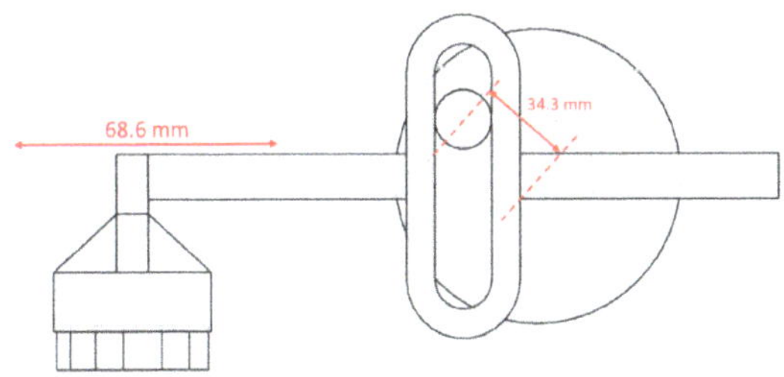

Figure 4. Dried longan pulp cutter reciprocating scotch yoke mechanism.

2.6. Dried Longan Pulp Peeling and Seed Removing Mechanisms

The pulp peeling and seed removing mechanisms are designed based on a wine opener's linkage mechanism. The main structure fixes the aluminum extruder on the aluminum plate. When the stepping motor rotates, it will drive the eccentric shaft disk. The eccentric shaft turntable will drive the row of teeth on the lifting mechanism to synchronously drive the ratchet to open and close the clamping arm. The eccentricity of the shaft on the turntable is 34.3 mm, and the maximum displacement stroke is 68.6 mm. There is a material holder above the lifting rod and a ball socket in this holder in which to place the dried longan. The side of the lifting rod is equipped with a row of teeth to drive the clamping arm. This arm's lower end has a semicircular ratchet for gear meshing transmission. The upper end of the clamping arm is equipped with clamping jaws to clamp and fix the dried longan. After the dried longan is clamped and fixed, a hole is cut in its pulp to remove its seed. Given that the dried longan's average height is approximately 16.8 mm, the clamping height is set to 11.2 mm to stop the longan from turning during the opening process, as shown in Figure 5. Two-thirds of the dried longan's average length must pass through the silicone soft board to achieve good peeling. Considering that the thickness of the silicone soft board is 5 mm, the mechanism's stroke is at least 16.2 mm, and the total mechanical stroke of the peeling operation is at least 27.4 mm, as shown in Figure 5.

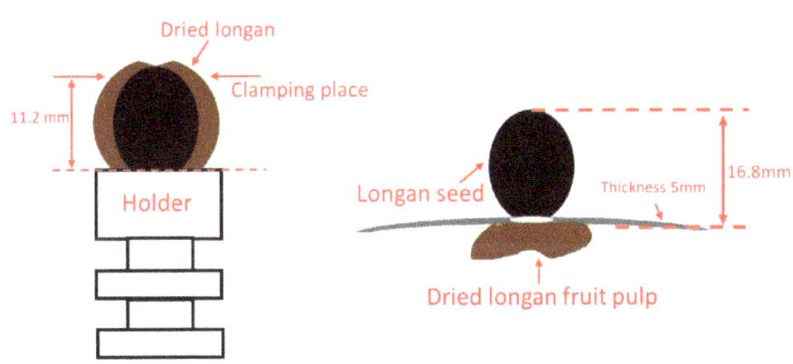

Figure 5. Schematic of the dried longan's clamping position.

After the hole is cut, the eccentric shaft turntable is driven by the stepping motor, driving the lifting rod upward. Simultaneously, the row of teeth synchronously drives the clamping arm's ratchet wheel to rotate downward, the clamping arms open, and the material base is shifted upward. The dried longan in the material holder is lifted by the lifting mechanism. The average diameter of the dried longan is 17.2 mm. The dried longan and the silicone soft plate are squeezed with a 6 mm through-hole silicone soft plate. This plate will deform, and its elasticity will be used to remove the dried longan pulp along the seed's surface. The longan seeds will be separated from the pulp through the hole in the plate and emerge on the other side, as shown in Figure 6.

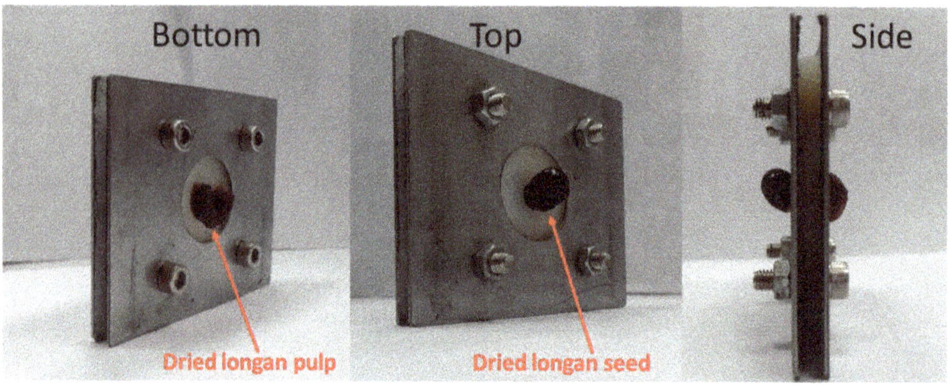

Figure 6. After squeezing the dried longan through the 6 mm hole of the silicone soft board, the pulp and seed are separated. The dried longan pulp will be below the plate while the seed will be above.

3. Results and Discussion

3.1. Success Rate of Cutting with the Designed Machine

Longan pulp cutting knives will affect hole size, affecting the success of subsequent peeling and seeding operations. In cutting the longan pulp, the dried seeds may be broken because of an uneven force applied by the cutter. The pulp also has a high sugar content, making it sticky. If the longan seeds are broken and shredded, they will stick to the pulp and seeding cannot be carried out. Figure 7 shows the definition of success and failure in this context. The hole cut in the dried longan pulp by the knife either (a) succeeds or (b) fails. If a hole is successfully cut in the dried longan pulp using the knife, the seed is clearly visible, as shown in Figure 7a; otherwise, it is not visible, as shown in Figure 7b.

Two different cutting tools were designed in this research, as described in Section 2.3 and shown in Figure 3. A DC motor drives the longan pulp cutting tool. This tool's speed is set to 508 mm/s, and 60 dried longans are used for the test. Table 1 shows the test results. The average width of the longan pulp breach formed by the cutter with a cutting surface width of 20 mm and a serrated bending angle of 30° is 7.4 mm; the success rate is 85%, and the seed breakage rate is 11.7%. For the cutter with a cutting surface width of 10 mm and a serrated bending angle of 110°, the average breach width is 5.5 mm, the success rate is 63.3%, and the seed breakage rate is 31.5%.

The cutting range of the cutter with a 30° serrated bending angle is larger than that of the one with a 110° serrated bending angle. The former tool sticks easily to the pulp, but this does not affect the pulp opening operation. The 30°-serrated-bending-angle tool cuts holes with a high success rate and avoids breaking longan seeds. Experience with manual pulp peeling and seed removing shows that larger breaches in the pulp allow the smooth surface of the longan seeds to be clearly seen, enabling the pulp to be easily peeled and the seed to be removed. Therefore, the 30°-serrated knife was used in follow-up experiments.

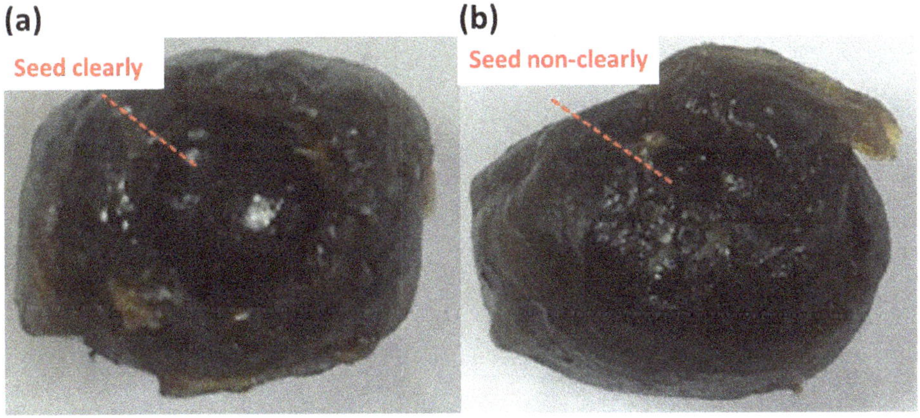

Figure 7. The hole cut in the dried longan pulp: (**a**) successfully; (**b**) fails.

Table 1. Cutting tool test results for dried longan.

	Average Width of the Breach (mm)	Cutting Success Rate (%)	Seed Damage Rate (%)
Blade width is 20 mm; Serrated bending angle is 30°	7.4	85.0 (51/60)	11.7 (7/60)
Blade width is 10 mm; Serrated bending angle is 110°	5.5	63.3 (38/60)	31.7 (19/60)

3.2. Success Rate of the Cutting Tool with Operating Speed

The Arduino UNO controller adjusts the longan pulp cutter's speed through pulse width modulation (PWM) to adjust the DC motor's power supply voltage. The greater the PWM duty cycle, the faster the DC motor's speed and the greater the force of the cutter. A cutter with a width of 20 mm and a 30° serrated bending angle was used to find the optimal cutting speed for dried longan pulp. The tests were carried out at speeds of 292.0, 330.0, 380.0 and 508.0 mm/s. Each speed was tested with 60 dried longans. Figure 8 shows the results.

When the cutting speed is 292.0 mm/s, the average width of the breach is approximately 7.5 mm and the success rate is approximately 95.0% (57/60). The test sample has no broken seeds. When the cutting speed is 330.0 mm/s, the breach's average width is approximately 7.3 mm, the success rate is approximately 85.0% (51/60), and the seed breaking rate is approximately 10.0% (6/60). When the cutting speed is 380.0 mm/s, the average width of the breach is approximately 7.5 mm, and the cutting success rate is approximately 80.0% (48/60), with a seed breakage rate of 15.0% (9/60). When the cutting speed is 508.0 mm/s, the average width of the breach is approximately 7.4 mm, the success rate is approximately 85.0% (51/60), and the seed breakage rate is approximately 11.7% (7/60). The test results suggest that reducing the cutting speed helps to increase the cutting success rate and reduce the seed breakage rate. A cutting speed of 292.0 mm/s yields the highest success rate without causing longan seed breakage.

3.3. Test Analysis of Peeling Dried Longan Pulp

A stepping motor is used to drive the lifting and clamping mechanism. The dried longan pulp is cut with a serrated cutter with a bending angle of 30° at a speed of 292 mm/s. The lifting mechanism pushes the dried longan upward against the 6 mm hole in the C-25 silicone soft plate such that the seed is separated, as shown in Figure 2.

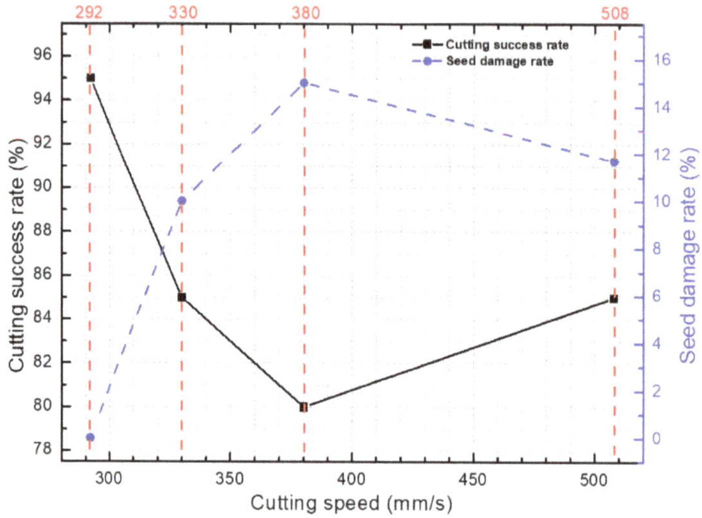

Figure 8. Success rates of the tool with a width of 20 mm and a sawtooth bending angle of 30° at various cutting speeds.

The lifting mechanism's upward pushing speed is varied, and the relationship between this speed and the peeling success rate is analyzed. Tests are performed using 60 longans per group, and complete separation of the dried longan pulp and seed is defined as the successful peeling of the pulp. If the longan seeds have residual pulp, are broken due to pushing or cannot be separated from the pulp, the experiment is classified as a failure.

As shown in Figure 9, when the pushing speed is 70.0 mm/s, the peeling success rate is approximately 66.7% (40/60) and the failure rate is approximately 33.3% (20/60). The pulp peeling processes of 20 dried longans failed. Of these, 12 failed because pulp remained on the seeds, two failed because the seeds were broken, and six seeds did not smoothly pass through the silicone plate. When the pushing speed is 56.0 mm/s, the peeling success rate is 70.0% (42/60) and the failure rate is approximately 30.0% (18/60). Of the failures, 12 seeds had pulp remaining, two were broken, and four could not pass through the silicone soft board. When the pushing speed is 42.0 mm/s, the peeling success rate is 80.0% (48/60) and the failure rate is 20.0% (12/60). Of the failures, six seeds had pulp remaining, two were crushed, and four did not pass through the plate. For a pushing speed of 28.0 mm/s, the peeling success rate is approximately 86.7% (52/60) and the failure rate is 13.3% (8/60). Of the failures, four seeds had pulp remaining, two were broken, and two did not pass through the plate. For a pushing speed of 14.0 mm/s, the peeling success rate is approximately 86.7% (52/60) and the failure rate is approximately 13.3% (8/60). No samples failed because they did not pass through the soft silicone plate; however, six seeds had pulp remaining, and two were broken because of pushing.

The experiment shows that reducing the pushing speed will help to improve pulp peeling success rate. The dried longan may fall off the holder when passing through the plate if the lifting speed is too fast. Because there is not enough time for the seed to pass through the holes, some will fail to be removed. The dried longan peeling and seeding operation is equally successful at 28 and 14 mm/s; therefore, a speed of 28 mm/s is preferable for efficiency.

Based on our results, we recommend using a tool with a cutting surface width of 20 mm and a sawtooth bending angle of 30° to breach the pulp at a speed of 292.0 mm/s. After cutting, the dried longan should be squeezed with a silicone soft plate at a speed of 28 mm/s for seed removing. Another 100 dried longan samples were prepared for testing to verify the mechanical stability. The peeling success rate was found to be 86.0%, with a failure rate of 14%. Of the failures, three seeds had pulp remaining and 11 seeds failed to pass through the plate. After the dried longan pulp was cut, if the dried longan seeds were broken or unable to be seen clearly, as shown in Figure 7, then it was judged that the cut had failed. After cutting, the dried longan and silicone soft plate could not pass through the through hole in the silicone soft plate, with residual pulp on the seed surface and dried longan seed breaks judged as failed peel, as shown in Table 2. The speed parameters of the semiautomatic flesh peeling in Table 2 and the seed removal mechanism of the dried longan are provided as references for future research.

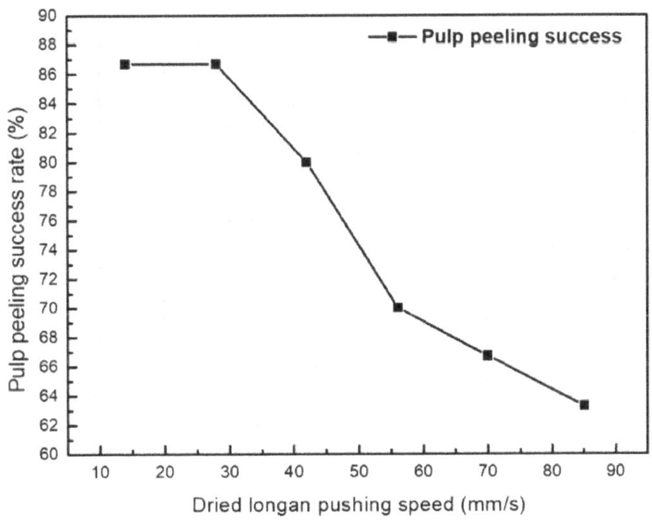

Figure 9. Pulp peeling success rate vs. pushing speed of the longan against the soft silicone plate.

Table 2. Parameters of the semiautomatic flesh peeling and seed removal mechanism for the dried longan design. Each parameter was tested using 60 dried longans.

Blade width of 20 mm and sawtooth bending angle of 30°						
Cutting speed (mm/s)	**292.0**	330.0		380.0		508.0
Cutting success rate (%)	**95.0**	85.0		80.0		85.0
Seed nonclearly visible (%)	**5.0**	5.0		5.0		3.3
Seed damaging rate (%)	**0.0**	10.0		15.0		11.7
The hole cut in the dried longan pulp and C-25 silicone soft plate peeling						
Dried longan pushing speed (mm/s)	14.0	**28.0**	42.0	56.0	70.0	85.0
Peeling success rate (%)	86.7	**86.7**	80	70	66.7	63.3
Squeeze failure rate (%)	0.0	**3.3**	6.7	6.7	10.0	20.0
Residual pulp on the seed surface (%)	10.0	**6.7**	10.0	20.0	20.0	16.7
Seed damaging rate (%)	3.3	**3.3**	3.3	3.3	3.3	0.0
Cutting speed of 292.0 mm/s and pushing speed of 28 mm/s. (100 dried longan samples)						
Peeling success rate (%)				**86.0**		
Squeeze failure rate (%)				**11.0**		
Residual pulp on the seed surface (%)				**3.0**		
Seed damaging rate (%)				**0.0**		

4. Conclusions

This study shows that the use of serrated tools can increase the success rate of breaching dried longan pulp while decreasing the rate of seed breakage. The cutting tool's surface width is 20.0 mm, and its bending angle is 30°. The use of a C-25 silicone soft plate with high elongation and high tensile strength can increase the success rate of pulp peeling and seed removing and extend the service lifetime of the pulp peeling parts. Reducing the cutting speed can increase the success rate. The optimal cutting speed is 292.0 mm/s, yielding a success rate of approximately 85%, which is the best among the samples. Reducing the pushing speed helps to increase the success rate of pulp peeling and seed removing. The optimal pushing speed is 28.0 mm/s, yielding a success rate of approximately 86.7%, which is the highest among the tests. Our pulp peeling mechanism was based on previous research by Pan et al. (2012) on squeezing down the thumbs of both hands [11]. The advantage of our mechanism is that the dried longan pulp cutting tool cannot be easily damaged and the overall structure is simple. Presently, if the peeling silicone soft plate is not manually installed in the machine, it takes around 5–6 s for each dried longan pulp to be cut and peeled, thereby showing that the average work efficiency of this design is lower than that of manual labor. Therefore, if this mechanical process can be converted into a workstation-based automatic pulp peeling and seed removal system, it is expected to be much faster and more hygienic than manual labor. The "Peeling and seed removal method" (patent number: TWI632863B) and "Pulp peeling device for dried longan" (patent number: TWM554709U) developed in the work have been patented in Taiwan.

5. Patents

We successfully obtained Taiwanese invention patents for the "Peeling and seed removing method" (patent number: TWI632863B) and the "Pulp peeling device for dried longan" (patent number: TWM554709U).

Author Contributions: C.-W.C. conceived and designed the experiments; Y.-T.H. performed the experiments; C.-H.L. and C.-W.C. wrote the paper; C.-H.L., C.-H.H. and C.-W.C. reviewed and edited the paper. All authors have read and agreed to the published version of the manuscript.

Funding: The authors thank the Agriculture and Food Agency, Council of Agriculture, Executive Yuan of the Republic of China for financially supporting, Taiwan, grant number 109AS-11.3.1-FD-Z1.

Acknowledgments: The authors thank the Agriculture and Food Agency, Council of Agriculture, Executive Yuan of the Republic of China for financially supporting.

Conflicts of Interest: The authors declare no conflict of interest.

References

1. Chen, Y.; Sun, J.; Lin, H.; Lin, M.; Lin, Y.; Wang, H.; Hung, Y.-C. Salicylic acid treatment suppresses Phomopsis longanae Chi-induced disease development of postharvest longan fruit by modulating membrane lipid metabolism. *Postharvest Biol. Technol.* **2020**, *164*, 111168. [CrossRef]
2. Jiang, Y.; Zhang, Z.; Joyce, D.C.; Ketsa, S. Postharvest biology and handling of longan fruit (*Dimocarpus longan* Lour.). *Postharvest Biol. Technol.* **2002**, *26*, 241–252. [CrossRef]
3. Lin, Y.; Lin, Y.; Lin, Y.; Lin, M.; Chen, Y.; Wang, H.; Lin, H. A novel chitosan alleviates pulp breakdown of harvested longan fruit by suppressing disassembly of cell wall polysaccharides. *Carbohydr. Polym.* **2019**, *217*, 126–134. [CrossRef] [PubMed]
4. Chen, Y.; Xie, H.; Tang, J.; Lin, M.; Hung, Y.-C.; Lin, H. Effects of acidic electrolyzed water treatment on storability, quality attributes and nutritive properties of longan fruit during storage. *Food Chem.* **2020**, *320*, 126641. [CrossRef] [PubMed]
5. Huang, F.; Hong, R.; Yi, Y.; Bai, Y.; Dong, L.; Jia, X.; Zhang, R.; Wang, G.; Zhang, M.; Wu, J. In vitro digestion and human gut microbiota fermentation of longan pulp polysaccharides as affected by Lactobacillus fermentum fermentation. *Int. J. Biol. Macromol.* **2020**, *147*, 363–368. [CrossRef] [PubMed]
6. Lin, Y.; Lin, Y.; Lin, H.; Shi, J.; Chen, Y.; Wang, H. Inhibitory effects of propyl gallate on membrane lipids metabolism and its relation to increasing storability of harvested longan fruit. *Food Chem.* **2017**, *217*, 133–138. [CrossRef] [PubMed]
7. Wang, D.; Zhang, H.; Wu, F.; Li, T.; Liang, Y.; Duan, X. Modification of Pectin and Hemicellulose Polysaccharides in Relation to Aril Breakdown of Harvested Longan Fruit. *Int. J. Mol. Sci.* **2013**, *14*, 23356–23368. [CrossRef] [PubMed]
8. Tippayawong, N.; Tantakitti, C.; Thavornun, S. Energy efficiency improvements in longan drying practice. *Energy* **2008**, *33*, 1137–1143. [CrossRef]

9. Yang, K.-M.; Chiang, P.-Y. Effects of smoking process on the aroma characteristics and sensory qualities of dried longan. *Food Chem.* **2019**, *287*, 133–138. [CrossRef] [PubMed]
10. Xie, H.J.; Qing, Y.M.; Zheng, D.; Lin, C.F.; Li, Z.W. Design and experiment of a longan core-removing mechanism. *Trans. CSAE* **2006**, *22*, 211–213.
11. Pan, Y.T.; Lee, Y.N.; Chen, Y.H.; Cheng, C.W. The designation and test of dried longan pneumatic flesh-peeling machine. *J. Agric. For.* **2012**, *61*, 337–350.

agriculture

MDPI

Article

Longitudinal Axial Flow Rice Thresher Performance Optimization Using the Taguchi Technique

Mohamed Anwer Abdeen [1,2], Abouelnadar Elsayed Salem [1,3] and Guozhong Zhang [1,*]

1 College of Engineering, Huazhong Agricultural University, Wuhan 430070, China; mohamed.anwer2010@yahoo.com (M.A.A.); abouelnadar@webmail.hzau.edu.cn (A.E.S.)
2 Agricultural Engineering Department, Zagazig University, Zagazig 44519, Egypt
3 Desert Research Center, Mataria 11753, Egypt
* Correspondence: zhanggz@mail.hzau.edu.cn

Abstract: Combine harvesters are widely used worldwide in harvesting many crops, and they have many functions that cover the entire harvesting process, such as cutting, threshing, separating, and cleaning. The threshing drum is the core working device of the combine harvester and plays an influential role in rice threshing efficiency, threshing power requirement, and seed loss. In this study, two structures of rice threshers (conical-shaped and cylindrical-shaped) were tested and evaluated for performance under different thresher rotating speeds of 1100, 1300, and 1500 rpm and different feeding rates of 0.8, 1.1, and 1.4 kg/s. The experiment was designed using the Taguchi method, and the obtained results were evaluated using the same technique. The thresher structure and operating parameters were assessed and optimized with reference to threshing efficiency, required power, and productivity. The obtained results revealed that increasing thresher rotating speed and the feeding rate positively related to threshing efficiency, power, and productivity. The highest efficiency of 98% and the maximum productivity of 0.64 kg/s were obtained using the conical-shaped thresher under a 1500 rpm rotating speed and a feed rate of 1.4 kg/s, whereas the minimum required power of 5.45 kW was obtained using the conical thresher under a rotating speed of 1100 rpm and a feed rate of 0.8 kg/s.

Keywords: rice threshing; rice combine; longitudinal axial flow thresher; Taguchi method

Citation: Abdeen, M.A.; Salem, A.E.; Zhang, G. Longitudinal Axial Flow Rice Thresher Performance Optimization Using the Taguchi Technique. *Agriculture* **2021**, *11*, 88. https://doi.org/10.3390/agriculture11020088

Received: 24 November 2020
Accepted: 18 January 2021
Published: 20 January 2021

1. Introduction

Rice is the second most important cereal after wheat, which together supply 95% of the world's population's whole staple food [1].

China is the biggest grain producer globally, with a planting area of 94,370.8 km^2 and accounting for 21.98% of the world's grain production [2].

The grain harvester is the most essential piece of agricultural machinery that improves harvesting efficiency and reduces labor costs [3–6].

Combine harvesters are widely used worldwide to harvest different crops under different environmental and operating conditions. They have many functions such as cutting, threshing, separating, cleaning, and sometimes storing crops.

Many small, medium, and giant threshers have existed for a long time, but because of their low performance compared to traditional threshing methods, they have never been adapted to a significant extent. Some of these threshers are hand-held, and others are pedal-operated [7].

Threshing is considered one of the most vital crop processing operations for separating grains from the ears and preparing them for the market [8].

Threshing is the process of separating the edible part of the cereal grain from the chaff that surrounds it, and it is done after harvesting the crop and before winnowing it [9].

The simplest threshing system is picking up rice stalks and trampling the panicles underfoot or beating them against a hard surface such as a rack, threshing board, or tub [10].

In several countries in Asia, Madagascar, and Africa, cereal crop is being trodden by humans or animals underfoot for threshing; this method results in a high loss due to grain breakage or buried under the earth [11].

Many researchers have done many experiments to examine grain threshing devices, and many kinds of grain threshers have been developed since the 1820s [12–17].

The mechanical threshing concept entails providing energy for turning materials, drawing in materials to be threshed, and creating different layers of velocity to rub grain heads together. The grain separating process directly acts on the linkage of grain and stalk. The threshing chamber is made of a drum and concave. The drum consists of a long cylinder mounted on bearings with spikes or rasp bars attached to its surface, and the concave is perforated to enable the threshed product to drop by gravity into a collector [18].

There are many types of threshing cylinders: the spike tooth, which threshes by striking action, and the rasp bar, which thresh by rubbing and friction. A spike teeth threshing drum gives higher throughput values at higher speeds than a rasp bar threshing drum due to the greater impact of the spikes against the stalk, even at lower speeds [19].

In axial flow threshers, the crop spirally moves between the concave and the rotating drum for several complete turns, which allows for multiple impacts between the concave and drum as the crop moves along the drum length [11].

In axial-flow threshers, around 80% of grains are separated in the drum's first half, whereas just 20% of grains are separated in the other half [20].

The factors influencing thresher performance are classified as crop factors (crop variety and crop moisture content), machine factors (cylinder diameter, cylinder type, feeding chute angle, spike shape, spike number and size, concave clearance, shape, and size), and operational factors (feed rate, cylinder speed, and machine adjustment) [21].

An optimum drum speed is necessary for improving thresher performance because extreme speed can cause grain crack but a low speed can result in an un-threshed head. The impact is the main threshing action for grain separating from the ear. In all threshers, this impact force is controlled by the thresher rotating speed [22].

A multi-crop thresher was developed and evaluated by Singh et al. (2015) [23], who found that increasing the drum speed highly affected the thresher threshing efficiency.

A paddy thresher was fabricated and evaluated at three threshing drum speeds (15.5, 17.3, and 19.0 m s^{-1}) and three feed rates (44, 720, and 1163 kg h^{-1}). The results revealed that the percentage of threshing efficiency and damaged grain increased with increasing drum speed for all feed rates [24]. The impact of drum speed, crop moisture content, and crop variety on grain damage, threshing loss, and power was studied. It was revealed that the increase in drum speed reduced threshing losses but increased the damaged grain due to spikes' greater impact against the crop stalks [25].

Increasing the drum speed increased the threshing efficiency because of high level of impacting to the plant spikes. The highest threshing efficiency was 99.76% at a feed rate of 15 kg/min and a drum speed of 1400 rpm (21.25 m/s) [26].

The threshing efficiency was positively affected by the cylinder speed, the concave clearance, the crop feeding rate, and the crop variety [27].

In this paper, a testing platform for a longitudinal axial flow rice thresher was constructed. Two kinds of threshers (cylindrical and conical) were tested and evaluated for performance under different rotating speeds and feeding rates using Taguchi techniques to get the highest possible threshing efficiency and productivity and the lowest required power.

2. Materials and Methods

2.1. Testing Platform

To simulate the rice threshing process, a longitudinal axial flow threshing platform was constructed in a factory. The platform comprised a conveying belt, a longitudinal axial flow thresher, a concave thresher cover, receiving boxes, a diesel engine, a feeding device, a frequency convertor, an electric motor, and a torque sensor. The platform is shown in Figure 1, and the characteristics of it are shown in Table 1.

Figure 1. Testing platform.

Table 1. Characteristics of the platform.

Part	Parameters	Value
platform	Length	3700 mm
	Width	1460 mm
	Height	1540 mm
Thresher	Length	1360 mm
	Diameter	370 mm
	No. of threshing bars	6
	No. of threshing teeth	87
	Tooth height	50–70 mm
	Tooth diameter	20 mm
	Distance between two adjacent teeth	80 mm
Concave	Length	1000 mm
	Wrap angle	180°
	Meshing size	20 × 36.6 mm
	Clearance	20 mm
Top cover	Deflector helical angle	24–30°
	The gap between the two deflectors	160°
Feeding auger	Length	193 mm
	Front diameter	230 mm
	rear diameter	270 mm

The conveying mechanism composed of a rotating belt with dimensions of 6 × 0.5 m, and it was driven by an electric motor. Its speed was controlled using a frequency converter. It was used to transport rice to the feeding auger, which consisted of a rotating auger and a rotating chain with steel bars. The power was conveyed from the diesel engine to the auger using a belt and pulley. Its function was to feed the rice from the conveying mechanism to the threshing unit.

The threshing unit consisted of a longitudinal axial flow thresher with spike teeth, a thresher cover with helical blades, and a stationary concave. The thresher composed of 6 bars with spike teeth, and the rotational speed was conveyed from the engine to the thresher using a pulley and belt. Two kinds of thresher structures (cylindrical and conical), and three thresher rotating speeds of (1100, 1300, and 1500 rpm) were tested for the experiment. The threshers are shown in Figure 2.

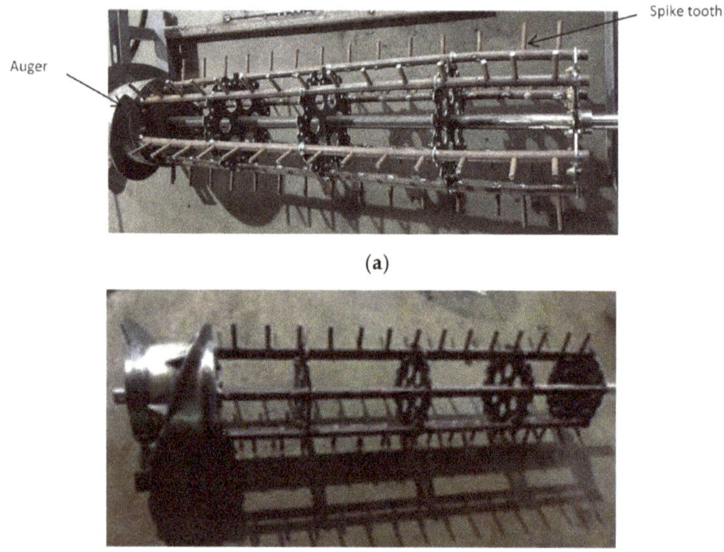

Figure 2. Longitudinal axial flow threshers: (**a**) Conical thresher and (**b**) cylindrical thresher.

2.2. Torque Sensor

The torque sensor (Figure 3) was installed on the thresher shaft to measure thresher rotating speed, threshing torque, and required power. The torque sensor's measuring range was 0–10,000 N.m and 0–12,000 rpm.

Figure 3. Torque sensor.

2.3. Testing Instruments

An TD 1001 electronic digital balance with an accuracy of 0.01 g produced by the Chengdu Cheng Sheng tools group company, LTD; an Sdh-1202 rapid halogen moisture meter produced by the same instruments company, LTD; a TMS-PRO type texture analyzer produced by the FTC USA company; an MB45 moisture meter produced by the OHAUS

USA company; an electronic digital display Vernier caliper; a tape; scissors; some sealing bags; a tachometer; and a frequency converter were used for the test.

2.4. *Rice Cultivar*

The Huanghuazhan rice variety was used for the experiment. It was planted in a field at Huazhong agricultural university, Wuhan, China. The planting method was artificial transplanting. Manual harvesting was used, with a stubble height of about 150 mm, and then the rice was transported to the university factory for testing.

Each rice stem was subjected to a three-point bending test and a shearing test using the TMS-PRO type texture analyzer at the engineering college's agricultural equipment laboratory, as shown in Figure 4. The properties of rice stalks and grains are shown in Table 2.

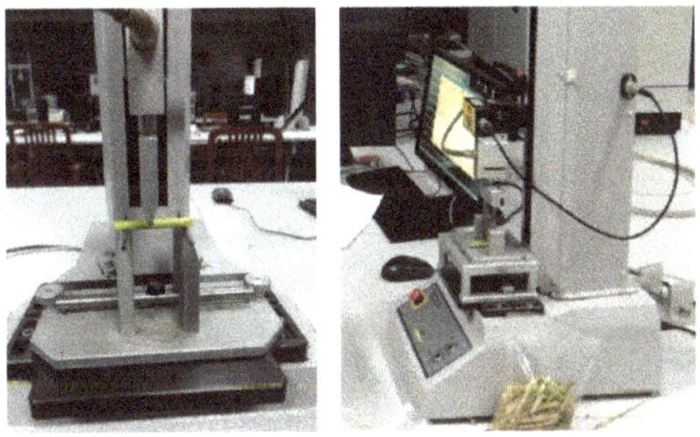

Bending test Shearing test

Figure 4. Bending and shearing test.

Table 2. Rice stalks and grain properties.

Properties	Value
Average Grain Length; mm	9.75
Average Grain Width; mm	2.75
Average Grain Thickness; mm	2.02
Grain Moisture Content; %	13.32
1000 Grains Weight; g	30.43
Average Stalk Length; mm	964.08
Stalk Moisture Content; %	63.21
Max Shearing Force; N	249.6
Max Bending Force; N	9

2.5. *Taguchi Method and Experiment Design*

The Taguchi method is used widely in engineering analysis. It is a dominant design that reduces the number of tests and minimizes the effects of factors that cannot be controlled [28,29]. It uses a loss function to calculate the deviation between the desired values and the experimental values. This loss function is converted into a signal–noise (S/N) ratio [29,30].

The S/N ratio can be divided into three categories given by Equations (1)–(3) [31]:

The nominal is the best:

$$\frac{S}{N} = 10 \log \frac{\overline{y}}{s_y^2} \quad (1)$$

The lower is better:

$$\frac{S}{N} = -10\ \log\frac{1}{n}\left(\sum y2\right) \quad (2)$$

The higher is better:

$$\frac{S}{N} = -\log\frac{1}{n}\left(\sum\frac{1}{y2}\right) \quad (3)$$

where $\overline{y}$ is the average of observed data, s_y^2 is the variation of y, n is the number of observations, and y is the observed data or each type of the characteristics.

2.5.1. Threshing Parameters and Their Levels

For the test, we used the cylindrical and conical types of threshers (A); three rotational speeds (B) of 1100, 1300, and 1500 rpm; and three feeding rates (C) of 0.8, 1.1, and 1.4 kg/s, as shown in Table 3.

Table 3. Threshing parameters levels.

Parameters	Symbol	Level 1	Level 2	Level 3
Thresher type (T)	A	Cylindrical	Conical	
Thresher rotating speed (N), rpm	B	1100	1300	1500
Feeding rate (F), Kg/s	C	0.8	1.1	1.4

2.5.2. Taguchi Full Factorial Design L_{18} ($2^1 \times 3^2$)

In this study, the Taguchi method was used to assess threshing performance and to compare two threshers' structures (cylindrical and conical). Taguchi's L_{18} arrangement was used for experimenting. To determine the optimal threshing conditions and the best operating parameters, the S/N ratio was calculated. The lower is better was used to determine the S/N ratio for power requirement, and the higher is better was used for efficiency and productivity. The experiment results and S/N ratios are shown in Table 4.

Table 4. The results of experiment and the signal–noise (S/N) ratios.

No	Thresher Type	Thresher Speed	Feeding Rate	Threshing Efficiency	S/N Efficiency	Productivity	S/N Productivity	Power	S/N Power
1	Cylindrical	1100	0.8	96.98	39.73	0.27	−11.37	6.20	−15.85
2	Cylindrical	1100	1.1	97.77	39.80	0.30	−10.46	8.27	−18.35
3	Cylindrical	1100	1.4	98.16	39.84	0.35	−9.12	10.37	−20.32
4	Cylindrical	1300	0.8	97.07	39.74	0.33	−9.63	6.31	−16.00
5	Cylindrical	1300	1.1	97.96	39.82	0.38	−8.40	8.96	−19.05
6	Cylindrical	1300	1.4	98.28	39.85	0.39	−8.18	10.43	−20.37
7	Cylindrical	1500	0.8	97.40	39.77	0.40	−7.96	6.57	−16.35
8	Cylindrical	1500	1.1	98.05	39.83	0.45	−6.94	9.04	−19.12
9	Cylindrical	1500	1.4	98.41	39.86	0.51	−5.85	10.63	−20.53
10	Conical	1100	0.8	97.21	39.75	0.33	−9.63	5.45	−14.73
11	Conical	1100	1.1	97.90	39.82	0.38	−8.40	7.22	−17.17
12	Conical	1100	1.4	98.21	39.84	0.39	−8.18	10.06	−20.05
13	Conical	1300	0.8	97.48	39.78	0.38	−8.40	5.89	−15.40
14	Conical	1300	1.1	98.09	39.83	0.42	−7.54	7.66	−17.68
15	Conical	1300	1.4	98.34	39.85	0.44	−7.13	10.22	−20.19
16	Conical	1500	0.8	97.92	39.82	0.48	−6.38	6.44	−16.18
17	Conical	1500	1.1	98.15	39.84	0.59	−4.58	8.90	−18.99
18	Conical	1500	1.4	98.60	39.88	0.64	−3.88	10.30	−20.26

2.6. Testing Procedure

The platform of the thresher was established in the factory. The conveyor belt's total length was 6 m, the first meter was left empty, and the rice straw was evenly spread on the last 5 m to ensure that it would be fed at a stable speed. The conveyor belt's speed was kept to 1 m/s, and different feeding rates of 0.8, 1.1, and 1.4 kg/s were tested. The drum speeds were 1100, 1300, and 1500 rpm. After the experiment, the rice grains were collected from the boxes under the concave and from the straw outlet, cleaned using a cleaning machine, and then weighed to measure the threshing efficiency and productivity. Additionally, the required power was measured using the torque sensor mounted on the thresher shaft.

2.7. Threshing Performance Indicators

Performance evaluation is a scientific method of ascertaining the working conditions of a system's main components to establish how the components contribute to the system's overall efficiency [32].

The criteria for evaluating threshing mechanisms' performance include threshing efficiency, grain loss, grain damage, output capacity, cleaning efficiency, and power requirement [33].

The crop's feed rate into the thresher and operating parameters such as drum speed significantly affected the threshing performance [34].

2.7.1. Threshing Efficiency

Threshing efficiency is the ratio between the mass of threshed grains received from thresher outlets and the total grain input per time unit expressed in percentage [25].

It was calculated regarding the following equation:

$$TE = \text{weight of threshed seed (g)}/\text{total weight of seed (g)} \times 100$$

2.7.2. Thresher Productivity

The throughput of a thresher is the mass of materials passing through the thresher per time unit [35].

$$\text{Throughput} = \text{total weight of seed}/\text{threshing time}$$

2.7.3. Power Requirement

The required power was calculated after analyzing the obtained data from the torque sensor.

3. Results and Analysis

3.1. Taguchi Technique Analysis

3.1.1. Analysis of the Signal-to-Noise (S/N) Ratio

Threshing efficiency, power, and productivity were measured using Taguchi techniques, and the optimization of the control factors was provided by signal-to-noise ratios using the Minitab software. The lowest value of power was effective on threshing performance enhancing, so the lower is better equation was used to determine its S/N ratio. Additionally, the highest values of threshing efficiency and productivity were very effective on threshing performance, so the higher is better was used. The values of the S/N ratios are shown in Tables 5–7 and show the optimal levels of control factors for optimal threshing efficiency, power, and productivity. These levels are also shown in graph forms in Figures 5–7.

The optimum level for each control factor was found regarding the highest S/N ratio in the levels of that control factor. The levels of the factors giving the best efficiency and productivity were specified as $A_2B_3C_3$. This means that the optimum efficiency and productivity were obtained using the conical shaped thresher (A_2), a rotating speed of 1500 (B_3), and a feed rate of 1.8 (C_3). On the other hand, the lowest power requirement was obtained with a thresher type (A_2), at rotating speed (B_1), and feeding rate (C_1).

Table 5. Response table for signal to noise ratios for threshing efficiency (larger is better).

Level	A	B	C
1	39.81	39.80	39.77
2	39.82	39.81	39.82
3		39.83	39.85
Delta	0.02	0.03	0.09
Rank	3	2	1

Table 6. Response table for signal to noise ratios for productivity (larger is better).

Level	A	B	C
1	−8.656	−9.527	−8.895
2	−7.124	−8.214	−7.720
3		−5.930	−7.055
Delta	1.532	3.597	1.840
Rank	3	1	2

Table 7. Response table for signal to noise ratios for power (smaller is better).

Level	A	B	C
1	−18.44	−17.74	−15.75
2	−17.85	−18.11	−18.39
3		−18.57	−20.28
Delta	0.59	0.83	4.53
Rank	3	2	1

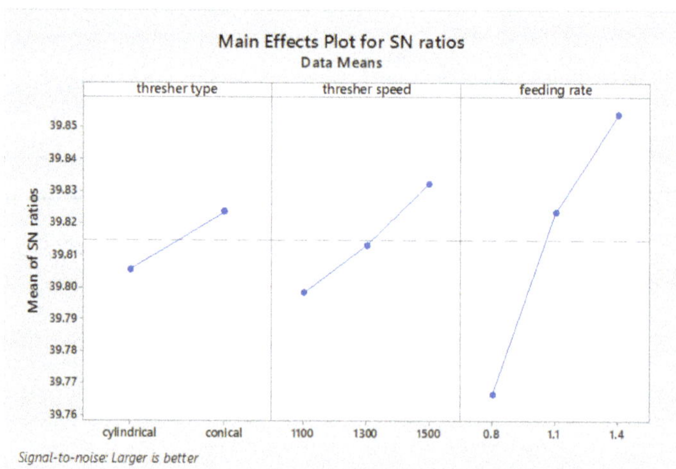

Figure 5. Effect of operating parameters on S/N ratio for threshing efficiency.

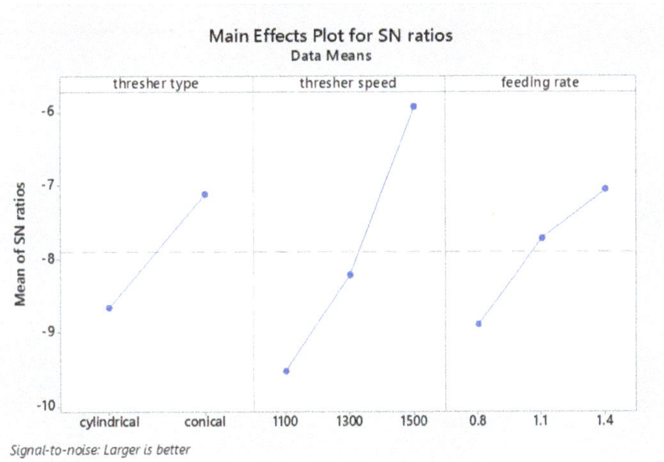

Figure 6. Effect of operating parameters on S/N ratio for productivity.

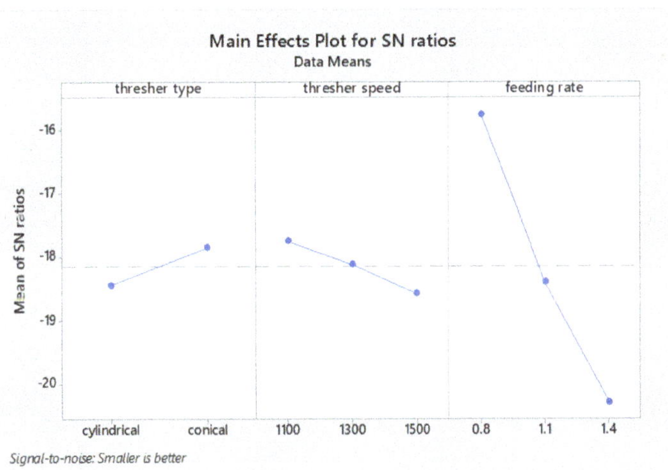

Figure 7. Effect of operating parameters on S/N ratio for power.

3.1.2. Analysis of Variance

ANOVA was used to determine the individual interaction of all of the control factors in the test. In this study, ANOVA was used to analyze the effects of thresher type, rotating speeds, and feeding rates on threshing performance. The ANOVA results are shown in Tables 7–9. This analysis was carried out at a 5% significance level and a 95% confidence level. The last column of the table shows the percentage value of each parameter contribution, which indicates the degree of influence on threshing performance.

According to Table 8, the percent contributions of the A, B, and C factors on the threshing efficiency were found to be 4.87, 11.63, and 79%, respectively. Thus, the most important factor affecting the threshing efficiency was feeding rate.

Table 8. ANOVA for threshing efficiency.

Source	DF	Adj SS	Adj MS	F-Value	p-Value	Contribution Rate (%)
Thresher Type (A)	1	0.1860	0.18600	12.99	0.004	4.87
Thresher Speed (B)	2	0.4440	0.22200	15.50	0.000	11.63
Feeding Rate (C)	2	3.0160	1.50799	105.29	0.000	79.00
Error	12	0.1719	0.01432			4.50
Total	17	3.8178				100.00

Table 9. ANOVA for productivity.

Source	DF	Adj SS	Adj MS	F-Value	p-Value	Contribution Rate (%)
Thresher Type (A)	1	0.024939	0.024939	33.25	0.000	16.16
Thresher Speed (B)	2	0.096544	0.048272	64.36	0.000	62.54
Feeding Rate (C)	2	0.023878	0.011939	15.92	0.000	15.47
Error	12	0.009000	0.000750			5.83
Total	17	0.154361				100.00

Referring to Table 9, the percent contributions of the A, B, and C factors in productivity were 16.16, 62.54, and 15.47%, respectively. Thus, the most effective factor on the productivity was thresher speed.

Regarding Table 10, the contributions percentage of the A, B, and C factors on the power were 2.11, 2.74, and 93.05%. Thus, the most effective factor was the feeding rate.

Table 10. ANOVA for power.

Source	DF	Adj SS	Adj MS	F-Value	p-Value	Contribution Rate (%)
Thresher Type (A)	1	1.196	1.1961	12.04	0.005	2.11
Thresher Speed (B)	2	1.555	0.7776	7.83	0.007	2.74
Feeding Rate (C)	2	52.752	26.3761	265.61	0.000	93.05
Error	12	1.192	0.0993			2.10
Total	17	56.695				100.00

3.2. *Results Evaluation and Discussion*

After the test was carried out, and after the collected data were analyzed according to Taguchi techniques, some graphs were drawn using the Origin software in order to assure the former obtained results.

3.2.1. Effect of Feed Rate on Threshing Efficiency under Different Rotating Speeds for the Cylindrical and Conical Thresher

It was concluded that increasing the thresher's feeding rate and rotating speed increased the threshing efficiency from 96.98% to 98.41% for the cylindrical thresher and from 97.21 to 98.6% for the conical thresher, as shown in Figure 8. These results were in agreement with the results of Osueke, 2013 [36], and Ahuja et al., 2017 [37]. The increase in threshing efficiency with drum speed could be attributed to the high frequency of collisions and impacts between spikes and grain heads, resulting in more grain threshing and separating, and it could also be attributed to the increased friction between the concave and grain heads. The efficiency increased with the increase of feeding rate due to greater amount of mass of the crop fed to the thresher per the time unit.

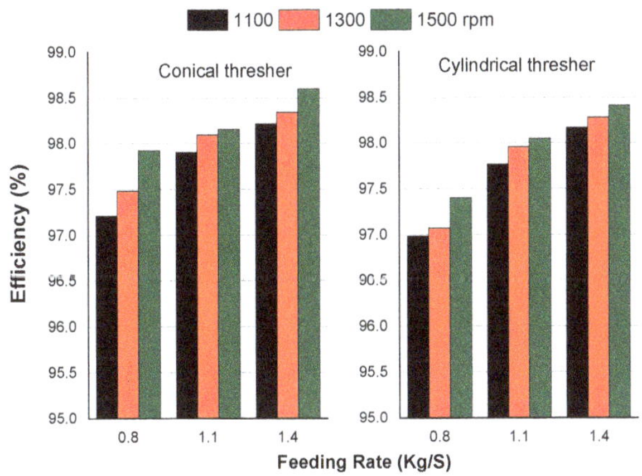

Figure 8. Effect of operating parameters on threshing efficiency.

3.2.2. Effect of Feed Rate on Power Requirement under Different Rotating Speeds for the Cylindrical and Conical Thresher

The required power increased with the increase in feeding rate and rotational speed. This may be attributed to a high load on the thresher because of the excessive stalks passing through the threshing gap. These results were the same as the obtained results by Ezzatollah et al., 2009 [8], who noticed that drum speed significantly affected the power requirements.

Using the conical-shaped thresher resulted in low power requirements when compared to the cylindrical thresher, as shown in Figure 9.

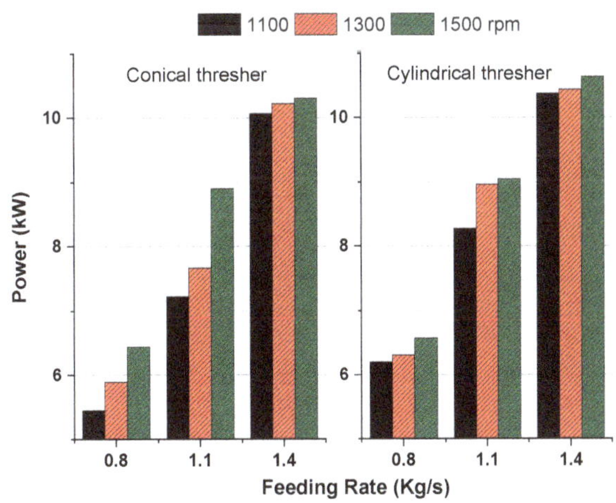

Figure 9. Effect of operating parameters on threshing power.

3.2.3. Effect of Feed Rate on Productivity under Different Rotating Speeds for the Cylindrical and Conical Thresher

Increasing the feed rate increased the productivity of the thresher from 0.27 to 0.51 kg/s for the cylindrical thresher and from 0.33 to 0.64 kg/s for the conical thresher, as illustrated in Figure 10. These results were in agreement with the work of Osueke, 2013 [36]. This may be attributed to the higher mass of rice passing through the thresher per time unit.

Additionally, the increase of the rotational speed increased productivity because the higher speed resulted in a low threshing time, which increased the threshed crop per the time unit. The conical thresher gave a higher productivity than the cylindrical thresher.

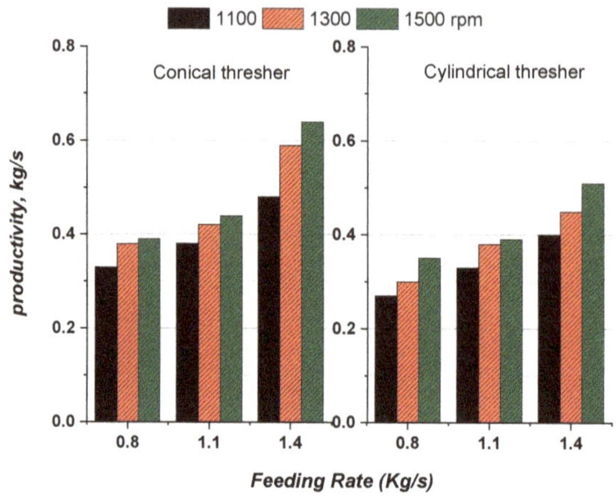

Figure 10. Effect of operating parameters on threshing productivity.

4. Conclusions

1. In this paper, two thresher structures were tested and evaluated for performance under different operating parameters such as thresher speed and feeding rate. The Taguchi technique was used to reduce the testing time and number and to analyze the data for experimental variable optimization. The obtained results revealed that increasing the feed rate and rotating speed positively correlated with threshing efficiency, productivity, and power requirements.
2. The highest threshing efficiency and highest productivity of 98.6% and 0.64 kg/s, respectively, were achieved using the conical thresher under a rotating speed of 1500 rpm and a feeding rate of 1.4 kg/s.
3. The lowest required power of 5.45 kW was obtained using the conical thresher under a rotational speed of 1100 rpm and a feeding rate of 0.8 kg/s.
4. It was concluded that the conical thresher was more effective than the cylindrical thresher because it achieved a higher efficiency, a higher productivity, and a lower power requirement.
5. This research provides a new method for assessing rice thresher performance and presents a new threshing drum structure that will be more efficient for rice threshing with a combine harvester.

Author Contributions: Conceptualization, M.A.A. and G.Z.; methodology, M.A.A.; software, M.A.A.; validation, M.A.A., G.Z., and A.E.S.; formal analysis, M.A.A.; investigation, M.A.A.; resources, M.A.A.; data curation, G.Z.; writing—original draft preparation, M.A.A.; writing—review and editing, M.A.A.; visualization, M.A.A.; supervision, G.Z.; project administration, M.A.A.; funding acquisition, G.Z. All authors have read and agreed to the published version of the manuscript.

Funding: This research was financially supported by the National Key Research and Development Program of China (Technological Innovation of Regenerated Rice Mechanization in the North of the Middle and Lower Reaches of the Yangtze River, 2017YFD0301404-05); and the Fundamental Research Funds for the Central Universities (2662018PY038).

Institutional Review Board Statement: Not applicable.

Informed Consent Statement: Not applicable.

Data Availability Statement: The data presented in this study are available on demand from the first author at (mohamed.anwer2010@yahoo.com).

Acknowledgments: G.Z. and National Key Research and Development Program of China (Technological Innovation of Regenerated Rice Mechanization in the North of the Middle and Lower Reaches of the Yangtze River, 2017YFD0301404-05); and the Fundamental Research Funds for the Central Universities (2662018PY038).

Conflicts of Interest: Authors declare no conflict of interest.

References

1. Center American Trends Panel. The forecast of world agricultural supply and demand forecast for February. *World Agric.* **2017**, *4*, 215–219.
2. Callaway, E.J.N. The birth of rice. *Nature* **2014**, *514*, S58. [CrossRef] [PubMed]
3. Qi, W. Impacts of nitrogen and zeolite managements on yield and physicochemical properties of rice grain. *Int. J. Agric. Biol. Eng.* **2016**, *9*, 93–100.
4. Cerquitelli, T.J.E.P. Predicting large scale fine grain energy consumption. *Energy Proc.* **2017**, *111*, 1079–1088. [CrossRef]
5. Zami, M.A. Performance evaluation of the BRRI reaper and Chinese reaper compared to manual harvesting of rice (Oryza sativa L.). *Agriculturists* **2014**, *12*, 142–150. [CrossRef]
6. Unakıtan, G.; Aydın, B.E.J. A comparison of energy use efficiency and economic analysis of wheat and sunflower production in Turkey: A case study in Thrace Region. *Energy* **2018**, *149*, 279–285. [CrossRef]
7. Chabrol, D. The Technical Centre for Agricultural and Rural Cooperation. *ACP-EU Lome* **1996**, *6*, 12–15.
8. Ezzatollah, A.A.-A.; Yousef, A.-G.; Saeid, A. Study of Performance Parameters of Threshing Unit in a Single Plant Thresher. *Ardabil* **2009**, *4*, 92–96.
9. Asundep, N.N. Determinants of access to antenatal care and birth outcomes in Kumasi, Ghana. *J. Epid. Glob. Health* **2013**, *3*, 279–288. [CrossRef]
10. Olayanju, T. An assessment of the status of small rice threshers in Nigeria. *J. Nat. Sci. Eng. Technol.* **2009**, *8*, 84–94.
11. Azouma, O.Y. Design of throw-in type rice thresher for small scale farmers. *Indian J. Sci. Tehnol.* **2009**, *2*, 10–14. [CrossRef]
12. Gbabo, A.; Gana, I.M.; Amoto, M.S. Design, fabrication and testing of a millet thresher. *Net J. Agric. Sci.* **2013**, *1*, 100–106.
13. Singh, B.; Linvill, D.E. Determining the effect of pod and grain moisture content on threshing loss and damage of navy beans. *Trans. ASAE* **1977**, *20*, 226–0227. [CrossRef]
14. Ndirika, V.J. A mathematical model for predicting output capacity of selected stationary spike-tooth grain threshers. *Appl. Eng. Agric.* **2006**, *22*, 195–200. [CrossRef]
15. Gang, W. Design of variable screw pitch rib snapping roller and residue cutter for corn harvesters. *Int. J. Agric. Biol. Eng.* **2016**, *9*, 27–34.
16. Li, H. CFD–DEM simulation of material motion in air-and-screen cleaning device. *Comp. Electron. Agric.* **2012**, *88*, 111–119. [CrossRef]
17. Li, Y. Development and application of mechanized maize harvesters. *Int. J. Agric. Biol. Eng.* **2016**, *9*, 15–28.
18. Tang, Z.; Li, Y.; Xu, L. Design and optimization for length of longitudinal-flow threshing cylinder of combine harvester. *Trans. Chin. Sci. Agric. Eng.* **2014**, *30*, 28–34.
19. Addo, A. Effect of different threshing cylinders on soybean quality. *J. Sci. Tehnol.* **2004**, *24*, 121–125. [CrossRef]
20. Alizadeh, M.R.; Bagheri, I. Field performance evaluation of different rice threshing methods. *Int. J. Nat. Sci. Eng.* **2009**, *3*, 139–143.
21. Osueke, C.O. Frictional impact modeling of a cereal thresher. *Am. J. Eng. Appl. Sci.* **2011**, *4*, 405–412. [CrossRef]
22. Khazaei, J. Force Requirement for Pulling off Chickpea Pods as Well as Fracture Resistance of Chickpea Pods and Grains. Ph.D. Thesis, Tehran University, Tehran, Iran, 2003.
23. Singh, K. Development and evaluation of multi millet thresher. *J. Appl. Nat. Sci.* **2015**, *7*, 939–948. [CrossRef]
24. Azhar Saeed, M. Testing and evaluation of hold-on paddy thresher. *Agric. Mech. Asia Africa* **1995**, *26*, 47.
25. Asli-Ardeh, E.A.; Abbaspour-Gilandeh, Y. Investigation of the effective factors on threshing loss, damaged grains percent and material other than grain to grain ratio on an auto head feed threshing unit. *Am. J. Agric. Biol. Sci.* **2008**, *3*, 699–705.
26. El-Haddad, W.Z. A simplified design and performance study of threshing and winnowing machine suitable for sample holdings. Master's Thesis, University Egypt, Tanta, Egypt, 2000.
27. Helmy, M.; Yousef, I.; Badawy, M.A. Performance evaluation of some sunflower threshers. *Egypt J. Agric. Res.* **2000**, *78*, 959–975.
28. Palanikumar, K.J.M. Experimental investigation and optimisation in drilling of GFRP composites. *Measurement* **2011**, *44*, 2138–2148. [CrossRef]
29. Asiltürk, I.; Akkuş, H. Determining the effect of cutting parameters on surface roughness in hard turning using the Taguchi method. *Measurement* **2011**, *44*, 1697–1704. [CrossRef]
30. Köksoy, O.; Muluk, F.Z. Solution to the taguchiâs problem with correlated responses. *Gazi Univ. J. Sci.* **2004**, *17*, 59–70.
31. Shetty, R.; Pai, R.B.; Rao, S.S.; Nayak, R. Taguchi's technique in machining of metal matrix composites. *J. Braz. Soc. Mech. Sci. Eng.* **2009**, *31*, 12–20. [CrossRef]
32. Chukwu, O. Performance evaluation of locally manufactured rice threshers in Niger state. *J. Eng. Appl. Sci.* **2008**, *3*, 602–606.

33. Behera, B.K. Development and testing of a power-operated wheat thresher. *Agric. Mech. Asia Africa Lat. Am. Jpn.* **1990**, *21*, 15–21.
34. Špokas, L.; Steponavičius, D.; Petkevičius, S. Impact of technological parameters of threshing apparatus on grain damage. *Agron. Res.* **2008**, *6*, 367–376.
35. Ndirika, V.I.O. Modeling the Performance of Selected Stationary Grain Threshers. Ph.D. Thesis, Department of Agricultural Engineering Ahmadu, Bello University, Zaria, Nigeria, 1997.
36. Osueke, E.C.O. Study of the influence of crop, machine and operating parameters on performance of cereal threshers. *Int. J. Eng. Res. Dev.* **2014**, *7*, 1–9.
37. Ahuja, M.; Dogra, B.; Narang, M.K.; Dogra, R. Development and evaluation of axial flow paddy thresher equipped with feeder chain type mechanical feeding system. *Curr. J. Appl. Sci. Technol.* **2017**, *23*, 1–10. [CrossRef]

agriculture

MDPI

Article

Deviation Tolerance Performance Evaluation and Experiment of Picking End Effector for Famous Tea

Yingpeng Zhu [1], Chuanyu Wu [1,2], Junhua Tong [1,2], Jianneng Chen [1,2], Leiying He [1,2], Rongyang Wang [3] and Jiangming Jia [1,*]

[1] Faculty of Mechanical Engineering & Automation, Zhejiang Sci-Tech University, Hangzhou 310018, China; 201910501018@mails.zstu.edu.cn (Y.Z.); cywu@zstu.edu.cn (C.W.); jhtong@zstu.edu.cn (J.T.); jiannengchen@zstu.edu.cn (J.C.); hlying@zstu.edu.cn (L.H.)
[2] Key Laboratory of Transplanting Equipment and Technology of Zhejiang Province, Hangzhou 310018, China
[3] College of Mechanical and Electrical Engineering, Huzhou Vocational and Technical College, Huzhou 313000, China; 2012005@hzvtc.net.cn
* Correspondence: jarky@zstu.edu.cn; Tel.: +86-0571-86843741

Abstract: Accurately obtaining the posture and spatial position of tea buds through machine vision and other technologies is difficult due to the small size, different shapes, and complex growth environment of tea buds. Therefore, end effectors are prone to problems, such as picking omission and picking error. This study designs a picking end effector based on negative pressure guidance for famous tea. This end effector uses negative pressure to guide tea buds in a top-down manner, thereby correcting their posture and spatial position. Therefore, the designed end effector has deviation tolerance performance that can improve the picking success rate. The pre-experiment is designed, the tip of apical bud is referred to as the descent position, and the negative pressure range is determined to be 0.6 to 0.9 kPa. A deviation tolerance orthogonal experiment is designed. Experimental results show that various experimental factors are ranked in terms of the significance level of the effect on the average success rate, and the significance ranking is as follows: negative pressure (P) > pipe diameter (D) > descent speed (V). An evaluation method of deviation tolerance performance is presented, and the optimal experiment factor-level combination is determined as: P = 0.9 kPa, D = 34 mm, V = 20 mm/s. Within the deviation range of a 10 mm radius, the average success rate of the negative pressure guidance of the end effector is 97.36%. The designed end effector can be applied to the intelligent picking of famous tea. This study can provide a reference for the design of similar picking end effectors for famous tea.

Keywords: famous tea; end effector; deviation tolerance; orthogonal experiment; evaluation method

Citation: Zhu, Y.; Wu, C.; Tong, J.; Chen, J.; He, L.; Wang, R.; Jia, J. Deviation Tolerance Performance Evaluation and Experiment of Picking End Effector for Famous Tea. *Agriculture* **2021**, *11*, 128. https://doi.org/10.3390/agriculture11020128

Academic Editor: José Pérez-Alonso
Received: 19 December 2020
Accepted: 1 February 2021
Published: 5 February 2021

1. Introduction

Tea is a natural health drink loved by people from all over the world [1,2]. Famous tea has high drinking value and economic benefits and is the pillar of the tea industry. At present, the picking of ordinary tea has been mechanized [3,4], whereas the picking of famous tea still relies on manual labor. With the increasing shortage of the tea picking labor force, the phenomenon of a "tea picker shortage" is becoming increasingly serious, thereby limiting the development of the tea industry [5,6]. Realizing the mechanized picking of famous tea is an objective requirement and inevitable trend for the sustainable development of the tea industry.

At present, research on mechanized picking of famous and excellent tea is still in the exploration stage, and the research direction mainly focuses on the identification and location of tea buds. Early research methods are mainly based on the color, shape, texture, and other characteristics of tea leaves to identify the tea buds [7–9]. Such methods have poor robustness and low accuracy. Deep learning has developed rapidly. Many researchers have begun to use network models, such as Faster R-CNN [10] and YOLO [11], to identify the tea buds and locate them on 2D images. However, many problems remain to be solved due to the small size, different shapes, and complex growth environment of tea buds, and the robustness, accuracy, and efficiency of algorithms need to be improved.

In the actual picking conditions, end effectors are prone to problems, such as picking omission and picking error [12–14]. Thus, error compensation is required. Mehta and Burks [15] proposed a hybrid translation controller based on pursuit guidance, with control accuracy of approximately 15 mm. Wang et al. [16] designed an efficient strawberry harvesting end effector with large misalignment tolerance. This end effector can complete picking within the positioning error of ± 7 mm, and the success rate of the indoor picking test was 97.7%. Ye et al. [17] proposed a dynamic positioning error analysis method to guide the fault-tolerant design of an end effector, and the success rate of the experiment was more than 90%. Zou et al. [18] designed a limited universal fruit-picking end effector, and the success rate of indoor and outdoor picking experiments was more than 84% and 78%, respectively. Xiong et al. [19] designed a novel cable-driven gripper with perception capabilities for strawberry-picking robots. The gripper is equipped with three infrared sensors to correct the position error. In the field test, the success rate of picking isolated strawberries was 96.77%. The abovementioned error compensation methods mainly include visual servo to expand the picking range of end effector and multisensor cooperative positioning. However, they cannot meet the requirements of picking tea buds in terms of the positioning accuracy and space occupation.

Different from many crops, the picking objects of famous tea are branches and leaves rather than fruits. Traditional end effectors are difficult to apply. Thus, developing new picking end effectors is necessary. Qin et al. [20] developed a picking end effector, and the intact rate of tea buds harvested was approximately 76.6%. Hao et al. [21] developed a bionic picking finger, and the picking success rate in the preliminary indoor experiment was approximately 70%. Motokura et al. [22] used a three-finger gripper on the end of a Kinova Jaco robotic arm to complete the tea picking action. Most of the existing picking end effectors for famous tea adopt some simple mechanical structures and have poor error compensation ability, which cannot ensure the success rate of picking and the intact rate of tea buds. Therefore, a deviation tolerance design of picking end effectors for famous tea is necessary.

This study aims to explore the picking technology of famous tea. The specific objectives are: (1) to study the harvesting mechanism of famous tea and develop a picking end effector with deviation tolerance performance based on negative pressure guidance; and (2) to establish a set of deviation tolerance performance evaluation methods to provide a reference for the design of similar end effectors.

2. Materials and Methods

2.1. Physical Properties of Tea

As shown in Figure 1, the picking object in this study was one-bud double-leaves, namely, the tea bud above the third leaf node. Determining the relevant physical characteristics of tea, including the overall dimension, shear force F, and average growth region area S_A, is necessary to make the end effector succeed in picking tea buds.

Figure 1. Schematic of tea characteristic parameters. Overall dimension: a—first leaf span, b—second leaf span, c—leaf width, d—node spacing.

The physical properties of tea were measured at the Lishui Comprehensive Test Station of the National Tea Industry Technology System on 24 October 2020. As shown in Figures 2 and 3, the measurement samples of the first five physical characteristics are 100 fresh tea leaves picked randomly, and the tea variety is tulip. The measuring tools are an AIRAJ brand digital caliper and a stem shear characteristic measuring instrument. As shown in Figure 4, the measuring tool of S_A is a self-made square frame with a side length of 400 mm. The measurement method is to place the frame on the canopy of the tea tree, and the number of tea buds in the frame is x. S_A can be calculated by using Formula (1), and a total of 20 measurements are made.

$$S_A = \frac{400 \times 400}{x}, \quad (1)$$

x—the number of tea buds in the frame.

Figure 2. Source of measurement samples.

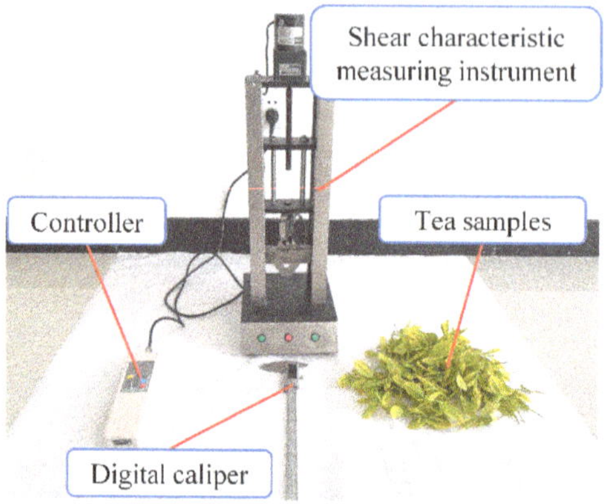

Figure 3. Measurement of the physical properties of tea.

Figure 4. Measurement of average growth region area.

The measurement results are statistically sorted, as shown in Figure 5 and Table 1. The results show that large individual differences are found in the tea buds, thereby affecting the effect of negative pressure guidance. Therefore, the end effector is required to have strong adaptability. The growth density of tea buds is different. The average growth region area of tea buds in the densest place is 3200 mm^2, thereby requiring the end effector to avoid the nontarget objects.

Table 1. Statistical table of the physical properties of tea.

Parameter Name	Statistical Parameters			
	Average Value	Maximum Value	Minimum Value	Standard Deviation
a (mm)	26.90	43.70	14.70	6.13
b (mm)	59.78	89.50	38.80	10.30
c (mm)	11.39	20.80	7.00	2.97
d (mm)	13.55	22.80	7.10	3.17
F (N)	7.64	10.56	5.15	1.17
S_A (mm^2)	4017.51	4571.43	3200.00	360.68

a—first leaf span, b—second leaf span, c—leaf width, d—node spacing, F—shear force, S_A—average growth region area.

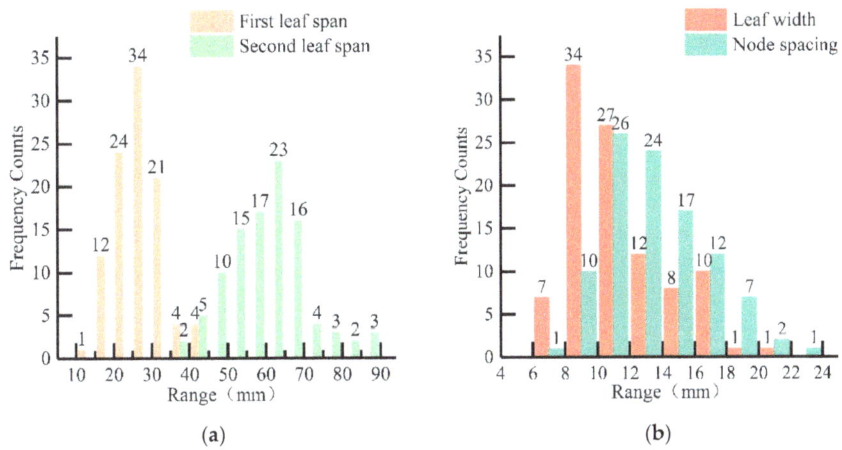

Figure 5. Frequency distribution of dimension parameters. (**a**) Frequency distribution of first leaf span a and second leaf span b, (**b**) Frequency distribution of leaf width c and node spacing d.

2.2. End Effector Structure and Principle

In accordance with the growth characteristics of tea leaves, the tea buds are mainly scattered on the canopy surface. Thus, the top-down picking method can reduce interference. A negative pressure guidance method was adopted in this study to solve the positioning and random errors. Its principle is to attract the tea bud with the airflow generated by the negative pressure. This condition can effectively improve the picking success rate of the end effector.

The end effector was designed in combination with lightweight design requirements, as shown in Figure 6, including the picking pipe, rope drive mechanism, and shearing mechanism. As shown in Figure 7, the picking pipe was formed by 3D printing, and the upper pipe opening was connected to a vacuum suction machine. The flange was fixed on the moving platform of the parallel manipulator, one end of the wire rope pipe of the rope drive mechanism was fixed on the flange, and the shear mechanism was fixed on the boss. As shown in Figure 8, the rope drive mechanism included a mounting support, steering gear, Arduino Mega 2560, wire rope pipe, and wire rope. Its function is to drive the shear mechanism for completing the shearing work. Its advantage is that the driving source is set on the frame, thereby reducing the burden of the parallel manipulator and avoiding limiting the movement speed of the manipulator. As shown in Figure 9, the shearing mechanism adopted a gear transmission, and two specially shaped blades were driven by the wire rope to complete the shearing. Its advantages are simple driving, small volume, and the blade being located directly below the lower pipe opening. Thus, interfering in the initial and shear states is difficult.

During operation, the end effector first reaches above the specified shear point to avoid interference. At this time, the vacuum suction machine begins to work, and negative pressure is generated at the lower pipe opening of the picking pipe. The parallel manipulator is driven to descend vertically so that the end effector reaches the shear point position. In this process, the tea bud is guided into the picking pipe with the airflow generated by negative pressure. The steering gear is controlled to rotate, and the wire rope is driven to tighten so that the shearing mechanism can complete the shearing. The tea buds are collected by negative pressure. At the same time, the steering gear is controlled to reset, and the shear mechanism returns to the initial position under the action of the spring.

The working principle of the end effector was analyzed. When the negative pressure guidance is successful, the probability of shear failure is minimal. Therefore, the success rate of picking depends mainly on the success rate of the negative pressure guidance.

Without considering the positioning error, the negative pressure of the vacuum suction machine, the pipe diameter of the lower pipe opening, and the descent speed of the picking pipe are the influencing factors of the negative pressure guidance.

The tea buds with a first leaf span less than 35 mm accounted for 92% of the samples, in accordance with the statistical results. Considering that negative pressure guidance can reduce the leaf span, the diameter of the lower pipe opening can be slightly smaller than the first leaf span. Thus, the minimum diameter of the lower pipe opening is 30 mm. The minimum value of the second leaf span was 38.8 mm. The maximum diameter of the lower pipe opening was set to 38 mm to reduce the inhalation of nontarget objects, such as the third and fourth leaves. Therefore, the pipe diameter range of the lower pipe opening was determined as 30–38 mm, and the three levels were set as 30, 34, and 38 mm, respectively.

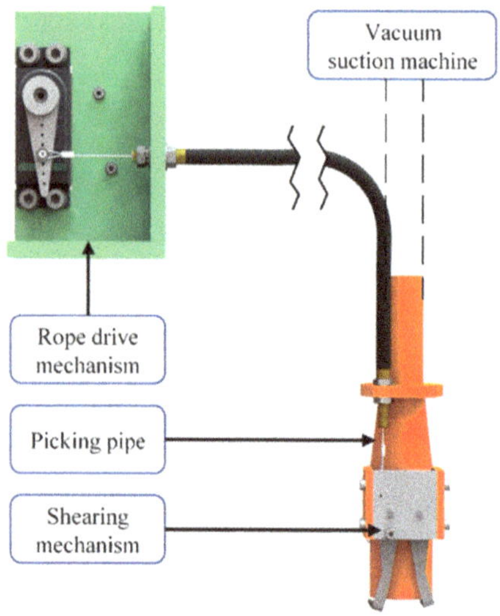

Figure 6. End effector.

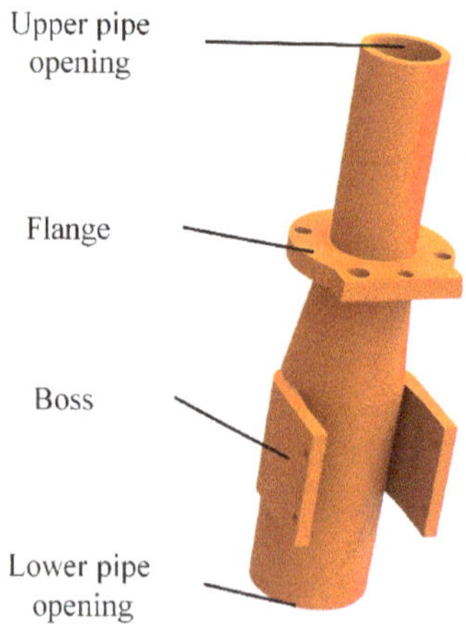

Figure 7. Picking pipe.

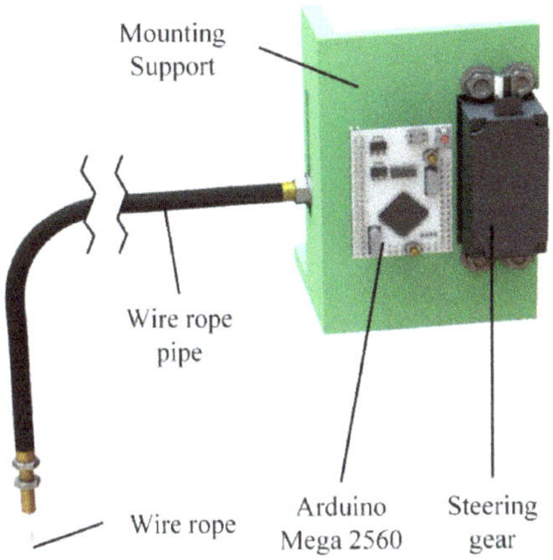

Figure 8. Rope drive mechanism.

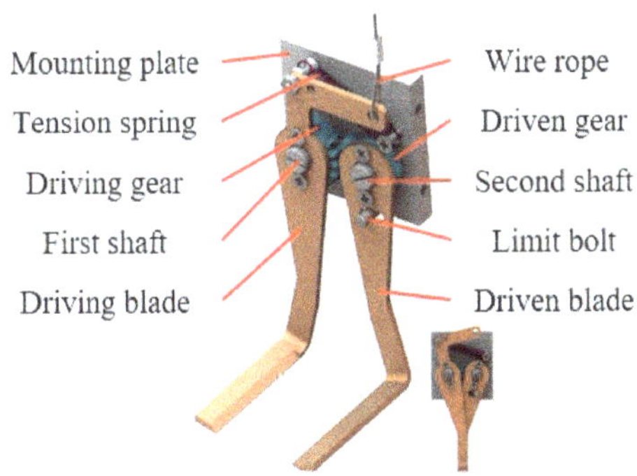

Figure 9. Shearing mechanism.

2.3. Design of Experiment

2.3.1. Experimental Conditions

The experimental platform was set up, as shown in Figure 10. The experimental equipment included the vacuum suction machine, parallel manipulator, digital vacuum gauge, laptop, high-speed camera, and light source. The high-speed camera used in this study was the Phantom VEO 340L (York Technologies Co., Ltd., Hong Kong, China). When the resolution is 2560 × 1600, the FPS is 800. The vacuum suction machine utilized was the 6281D model (Shanghai Yili Electric Co., Ltd., Shanghai, China). After modification, the maximum internal negative pressure is 0.9 kPa, which can be adjusted in a stepless manner with a knob. The brand of the digital vacuum gauge is Japanese Sanliang and the model is DP360. Its measuring range is ±10 kPa and the accuracy is 0.3%.

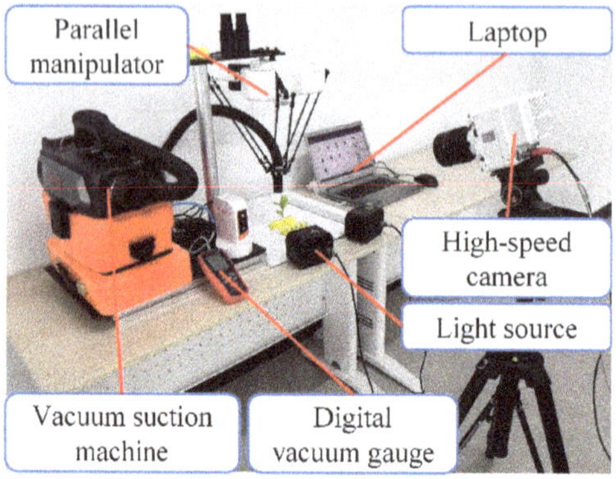

Figure 10. Experimental platform.

As shown in Figure 11, a ruler, a guide pipe with pipe diameters of 30, 34, and 38 mm (referred to as D30, D34, and D38, respectively), a fixed seat, end cap 1, end cap 2, and a locating pin were required. The fixed seat was used to fix the tea leaves. As shown in Figure 12, end cap 1 and the locating pin constitute location component 1, which can be used to locate the center position of D30 and D38. End cap 2 and the locating pin constitute location component 2, which can be used to locate the center position of D34.

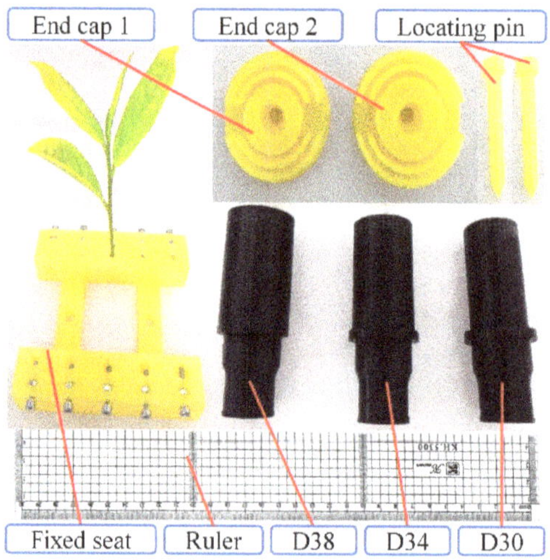

Figure 11. Experimental supplies.

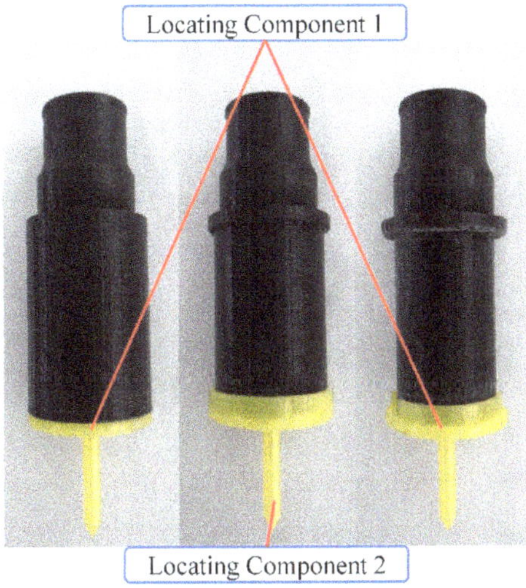

Figure 12. Assembly diagram of locating components.

2.3.2. Pre-Experiment

Multiple groups of experiments should be conducted to avoid losing generality due to the large individual differences of the tea buds. At the same time, uniformly regulating the placement of tea leaves is necessary to avoid the influence of placement errors on the test results. As shown in Figure 13, O-XYZ was the coordinate system of the parallel manipulator. The tea placement principle is that the second leaf is located in the negative direction of the X-axis, and the first leaf is located in the positive direction of the X-axis. The line between the tip of the first leaf G_1 and the tip of the second leaf G_2 is parallel to the XOZ plane.

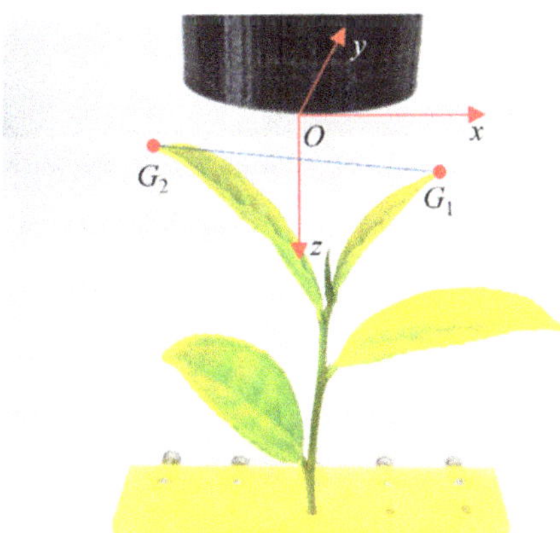

Figure 13. Diagram of tea placement.

The descent position of the guide pipe affects the negative pressure guidance. The first step is to determine the optimal descent position through the pre-experiment and unify it in the subsequent experiment for obtaining a better experiment effect. Considering the reliability of visual positioning and negative pressure guidance, the optimal descent position should be located at the center of the first leaf span or the tip of the apical bud. In other words, when tea leaves are placed, the tip of the locating pin should be located at the center of the first leaf span or the tip of the apical bud. The center of the first leaf span can be determined with a ruler. The test steps are as follows:

1. Randomly picked tea leaves are placed in accordance with the placement principle of tea leaves. The locating pin is used to align the center of the guide pipe with the center of the first leaf span.
2. Turn on the vacuum suction machine, control the parallel manipulator to descent, guide the tea buds with negative pressure, and use the high-speed camera to record the guiding process.
3. After reposition, use the locating pin to align the center of the guide pipe with the tip of the apical bud, and repeat step 2.

The negative pressure guidance process of two descent positions taken by the high-speed camera is shown in Figures 14 and 15, and the image evaluation system used was the Phantom Camera Control Application version 3.1. Whether the descent position is the center of the first leaf span or the tip of the apical bud when the second leaf is guided into the guide pipe, the first leaf moves some distances to the positive direction of the X-axis. As shown in Figure 14d, when the descent position is the center of the first leaf span, the first leaf is far from the guide pipe center. Thus, escaping from the range of negative pressure guidance is easier than when the descent position is the tip of the apical bud, resulting in the failure of guidance. Therefore, the tip of the apical bud was chosen as the descent position, and the descent position was unified as the tip of the apical bud in the follow-up experiment.

Figure 14. Descent position is the center of the first leaf span. (**a**) initial state, (**b**) Guidance start, (**c**) During guidance, (**d**) End of guidance.

Figure 15. Descent position is the tip of the apical bud. (**a**) initial state, (**b**) Guidance start, (**c**) During guidance, (**d**) End of guidance.

The size of negative pressure directly affects the effect of negative pressure guidance. Determining the range of negative pressure through a pre-experiment is necessary to ensure the reliability of a deviation tolerance orthogonal experiment. The experiment samples were 60 fresh tea leaves (one-bud four-leaves) randomly picked from the tea garden, and the variety was tulip. The maximum internal negative pressure of the vacuum tea suction machine used in this study was 0.9 kPa for a long time and could be increased to 1.1 kPa for a short time. When the negative pressure exceeded 1.0 kPa, some tea leaves shook rapidly during the negative pressure guidance, thereby damaging the tea buds. This condition is attributed to turbulence formed by the fast-flowing air at the lower pipe opening. The damage rates of tea buds were 5% and 20% when the negative pressures were 1.0 and 1.1 kPa, respectively. Therefore, the maximum negative pressure was set to 0.9 kPa by considering the quality of tea bud picking and the stability of the equipment operation. The function of negative pressure guidance is to guide the tea buds into the guide pipe. The designed experiment steps to determine the minimum negative pressure are as follows:

1. The 60 randomly picked fresh tea leaves are divided into groups A, B, and C, which are used as experimental samples of D30, D34, and D38, respectively. D30 is then installed on the parallel manipulator.
2. Take the experiment samples of group A, and place the tea leaves in accordance with the tea placement principle. Set the descent speed of the parallel manipulator to 20 mm/s (low speed is good for negative pressure guidance). Turn on the vacuum suction machine, set the negative pressure to 0.2 kPa (the guiding effect is extremely weak when negative pressure <0.2 kPa), and then control the parallel manipulator to descent.
3. If the guidance is successful, the negative pressure is recorded to be 0.2 kPa. If it fails, the negative pressure increases by 0.05 kPa each time until it is successful. The negative pressure is recorded to be ($0.2 + 0.05\,q$) kPa, and q is the times of negative pressure increase.
4. Replace the tea leaves, then repeat steps 2 and 3, and complete 20 data records.
5. Replace the guide pipe and corresponding experimental sample, and then repeat steps 2, 3, and 4.

Figure 16 shows the frequency distribution of the experiment results of the three groups. All the negative pressure values in the D30 experiment group were less than or equal to 0.6 kPa, and 95% of the negative pressure values in the D34 and D38 experiment groups were less than or equal to 0.6 kPa. Therefore, 0.6 kPa was selected as the minimum negative pressure, and the negative pressure range was 0.6–0.9 kPa.

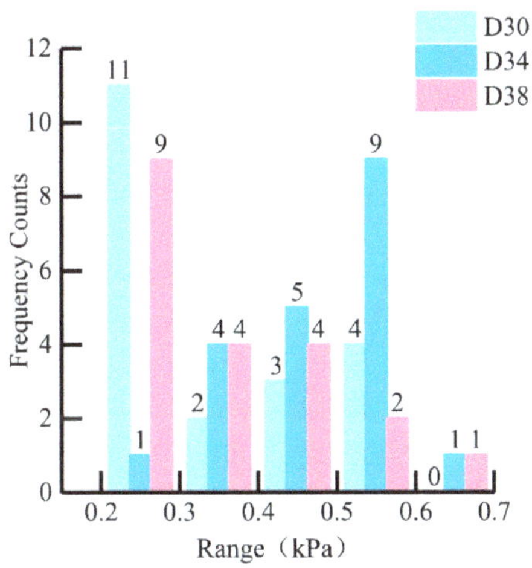

Figure 16. Frequency distribution of minimum negative pressure.

2.3.3. Design of Deviation Tolerance Orthogonal Experiment

The purpose of a deviation tolerance orthogonal experiment is to evaluate the deviation tolerance performance of the end effector under different design parameters, that is, the ability to guide the tea buds successfully when a deviation occurs in the descent position of the guide pipe. The deviation tolerance orthogonal experiment was conducted at the Lishui Comprehensive Test Station of the National Tea Industry Technology System from 2 November to 5 November 2020. The experiment samples were 180 fresh tea leaves (one-bud four-leaves) randomly picked from the tea garden, and the variety was tulip. Only 5 fresh tea leaves were picked each time and collected again after the experiment to avoid the water loss of fresh tea leaves, leading to great changes in physical properties and affecting the experimental results. This process was performed due to the large amount and time consumption of this experiment.

The deviation tolerance orthogonal experiment was conducted by using an $L9(3^4)$ orthogonal table array and performing 20 tests for each combination of test conditions. The negative pressure range was 0.6–0.9 kPa, and the pipe diameter range was 30–38 mm. The speed range of the parallel manipulator was 0–100 mm/s. Considering the picking efficiency, the descent speed range was limited to 20–100 mm/s, and the factor level is shown in Table 2. The descent position was arranged in accordance with Figure 17 in the experiment, where origin O is the tip of the apical bud, the evenly distributed blue points are the descent position, and the distance between adjacent points is 4 mm. If the descent position of the guide pipe shifts to the point (4,4), then the lower pipe opening of D30, D34, and D38 shifts to the position shown in the figure, which is unconducive to negative pressure guidance. In the same direction, the farther from origin O, the higher the failure probability of negative pressure guidance. The descent position that can be successfully guided during each experiment is marked. The success rate of guidance at different descent positions can be obtained for each group of experiments. This rate can be used to analyze and evaluate the deviation tolerance performance of the end effector.

Table 2. Factor level of deviation tolerance orthogonal experiment.

Factors Level	Experimental Factors		
	Negative Pressure P/(kPa)	Pipe Diameter D/(mm)	Descent Speed V/(mm/s)
1	0.60	30	20
2	0.75	34	60
3	0.90	38	100

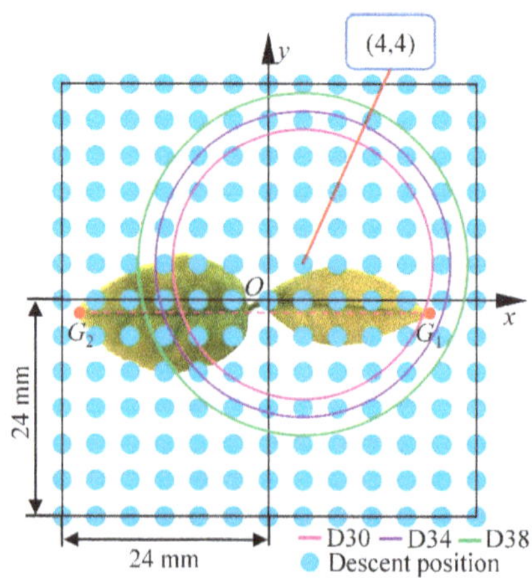

Figure 17. Layout diagram of descent position.

This study presents a deviation tolerance performance evaluation method that uses contour maps to statistically analyze the test results. The evaluation indicators included maximal guide area S_M and average success rate R_S. On this basis, the level combination with less energy and time consumptions can be taken as the optimal choice. As shown in Figure 18, the range with the marginal guidance success rate of zero is defined as the maximal guidance range, and the area is S_M. The set of end effector deviation positions is defined as the deviation range, the deviation range center is located at the origin, the radius is the maximum positioning deviation, and the area is S_D. Four contour areas were assumed to intersect with the deviation range in the contour map, which are A_1, A_2, A_3, and A_4 from the outside to the inside, and the corresponding guidance success rates are E_1, E_2, E_3, and E_4. The overlapping areas of each contour area and deviation range are B_1, B_2, B_3, and B_4, and the corresponding areas are S_1, S_2, S_3, and S_4. The average success rate R_S can be calculated using Formula (2).

$$R_s = (E_1\, S_1 + E_2\, S_2 + E_3\, S_3 + E_4\, S_4)/S_D, \tag{2}$$

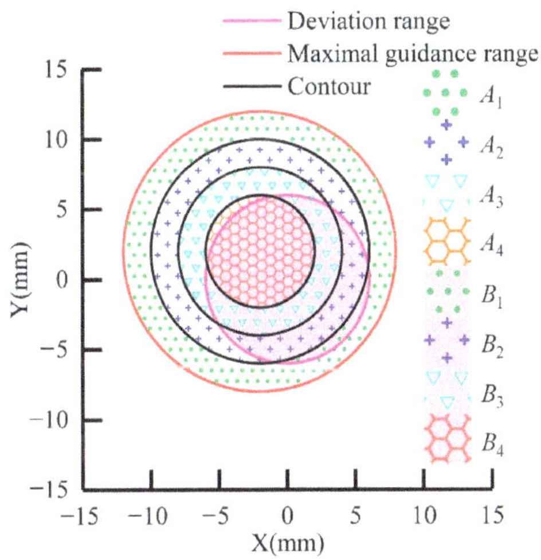

Figure 18. Schematic of evaluation method. A_1, A_2, A_3, and A_4—contour area, B_1, B_2, B_3, and B_4—overlapping areas of each contour area (A_1, A_2, A_3, and A_4) and deviation range.

Similarly, when the number of contour areas intersecting with the deviation range is n, the average success rate R_S can be calculated using Formula (3).

$$R_s = \frac{\sum_{j=1}^{n} E_j S_j}{S_D}, \tag{3}$$

In field picking conditions, the machine vision positioning system used in this study had a maximum positioning deviation of 10 mm in the XY direction. Thus, S_D is equal to 314.16 mm^2. In accordance with the definition, the larger the S_M is, the larger the influence range of negative pressure, and the easier it is to attract nontarget objects. The larger the R_S is, the higher the average success rate of negative pressure guidance within the deviation range. S_M should be determined and limited in accordance with the average growth region area of tea buds to ensure the quality and efficiency of picking. When S_M satisfies the condition, the level combination of experimental factors with large R_S is the best.

3. Results and Discussion

Figure 19 shows the deviation tolerance effect of nine experimental groups based on the contour map, verifying the influence of different design parameters on the deviation tolerance performance of the end effector. The range with the marginal guidance success rate of 95% is defined as the optimal guidance range. As clearly shown in Figure 19, when P = 0.9 kPa, D = 34 mm, and V = 20 mm/s, the area of the optimal guidance range of the end effector is the largest. When P = 0.6 kPa, D = 38 mm, and V = 100 mm/s, the area of the optimal guidance range of the end effector is the smallest.

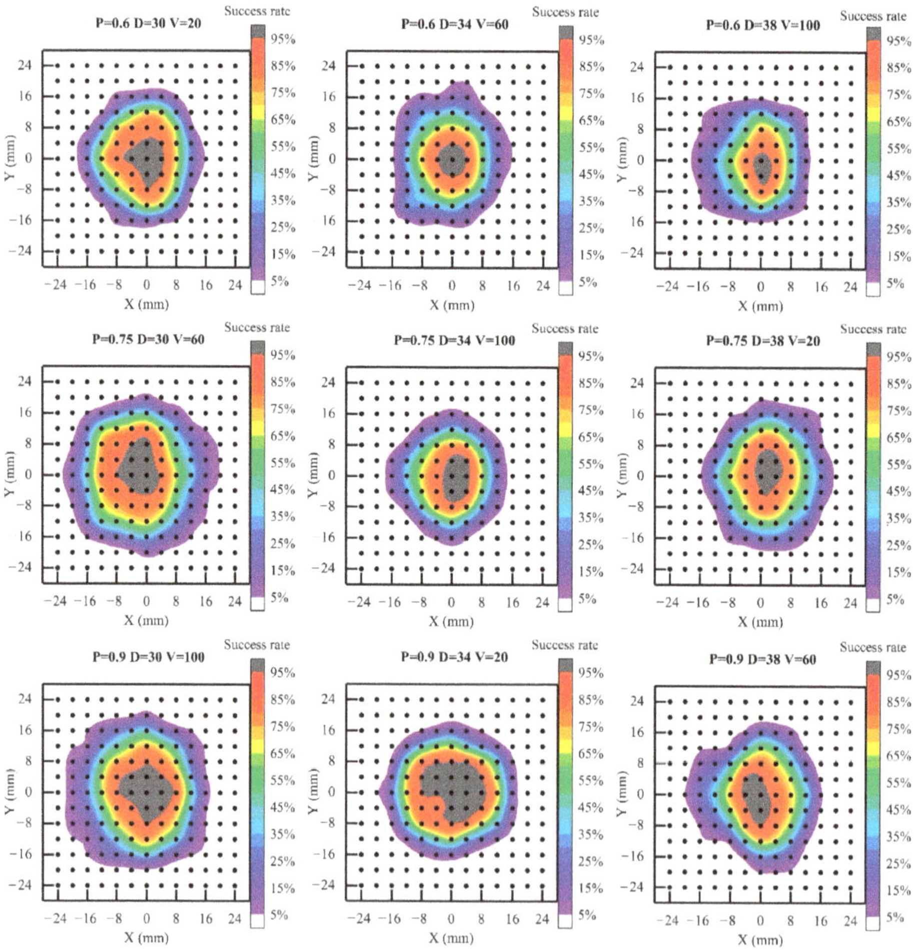

Figure 19. Diagram of deviation tolerance effect.

The results of the deviation tolerance orthogonal experiment are shown in Table 3, where k_i represents the average value of S_M when the experimental factor is at the i level, and t_i represents the average value of R_S when the experimental factor is at the i level, where i = 1, 2, 3. Range R_1 represents the difference between the maximum and minimum values in k_1, k_2, and k_3, and range R_2 represents the difference between maximum and minimum values in t_1, t_2, and t_3. Their formulas are expressed as follows:

$$R_1 = Max\{k_1, k_2, k_3\} - Min\{k_1, k_2, k_3\}, \tag{4}$$

$$R_2 = Max\{t_1, t_2, t_3\} - Min\{t_1, t_2, t_3\}, \quad (5)$$

Table 3. Results of deviation tolerance orthogonal experiment.

Experimental Number	Experimental Factors			Evaluating Indicator	
	P	D	V	S_M	R_S
1	0.60	0.30	20	945.29	88.84%
2	0.60	0.34	60	930.24	79.31%
3	0.60	0.38	100	859.96	73.96%
4	0.75	0.30	60	1345.64	89.72%
5	0.75	0.34	100	832.16	82.64%
6	0.75	0.38	20	1055.53	84.39%
7	0.90	0.30	100	1316.43	94.15%
8	0.90	0.34	20	1023.58	97.36%
9	0.90	0.38	60	1018.39	85.32%
k_1	911.83	1202.45	1008.13		
k_2	1077.78	928.66	1098.09		
k_3	1119.47	977.96	1002.85		
t_1	80.70%	90.90%	90.20%		
t_2	85.58%	86.44%	84.78%		
t_3	92.28%	81.22%	83.58%		
R_1	207.64	273.79	95.24		
R_2	11.57%	9.68%	6.61%		

P—negative pressure, D—pipe diameter, V—descent speed, S_M—maximal guidance area, R_S—average success rate, k_i—average value of S_M when the experimental factor is at the i level, t_i—average value of R_S when the experimental factor is at the i level, R_1—difference between the maximum and minimum values in k_1, k_2, and k_3, R_2—difference between the maximum and minimum values in t_1, t_2, and t_3.

The range R_1 values of the three experimental factors are 207.64, 273.79, and 95.24, respectively. Therefore, the order of primary and secondary factors affecting S_M is pipe diameter, negative pressure, and descent speed. As shown in column D of Table 3, $k_2 < k_3 < k_1$ indicates that S_M decreases first and then increases as the pipe diameter increases in most of the cases observed. The main reason is that the air velocity at the lower pipe opening decreases with the increase in pipe diameter. This condition leads to the weakening of the guiding ability on the tea bud and reduces the maximal guidance range. However, the coverage range of the pipe increases with the further increase in pipe diameter, so that the maximal guidance range increases. As shown in column P of Table 3, $k_1 < k_2 < k_3$ indicates that S_M increases as the negative pressure increases in most of the cases observed. The main reason is that the air velocity at the lower pipe opening increases with the increase in negative pressure. This condition enhances the guiding ability on the tea bud and attracts further tea buds, thereby increasing the maximal guidance range. As shown in column V of Table 3, $k_3 < k_1 < k_2$ indicates that S_M increases first and then decreases as the descent speed increases in most of the cases observed. The main reason is that when the descent speed is slow, the guidance time is sufficient, and the first leaf is prone to a large offset, leading to the failure of guidance. With the increase in descent speed, the guiding time decreases, and the first leaf of the tea bud is inhaled before it moves out of the guidance range, slightly increasing the maximal guidance range. However, when the descent speed continues to increase, the second leaf of the tea bud cannot deflect in time, probably leading to the failure of the guidance and slightly reducing the maximal guidance range.

The range R_2 values of the three experimental factors are 11.57%, 9.68%, and 6.61%, respectively. Therefore, the order of primary and secondary factors affecting R_S is negative pressure, pipe diameter, and descent speed. As shown in column P of Table 3, $t_1 < t_2 < t_3$ indicates that R_S increases as the negative pressure increases in most of the cases observed. As shown in column D of Table 3, $t_3 < t_2 < t_1$ indicates that R_S decreases as the pipe diameter increases in most of the cases observed. As shown in column V of Table 3, $t_3 < t_2 < t_1$ indicates that R_S decreases as the descent speed increases in most of the cases observed.

The main reason is that increasing the negative pressure or decreasing the pipe diameter can increase the air velocity at the lower pipe opening. This condition can effectively improve the guidance success rate of the tea buds near the lower pipe opening, thereby increasing R_S. Considering that guiding the tea buds requires time, reducing the descent speed can allow the tea buds near the lower pipe opening to have sufficient time to be guided, without excessive deviation. This condition can improve the guidance success rate of the tea buds near the lower pipe opening, thereby increasing R_S.

The optimal level combination of the end effector should be determined in accordance with different tea picking conditions. In accordance with Section 2.1, the average growth region area is 3200 mm^2 in the place where the tea buds grow most densely. The boundary is divided by a square, and the side length is 56.57 mm. As shown in Figure 20, the distance of O_1O_2 is 10 mm. The maximal guidance range should be controlled within the boundary to reduce the effect of negative pressure on nontarget objects. The maximal guidance range is approximated as a circle, and the maximal radius of the maximal guidance range in the figure is up to 18.285 mm, that is, S_M needs to meet the following formula.

$$S_M \le \pi \times 18.285^2 = 1050.36\text{mm}^2 \tag{6}$$

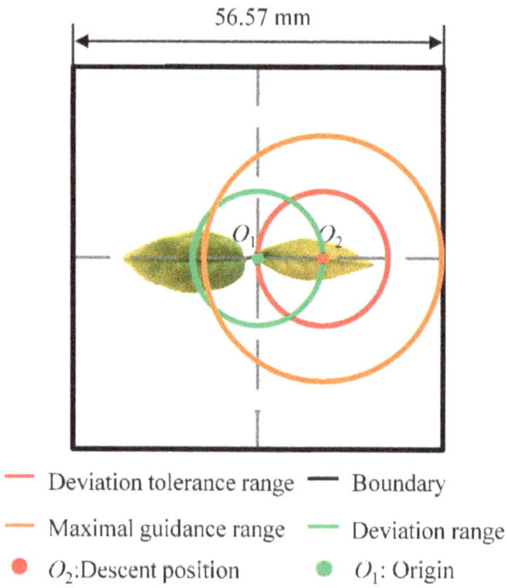

Figure 20. Limitation of maximal guidance range.

If the above conditions are satisfied, the level combined with the highest average success rate is optimal. Therefore, the optimal experimental level combination is $P_3D_2V_1$, which is the eighth group. At this time, the maximal guidance area S_M is 1023.58 mm^2, and the average success rate R_S is 97.36%.

The level combinations that may be better than the eighth group in terms of R_S are $P_3D_1V_1$ and $P_3D_1V_2$. However, the S_M of the two combinations should be greater than 1050.36 mm^2, which does not meet the design requirements. The verification experiment of level combination $P_3D_2V_1$ was conducted to verify the stability of the orthogonal experiment. The results are consistent with the orthogonal experiment. S_M is 1009.51 mm^2, R_S is 97.04%, and the error rates of the two indicators are 1.37% and 0.33%, respectively. The reason for the error should be the different samples of tea buds. The verification test shows that the results of the orthogonal experiment are stable and reliable.

4. Conclusions

In this study, a picking end effector based on negative pressure guidance was developed to solve the visual positioning error of tea buds. On this basis, a deviation tolerance performance evaluation method was presented to optimize the design parameters of the end effector. The deviation tolerance performance of the end effector was verified through experiments. The conclusions are as follows:

1. The developed picking end effector adopts a lightweight design to meet the requirements of the rapid movement of parallel manipulators and is suitable for intensive picking operations. It also has the advantages of being simple, reliable, and low cost, which is conducive to popularization and application.
2. The presented deviation tolerance performance evaluation method can optimize the design of the corresponding end effector for different tea picking conditions. This method can effectively improve the picking success rate of the tea buds and has guiding importance.
3. The picking end effector based on negative pressure guidance has deviation tolerance performance and can pick the tea buds in the case of a positioning deviation in the XY direction. Under the optimal level combination condition, the area of the maximal guidance range is 1023.58 mm^2, and the average success rate of negative pressure guidance within the deviation range with a 10 mm radius is 97.36%. These findings can effectively improve the success rate of picking tea buds, reduce the visual positioning requirements, and accelerate the development process of the tea industry.

Author Contributions: Conceptualization, C.W. and J.C.; data curation, J.T., R.W., and J.J.; writing—original draft, Y.Z. and L.H. All authors have read and agreed to the published version of the manuscript.

Funding: This work was funded by the Post Scientist of the Modern Agriculture Industry System (Intelligent Tea Picking) of the Ministry of Agriculture and Rural Affairs of the People's Republic of China (CARS-19) and Zhejiang Provincial Natural Science Foundation of the People's Republic of China (Grant No. LY18E050025).

Institutional Review Board Statement: Not applicable.

Informed Consent Statement: Not applicable.

Data Availability Statement: The data presented in this study are available on request from the corresponding author.

Acknowledgments: We would also like to thank the samples provided by the Lishui Comprehensive Test Station of The National Tea Industry Technology System.

Conflicts of Interest: The authors declare no conflict of interest.

References

1. Firouzi, S.; Azarian, F. Propellants of mechanical pruning and plucking of tea (a case of developing countries). *Inf. Process. Agric.* **2019**, 6. [CrossRef]
2. Thangavel, S.K.; Murthi, M. A semi automated system for smart harvesting of tea leaves. In Proceedings of the 2017 4th International Conference on Advanced Computing and Communication Systems, ICACCS, Coimbatore, India, 6–7 January 2017; IEEE: New York, NY, USA, 2017. [CrossRef]
3. Han, Y.; Xiao, H.; Qin, G.; Song, Z.; Ding, W.; Mei, S. Developing Situations of Tea Plucking Machine. *Engineering* **2014**, 6. [CrossRef]
4. Han, Y.; Xiao, R.H.; Song, Y.Z.; Ding, Q.W. Design and Evaluation of Tea-Plucking Machine for Improving Quality of Tea. *Appl. Eng. Agric.* **2019**, 35. [CrossRef]
5. Wu, C.C. Developing Situation of Tea Harvesting Machines in Taiwan. *Eng. Technol. Appl. Sci. Res.* **2015**, *5*, 871–875. [CrossRef]
6. Du, Z.; Hu, Y.; Wang, S. Simulation and Experiment of Reciprocating Cutter Kinematic of Portable Tea Picking Machine. *Nongye Jixie Xuebao/Trans. Chin. Soc. Agric. Mach.* **2018**, *49*. [CrossRef]
7. Chen, J.; Chen, Y.; Jin, X.; Che, J.; Gao, F.; Li, N. Research on a Parallel Robot for Tea Flushes Plucking. In Proceedings of the 2015 International Conference on Education, Management, Information and Medicine, Shenzhen, China, 24–26 April 2015. [CrossRef]
8. Bakhshipour, A.; Sanaeifar, A.; Payman, S.H.; de la Guardia, M. Evaluation of Data Mining Strategies for Classification of Black Tea Based on Image-Based Features. *Food Anal. Methods* **2018**, 11. [CrossRef]

9. Zhang, L.; Zhang, H.; Chen, Y.; Dai, S.; Li, X.; Imou, K.; Liu, Z.; Li, M. Real-time monitoring of optimum timing for harvesting fresh tea leaves based on machine vision. *Int. J. Agric. Biol. Eng.* **2019**, 12. [CrossRef]
10. Chen, Y.T.; Chen, S.F. Localizing plucking points of tea leaves using deep convolutional neural networks. *Comput. Electron. Agric.* **2020**, 171. [CrossRef]
11. Yang, H.; Chen, L.; Chen, M.; Ma, Z.; Deng, F.; Li, M.; Li, X. Tender Tea Shoots Recognition and Positioning for Picking Robot Using Improved YOLO-V3 Model. *IEEE Access* **2019**, 7. [CrossRef]
12. Zou, X.; Zou, H.; Lu, J. Virtual manipulator-based binocular stereo vision positioning system and errors modelling. *Mach. Vis. Appl.* **2012**, 23. [CrossRef]
13. Bac, C.W.; Hemming, J.; Van Henten, E.J. Stem localization of sweet-pepper plants using the support wire as a visual cue. *Comput. Electron. Agric.* **2014**, 105. [CrossRef]
14. Font, D.; Pallejà, T.; Tresanchez, M.; Runcan, D.; Moreno, J.; Martínez, D.; Teixidó, M.; Palacín, J. A proposal for automatic fruit harvesting by combining a low cost stereovision camera and a robotic arm. *Sensors* **2014**, 14. [CrossRef] [PubMed]
15. Mehta, S.S.; Burks, T.F. Vision-based control of robotic manipulator for citrus harvesting. *Comput. Electron. Agric.* **2014**, 102. [CrossRef]
16. Wang, L.; Zhang, T.; Chu, J.; Yang, L.; Zhang, J.; Duan, Y. An efficient strawberry harvesting end-effector with large misalignment tolerance. *Nongye Jixie Xuebao/Trans. Chin. Soc. Agric. Mach.* **2014**, *45*. [CrossRef]
17. Ye, M.; Zou, X.; Luo, L.; Liu, N.; Mo, Y.; Chen, M.; Wang, C. Error analysis of dynamic localization tests based on binocular stereo vision on litchi harvesting manipulator. *Nongye Gongcheng Xuebao/Trans. Chin. Soc. Agric. Eng.* **2016**, *32*. [CrossRef]
18. Zou, X.; Ye, M.; Luo, C.; Xiong, J.; Luo, L.; Wang, H.; Chen, Y. Fault-tolerant design of a limited universal fruit-picking end-effector based on vision-positioning error. *Appl. Eng. Agric.* **2016**, 32. [CrossRef]
19. Xiong, Y.; From, P.J.; Isler, V. Design and Evaluation of a Novel Cable-Driven Gripper with Perception Capabilities for Strawberry Picking Robots. In Proceedings of the IEEE International Conference on Robotics and Automation, Brisbane, Australia, 21–25 May 2018. [CrossRef]
20. Qin, G.; Zhao, Y.; Xiao, H.; Jin, Y.; Xu, L. 4CZ-12 intelligent tea harvest robot design and field experiment. *J. Chin. Agric. Mech.* **2014**, 35. [CrossRef]
21. Hao, M.; Chen, Y.; Pan, Z.; Sun, Y. Development of bionic plucking finger for high-quality green tea. *Food Mach.* **2018**, 34. [CrossRef]
22. Motokura, K.; Takahashi, M.; Ewerton, M.; Peters, J. Plucking motions for tea harvesting robots using probabilistic movement primitives. *IEEE Robot. Autom Let* **2020**, 5. [CrossRef]

agriculture

MDPI

Article

Structural Optimization and Performance Evaluation of Blocking Wheel-Type Screw Fertilizer Distributor

Xiantao Zha [1,2], Guozhong Zhang [1,2,*], Yuhang Han [1,2], Abouelnadar Elsayed Salem [1,3], Jianwei Fu [1,2] and Yong Zhou [1,2]

1 College of Engineering, Huazhong Agricultural University, Wuhan 430070, China; zhaxiantao@webmail.hzau.edu.cn (X.Z.); hanyh@webmail.hzau.edu.cn (Y.H.); abouelnadar@webmail.hzau.edu.cn (A.E.S.); fjwtap@mail.hzau.edu.cn (J.F.); zhyong@mail.hzau.edu.cn (Y.Z.)

2 Key Laboratory of Agricultural Equipment in Mid-Lower Yangtze River, Ministry of Agriculture and Rural Affairs, Wuhan 430070, China

3 Soil Conservation Department, Desert Research Center, Matraia 11753, Egypt

* Correspondence: zhanggz@mail.hzau.edu.cn

Abstract: In order to solve the problem where the amount of screw fertilizer distributor can only be adjusted by rotating speed and poor fertilization uniformity at low rotational speeds, a blocking wheel-type screw fertilizer distributor was designed. Single factor and $L_9(3^4)$ orthogonal simulation tests based on EDEM software were carried out to optimize the distributor variables at a speed of 20 r/min. The bench verification test was built under the same conditions as the simulation tests to verify the results of the simulation. Finally, the bench performance tests were carried out to evaluate distributor performance. The results of simulation tests revealed that the minimum coefficient of variation of fertilization uniformity (CVFU) was 19.27%, with the structural parameter combination of the inner diameter (17 mm), pitch (45 mm), outlet distance (40 mm), and number of screw heads (1). The verification test results showed that the changing trend and values of the CVFU were almost the same as the simulation tests. The results of the performance test revealed that when the opening width of the blocking wheel was 10–30 mm and the rotation speed was 20–60 r/min, the amount of fertilizer per lap (FAPL) was in the range of 27.74–38.15 g/r; the maximum CVFU and the coefficient of variation of fertilization stability (CVFS) were 29.43% and 2.18%, respectively, which met the requirements of the industry standard. This research provides a good reference for optimizing the screw fertilizer distribution and for researchers in the field of precision fertilization.

Keywords: screw fertilizer distributor; blocking wheel; discrete element method (DEM); fertilizing performance; fertilization uniformity

Citation: Zha, X.; Zhang, G.; Han, Y.; Salem, A.E.; Fu, J.; Zhou, Y. Structural Optimization and Performance Evaluation of Blocking Wheel-Type Screw Fertilizer Distributor. *Agriculture* **2021**, *11*, 248. http://doi.org/10.3390/agriculture11030248

Academic Editor: José Pérez-Alonso

Received: 10 February 2021
Accepted: 11 March 2021
Published: 14 March 2021

1. Introduction

Mechanized precision fertilization is an effective way to improve the fertilizer utilization rate and increase cost savings and efficiency during farming [1–4]. In order to solve the impact of the variable forward speed of machines on fertilization accuracy, researchers have developed machines using rotary encoders and Global Position System (GPS) to measure the forward speed of the machines and adjust the fertilizer quantity rate to the forward speed [5–7]. To meet the requirements of the different fertilizer amounts caused by different soil fertilities, Meng et al. [8,9] and Jeong et al. [10] carried out a map-based variable rate fertilizer application system. They used GPS to obtain the position information and ground wheels to obtain the forward speed of the machine. Then, the fertilization rate would be adjusted according to the forward speed, position information, and pre-stored prescription map. Kim et al. [11], Chen et al. [12], and Cho et al. [13] used sensors to obtain crop growth status or soil organic matter information to determine the fertilizer requirement. The fertilization rate was then be adjusted according to the fertilization requirements and the forward speed of the machine. To improve the fertilizer discharging precision of the fertilizer distributor, Yuan et al. [14] used a controller based on Advanced RISC Machines/Digital Signal Processor (ARM/DSP) to control the electric push rod to change the effective working length of the groove wheel fertilizer distributor and to control the speed of the direct-current (DC) motor to adjust the speed of the screw shaft. The research referred to above is primarily concerned with speed control of the fertilizer distributor and coordination with forward speed and other decision-making information. However, the distributors were the basic components of the machines, which also had a significant impact on the performance of the machines.

In recent years, researchers have continued to innovate and optimize the structure of fertilizer distributor to improve their fertilization performance [15–18]. For example, to improve the stability and uniformity of fertilizer distributor, researchers optimized the structure of groove wheel [19], fertilizer tongue [20], and fertilizer guide devices [21]. Nevertheless, this type of distributor was only suitable for granular fertilizers with good fluidity and dryness, so its application has strong limitations. Because of its strong adaptability to fertilizers with different fluidity, the screw fertilizer distributor has received more attention. Nukeshev et al. [22] added a segment reflector and determined the screw diameter, the number of screw heads, and the number of ripples of reflector to screw distributor to solve the problem of unevenly discharging fertilizers. Chen et al. [23] used different kinds of granular fertilizer and powder fertilizer to evaluate the performance of double-level horizontal screw fertilizer distributor, which confirmed that the screw fertilizer distributor had high fertilization stability and adaptability to fertilizers with different fluidity. Cao [24] optimized the structural parameters of screw fertilizer distributor and analyzed the influence of screw speed on fertilizer distributing performance by DEM and bench test, determining the optimal screw structure parameters and working speed. Dong [25] studied the movement law of fertilizer in the vertical screw fertilizer distributor utilizing theoretical analysis, numerical simulation, and tests, determined its optimal structural parameters and working parameters, and proved that the vertical screw fertilizer distributor had good fertilization stability and uniform rotating speed. In particular, the above research examined the adaptability of the structure and working parameters of the screw fertilizer distributor of different fertilizers. The analysis of the influence of law and mechanism on the structure and working parameters of the screw fertilizer distributor on the fertilization uniformity was not sufficient. The fertilizer amount of existing screw fertilizer distributors can be only adjusted by the screw shaft rotation speed, and the uniformity of fertilizer discharge is poor when the rotation speed is low, so it is difficult to meet the requirements of precision fertilizer application.

Regarding the above-mentioned problems, this paper has designed a blocking wheel-type screw fertilizer distributor. (1) Structural parameters of the distributor were optimized by single factor and $L_9(3^4)$ orthogonal simulation tests at a low speed of 20 r/min based on EDEM software. (2) Bench verification tests under the same conditions as the simulation tests were carried out to verify the simulation results. (3) The performance of the distributor was evaluated by bench performance tests. This research provides a useful reference for the optimization of the screw fertilizer distributor and the research of precision fertilization technology.

2. Materials and Methods

2.1. Structural Design

The structure of the blocking wheel-type screw fertilizer distributor is shown in Figure 1. It consisted of a screw fertilizer transporting axis, a sleeve of screw, a pyramidal feeding inlet, a spiral blocking wheel, and a fertilizer outlet. In the figure, D is the outer diameter of the screw blade, d is the inner diameter of the screw blade, P_t is the pitch, b is the average thickness of the screw blade, K is the spiral blocking wheel opening width, and S is the outlet distance (short for distance between the end of screw blade and outlet).

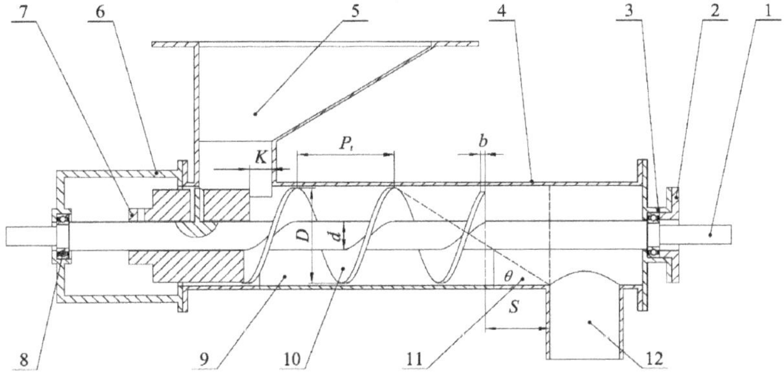

Figure 1. Structure of blocking wheel-type screw fertilizer distributor. 1. Screw shaft 2. Bearing cover 3. End cover of sleeve 4. Sleeve of screw 5. Pyramidal feeding inlet 6. Bearing seat 7. Spiral blocking wheel 8. Bearing 9. Fertilizer cleaning outlet and its cover 10. Screw blade 11. Fertilizer buffering zone 12. Fertilizer outlet. Note: D is the outer diameter of the screw blade, mm; d is the inner diameter of the screw blade, mm; P_t is the pitch, mm; b is the average thickness of the screw blade, mm; K is the spiral blocking wheel opening width, mm; S is the outlet distance, mm.

2.2. Operation Principle

When working, the fertilizer particles fill the beginning screw section through the feeding inlet under the action of their own gravity. With the rotation of the screw shaft, fertilizers in the beginning section of the screw are moved to the end of the screw under the action of the screw blade. Then they are discharged from the fertilizer outlet through the fertilizer buffer zone between the end of the screw blade and the fertilizer outlet.

Compared with the existing screw fertilizer distributor, a fertilizer buffer zone is set between the end of the screw blade and the fertilizer outlet, so that the fertilizer transported to the end of the distributor can be slowly discharged from the fertilizer outlet, which can improve the fertilization uniformity. The blocking wheel can be used to change the fertilizer amount per lap (FAPL) of the screw fertilizer distributor. As a result of the low target fertilization rate, the problem of low uniformity of fertilizer distributor can be resolved by reducing the fertilizer amount per lap and increasing the working speed of the screw shaft.

2.3. Force Analysis of Fertilizer Particles

The fertilizer particles in the fertilizer distributor are subjected to the combined effect of gravity G, the friction force f_w generated by the outer shell wall of the fertilizer distributor, and the force of the screw blade on the fertilizer particles.

The force of the screw blade on the fertilizer particles is shown in Figure 2; the particles are subjected to the tangential friction force f_s and the normal pressure F_N. Their resultant force is F. Force F can be decomposed into an axial force P_1 and a circumferential force P_2. The axial force P_1 makes the fertilizer move to the outlet axially, and the circumferential force P_2 makes the fertilizer particles move in a circumferential direction. However, gravity G and friction force f_w overcomes this movement to make the fertilizer particles slide axially in the fertilizer distributor.

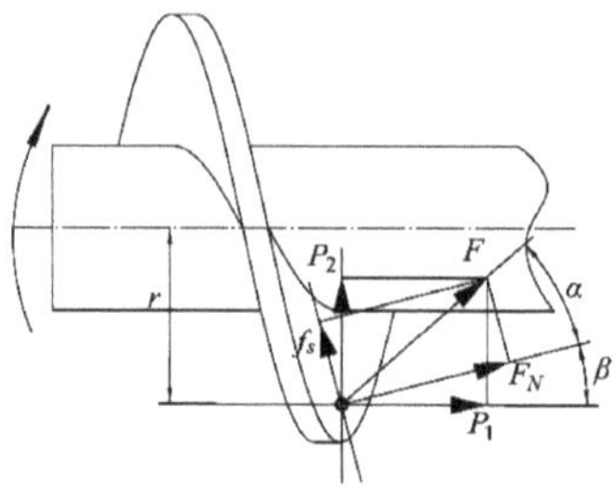

Figure 2. Force of screw blade on fertilizer particles. Note: F is the total force of screw blade acting on fertilizer particles, N; F_N is the normal pressure of screw blade on fertilizer particles, N; f_s is the friction force of screw blade on fertilizer particles, N; P_1 is the axial component of F, N; P_2 is the circumferential component of F, N; α is the friction angle between blade and fertilizer, (°); β is the lift angle of screw blade, (°); r is the distance between particles and axis of screw shaft, mm.

According to Figure 2, the force F is decomposed by Equation (1):

$$\begin{cases} P_1 = F\cos(\alpha+\beta) \\ P_2 = F\sin(\alpha+\beta) \\ \alpha = \tan^{-1}\mu \\ \beta = \tan^{-1}\frac{P_t}{2\pi r} \end{cases} \tag{1}$$

where F is the total force of screw blade acting on fertilizer particles, N; P_1 is the axial component of F, N; P_2 is the circumferential component of F, N; α is the friction angle between blade and fertilizer, (°); β is the lift angle of screw blade, (°); μ is the friction coefficient between fertilizer particles and screw blade; P_t is the pitch of screw blade, mm; r is the distance between particles and axis of screw shaft, mm.

2.4. *Key Performance Parameters of Screw Fertilizer Distributor*

2.4.1. Fertilizer Amount Per Lap (FAPL)

The FAPL is an important parameter to scale the fertilizer discharge capacity of the fertilizer distributor. It is generally believed that under ideal conditions, fertilizer does not move in a circle with the fertilizer screw, but only moves horizontally in the axial direction. Then the FAPL can be calculated by Equation (2) [21]:

$$q = \left[\pi\left(D^2 - d^2\right)P_t/4 - ZbHL_p\right]\rho\varphi \times 10^{-3} \tag{2}$$

where q is FAPL, g; D is outer diameter of screw blade, mm; d is inner diameter of screw blade, mm; P_t is pitch of screw blade, mm; Z is the number of screw heads; b is the average thickness of screw teeth, mm; H is the depth of screw teeth, $H = \frac{D-d}{2}$, mm; L_p is the average screw length of single circle, $L_p = \sqrt{\left[\frac{\pi(D+d)}{2}\right]^2 + t^2}$, mm; ρ is the density of fertilizer, g/cm^3; φ is the fertilizer filling rate of the fertilizer screw.

According to Equation (2), the main parameters affecting FAPL of the screw fertilizer distributor are the outer diameter of screw blade D, inner diameter of screw blade d, pitch P_t, and filling rate of the screw φ. Among them, the filling rate φ is the ratio of the volume of fertilizer pushed by the blade to the volume formed by the blade for each revolution of the screw, which is mainly related to the material characteristics of fertilizer (such as particle shape, particle size, particle size distribution, and fluidity), blocking wheel opening width, vibration of the distributor, rotation speed of screw shaft, and other factors. When designing the structure parameters of the screw distributor according to FAPL, the outer diameter of the screw blade D, inner diameter of the screw blade d, and pitch of the blade P_t should be mainly considered.

In actual operation, when the fertilizer exhauster works stably, the equation of FAPL is as follows:

$$q = \frac{60 \cdot \Delta m}{n \cdot \Delta t} \tag{3}$$

where q is FAPL, g/r; Δt is the time interval between the beginning and the end of a measurement, s; Δm is the total mass of discharged fertilizer during the period of Δt, g; n is the rotation speed of fertilizer screw, r/min.

2.4.2. Coefficient of Variation of Fertilization Uniformity and Stability (CVFU and CVFS)

CVFU and CVFU are important indices for assessing the uniformity and stability of the fertilizer distributor. Their test method in this research followed the China Industry Standard NY/T 1003-2006 (Technical specification of quality evaluation for fertilization machinery) [26].

The CVFU should be tested according to the following steps. First, ensure that the fertilization machine passes through a working length of more than 10 m at normal operating speed smoothly. Then take a length of 3 m from it, where the machine is stably forward. Thirdly, divide the 3 m length of fertilizers into 30 sections continuously. Fourth, the fertilizer quantity of each section should be weighed with an electronic balance (measurement accuracy is not less than 0.1 g). Finally, the CVFU could be calculated by Equation (4):

$$\begin{cases} m = \frac{1}{a}\sum\limits_{i=1}^{a} m_i \\ S = \sqrt{\frac{1}{a-1}\sum\limits_{i=1}^{a}(m_i - m)^2} \\ V = \frac{S}{m} \times 100 \end{cases} \tag{4}$$

where m_i is the quantity of fertilizer in the i-th section (i = 1, 2, 3, ... , 30), g; m is the average value of fertilizer quantity of the 30 section, g; a is the total number of the sections, which is equal to 30 when measuring the CVFU; S is the standard deviation of fertilizer quantity, g; V is the CVFU, %.

The CVFS is also calculated by Equation (4), but the testing method is not the same. It should be tested as follows. Firstly, take the fertilizer distributed by the distributor within a period (not less than 2 min). Secondly, weigh the fertilizer quantity with an electronic balance (measurement accuracy is not less than 0.5 g). Thirdly, repeat the above two steps five times, and each test should last for a same period. Then, m_i is the quantity of fertilizer in the i-th test (i = 1, 2, 3, 4, 5), g; m is the average value of fertilizer quantity of the 5 tests, g; a is the total number of the tests, which is set to 5 when measuring the CVFS; S is the standard deviation of fertilizer quantity, g; V is the CVFS, %.

The equation for the calculation of CVFU and CVFS is the same. However, according to the standard (NY/T 1003–2006), the number of samples required for calculating CVFU is 30, and the fertilization distance of each sample is 0.1 m. Thus, CVFU reflects the uniformity and stability of the fertilization quantity over a short time and distance. On the contrary, the number of samples required for calculating the CVFS is 5, and the measurement time of each sample is not less than 2 min (if the machine moves forward at a speed of 0.5 m/s, the fertilization distance is 60 m within 2 min). The CVFS therefore reflects the uniformity and stability of the fertilization quantity over a long time and distance.

2.5. Methods of DEM Simulation Tests

In order to facilitate the research, this paper has fixed the outer diameter of the screw blade to D = 45 mm and analyzed the influence of the parameters (inner diameter of the screw blade d, pitch P_t, number of screw heads Z, outlet distance S, and the opening width of the locking wheel K) on FAPL and CVFU. All the simulations were carried out at a low screw shaft rotation speed of 20 r/min.

2.5.1. Simulation Model and Parameter Setting

Studies have shown that the particles of compound fertilizer are mostly spherical or ellipsoidal discrete elements, with high sphericity and no adhesion between the particles [27–29], so this paper chose the "Hertz-Mindlin (no slip)" contact model to simulate the composite fertilizer particles. Referring to the existing research, the average particle size fertilizer particles d_s was set to 3.3 mm, the generation method of the particle size range as "random", with a setting range of 0.75–1.25. Then, the diameter of generated particles was 2.475–4.125 mm [28,29]. The other simulation parameters are shown in Table 1.

Table 1. Simulation parameters.

Parameters	Value
Poisson's ratio of fertilizer	0.28
Density of fertilizer (kg/m^3)	1511
Shear modulus of fertilizer (Pa)	1×10^7
Poisson's ratio of ABS	0.394
Density of ABS (kg/m^3)	1060
Shear modulus of ABS (Pa)	8.9×10^8
Coefficient of Restitution between fertilizers	0.35
Coefficient of Static Friction between fertilizers	0.4774
Coefficient of Rolling friction between fertilizers	0.21
Coefficient of restitution between fertilizer and ABS	0.359
Coefficient of Static Friction between fertilizer and ABS	0.1962
Coefficient of restitution between fertilizer and ABS	0.15

2.5.2. Calculation of CVFU in Simulation Tests

The test method in Section 2.4.2 made the simulation domain too large, which led to low efficiency of the simulation test. To increase the simulation speed, this paper carried out the simulation test according to the following method.

In the preparation step, a 3D model of the fertilizer distributor with a needed blocking wheel width K was built in Solidworks software.

During the simulation model set-up step, the 3D model was imported into EDEM and the rotation speed of screw shaft was set to 20 r/min. Other simulation parameters were set referring to Table 1. Then a particle factory was built to add 1 kg of fertilizer particles to the fertilizer box, and all the particles were generated within 1 s. The fertilizer screw was rotated after the particles in the fertilizer box were steady.

After the simulation was done, the FAPL of the screw fertilizer distributor could be calculated as follows. Firstly, a total mass sensor for remained fertilizer was built as shown in Figure 3. Secondly, after the fertilizer had been distributed steadily, the timeline was two different times and two total mass values and were obtained by the sensor, then the FAPL could be calculated with Equation (3).

The CVFU of the screw fertilizer distributor could be calculated as follows. Firstly, a cylindrical mass sensor for the fertilizers was built at the fertilizer outlet as shown in Figure 3. Secondly, the mass of the fertilizer discharged from the fertilizer outlet was obtained by the sensor for every 0.2 s. Thirdly, a continuous 6 s with 30 mass values could be selected after the fertilizer was being discharged steadily, and Equation (4) could be used to calculate the CVFU. This testing method was equivalent to the distributor passing a 3 m discharging length at a steady forward speed of 0.5 m/s.

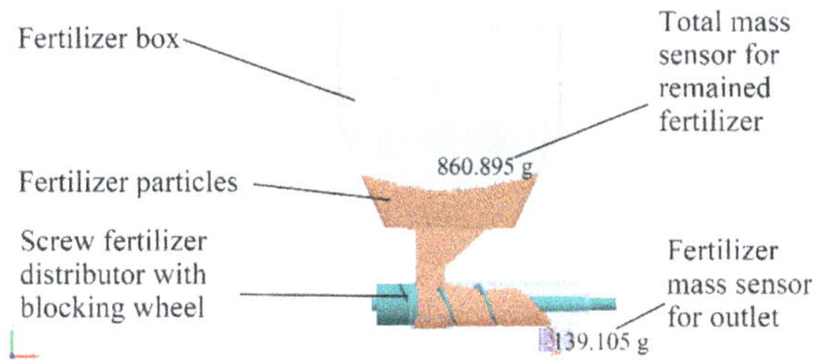

Figure 3. Simulation model and mass sensors.

2.5.3. Single Factor Tests Method

According to the design requirements of the horizontal screw conveyor, the relationship between screw pitch and its outer diameter was as follows:

$$P_t = k \cdot D \tag{5}$$

where P_t is the pitch, mm; D is the outer diameter of screw blade, mm; k is the pitch coefficient, usually k is 0.5–1.

Therefore, the range of P_t was 22.5–45 mm calculated by Equation (5). As a result, the pitch was set to 25, 35 and 45 mm, respectively.

When the fertilizer distributor is in a static state, to ensure the fertilizer particles are fully accumulated in the fertilizer distributor, the outlet distance S should be calculated as in Equation (6):

$$\begin{cases} S = r - P_t \\ r = \frac{D}{\sin\theta} \end{cases} \tag{6}$$

where S is the outlet distance, mm; P_t is the pitch, mm; r is the stacking radius, mm; D is the outer diameter of screw blade, mm; θ is the fertilizer stacking angle, °.

The stacking angle of commonly used fertilizer is about 30° [30]. If the stacking height is approximately equal to outer diameter D = 45 mm, then the outlet distance S should be about 33 mm to ensure the full accumulation of fertilizer particles under static conditions when the pitch P_t is 45 m. Considering the influence on the fertilizer accumulation of the rotation of the screw shaft and the vibration of the fertilizer distributor, the outlet distance S was set to a maximum value of 40 mm in the single simulation factor test.

Only one factor was changed for each test group, while the other factors were fixed, as shown in Table 2. The structure of the screw shaft with different number of screw heads is shown in Figure 4, and their pitch was 45 mm.

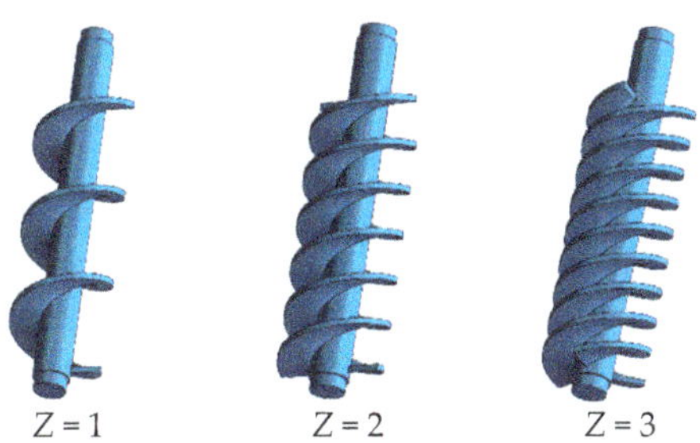

Figure 4. Screw shaft with different number of heads.

Table 2. Single factor simulation test table.

Test Group	Inner Diameter of Screw Blade d/mm	Pitch of Screw P_t/mm	Number of Screw Heads Z	Outlet Distance S/mm	Blocking Wheel Opening Width K/mm
1	17, 24, 31	45	1	0	30
2	17	25, 35, 45	1	0	30
3	17	45	1, 2, 3	0	30
4	17	45	1	0, 20, 40	30
5	17	45	1	40	10, 15, 20, 25, 30

2.5.4. Orthogonal Simulation Tests Method

To further clarify the effect of outlet distance S, pitch P_t, and the blocking wheel opening width K on fertilization uniformity of the screw fertilizer distributor, orthogonal simulation tests of the above three factors were carried out. In the tests, the outer diameter of screw blade D was set to 45 mm, the inner diameter of screw blade d was set to 17 mm, the number of screw heads was set 1, and the rotation speed of screw shaft n was set to 20 r/min. The influence factors of the orthogonal test are shown in Table 3.

Table 3. Factors and levels of influence of the orthogonal test.

Level	Blocking Wheel Opening Width K/mm	Pitch of Screw P_t/mm	Outlet Distance S/mm
1	30	25	0
2	20	35	20
3	10	45	40

The CVFU was used as an evaluating index. Table $L_9(3^4)$ was selected for the orthogonal test, and Minitab was used to deal with the results of the orthogonal test, test statistical assumptions, and variance analysis. The experimental arrangement and results are shown in Table 4.

Table 4. Orthogonal test results.

Test NO.	Factors			CVFU CV/%
	K/mm	P_t/mm	S/mm	
1	1	1	1	53.41
2	1	2	2	41.13
3	1	3	3	19.27
4	2	1	2	34.49
5	2	2	3	28.10
6	2	3	1	38.72
7	3	1	3	37.60
8	3	2	1	47.65
9	3	3	2	29.95
k_1	37.94	41.83	46.59	
k_2	33.77	38.96	35.19	
k_3	38.40	29.31	28.32	
R	4.63	12.52	18.27	
Order	$S > P_t > K$			

2.6. Method and Metarial of Bench Tests

2.6.1. Test Bench and Fertilizer

According to the optimal structural parameters obtained from the simulation test results (D = 45 mm, d = 17 mm, P_t = 45 mm, S = 40 mm), the FDM rapid prototyping technology was used to trial-produce the distributor, and the test bench shown in Figure 5 was built in the Engineering Training Base in Huazhong Agricultural University, China. The bench consisted of a blocking wheel-type screw fertilizer distributor, a driving moto (12V 60W), a moto driver (based on L298N), a photoelectric sensor (based on LM393 and H2010 Optocoupler), a Raspberry Pi 3B, 5V and 12V batteries.

Figure 5. Testing bench.

The fertilizer used in bench test was Xiangyun high-sulfur-based compound fertilizer ($N-P_2O_5-K_2O$ content was 15%-15%-15%, bulk density was 1.20 g/cm^3 and moisture content was about 7.35%).

2.6.2. Verification Test

In order to verify the results of the simulation tests, the CVFU of the optimized distributor was tested under the same conditions of the simulation tests. In the test, the rotation speed of screw shaft was set at 20 r/min, while the forward speed of the conveyor belt was set to 0.5 m/s. The blocking wheel opening width K was set to 10, 20, and 30 mm, respectively, and each test was repeated three times. After carrying out the test, the CVFUs of the fertilizer distributor were calculated according to the method in Section 2.4.2.

2.6.3. Performance Tests of the Screw Fertilizer Distributor

When evaluating the performance of the optimized screw fertilizer distributor, the CVFS and the FAPL were selected as the evaluating indexes, and blocking wheel opening width K (which was set to 10, 20, 30 mm) and rotating speed of the screw shaft (which was set to 20, 30, 40, 50, 60 r/min) were taken as the influence factors.

In the test, the distributor was operated according to the above specified parameters first. After the fertilizer particles were discharged steadily, a collecting bucket was placed under the fertilizer outlet. When the fertilizer was collected for 2 min, the bucket was removed. Then the total mass of the collected fertilizer was weighted by an electronic balance. Each step was repeated for five times. After the tests, FAPL and CVFS were calculated according to the method in Sections 2.4.1 and 2.4.2, respectively.

2.7. Field Tests

Five screw fertilizer distributors were installed to a DAIDO DXZ830 rice seeding machine (10 rice seeding rows, with a working width of 2.55 m), as shown in Figure 6. Each screw fertilizer distributor corresponded to two rice seeding rows. The distributors were driven by a motor (12 V 100 W), and the motor speed controlling system was the same as the [31]. The speed of screw shaft could be synchronized with that of the (power-take-off) PTO at a set transmission ratio (PTO: screw) by the controlling system, which was set according to the target fertilizer amount of per unit area and the FAPL. In this research, it was set to 11.7:1 (calculated with a target fertilizer amount of per unit area of 400 kg/hm^2 and a FAPL of 39 g/r).

Figure 6. Screw fertilizer distributors with DAIDO DXZ830 rice seeding machine.

The fertilizer used in field test was Guihu compound fertilizer (N-P_2O_5-K_2O content was 15%-15%-15%, bulk density was 1.11 g/cm^3, and moisture content was about 3.71%).

The paddy field was divided into six experiment plots, each plot had a length of 40 m and a width of 2.55 m. The field test was repeated three times, each test fertilized in two plots. After the test, the CVFS, variation coefficient of the fertilization consistency in different rows (CVODR), and the relative error of fertilization amount were measured.

The CVODR was also calculated by Equation (4), where m_i is the total quantity of the i-th distributor during the test (i = 1, 2, 3, 4, 5), g; m is the average value of fertilizer quantity of the 5 distributors, g; a is the total number of the distributors; S is the standard deviation of fertilizer quantity of the 5 distributors, g; V is the CVODR, %.

The relative error of fertilization amount was calculated according to Equation (7)

$$\gamma_s = [10(W_0 - W_1)/A - T]/\mathrm{T} \times 100 \tag{7}$$

where γ_s is the relative error of fertilization amount, %; W_0 is the fertilizer amount added into fertilizer box before the test, g; W_1 is the fertilizer amount remaining in the fertilizer box after the test, g; A is the working area; m^2; T is the target fertilizer amount of per unit area, kg/hm^2.

3. Results

3.1. The Single Factor Tests

3.1.1. Inner Diameter of Screw Blade

The simulation test results showed that the FAPL of the screw fertilizer distributor was 39.37, 32.86 and 24.05 g/r, and the CVFU was 36.65%, 40.44% and 38.77% respectively, for the three different inner diameters of 17, 24, and 31 mm. When other factors were fixed, the FAPL of the distributor decreased significantly, and CVFU did change within the error tolerance with the increase of the inner diameter of the screw blade.

3.1.2. Pitch of the Screw

When the pitch was set to 25, 35, and 45 mm, the results showed that the FAPL was 20.08, 29.32, and 37.41 g/r and the CVFU was 53.41%, 47.27%, and 36.65%, respectively. The results indicated that by keeping the screw diameter, the outlet distance, the rotation speed, and the blocking wheel opening width constant and increasing pitch within the range of 25–45 mm, the FAPL of the screw fertilizer distributor increased, but the CVFU decreased. Therefore, when the outer diameter of the screw blade is 45 mm, the pitch should be set to 45 mm, which could improve the fertilization uniformity of the screw fertilizer distributor.

3.1.3. Number of Screw Heads

The results showed that when the number of screw heads was 1, 2, and 3, respectively; the corresponding FAPL of fertilizer distributor was 39.37, 32.62 and 23.48 g/r, and the CVFU was 36.65%, 53.83%, and 62.45%. The number of screw heads had a significant effect on both the FAPL and the CVFU. With the increase of the number of screw heads, the FAPL decreased while the CVFU increased. The results showed that the best number of screw heads was $Z = 1$ under the parameter setting of this study.

3.1.4. Outlet Distance

The simulation results showed that when the outlet distance S was 0, 20, and 40 mm, respectively. The corresponding FAPL was 39.37, 38.34, 37.41 g/r, and the CVFU was 36.65%, 26.94%, and 19.27%, respectively. With the increase of the outlet distances, the change of FAPL was not obvious, but the fertilization uniformity was significantly improved. Therefore, the optimal fertilizer outlet distance is 40 mm.

3.1.5. Blocking Wheel Opening Width

Figure 7a showed the influence of the blocking wheel opening width K on the FAPL within the range of the blocking wheel opening width K of 10–30 mm. There was a good positive linear relationship between the FAPL and the blocking wheel opening width K. The regression equation was $q = 0.48K + 24.38$, and the coefficient of determination (R^2) was 0.9557.

Figure 7b showed that when the range of the blocking wheel opening width K was set to 10–30 mm, the CVFU decreased with the increase of the blocking wheel opening width. However, the linear fitting degree was not very high, and the coefficient of determination (R^2) was 0.8932.

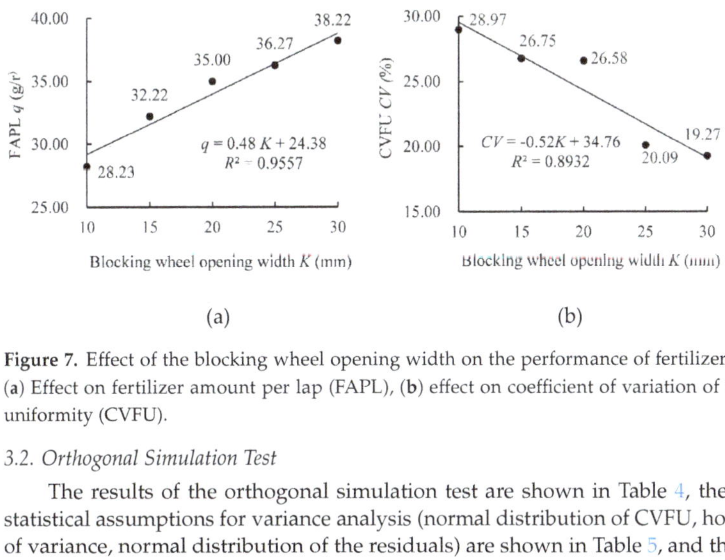

Figure 7. Effect of the blocking wheel opening width on the performance of fertilizer distributor. (**a**) Effect on fertilizer amount per lap (FAPL), (**b**) effect on coefficient of variation of fertilization uniformity (CVFU).

3.2. Orthogonal Simulation Test

The results of the orthogonal simulation test are shown in Table 4, the results of statistical assumptions for variance analysis (normal distribution of CVFU, homogeneity of variance, normal distribution of the residuals) are shown in Table 5, and the variance analysis results are shown in Table 6.

Table 5. Statistical assumptions for variance analysis.

Item		Test Method	p-Value
Normal distribution of CV		Anderson-Darling test	0.973
Homogeneity of variance	K and CV	Bartlett's Test	0.344
	P_t and CV		0.999
	S and CV		0.828
Normal distribution of the residuals		Anderson-Darling test	0.792

Table 6. Variance analysis performed on the orthogonal test results of CVFU (CV%).

Source	SS	df	MS	F Value	Significant
P_t	258.06	2	129.03	6.23	*
S	510.98	2	255.49	12.34	**
K } e^{Δ}	39.01 } 82.81	2 } 4	19.51		
e	43.80	2			
Sum	851.85	8			

Critical value of F-test: $F_{0.1}$ (2,4) = 4.32; $F_{0.05}$ (2,4) = 6.94. * means the factor is significant when α = 0.1, ** means the factor is significant when α = 0.05.

The results of statistical assumptions for variance analysis showed that the p value of all the items was higher than 0.05. The data of CVFUs were normally distributed, the variances were homogenous between factors and CVFU (K and CV, P_t and CV, S and CV), and the residuals were also normally distributed.

In the variance analysis, since $MS_K < MS_e$, when the blocking wheel opening width K was included in the error term, then the corrected error e^{Δ} was obtained.

Orthogonal simulation test and variance analysis results showed that the order of each factors on CVFU was ranked as: $S > P_t > K$. Among them, outlet distance S had an extremely significant effect on CVFU, the pitch P_t had a significant effect on CVFU, and the blocking wheel opening width K had no significant effect.

The best combination of orthogonal test was $K_xP_{t3}S_3$ at a pitch P_t of 45 mm and an outlet distance S of 40 mm. Combined with the single factor test results of the blocking wheel opening width K, when the blocking wheel opening width K was 30 mm at a rotating speed of 20 r/min, the CVFU was the smallest, which was 19.27%.

3.3. Bench Tests

3.3.1. Verification Tests

The results of verification tests showed that the corresponding CVFU was 29.43%, 20.40%, and 19.20% when the blocking wheel opening width K was 10, 20, and 30 mm, respectively. The CVFU decreased with the increase of the blocking wheel opening width K. Within the allowable error range, the simulation test results were consistent with the bench test results.

3.3.2. Performance Tests of the Screw Fertilizer Distributor

The performance testing results of the blocking wheel-type screw fertilizer distributor are shown in Tables 7 and 8 and Figure 8.

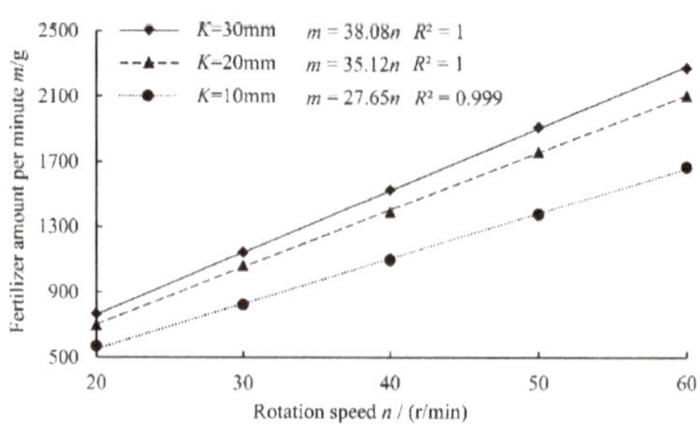

Figure 8. Effect of the rotation speed on the fertilizer amount per minute.

Table 7. Results of coefficient of variation of fertilization stability (CVFS) determination tests.

CVFS ***CV*/%**		**Blocking Wheel Opening Width** ***K*/mm**		
		30	**20**	**10**
Rotation speed n/(r/min)	20	0.81	1.15	1.24
	30	0.76	0.85	0.84
	40	1.4	0.82	1.29
	50	0.67	1.72	0.49
	60	0.47	0.24	2.18

Table 8. Results of FAPL determination tests.

FAPL q/g·r^{-1}		**Blocking Wheel Opening Width** ***K*/mm**		
		30	**20**	**10**
Rotation speed n/(r/min)	20	38.33	35.03	28.49
	30	38.11	35.45	27.44
	40	38.19	34.81	27.43
	50	38.23	35.21	27.54
	60	37.90	35.13	27.79
Average		38.15	35.13	27.74
Variation coefficient of FAPL/%		0.42	0.67	1.60

The results of the CVFS determination tests showed that within the range of the blocking wheel opening width K of 10–30 mm and the rotating speed of screw shaft of 20–60 r/min, the maximum CVFS of the screw fertilizer distributor was only 2.18%.

The results of FAPL determination tests showed that the FAPL was not significantly affected by the rotating speed when the rotating speed of the screw shaft changed within the range of 20–60 r/min. This result indicated that there was a strict linear proportional relation between the fertilizer discharging rate and the speed of screw shaft.

In the range of 10–30 mm of the blocking wheel opening width K, the adjusting range of FAPL was 27.74–38.15 g/r and the FAPL increased with the increase of the blocking wheel opening width K. It realized the design goal of adjusting the FAPL by the blocking wheel.

3.4. Field Test

The results of field tests were shown in Table 9. It showed that the maximum CVODR was 3.13%, the maximum relative error of fertilization amount was 3.95%, and the maximum CVFS of five screw distributors was 2.36%. Figure 9 showed that the fertilizers were uniformly distributed in the fertilizer ditch by the optimized screw distributor.

Figure 9. Presentation of paddy field after fertilization.

Table 9. Results of field tests.

Test NO.	Fertilization Amount of Each Screw Distributor ($W_0 - W_1$)/kg					CVODR	Relative Error
	1	2	3	4	5	CV/%	γ_s/%
1	1562.34	1587.45	1545.34	1530.98	1611.34	2.06	3.95
2	1624.15	1593.56	1538.64	1520.54	1580.12	2.66	3.71
3	1557.22	1650.97	1517.40	1555.22	1569.48	3.13	3.80
CVFS CV/%	2.36	2.18	0.95	1.16	1.37		

4. Discussion

4.1. Effect of Inner Diameter of Screw Blade on FAPL and CVFU

Figure 2 showed that when the fertilizer screw rotates, the change in the inner diameter of the screw has little effect on the stress of the fertilizer particles that have been filled into the distributer, and the filling rate will not change. The screw blade inner diameter has little impact on the CVFU, but the volume formed by the blade for each revolution of the screw will change and the volume of particles filled into the distributor will be changed, so the FAPL will be strongly affected by the screw inner diameter d.

It has been concluded that when designing a screw distributor, the inner diameter of the screw blade is a useful parameter to adjust the FAPL to the required value.

4.2. Effect of Pitch of Screw on FAPL and CVFU

According to Figure 2 and Equation (1), when the outer diameter of screw blade D was fixed, the lift angle of screw blade β increased with the increase of screw pitch P_t, the friction angle α did not change, the axial component force P_1 decreased, and the circumferential component P_2 increased as well. Within a certain reasonable range, fertilizer particles moved along the axis and the of FAPL increased. However, the screw pitch P_t increased to gravity G and the friction force f_w generated by the outer shell wall of the fertilizer distributor was not enough to overcome the trend of fertilizer particles moving in a circle with the blade. In this situation, the efficiency of fertilizer particle axial transportation decreased gradually, until they were moving completely in the circumferential direction without axial movement. It is the reason why the pitch coefficient k was assigned a maximum value of 1 in Equation (5).

In addition, when the outer diameter D was fixed, increasing the pitch P_t within a reasonable range, the angle between the horizontal plane and the cut plane of the screw blade decreased and the sliding speed of fertilizer particles from the screw surface decreased. At the same time the FAPL increased, so that in a rotating cycle, the time for fertilizer flowing out of the fertilizer outlet was prolonged and the time of no fertilizer discharging was reduced to improve the fertilization uniformity.

4.3. Effect of Number of Screw Heads on FAPL and CVFU

When the outer diameter D, inner diameter d, pitch P_t, outlet distance S, and the rotation speed n were fixed, with the increase of the number of screw heads Z there would be two main changes. On one hand, the axial force P_1 of screw blades on fertilizer particles did not change, but the circumferential force P_2 increased. On the other hand, the fertilizer transported in one cycle was discharged several times. All these changes would affect the CVFU of the fertilizer distributor.

In addition, with the increase of the number of screw heads Z, the total volume of screw blade increased and the proportion of fertilizer particles moving along the circumferential direction also increased, which led to a decrease of FAPL.

4.4. Effect of Outlet Distance on FAPL and CVFU

The outlet distance S does not affect the FAPL of a screw fertilizer distributor but affects the CVFU of it. When the outlet distance S is enough, fertilizer can be fully accumulated in the buffer zone. The friction between the fertilizer particles and the shell of the fertilizer distributor, the screw shaft, and other particles will form a repose angle θ, as shown in Figure 10. At this time, the fertilizer will move axially with the rotation of the screw shaft. When the overall transport distance of the fertilizer reaches δ, the tangential component force F_σ produced by the gravity G of fertilizer particles in the shadow area outside the fertilizer pile will be greater than the friction force f, and the fertilizer within the scope of δ area will collapse freely from the bottom and flow out of the fertilizer outlet. The steadiness of the collapse will improve the CVFU.

If outlet distance S is not enough, the accumulation of fertilizer in the buffer zone will be not sufficient, the unstable accumulation will be formed in the area outside the angle of θ, and the fertilizer will slide unsteadily, which will reduce the fertilization uniformity.

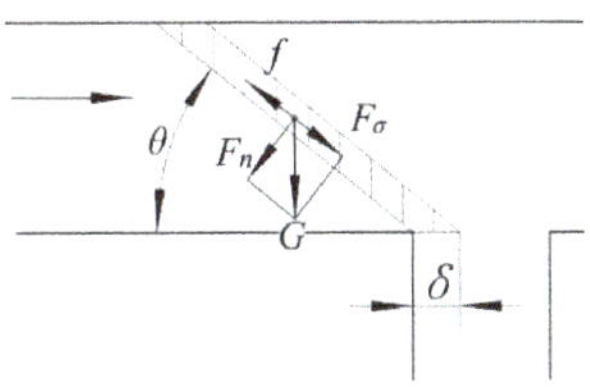

Figure 10. Effect of outlet distance on fertilizer amount uniformity.

4.5. Effect of Blocking Wheel Opening Width on FAPL and CVFU

When other parameters were kept constant, by reducing the blocking wheel opening width K the space of the filling section of the screw fertilizer distributor became smaller, therefore the fertilizer filling rate φ of the fertilizer conveying section was reduced. The FAPL of the screw fertilizer distributor q decreased according to Equation (2).

With the increasing of the blocking wheel opening width K, the fertilizer filling rate φ in the conveying section was increased, the fertilizer accumulated more fully, and the stability of fertilizer free collapsing was improved, so the fertilization uniformity was improved.

4.6. Effect of Rotation Speed of Screw Shaft on FAPL and CVFU

In theory, there is a critical speed n_0 of the screw shaft when the screw fertilizer distributor is working; that is, the highest speed at which the filling rate can maintain the maximum value when the fertilizer enters into the fertilizer distributor by gravity. If the rotation speed of the screw shaft is higher than n_0, the fertilizer filling rate will decrease with the increase of the rotation speed, which will lead to the decrease of the FAPL of the distributor. To ensure that the FAPL is less affected by the rotation speed, its working rotation speed should be less than the critical speed n_0. According to the structural characteristics of the screw fertilizer distributor, when the rotating speed is lower than the critical speed n_0, the higher the rotating speed, the higher the uniformity of fertilizer quantity. This research only optimized the fertilizer distributor at a low speed of 20 r/min to make the CVFU meet the requirements of the industry standard (NY/T 1003-2006).

5. Conclusions

(1) A blocking wheel-type screw fertilizer distributor was designed. Its working principle and the affecting force of fertilizer particles were explored in this research.

(2) Single factor and $L_9(3^3)$ orthogonal simulation tests based on EDEM software were carried out to optimize the parameters of the distributor at a low speed of 20 r/min. The results showed that when the outer diameter of the screw blade was set to D = 45 mm, the optimal parameters were: inner diameter of screw blade d = 17 mm, screw pitch P_t = 45 mm, outlet distance S = 40 mm, number of screw heads Z = 1, and the blocking wheel opening width K = 30 mm. The minimum CVFU value was 19.27% at the optimal parameters. The blocking wheel opening width K had no significant effect on CVFU but had a significant effect on FAPL.

(3) According to the optimal structure parameters obtained from the simulation test, the FDM rapid prototyping technology was used to trial-product the fertilizer distributor, and the verification bench test was carried out under the same conditions as the simulation. The results showed that the structural parameters of the distributor were the same as those obtained by EDEM.
(4) Furthermore, bench and field tests were carried out to evaluate the performance of the distributor. The bench test results showed that the range of FAPL was 27.73–38.15 g/r when the blocking wheel opening width K was 10–30 mm and the screw shaft rotation speed was 20–60 r/min. The FAPL was significantly affected by the blocking wheel opening width K, but less regulated by rotation speed of screw shaft. More importantly, the maximum CVFU and CVFS was 29.43% and 2.18%, respectively. The field test results showed that when setting the target fertilizer amount of per unit area to 400 kg/hm^2, the maximum CVODR was 3.13%, the maximum relative error of fertilization amount was 3.95%, and the maximum CVFS of five screw distributors was 2.36%. All these indexes are significantly better than the requirement in the "NY/T 1003–2006". The standard stipulated that the CVFU, CVFS, and CVODR should not be higher than 40%, 7.8% and 13%, respectively. This research solved the problems of the traditional screw fertilizer distributor only adjusting the fertilizer amount by rotating speed, and the fertilizer uniformity being poor at a low speed. It provided a meaningful reference for the optimization of the screw fertilizer distributor.

6. Patents

Two patents have been applied in China for the blocking wheel-type screw fertilizer distributor reported in this manuscript (Patent No. CN212436347U and Application No. CN111837549A).

Author Contributions: Conceptualization, X.Z. and G.Z.; methodology, X.Z.; software, X.Z.; validation, X.Z., G.Z., Y.H., A.E.S., J.F. and Y.Z.; formal analysis, X.Z.; investigation, X.Z., Y.H, A.E.S., J.F., and Y.Z.; resources, X.Z.; data curation, X.Z.; writing—original draft preparation, X.Z.; writing—review and editing, X.Z., G.Z.; visualization, X.Z.; supervision, G.Z.; project administration, G.Z.; funding acquisition, G.Z. All authors have read and agreed to the published version of the manuscript.

Funding: This work was financially supported by the Integrated Innovation of Mechanized Precision Fertilization Technology for Japonica Rice (Grant No. 2018YFD0301304-03) and Mechanization Technology Innovation of Ratoon Rice in the North of the Middle and Lower Reaches of the Yangtze River (Grant No. 2017YFD0301404-05), which are the sub-projects of the National Key Research and Development Project of China.

Institutional Review Board Statement: Not applicable.

Data Availability Statement: The data presented in this study are available on demand from the first author at (zhaxiantao@163.com).

Acknowledgments: This work was financially supported by the National Key Research and Development Program of China (Integrated Innovation of Mechanized Precision Fertilization Technology for Japonica Rice, 2018YFD0301304-03; Technological Innovation of Regenerated Rice Mechanization in the North of the Middle and Lower Reaches of the Yangtze River, 2017YFD0301404-05). We thank Mohamed Anwer for his linguistic assistance during the preparation of this manuscript. All supports and assistance are sincerely appreciated.

Conflicts of Interest: The authors declare no conflict of interest.

References

1. Wu, M.; Li, G.; Li, W.; Liu, J.; Liu, M.; Jiang, C.; Li, Z. Nitrogen fertilizer deep placement for increased grain yield and nitrogen recovery efficiency in rice grown in subtropical China. *Front. Plant Sci.* **2017**, *8*, 1227. [CrossRef] [PubMed]
2. Bandaogo, A.; Bidjokazo, F.; Youl, S.; Safo, E.; Abaidoo, R.; Andrews, O. Effect of fertilizer deep placement with urea supergranule on nitrogen use efficiency of irrigated rice in Sourou Valley (Burkina Faso). *Nutr. Cycl. Agroecosys.* **2015**, *102*, 79–89. [CrossRef]
3. Pan, S.G.; Wen, X.C.; Wang, Z.M.; Ashraf, U.; Tian, H.; Duan, M.Y.; Mo, Z.W.; Fan, P.S.; Tang, X.R. Benefits of mechanized deep placement of nitrogen fertilizer in direct-seeded rice in South China. *Field Crop. Res.* **2017**, *203*, 139–149. [CrossRef]

4. Chen, Y.Y.; Fan, P.S.; Mo, Z.W.; Kong, L.L.; Tian, H.; Duan, M.Y.; Li, L.; Wu, L.J.; Wang, Z.M.; Tang, X.R.; et al. Deep Placement of Nitrogen Fertilizer Affects Grain Yield, Nitrogen Recovery Efficiency, and Root Characteristics in Direct-Seeded Rice in South China. *J. Plant Growth Regul.* **2020**, *40*, 1–9. [CrossRef]
5. Schueller, J.K.; Wang, M. Spatially-variable fertilizer and pesticide application with GPS and DGPS. *Comput. Electron. Agric.* **1994**, *11*, 69–83. [CrossRef]
6. Fulton, J.P.; Shearer, S.A.; Chabra, G.; Higgins, S.F. Performance assessment and model development of a variable-rate, spinner-disc fertilizer applicator. *Trans. ASAE* **2001**, *44*, 1071–1081. [CrossRef]
7. Tola, E.; Kataoka, T.; Burce, M.; Okamoto, H.; Hata, S. Granular fertiliser application rate control system with integrated output volume measurement. *Biosys. Eng.* **2008**, *101*, 411–416. [CrossRef]
8. Meng, Z.; Zhao, C.; Liu, H.; Huang, W.; Fu, W.; Wang, X. Development and performance assessment of map-based variable rate granule application system. *J. Jiangsu Univ. (Nat. Sci. Ed.)* **2009**, *30*, 338–342.
9. Meng, Z.; Zhao, C.; Fu, W.; Ji, Y.; Wu, G. Prescription map identification and position lag calibration method for variable rate application of fertilizer. *Trans. Chin. Soc. Agric. Mach.* **2011**, *42*, 204–209.
10. Jeong, I.G.; Jeong, S.O.; Seong, J.H.; Lee, C.G. Development of map-based variable-rate applicator. In Proceedings of the Korean Society for Agricultural Machinery Conference, Korea, July, 2006; pp. 345–348. Available online: http://koreascience.kr/article/CFKO200636035485058.page (accessed on 30 January 2020).
11. Kim, Y.J.; Kim, H.J.; Ryu, K.H.; Rhee, J.Y. Fertiliser application performance of a variable-rate pneumatic granular applicator for rice production. *Biosys. Eng.* **2008**, *100*, 498–510. [CrossRef]
12. Chen, M.; Shi, Y.; Wang, X.; Sun, G.; Li, Y. Design and experiment of variable rate fertilizer applicator based on crop canopy spectral reflectance. *Trans. Chin. Soc. Agric. Mach.* **2015**, *46*, 26–32. [CrossRef]
13. Cho, S.I.; Choi, S.H.; Kim, Y.Y. Development of electronic mapping system for N-fertilizer dosage using real-time soil organic matter sensor and DGPS. *Biosys. Eng.* **2002**, *27*, 259–266.
14. Yuan, J.; Liu, C.L.; Li, Y.M.; Zeng, Q.B.; Zha, X.F. Gaussian processes based bivariate control parameters optimization of variable-rate granular fertilizer applicator. *Comput. Electron. Agric.* **2010**, *70*, 33–41. [CrossRef]
15. Zeng, S.; Tan, Y.P.; Wang, Y.; Luo, X.W.; Yao, L.M.; Huang, D.P.; Mo, Z.W. Structural design and parameter determination for fluted-roller fertilizer applicator. *Int. J. Agric. Biol. Eng.* **2020**, *13*, 101–110. [CrossRef]
16. Tang, H.; Jiang, Y.; Wang, J.; Wang, J.; Zhou, W. Numerical analysis and performance optimization of a spiral fertilizer distributor in side deep fertilization of a paddy field. *Proc. Inst. Mech. Eng. C J. Mech. Eng. Sci.* **2020**. [CrossRef]
17. Sun, J.F.; Chen, H.M.; Duan, J.L.; Liu, Z.; Zhu, Q.C. Mechanical properties of the grooved-wheel drilling particles under multivariate interaction influenced based on 3D printing and EDEM simulation. *Comput. Electron. Agric.* **2020**, *172*, 105329. [CrossRef]
18. Liu, X.; Ding, Y.; Shu, C.; Liu, W.; Wang, K.; Du, C.; Wang, X. Design and experiment of spiral disturbance cone centrifugal fertilizer apparatus. *Trans. Chin. Soc. Agric. Eng.* **2020**, *36*, 40–49. [CrossRef]
19. Zhang, L.; Zhang, L.; Zheng, W. Fertilizer Feeding Mechanism and Experimental Study of a Spiral Grooved-Wheel Fertilizer Feeder. *J. Eng. Technol.* **2018**, *11*, 107–115. [CrossRef]
20. Dun, G.; Yu, C.; Yang, Y.; Chen, H.; Ji, W.; Ye, J. Discharging characteristic test of outergroove wheel fertilizer and parameter optimization of fertilizer tongue by discrete element simulation. *J. Hunan Agric. Univ. Nat. Sci.* **2018**, *44*, 661–665.
21. Ding, H.; An, X.; Wu, G.; Li, L.; Zhu, Q. Optimization and Simulation of Fertilizer Guide Device Parameters Based on EDEM Software. In Proceedings of the International Conference on Computer and Computing Technologies in Agriculture, Jilin, China, 12–15 August 2017; pp. 377–388.
22. Nukeshev, S.; Romanyuk, N.; Yeskhozhin, K.; Zhunusova, A. Methods and results of the research of the fertilizer distributor of cutter-ridger-fertilizer. *H. Sci. S. Seifullin Kazakh Agro. Tech. Univ.* **2016**, *3*, 138–145.
23. Chen, X.; Luo, X.; Wang, Z.; Zhang, M.; Hu, L.; Yang, W.; Zeng, S.; Zang, Y.; Wei, H.; Zheng, L. Design and experiment of fertilizer distribution apparatus with double-level screws. *Trans. Chin. Soc. Agric. Eng.* **2015**, *31*, 10–16. [CrossRef]
24. Cao, Y. Simulation Analysis and Experimental Study of Screw Fertilizer Distributor Based on EDEM. Master's Thesis, Jilin Agricultural University, Changchun, China, 2017.
25. Dong, X. Mechanism and Parameters of Vertical Spiral Rice Deep-Side Fertilizing Device. Ph'D Thesis, Heilongjiang Bayi Agricultural University, Daqing, China, 2019.
26. *Technical Specification for Quality Evaluation of Fertilization Machinery*; Ministry of Agriculture of the People's Republic of China: Beijing, China, 2006; NY/T 1003-2006.
27. Liu, Z.; Wang, Q.; Li, H.; He, J.; Lu, C.; Yu, C. Fertilizer injecting route analysis and test for air-blowing seed-fertilizer hole-applicator via CFD-DEM coupling. *Trans. Chin. Soc. Agric. Eng.* **2019**, *35*, 18–25. [CrossRef]
28. Shi, Y.Y.; Chen, M.; Wang, X.C.; Odhiambo, M.O.; Ding, W.M. Numerical simulation of spreading performance and distribution pattern of centrifugal variable-rate fertilizer applicator based on DEM software. *Comput. Electron. Agric.* **2018**, *144*, 249–259. [CrossRef]
29. Liu, J.; Yuan, C.; Liu, J.; Li, Y.; Zhou, C.; Gu, Z. ARM and DSP-based Bivariable Fertilizing Control System Design and Implementation. *Trans. Chin. Soc. Agric. Mach.* **2010**, *41*, 233–238.
30. Wu, M. The design of Rice Side Deep Fertilizing Device. Master's Thesis, Zhejiang Sci-Tech University, Hangzhou, China, 2017.
31. Zha, X.; Zhang, G.; Zhang, S.; Hou, Q.; Wang, Y.; Zhou, Y. Design and experiment of centralized pneumatic deep precision fertilization device for rice transplanter. *Int. J. Agric. Biol. Eng.* **2020**, *13*, 109–117. [CrossRef]

agriculture

MDPI

Article

Design, DEM Simulation, and Field Experiments of a Novel Precision Seeder for Dry Direct-Seeded Rice with Film Mulching

Hui Li [1,2], Shan Zeng [1], Xiwen Luo [1,2], Longyu Fang [2], Zhanhao Liang [1] and Wenwu Yang [1,*]

1 Key Laboratory of Key Technology on Agricultural Machine and Equipment, Ministry of Education, South China Agricultural University, Guangzhou 510642, China; lihui17@mails.jlu.edu.cn (H.L.); shanzeng@scau.edu.cn (S.Z.); xwluo@scau.edu.cn (X.L.); lzkou@stu.scau.edu.cn (Z.L.)
2 College of Biological and Agricultural Engineering, Jilin University, Changchun 130022, China; fangly20@mails.jlu.edu.cn
* Correspondence: yangwenwu@scau.edu.cn

Abstract: Existing devices for dry direct-seeded rice with film mulching in northern China have limitations such as imprecise sowing, unadjustable sowing depth, and seeding device blocking. In this regard, this study proposes a combined seeding method of 'mini shovel + telescopic pipe' for dry direct-seeded rice with film mulching. A precision seeder for dry direct-seeded rice with film mulching was developed through theoretical calculations, discrete element modelling (DEM) simulations, and field experiments. The configuration and diameter of the rollers were obtained. Twelve telescopic pipes were evenly distributed on the circumference of the roller, with a contact ratio exceeding one. This ratio reduced the slip rate of the roller effectively. Subsequently, DEM was used to develop a 3^3 central composite design. The response surface was established with the sowing depth as the response value. According to agronomic requirements, the sowing depth was set to 20 mm. The optimal combination of working parameters was obtained by optimizing the regression equation. The field experiments showed that the performance of the precision seeder for dry direct-seeded rice with film mulching satisfied the requirements of agricultural production, working stably and reliably. The developed device represents a useful solution for dry direct-seeded rice with film mulching.

Keywords: film mulching; mini shovel; telescopic pipe; discrete element modelling; hill-drop sowing

Citation: Li, H.; Zeng, S.; Luo, X.; Fang, L.; Liang, Z.; Yang, W. Design, DEM Simulation, and Field Experiments of a Novel Precision Seeder for Dry Direct-Seeded Rice with Film Mulching. *Agriculture* **2021**, *11*, 378. https://doi.org/10.3390/agriculture11050378

Academic Editor: José Pérez-Alonso

Received: 28 March 2021
Accepted: 19 April 2021
Published: 21 April 2021

1. Introduction

Rice transplanting is a conventional rice cultivation technology in Asia consisting of four basic steps: nursery bed preparation, seedling raising, seedling uprooting, and transplanting the seedling into the field. All of these steps are highly labor-intensive and water-intensive [1,2]. However, with the rapid economic development in China, the phenomenon of labor shortage appears, leading to the rising of labor wages and production cost in agriculture. Thus, rice production is becoming less profitable [3]. Furthermore, water scarcity is becoming more severe and widespread for rice cultivation [4]. Direct-seeded rice is a feasible alternative method to deal with the shortages of water and labor [5]. Direct-seeded rice consists of directly sowing seeds into the field and represents a simplified cultivation technology, which is a trend toward the future of rice cultivation [6–8]. Xu et al. [9] compared the effects of rice transplanting and direct-seeding rice on yield. The results showed that direct-seeding rice could produce comparable yields to rice transplanting, but direct-seeded rice was exposed to more risks such as weed encroachment and extreme weather. Kakumanu et al. [10] reported that direct-seeded rice was efficient in terms of reducing impacts of climate, water use, increasing grain yield, and better returns to farmers.

In recent years, dry direct-seeded rice with film mulching technique has been adopted to improve water utilization rate, reduce herbicide usage, and improve soil temperature after sowing in rice areas of northern China, which could effectively solve and mitigate the risk of direct-seeding rice [11–13]. Jabran et al. [14] reported that plastic film mulching improved the soil moisture retention, and enhanced rice water productivity, grain yield, and rice quality. Li et al. [15] showed that plastic film mulching with no flooding could provide higher rice yield under appropriate water condition and change the soil nutrient cycle. Row drilling and hill-drop drilling are the main sowing methods for direct-seeded rice. Specifically, hill-drop drilling is beneficial for improving field permeability, leaf photosynthesis, and photosynthetic productivity [16,17]. In this regard, hill-drop drilling is the better sowing method for dry direct-seeded rice with the film mulching technique due to its low film-breaking rate.

Currently, the dry direct-seeded rice with film mulching devices are mainly based on the duckbill-type roller seeder, which is a modified design from seeders for other crops, such as wheat and corn seeders. However, these devices cannot obtain precise seeding because of the differences between rice and other crops in terms of the flowability and triaxial size of seeds. The duckbill-type seed-metering device has a disadvantage: the sowing depth cannot be adjusted. Moreover, stone and caking soil are present in seeding beds in some parts of northern China, causing obstructions in the duckbill-type seed-metering device, leading to miss-seeding, reseeding, or even sparse failures, which seriously affect the rice yield. Various studies have been conducted to solve or mitigate the risks of the duckbill-type seeder. A CAM mechanism was added into the duckbill-type seed-metering device to change the passive opening into the active aperture of the seed-metering device, which alleviated the miss-seeding caused by the failure of the duckbill-type seeder to open in the case of hard objects [18]. The length and shape of the duckbill seeder were optimized to reduce the size of the film hole and avoid the formation of plastic film hanging on the duckbill-type seeder [19]. The internal seed-taking structure of the duckbill-type seeder was optimized, improving the seeding accuracy [20]. All the above studies have played a positive role in improving the performance of the duckbill-type roller seeder, achieving good results. However, the problems such as blocking of the duckbill seeder and inability to adjust the sowing depth remain unsolved.

In this study, to overcome these limitations, the 'mini shovel + telescopic pipe' combination-seeding method was proposed, and a precision seeder for dry direct-seeded rice with film mulching was developed. The mechanism was developed based on the previous studies and through the analysis of the traditional artificial sowing process (including three steps: dig a hole with a shovel→seed pitching by reaching out hand→take back hand and shovel) and considering the technique of seeding in dry soil and drip irrigation for emergence.

The proposed rice precision seeder combines a type-hole seed-metering device to realise precision sowing and adjust the seeding quantities [21,22]. The mini shovel pushes aside the soil and hard materials within the seeding area to form a cavity. Subsequently, the telescopic pipe stretches out like a hand to cast seeds, which effectively prevent the telescopic pipe from being blocked. The extension length of the telescopic pipe controls the sowing depth adjustment by changing the upper and lower position of the drive slideway. Optimal design of the critical component was conducted, and experimental analysis of the working performance and operation effect was performed to obtain the optimal combination of working parameters. The field experiments contribute to providing a new machine for dry direct-seeded rice with film mulching of high quality, without blockage of the telescopic pipe, as well as the development of precision agriculture.

2. Structure and Working Principles of the Proposed Seeder

2.1. Machine Structure

Figure 1 illustrates the overall structure of the developed precision seeder for dry direct-seeded rice with film mulching. The rice precision seeder consists of frame, suspension system, control, drive systems, combined type-hole seed-metering device, roller, mini shovel, telescopic pipe, and other key components.

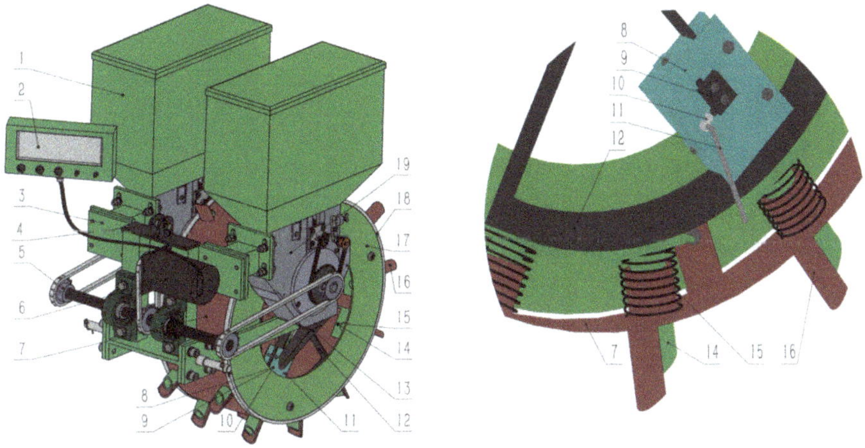

(**a**) (**b**)

Figure 1. 3D model of the novel precision seeder for dry direct-seeded rice with film mulching: (**a**) Axonometric drawing of the entire machine; (**b**) Partial enlarged view. 1. Seed tank 2. Control box 3. Frame 4. Suspension system 5. Chain drive system 6. Stepping motor 7. Roller 8. Dividing seeds box 9. Touch switch 10. Micro CAM 11. Crank 12. Drive slideway 13. Seed tube 14. Telescopic pipe 15. Reset spring 16. Mini shovel 17. Protection plate 18. Rubber ring 19. Combined type-hole seed-metering device.

2.2. Working Principle

The precision seeder for dry direct-seeded rice with film mulching has two rows, which are connected to the traction device using a trifilar suspension system. The movement of the traction device drives the roller rotation, and the mini shovel enters the soil as the roller rotates. The soil and hard materials within the seeding area are pushed aside to form a cavity simultaneously, and the telescopic pipe sticks out under the action of the drive slideway. Seeds are dropped into the cavity through the telescopic pipe after the top end of the telescopic pipe touches the crank. The crank drives the micro CAM to rotate so that the touch switch is on. The control system commands the stepping motor to drive the combined type-hole seed-metering device to rotate 45° to seed to the dividing seeds box for the next sowing. After sowing, the telescopic pipe returns quickly to detach the adhesive attachment under the action of the reset spring. The mini shovel leaves the soil, and the seeding operation finishes. The extension length of the telescopic pipe controls the sowing depth adjustment by changing the upper and lower position of the drive slideway.

2.3. Structural Design of Critical Component

2.3.1. Roller and Contact Ratio

The roller is the supporting part of the seed-throwing mechanism and provides power for moving components. According to the preliminary experiments and referring to previous studies, the diameter of the roller was selected as 410 mm [23–25]. The agronomic requirement of holes distance is 140 ± 10 mm. According to the preliminary simulated experiments, the length of the mini shovel is 70 mm, which can satisfy the seeding performance and increase the roller torque; thus,

$$N = \pi(D + 2H)/S \tag{1}$$

N—Number of mini shovels;

D—Diameter of the roller, mm;
H—Length of the mini shovel, mm;
S—Hole spacing, mm.

Substituting the parameters into Equation (1), considering N = 12.3, rounded to N = 12, and isolating the holes distance separation from Equation (1), a hole spacing of S = 143.9 mm is obtained, which satisfies the agronomic requirements of holes distance.

A certain sliding phenomenon exists in the operation of the roller; that is, the roller changes from rotating to sliding, which will affect the stability of the holes distance separation and reduce the performance of the device. The slip rate of the roller will reduce if a certain degree of contact ratio between the mini shovels and the soil is present. As shown in Figure 2., point A is the entry point, while point B is the exit point. The contact ratio (ε) is the ratio of the mini shovel angle of rotation (θ_1 + θ_2) to the center angle corresponding to two adjacent mini shovels (θ_0). If the contact ratio (ε) is higher than one, the slip rate will reduce, and the adverse effects will reduce.

$$\theta_1 = \theta_2 = \cos^{-1}\frac{R}{R+H} \tag{2}$$

$$\theta_0 = \frac{2\pi}{N} \tag{3}$$

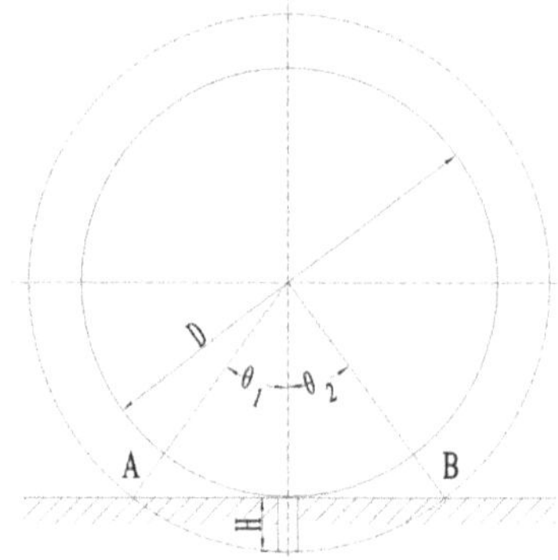

Figure 2. The contact ratio of the mini shovels.

Then:

$$\varepsilon = \frac{\theta_1+\theta_2}{\theta_0} = \frac{N}{\pi}\cos^{-1}\frac{R}{R+H} \tag{4}$$

θ_0—Center angle, °;
θ_1, θ_2—Angle of rotation of the mini shovel;
ε—Contact ratio;
N—Number of mini shovels;
R—Roller radius(D/2), mm;
H—Length of mini shovel, mm.

The parameters *R*, *H* and *N* are substituted into the above equation; thus, $\varepsilon = 2.7868 > 1$.

The contact ratio is ε = 2.7868 > 1 when the roller diameter is 410 mm, the number of mini shovels is 12, and the length of each mini shovel is 70 mm. These parameters provide a useful configuration to reduce the slip rate of the roller and ensure stable holes distance separation.

2.3.2. Drive Slideway

The drive slideway is the core part of the device that ensures the stable displacement of the telescopic pipe and adjusts the extension length of the telescopic pipe; that is, the sowing depth. According to its function, the drive slideway can be divided into two parts: the drive and holding sections.

When the telescopic pipe touches the drive slideway in the drive section, it gradually stretches out during the roller rotation. The telescopic pipe moves following the spiral of the law of Archimedes. The position equation of the contact point between the telescopic pipe and the drive slideway is:

$$\begin{cases} x = [85 + 0.5(\theta - 15)]\cos\theta \\ y = [85 + 0.5(\theta - 15)]\sin\theta + (35 - H_1) \end{cases} \tag{5}$$

θ—Angle of rotation of the telescopic pipe, °, $\theta \in [15, 220]$;
x, y—mm;
H_1—Telescopic pipe stretching out length, mm.

The telescopic pipe rotates with the roller and no longer extends in the holding section, thus, maintaining a fixed extended length. As shown in Figure 2, this section is mainly the interval between the entry point A and the exit point B; the position equation of the contact point between the telescopic pipe and the drive slideway is:

$$\begin{cases} x = 180\cos\theta \\ y = 180\sin\theta + (35 - H_1) \end{cases} \tag{6}$$

θ—Angle of rotation of the telescopic pipe, °, $\theta \in [220, 300]$;
x, y—mm;
H_1—Telescopic pipe stretching out length, mm.

The smooth extension and small vibration impact of the telescopic pipe driven by the drive slideway is an essential performance of the drive slideway, which guarantees the operation quality. In this study, motion simulation function in SolidWorks is used to perform kinematic analysis of the operation process. The preset extended length of the telescopic pipe is 35 mm, the forward speed of the traction device is 0.4 m/s, and the motion curve of the telescopic pipe is shown in Figure 3.

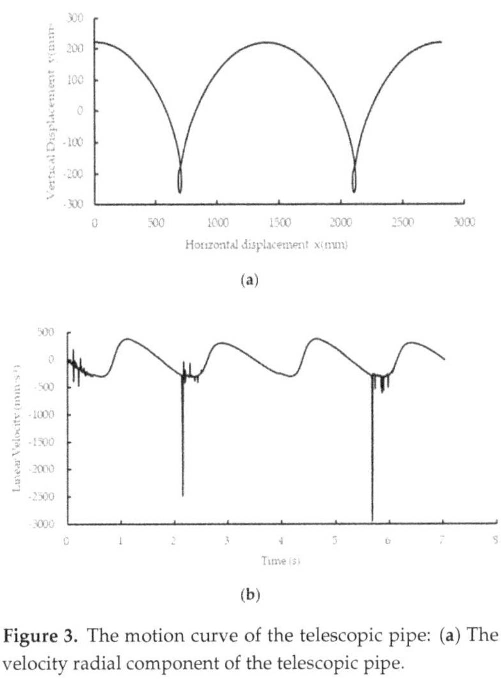

Figure 3. The motion curve of the telescopic pipe: (**a**) The path of the telescopic pipe; (**b**) Linear velocity radial component of the telescopic pipe.

As shown in Figure 3a, the motion path of the telescopic pipe under the preset motion condition is smooth, with a convex point in the return, which is basically consistent with the design goal. The convex point is due to the rapid return of the telescopic pipe under the action of the reset spring when it detaches from the drive slideway.

As shown in Figure 3b, velocity bulges are visible near the starting point and the point the telescopic pipe detaches from the drive slideway. Except for that, the change in velocity is flat without sudden impacts; thus, the vibration is reduced to a certain extent, and the seeder reliability and sowing quality are effectively guaranteed.

2.3.3. Mini Shovel and Telescopic Pipe

The mini shovel and telescopic pipe are crucial parts in contact with the soil, which directly determine the performance of the device. The mini shovel has two functions: pushing aside the soil and hard materials within the seeding area to form a cavity and increasing the torque of the roller. The structure and size of the mini shovel are obtained according to experience and pre-tests, as shown in Figure 4a. The arc at the top of the mini shovel helps to cut the mulch film and reduce the possibility of scraping the mulch film.

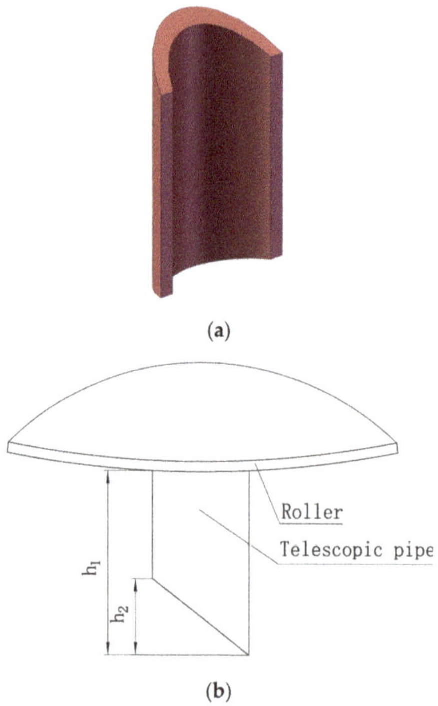

Figure 4. The seeds throwing mechanism: (**a**) Mini shovel; (**b**) The structure of telescopic pipe.

The telescopic pipe is the part performing the sowing, and the extended length of the telescopic pipe determines the sowing depth, as shown in Figure 4b. However, it is not equal to the sowing depth. The sowing depth is related to factors such as the forward speed of the traction device, the extended length of the telescopic pipe, the height of the lower oblique section of the telescopic pipe, and the soil characteristics. Thus, determining the parameters of the telescopic pipe is a complicated process. No accurate model can directly determine the optimal values of these parameters; therefore, the discrete element method (DEM) is used to simulate and analyze the operation process of the telescopic pipe. Section 3.1 presents the detailed information and procedures used in this study.

2.3.4. Electronic Control System

The electronic control system is an essential part of the device that directly affects the performance of the machine. Eight holes are evenly distributed on the wheel circumference of the combined type-hole seed-metering device. The average center angle is 45°, that is, rotating 45° to sow once. As shown in Figure 5, this study considers an MCU (single-chip microcomputer) as the control core to ensure the accuracy of the rotation angle. Moreover, a stepping motor is selected to drive the combined type-hole seed-metering device. Others external affecting factors are also reduced. In this regard, stepping motors able to self-lock are selected to prevent rotation when non-system instruction after the system starts and improve the sowing quality. The system has self-protection function; thus, when the stepping motor is overloaded, the system stops sending pulse instructions to the stepping motor and starts the alarm device. The control flowchart is shown in Figure 6.

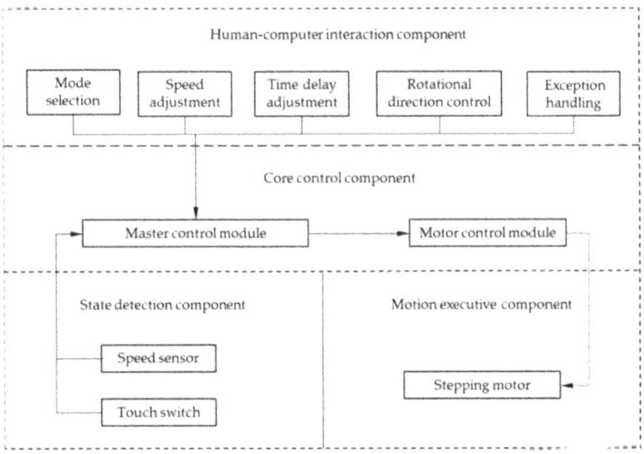

Figure 5. Schematic diagram of electronic control system.

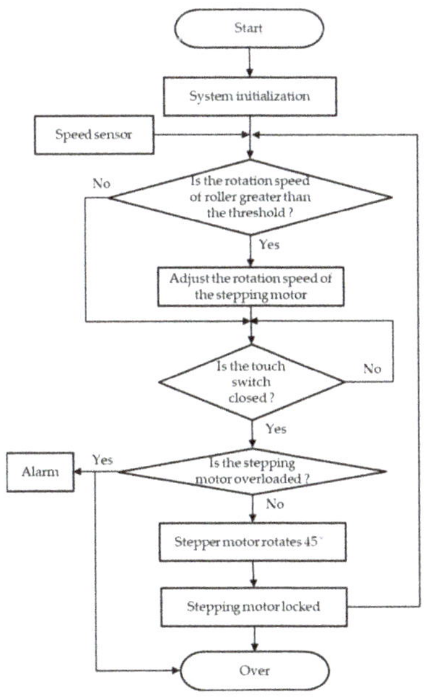

Figure 6. Control flow chart.

The system consists of a closed-loop control, where the rotation speed of the roller is obtained through the speed sensor in real-time. When the rotation speed exceeds or is lower than the preset range, the system adjusts the rotation speed of the stepping motor in real-time to arrange the sowing time and guarantee the filling performance of the combined type-hole seed-metering device.

A direct current voltage transducer was installed to provide power to the control box, stepping motor, and alert and speed sensors. The device obtains the direct current voltage from the battery of a tractor and converts it to the required voltage.

3. Materials and Methods

3.1. DEM Simulation

The kinematic and dynamic modelling of the device in the soil layer is complex; thus, it is difficult to perform a theoretical analysis. As a useful numerical computation approach, the DEM is increasingly being applied to the research between discrete units and machinery in agricultural mechanization production [26–28]. Therefore, this study adopted DEM to simulate the sowing process. In particular, the discrete element simulation software EDEM 2018 was used to perform the simulation experiments. The purpose of the DEM simulation is to establish the regression equation of sowing depth and determine the structure size of the telescopic pipe through the study of the sowing process.

3.1.1. Simulation Model and Parameters

The simplified solid assembly model of the dry direct-seeded rice with film mulching device was imported into the software preprocessing module, illustrated in Figure 7a. Multi-spherical polymerization was selected considering the characteristics of rice shapes, and a total of 25 sphere models with various diameters were connected to construct the rice particle model for the long and irregular shape of the rice (as shown in Figure 7b). The soil particle model considered spheres with diameters of 2 mm. The Hertz–Mindlin (no-slip) contact model included in EDEM was used for the simulation analysis, and the material physical and contact mechanical properties parameters are presented in Table 1.

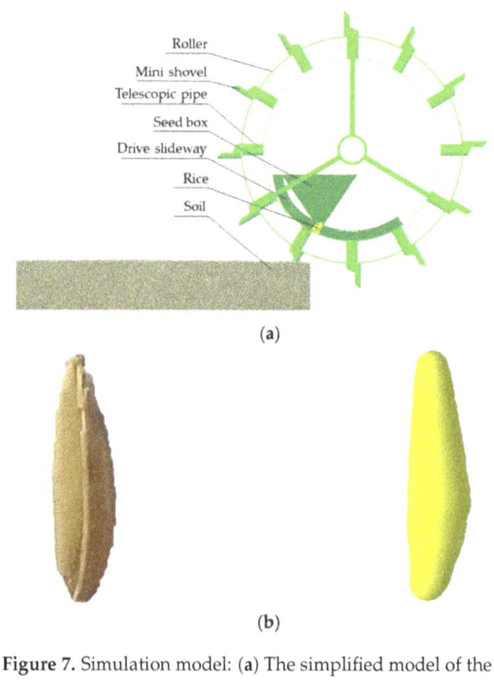

(**a**)

(**b**)

Figure 7. Simulation model: (**a**) The simplified model of the device; (**b**) Rice and rice particle model.

Table 1. Simulation experiments parameters.

Materials	Parameters	Value
Rice	Density (kg·m^{-3})	1147
	Poison ratio	0.25
	Shear modulus (Mpa)	108
Soil	Density (kg·m^{-3})	1914
	Poison ratio	0.38
	Shear modulus (Mpa)	1
Steel	Density (kg·m^{-3})	7850
	Poison ratio	0.3
	Shear modulus (Mpa)	79,000
Rice-Rice	Elastic restitution coefficient	0.13
	Static friction coefficient	0.52
	Rolling friction coefficient	0.1
Soil-Soil	Elastic restitution coefficient	0.11
	Static friction coefficient	0.6
	Rolling friction coefficient	0.4
Rice-Steel	Elastic restitution coefficient	0.41
	Static friction coefficient	0.48
	Rolling friction coefficient	0.01
Soil-Steel	Elastic restitution coefficient	0.12
	Static friction coefficient	0.3
	Rolling friction coefficient	0.05
Rice-Soil	Elastic restitution coefficient	0.08
	Static friction coefficient	0.9
	Rolling friction coefficient	0.7

3.1.2. Central Composite Design

According to the actual device shape and size, the particle factory was statically developed to form a soil tank of 600 × 300 × 150 mm (length × width × height), the seeder working ground speed ranged from 0.4 to 0.6 m/s, and the fixed time step was set to 20% of the Rayleigh time step (8.86 × 10^{-7} s). A single simulation was run for 15 s to ensure the consistency of the sowing process.

In the simulation experiments, the forward speed of traction device (V), the extended length of the telescopic pipe (h_1), and the height of the lower oblique section of the telescopic pipe (h_2) were considered as the design variables, while the sowing depth (Y) was considered as the objective function. Preliminary combined factor experiments were conducted to determine the range of the experimental factors. The central composite design simulation experiment with three factors and three levels was performed using Design-Expert software [29], and the 0-level experiment was repeated five times. The code value of simulated experimental factors is shown in Table 2, while the simulation experiment scheme is shown in Table 3.

Table 2. Code of simulation experiment factors.

Code Value	Forward Speed V (m·s^{-1})	Extended Length h_1 (mm)	Height of the Lower Oblique Section h_2 (mm)
1	0.6	25	15
0	0.5	20	10
−1	0.4	15	5

Table 3. Simulation experiment scheme and results.

No.	Forward Speed V (m·s^{-1})	Extended Length h_1 (mm)	Height of the Lower Oblique Section h_2 (mm)	Sowing Depth Y (mm)
1	−1	−1	−1	18.8
2	1	−1	−1	23.6
3	−1	1	−1	27.9
4	1	1	−1	29.4
5	−1	−1	1	17.7
6	1	−1	1	21.5
7	−1	1	1	31.7
8	1	1	1	32.1
9	−1	0	0	22.5
10	1	0	0	24.6
11	0	−1	0	21
12	0	1	0	34.2
13	0	0	−1	26.1
14	0	0	1	26.9
15	0	0	0	25.8
16	0	0	0	26.3
17	0	0	0	22.7
18	0	0	0	23.7
19	0	0	0	26.3

3.2. *Field Experiment*

The working performance of the device for dry direct-seeded rice with film mulching was tested on 12 September 2020, in Gexinzhuang Village, Yilunbu Township, Renqiu City, Cangzhou City, Hebei Province (38°42′30″ N, 116°10′30″ E, altitude 8 m). The qualified rate of sowing depth, the rate of empty seed, the qualified rate of seeds, the stagger rate, and the qualified rate of hole spacing, were investigated. In order to facilitate the observation of the operation effect of the device and the collection of data, the white polyethylene plastic film with a thickness of 0.01 mm was used in the field experiment. When sowing, the soil was soft, and the average soil compatibility of 100 mm soil layer was 283.5 kPa; the soil layer contained small quantities of soil blocks and stones; the soil moisture was 15.23%, which was slightly higher than the normal soil moisture during the seeding period. The tractor power was 29.42 kW. On average, there were 12 rice seeds in each hole. Before the field experiments, the rice precision seeder operating parameters were adjusted to achieve the optimal combination determined from the response surface optimization of the above simulation experiments: forward speed of tractor V= 0.494 m/s, the extended length of the telescopic pipe h_1 = 15 mm, and the height of the lower oblique section of the telescopic pipe h_2 = 15 mm. Three rows were sown, and a single row was 30 m in length. The field experiments were conducted in accordance with Chinese Agriculture Industrial Standard NY/T 987-2006 'Operating quality grain film-covering hill-drop drill' for data collection and device performance evaluation.

3.3. *Data Analysis*

The DEM simulation was focused on obtaining the sowing depth of different experimental combinations and establishing the response surface with the sowing depth as the response value. The images of the simulation results exported from the EDEM 2018 software were imported into CAD software to measure the sowing depth. The regression equation was analyzed through the variance and quadratic optimized. Data processing and analysis were performed using Microsoft Excel and Design Expert software.

4. Results and Discussion

4.1. *Simulation Experimental Results*

The results of the simulation experiments are shown in Table 3. The Quadratic model was used to establish the regression equation of the sowing depth. The results of the analysis of variance for the regression equation showed a p-value of less than 0.0001. As a p-value greater than 0.05 indicates a lack of fit, the obtained regression equation is extremely significant; thus, the regression equation is effective. Among the three experimental factors, V is very significant, h_1 is extremely significant, and h_2 has a tiny effect. Regarding the interaction terms, only h_1 and h_2 items have a significant effect on the sowing depth, while the interaction of other terms is not significant. Concerning the quadratic term of factors, V^2 has a significant effect on sowing depth, while $h_1{}^2$ has an effect. After removing the terms with no effect, the regression equation of sowing depth was obtained by the quadratic optimization of the regression equation, as shown Equation (7), and the determination coefficient of the regression equation $R^2 = 0.9433$, and the adjustment determination coefficient $R^2{}_{adj} = 0.9072$. The variance analysis of the optimized regression equation is shown in Table 4.

$$Y = -29.745 + 253.572V - 1.789h_1 - 0.882h_2 - 1.65Vh_1 + 0.048h_1h_2 - 207.972V^2 + 0.0797h_1{}^2, \tag{7}$$

Table 4. ANOVA of the modified regression model for the central composite design.

Source	Sum of Squares	df	Mean Square	F-Value	p-Value
Model	329.28	7	47.04	26.14	<0.0001 *
V	15.88	1	15.88	8.82	0.0127 *
h_1	278.31	1	278.31	154.66	<0.0001 *
h_2	1.52	1	1.52	0.8453	0.3776
Vh_1	5.44	1	5.44	3.03	0.1098
h_1h_2	11.52	1	11.52	6.4	0.028 *
V^2	13.65	1	13.65	7.58	0.0188 *
$h_1{}^2$	12.53	1	12.53	6.96	0.023 *
Residual	19.79	11	1.8		
Lack of Fit	8.84	7	1.26	0.4612	0.8251
Pure Error	10.95	4	2.74		
Cor Total	349.08	18			

$R^2 = 0.9433$; $R^2{}_{adj} = 0.9072$; CV = 5.28%; Adequate precision = 17.8366; Note: * shows the term is significant ($p < 0.05$).

According to the requirements of agronomy, the sowing depth should be 15 to 25 mm. If the mean sowing depth is 20 mm, the target value is Y=20 mm. Subsequently, the optimal combination of V = 0.494 m/s, h_1 = 15 mm and h_2 = 15 mm was obtained by optimizing the regression equation. Simulation verification experiments (as shown in Figure 8.) were performed for the optimal combination under the target value; the average sowing depth was 21.35 mm, which was consistent with the predicted value.

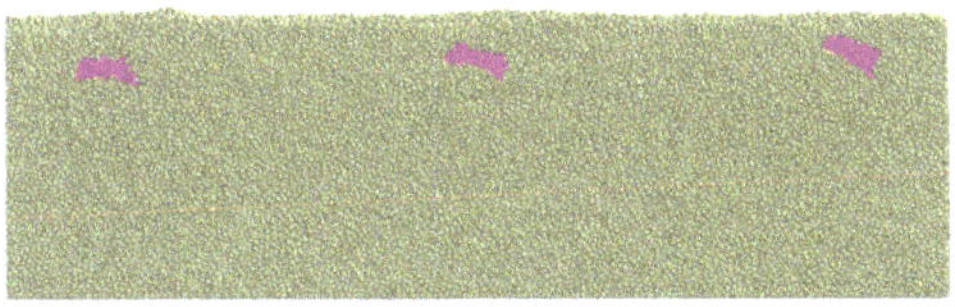

Figure 8. Simulation validation experiments of optimal combination.

4.2. Interactive Analysis and Discussion

According to the analysis of variance of the regression equation Equation (7), the following factors interact with each other: forward speed V, extended length h_1, and height of the lower oblique section h_2. The response surface analysis was conducted to account for the interaction between the factors. Three factors were considered; one factor was set to zero level to evaluate the interaction analysis between the two other factors. Figure 9 shows the response surface analysis.

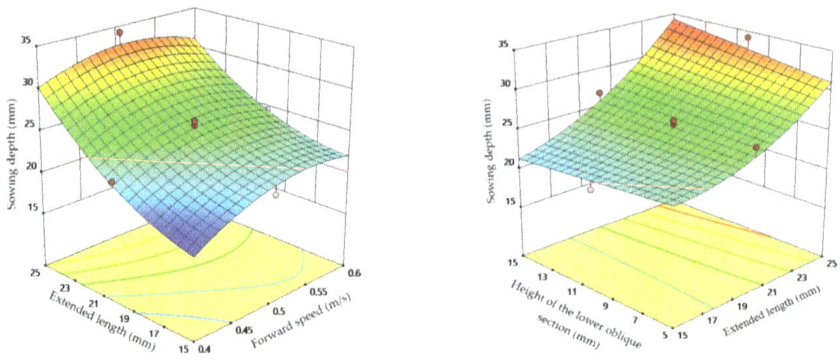

$Y = f(x_1, x_2, 0)$
(a)

$Y = f(0, x_2, x_3)$
(b)

Figure 9. Interaction effect between factors: (**a**) Interaction between the forward speed and the extended length; (**b**) Interaction between extended length and height of the lower oblique section.

The response surface analysis of the interaction between the forward speed V and extended length h_1 is shown in Figure 9a. When the extended length h_1 was constant, the sowing depth increases with the forward speed V, and then the sowing depth tends to decrease. The maximum sowing depth occurs when the current feeding speed ranges from 0.5 to 0.55 m/s. This is because, as the forward speed V increases, the time when the telescopic pipe is in the soil decreases. Meanwhile, when the forward speed V is excessively high, the rice seeds cannot fall into the soil timely, and they would fall into the soil from the telescopic pipe during the excavation process, thus reducing the sowing depth. The sowing depth increases with the extended length h_1. Furthermore, an increase in the sowing depth caused by the extended length h_1 is faster than that caused by the forward speed V. The extended length h_1 presents a more significant effect on the sowing depth than the other factors.

As shown in Figure 9b, when the extended length h_1 is constant, the sowing depth slightly increases with the height of the lower oblique section h_2. The sowing depth increases with the extended length h_1. The extended length h_1 increases rapidly and has a significant effect on the sowing depth, while the height of the lower oblique section h_2 does not present a significant effect on the sowing depth.

4.3. Field Experiment Results

On 12 September 2020, a field experiment was conducted with the experimental combination of $V = 0.494$ m/s, $h_1 = 15$ mm, and $h_2 = 15$. The experimental site was located at the experimental field of Hebei Zhengrong Agricultural machinery Manufacturing Company. The developed device and operation effects are shown in Figure 10.

As shown in Table 5, the experimental results indicate that the device has achieved the design objectives and the agronomic requirements of dry direct-seeded rice with film mulching.

Table 5. Experiment result statistics.

Parameters	Right Rate of Hill Distances/%	Stagger Rate of Film Hill and Hill/%	Right Rate of Depth of Sowing under Film/%	Right Rate of Seeds Per Hill/%	Rate of No Seed Hill/%
Value	100	4.8388	95	83.33	3.2258

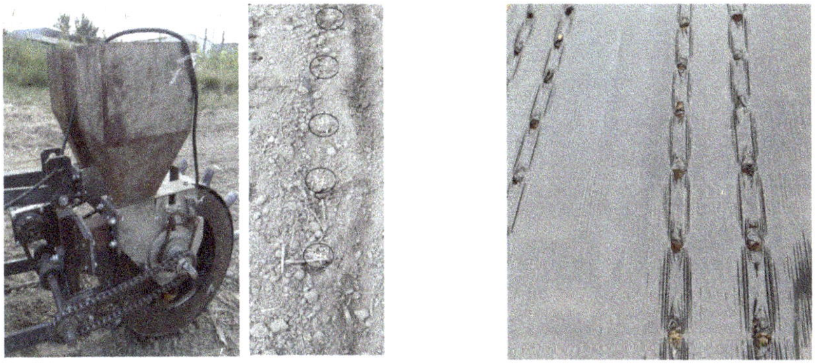

(a) (b)

Figure 10. Field experiments: (**a**) The developed device; (**b**) The effect of sowing.

5. Conclusions

Aimed at improving the technique of seeding in dry soil and drip irrigation for emergence, we proposed the combination seeding method of 'mini shovel + telescopic pipe' for dry direct-seeded rice with film mulching. The device for dry direct-seeded rice with film mulching was developed by theoretical calculation, DEM simulations, and field experiments. The research results showed that:

(1) The precise rice sowing was obtained by selecting the combined type-hole seed-metering device. Moreover, the sowing depth can be adjusted within the range of 15 to 40 mm to satisfy the agronomic requirements of different sowing depths. The drive slideway converted passive sowing to forced sowing, and the mini shovel pushed away from the soil and obstacles in the seeding area, thus reducing the possibility that the telescopic pipe would be clogged.

(2) DEM allowed the simulation of the device sowing process. Moreover, a regression equation was established with the sowing depth as the dependent variable, and the forward speed of the traction device, the extended length of the telescopic pipe, and the height of the lower oblique section of the telescopic pipe as the independent variables. The simulation experiments and field experiments showed that the regression equation could predict the sowing depth accurately.

(3) Field experiments were conducted according to agronomic requirements. The experiment results showed that the right rate of hill distances, stagger rate of film hill and hill, right rate of depth of sowing under film, right rate of seeds per hill, and rate of no seed hill were 100%, 4.8388%, 95%, 83.33%, and 3.2258%, respectively. The experimental results showed that the rice precision seeder with trepanning in the plastic film could achieve the design and agronomic requirements for actual production.

The field experiments of the proposed precision seeder achieved the desired effect; however, the sample size and scope of the field experiments are not substantial enough to understand the adaptability of the device to different regions and soil environments. The adaptability of the device should be verified in the future.

Author Contributions: Methodology, H.L. and W.Y.; software, H.L. and L.F.; validation, W.Y. and S.Z.; formal analysis, X.L.; investigation, H.L. and Z.L.; data curation, H.L.; writing—original draft preparation, H.L.; writing—review and editing, W.Y. and S.Z.; visualization, H.L.; supervision, W.Y.; project administration, W.Y.; funding acquisition, W.Y. and S.Z. All authors have read and agreed to the published version of the manuscript.

Funding: This research was funded by the National Key R & D Program (2018YFD0200700, 2017YFD0700704, and 2016YFD0200606) and Special Projects in Key Area of 'Artificial Intelligence' of General Colleges and Universities in Guangdong Province (2019KZDZX1003).

Institutional Review Board Statement: Not applicable.

Informed Consent Statement: Not applicable.

Data Availability Statement: The data presented in this study are available on demand from the first author at (lihui17@mails.jlu.edu.cn).

Acknowledgments: The authors would like to thank the technical support of the teacher and supervisor. We also appreciate the assistance provided by team members during the experiments. Moreover, we would like to thank Hebei Zhengrong Agricultural machinery Manufacturing Company for providing the device manufacturing. Additionally, we sincerely appreciate the work of the editor and the reviewers of the present paper.

Conflicts of Interest: The authors declare no conflict of interest.

References

1. Kumar, V.; Ladha, J.K. Direct Seeding of Rice: Recent Developments and Future Research Needs. *Adv. Agron.* **2011**, *111*, 297–413. [CrossRef]
2. Tuong, T.; Bouman, B. Rice production in water scarce environments. In *Water Productivity in Agriculture: Limits and Opportunities for Improvement*; Kijne, J., Barker, R., Molden, D., Eds.; CABI Publishing: Wallingford, UK, 2003; pp. 53–67.
3. Chakraborty, D.; Ladha, J.K.; Rana, D.S.; Jat, M.L.; Gathala, M.K.; Yadav, S.; Rao, A.N.; Ramesha, M.S.; Raman, A. A global analysis of alternative tillage and crop establishment practices for economically and environmentally efficient rice production. *Sci. Rep.* **2017**, *7*, 9342. [CrossRef] [PubMed]
4. Tuong, T.P.; Castillo, E.G.; Cabangon, R.C.; Boling, A.; Singh, U. The drought response of lowland rice to crop establishment practices and N-fertilizer sources. *Field Crop. Res.* **2002**, *74*, 243–257. [CrossRef]
5. Sun, L.M.; Hussain, S.; Liu, H.Y.; Peng, S.B.; Huang, J.L.; Cui, K.H.; Nie, L.X. Implications of low sowing rate for hybrid rice varieties under dry direct-seeded rice system in central China. *Field Crop. Res.* **2015**, *175*, 87–95. [CrossRef]
6. Farooq, M.; Siddique, K.H.; Rehman, H.; Aziz, T.; Lee, D.J.; Wahid, A. Rice direct seeding: Experiences, challenges and opportunities. *Soil Tillage Res.* **2011**, *111*, 87–98. [CrossRef]
7. Liu, H.Y.; Hussain, S.; Zheng, M.M.; Peng, S.B.; Huang, J.L.; Cui, K.H.; Nie, L.X. Dry direct-seeded rice as an alternative to transplanted-flooded rice in Central China. *Agron. Sustain. Dev.* **2015**, *35*, 285–294. [CrossRef]
8. Sandhu, N.; Subedi, S.R.; Singh, V.K.; Sinha, P.; Kumar, S.; Singh, S.P.; Ghimire, S.K.; Pandey, M.; Yadaw, R.B.; Varshney, R.K.; et al. Deciphering the genetic basis of root morphology, nutrient uptake, yield, and yield-related traits in rice under dry direct-seeded cultivation systems. *Sci. Rep.* **2019**, *9*, 9334. [CrossRef]
9. Xu, L.; Li, X.X.; Wang, X.Y.; Xiong, D.L.; Wang, F. Comparing the Grain Yields of Direct-Seeded and Transplanted Rice: A Meta-Analysis. *Agronomy* **2019**, *9*, 767. [CrossRef]
10. Kakumanu, K.R.; Kotapati, G.R.; Nagothu, U.S.; Kuppanan, P.; Kallam, S.R. Adaptation to climate change and variability: A case of direct seeded rice in Andhra Pradesh, India. *J. Water Clim. Chang.* **2019**, *10*, 419–430. [CrossRef]
11. Fawibe, O.O.; Hiramatsu, M.; Taguchi, Y.; Wang, J.; Isoda, A. Grain yield, water-use efficiency, and physiological characteristics of rice cultivars under drip irrigation with plastic-film-mulch. *J. Crop Improv.* **2020**, *34*, 414–436. [CrossRef]
12. Zhang, Z.C.; Zhang, S.F.; Yang, J.C.; Zhang, J.H. Yield, grain quality and water use efficiency of rice under non-flooded mulching cultivation. *Field Crop. Res.* **2008**, *108*, 71–81. [CrossRef]
13. He, H.B.; Ma, F.Y.; Yang, R.; Chen, L.; Jia, B.; Cui, J.; Fan, H.; Wang, X.; Li, L. Rice Performance and Water Use Efficiency under Plastic Mulching with Drip Irrigation. *PLoS ONE* **2013**, *8*, e83103. [CrossRef] [PubMed]
14. Jabran, K.; Ullah, E.; Hussain, M.; Farooq, M.; Zaman, U.; Yaseen, M.; Chauhan, B.S. Mulching Improves Water Productivity, Yield and Quality of Fine Rice under Water-saving Rice Production Systems. *J. Agron. Crop Sci.* **2015**, *201*, 389–400. [CrossRef]
15. Li, Y.S.; Wu, L.H.; Zhao, L.M.; Lu, X.H.; Fan, Q.L.; Zhang, F.S. Influence of continuous plastic film mulching on yield, water use efficiency and soil properties of rice fields under non-flooding condition. *Soil Tillage Res.* **2007**, *93*, 370–378. [CrossRef]
16. Wang, Z.M.; Luo, X.W.; Chen, X.F.; Mo, Z.W. Effects of precision rice hill-drop drilling on rice quality. *Trans. Chin. Soc. Agric. Eng.* **2015**, *31*, 16–21. [CrossRef]
17. Wang, Z.M.; Dai, Y.Z.; Wang, B.L.; Zhang, M.H.; Chen, X.F.; Mo, Z.W.; Luo, X.W. Research on Hill-drop Drilling and Row Drilling of Rice. *China Rice* **2016**, *22*, 19–20. [CrossRef]
18. Wang, J.P. Study of Framework and Parameter of Film Mulching Hill-Drop Planter. Master's Thesis, China Agriculture University, Beijing, China, 2005.
19. Zhang, S.J. Design and test of 2BMK-1/3 dibble filming planter machine. *J. Chin. Agric. Mech.* **2013**, *34*, 113–117. [CrossRef]
20. Niu, Q.; Wang, S.G.; Chen, X.G. Design of Rice Planter with Plastic Film Mulched Drip Irrigation. *Trans. Chin. Soc. Agric. Mach.* **2016**, *47*, 102. [CrossRef]
21. Luo, X.W.; Wang, Z.M.; Jiang, E.C.; Li, J.H.; Li, Q.; Chen, W.T. Design of Disassemble Rubber Guard Device for Cell Wheel Feed. *Trans. Chin. Soc. Agric. Mach.* **2008**, *39*, 60–63.
22. Zhang, M.H.; Luo, X.W.; Wang, Z.M.; Dai, Y.Z.; Wang, B.L.; Zheng, L. Design and Experiment of Combined Hole-type Metering Device of Rice Hill-drop Drilling Machine. *Trans. Chin. Soc. Agric. Mach.* **2016**, *47*, 29–36.

23. Liu, H.X.; Su, H.; Li, J.L.; Liu, Z.J. Interactive Optimal Design System of Drum-type No-till Planter Mechanism. *Trans. Chin. Soc. Agric. Mach.* **2019**, *50*, 58–68.
24. Liu, H.X.; Wen, H.N.; Gai, G.W.; Tang, S.F. Design and Experiment on Passive Drum-type No-till Planter Cavitation Mechanism. *Trans. Chin. Soc. Agric. Mach.* **2017**, *48*, 53–61. [CrossRef]
25. Zhang, X.J.; Yang, Y.; Zhou, L. Selection and definitiveness on main parameters of bunch planting cylinder. *Trans. Chin. Soc. Agric. Mach.* **1998**, *29*, 63–66.
26. Shi, Y.Y.; Sun, X.; Wang, X.C.; Hu, Z.C.; Newman, D.; Ding, W.M. Numerical simulation and field tests of minimum-tillage planter with straw smashing and strip laying based on EDEM software. *Comput. Electron. Agric.* **2019**, *166*, 105021. [CrossRef]
27. Barr, J.; Desbiolles, J.; Ucgul, M.; Fielke, J.M. Bentleg furrow opener performance analysis using the discrete element method. *Biosyst. Eng.* **2020**, *18*, 99–115. [CrossRef]
28. Li, M.S.; Mitsuoka, M.; Inoue, E.; Ye, J.; Liu, J.; Yang, S.; Zeng, B.G.; Song, X.X.; Okayasu, T.; Hirai, Y. Design and Performance Analysis of a Seed Metering Device for a Buckwheat Seeder Adopting Discrete Element Analysis. *J. Fac. Agric. Kyushu Univ.* **2020**, *65*, 123–129. [CrossRef]
29. Ge, Y. *Experimental Design Method and Application of Design-Expert Software*; Harbin Institute of Technology Press: Harbin, China, 2015; pp. 62–95.

agriculture

Review

Applications of Discrete Element Method in the Research of Agricultural Machinery: A Review

Hongbo Zhao [1], Yuxiang Huang [1], Zhengdao Liu [1], Wenzheng Liu [2] and Zhiqi Zheng [1,*]

[1] College of Mechanical and Electronic Engineering, Northwest A&F University, Yangling 712100, China; zhaohb@nwafu.edu.cn (H.Z.); hyx@nwsuaf.edu.cn (Y.H.); liuzd@nwafu.edu.cn (Z.L.)

[2] College of Enology, Northwest A&F University, Yangling 712100, China; lwzheng@nwafu.edu.cn

* Correspondence: zhiqizheng@nwsuaf.edu.cn

Abstract: As a promising and convenient numerical calculation approach, the discrete element method (DEM) has been increasingly adopted in the research of agricultural machinery. DEM is capable of monitoring and recording the dynamic and mechanical behavior of agricultural materials in the operational process of agricultural machinery, from both a macro-perspective and micro-perspective; which has been a tremendous help for the design and optimization of agricultural machines and their components. This paper reviewed the application research status of DEM in two aspects: First is the DEM model establishment of common agricultural materials such as soil, crop seed, and straw, etc. The other is the simulation of typical operational processes of agricultural machines or their components, such as rotary tillage, subsoiling, soil compaction, furrow opening, seed and fertilizer metering, crop harvesting, and so on. Finally, we evaluate the development prospects of the application of research on the DEM in agricultural machinery, and look forward to promoting its application in the field of the optimization and design of agricultural machinery.

Keywords: discrete element method (DEM); agricultural machinery; agricultural materials; model; interaction

Citation: Zhao, H.; Huang, Y.; Liu, Z.; Liu, W.; Zheng, Z. Applications of Discrete Element Method in the Research of Agricultural Machinery: A Review. *Agriculture* **2021**, *11*, 425. https://doi.org/10.3390/agriculture11050425

Academic Editor: José Pérez Alonso

Received: 19 April 2021
Accepted: 6 May 2021
Published: 8 May 2021

1. Introduction

In order to improve the operation performance and efficiency of agricultural machines and satisfy the production requirements of modern agriculture, it is necessary to upgrade, update, design, and optimize agricultural machines and their components [1,2]. Theoretical analysis, numerical calculation, and experiment are the three most commonly used approaches in the design and optimization process [3]. Computer simulation, which shows the great advantages of a short cycle, low cost, and being free from the farming season compared with experiments, has been popularly used in recent years in the research of agricultural machinery [4,5].

There are two generally adopted computer simulation methods, the finite element method (FEM) and the discrete element method (DEM). FEM has been mostly used in the research of continuous mediums but most agricultural materials are discrete, such as soil particles and crop seeds; therefore, FEM shows limitations in the study of soil flow and mixture, failure and deformation, and the motion of seed flow [6,7]. Whereas DEM, a numerical calculation approach for the analysis of complex dynamic discontinuous mechanical discrete systems [8,9], has shown its advantages in the research of agricultural machinery.

In DEM, research targets are divided into the discrete elements of particles and walls. In the simulation of operational process of agricultural machinery, agricultural materials can be modeled by particles or particle agglomerates, while agricultural machinery or its components can be modeled by a combination of walls. The trajectory, movement, and mechanical behavior of the agricultural materials and the resistance of the agricultural machinery and its components can be recorded in real time, so as to conduct an in-depth study of the interaction mechanism between agricultural materials and agricultural machinery, and offer guidance for the design and optimization of agricultural machinery.

Recently, many kinds of DEM software have been developed, such as EDEM, PFC, Pasimodo, Agridem, Yade-DEM, DEMeter++, and so on [10–13]. Using this software, many kinds of agricultural material models with high accuracy have been established, and with these established models, many simulation studies of the interaction processes between agricultural machinery and agricultural materials with high reliability and high fidelity have been carried out [14,15].

The aim of this review is to introduce the application research progress of DEM in agricultural machinery, including the model establishment of common agricultural materials and the typical operational process simulation of agricultural machinery and its components. The prospects of DEM's application in research on agricultural machinery are then proposed, while looking forward to promoting the application of DEM in the field of the optimization and design of agricultural machinery.

2. DEM Model Establishment and Parameter Calibration of Agricultural Materials

Accurate and reliable DEM models of agricultural materials are the foundation of the high fidelity simulation of agricultural machinery operation processes. Agricultural materials mainly include soil, crop seed, straw, and fertilizer, etc.

2.1. Soil Model Establishment and Parameter Calibration

Soil is one of the basic research objects in agricultural machinery. Affected by geographical location, climate, soil texture, and many other factors, the soil in different regions shows a great variance in physical and mechanical properties. As listed in Table 1, various contact models have been developed in DEM to represent the contacts between soil particles with different physical and mechanical properties, providing the basis for the establishment of a soil model. To establish a soil model, a suitable contact model needs to be established first, then the key parameters that affect the mechanical and dynamic behavior of the soil need to be calibrated.

2.1.1. Establishment of Soil Contact Model

A DEM soil model can be classified into a cohesive model or a cohesionless model, based on whether the soil has cohesion [16]. In the modeling of cohesionless soil, the Hertz–Mindlin contact model (HMCM), in which the deformation at the contact point is assumed to be non-linear elastic, can simulate the movement of cohesionless soil and the resistance of the agricultural machinery with a certain accuracy, but only the elasticity behavior of the soil can be modeled [17–19]. Ucgul et al. [20] developed a hysteretic spring contact model (HSCM), consisting of linear elasticity and cohesion, where the contact between particles of this model showed a linear elasticity when the stress was low, once the stress exceeded the threshold value, particle contact behaved in a plastic way. Experiment results showed that compared with HMCM, HSCM improved modeling accuracy in operation resistance under different travelling speeds, especially in vertical resistance [21]. Moreover, in order to simplify the calculation process, the contact between soil and machine mainly adopted a cohesionless contact model such as the Hertz–Mindlin (no slip) model [22,23].

Soil shows cohesion due to the existing liquid bridge. The parallel bond contact model (PBCM), which takes cohesion into consideration based on HMCM, is often adopted in DEM software PFC (Itasca Consulting Group, Inc., Minneapolis, MN, USA) [24,25]. The cohesion between soil particles was represented by a cylinder material, through which both force and moment can be delivered. When the distance between two adjacent particles was smaller than the preset value, the parallel bond was created spontaneously; when the stress exceeded a threshold value, the parallel bond broke off, and the two particles were separated, so as to mimic the soil breaking behavior. PBCM has mostly been adopted in sandy soil or sandy loam soil models, but only cohesions between soil particles can be simulated [26,27]. By adding cohesion and adhesion in a normal direction in the HSCM, the adhesion between soil and machinery can be modeled [28]. Experiments showed that the improved HSCM with LCM (liner cohesion model) could predict resistance more accurately under different moisture contents and pressures [29,30]. The Hertz–Mindlin with bonding model, one of the most used models in EDEM software (DEM Solutions Ltd., Edinburgh, UK), employed an "adhesive glue" to stick soil particles together. Shear and normal displacement could be carried with the "adhesive glue" until the shear stress reached the maximum value, then the "adhesive glue" breaks, in this way soil breaking behavior in rotary tillage, subsoiling, and separation of soil and potatoes and so on could be simulated [31–33]. However, this model was only appropriate for soils with low cohesion, because the soils with a high cohesion show more cohesive and plastic behavior in a macroscopic view [34]. The Hertz–Mindlin with JKR cohesion model uses JKR (Johnson–Kendall–Roberts) normal elasticity contact force to represent soil cohesions [35]; with this model, Xing et al. [36] established a model of lateritic soil in hot areas of Hainan Province. The soil breaking resistance error was 3.43%; Li et al. [34] simulated a clayey black soil in the Northeast Plain in China, and the soil aggregation phenomenon when moisture content exceeded 12% could be simulated.

All the above-mentioned soil models were constituted by spherical particles with the same physical and mechanical properties. However, soil composition is complex in real fields. Soil properties show distinctions at different soil depth, and are often mixed with straw, stubble, and other materials. Current studies have established a compound model of different soil layers and a compound root–soil model. In the establishment of a compound soil model of different soil layers, one method is to set different colors as different soil depth layers (Figure 1a) [37–39]; the other is determining the soil contact and constitutive parameters according to the soil structure, such as the tillage layer, plow-pan layer, and subsoil layer for subsoiling (Figure 1b) [23]. These models can be used to simulate the movement of soil particles at different depths and the resistance produced by specific soil layers on the tillage device. In the study of a root–soil model, Liu et al. [40] established a maize root model with bonded DEM particles and fixed the root model with soil particles. The model could be broken off by external forces (Figure 1c); the root–soil model established by Frank et al. [41] used sphere particles to model soil and flexible cylinders to model stubble, the stubble could bend and stretch, and the soil could be broken off. This root–soil model could be used in the study of tillage processes in a no-till field. These compound models offer new concepts for the modelling of complex soil.

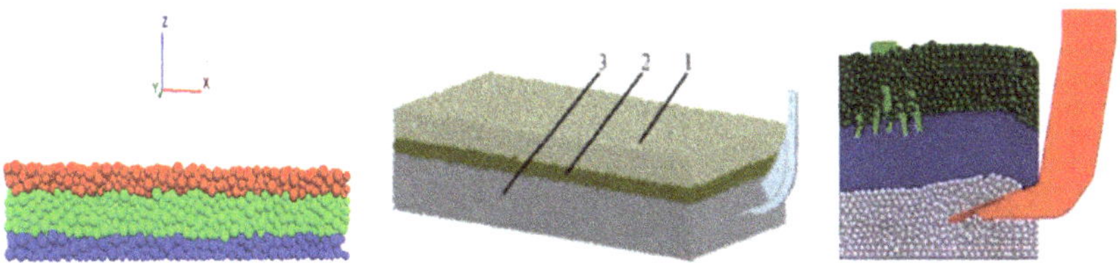

(a) (b) (c)

Figure 1. Compound soil model. (**a**) Compound soil model with different layers; (x,y,z reperesents the axes of the 3D DEM model) (**b**) Soil model for subsoiling [23]; (**c**) Root–soil model [40]. 1. Tillage layer 2. Plow-pan layer 3. Subsoil layer.

2.1.2. Calibration of Physical, Contact, and Constitutive Parameters of a Soil Model

In order to predict the movement of soil particles and resistance in the operation of agricultural machinery accurately, the DEM parameters of a soil model should be calibrated.

There are two kinds of DEM parameter, namely physical parameters, such as porosity and density, and contact and constitutive parameters, such as the friction coefficient, particle strength and stiffness, and bond strength and stiffness [19]. Most research has modeled soil particles with sphere particles, while Tekeste et al. [42] scanned the soil profile with a 3D scanner (Artec Space Spider model, Artec Studio, Luxembourg), then imported the profile data into a 3D modeling software ANSYS SpaceClaim (SpaceClaim Corp., Concord, MA, USA) and combined spherical particles to create a soil particle model that resembled the real profile. The radius of soil particles has a vital influence on the modeling accuracy and time spent. Many studies found that a decreased particle radius could improve modeling accuracy, with an increase in the time spent [21,28,43]. For the purpose of reducing the time spent, it was common to increase particle radius, so long as it could still ensure modeling accuracy [44–47]. When the particle radius is changed, other parameters need adjustment at the same time. Ucgul et al. [20] found that when soil particle radius was increased from 4 to 9 mm, the collision coefficient of restitution, static friction coefficient, rolling friction coefficient, and time step all increased; Thakur et al. [48] found that in order to match the macroscopic motion behavior with real experiments, particle stiffness should be increased after the increase of particle radius.

There are two common methods for calibrating the contact and constitutive parameters [15,49]. The first is determining the parameters directly with test apparatus; the second is comparing the simulated dynamic or mechanistic behaviors with the experimental ones through a repose angle test (Figure 2a), direct shear test (Figure 2b), triaxial pressure test (Figure 2c), and tillage test, by changing the parameter value until the error between the simulation and experimental results was controlled in a limited range, optimal parameter values could be obtained [17,27,29,50–52].

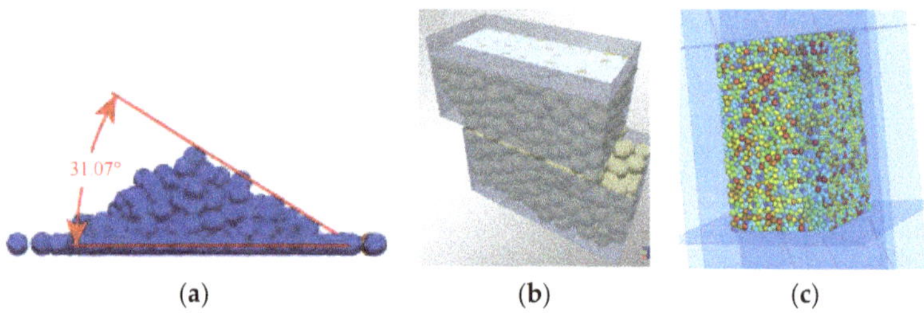

Figure 2. Typical calibration method for soil model parameters. (**a**) Repose angle test [29]; (**b**) Direct shear test [20]; (**c**) Triaxial compression test.

Friction coefficient is one of the most important parameters influencing soil dynamic behavior, including the static and rolling friction coefficient between soil particles, and the static and rolling friction coefficient between soil particles and the contact materials. The repose angle test was the most commonly used method to calibrate friction coefficient between soil particles. First, a repose angle experiment was conducted to obtain the real repose angle; second, the friction coefficient value of the DEM model was changed until the simulated repose angle was approximately that of the experiment value, then the friction coefficients could be calibrated [53,54]. Friction coefficients between soil particles and the contact material could be obtained directly with a sliding test on a slope or flat sliding test [53]. The collision coefficients of the restitution between soil particles and contact material were generally determined by collision test [55]. Soil cohesion was mostly affected by particle stiffness, particle strength, bond stiffness, and bond strength [56–60]. Mak et al. [22] found that the operation resistance of a simple soil engaging tool increased with the increase of particle normal and shear stiffness, the soil disturbance area increased with bond shear strength, and the optimal DEM parameter values varied with soil type. Chen et al. [53] calibrated particle stiffness for three kinds of soil by comparing the draft force, vertical resistance, and soil disturbance in a furrow opening test; the relative error was smaller than 10%. Pue et al. [58] calibrated the soil Young's modulus, Poisson's ratio, and bond shear and normal stiffness with a triaxial compression test.

Table 1 summarizes the commonly used soil contact models and key model parameters in the published literature.

Table 1. Common DEM models and key parameters of soil.

Soil Contact Model	Soil Type	Key DEM * Parameters					
		Static Friction Coefficient between Soil Particles	Rolling Friction Coefficient between Soil Particles	Static friction Coefficient between Soil Particles and Contact Material	Rolling Friction Coefficient between Soil Particles and Contact Material	Collision Coefficient of Restitution	Mechanical Parameters
HMCM *	sea sandy soil [16]	[18]			Steel [53]		Elasticity modulus [18] Particle stiffness [16]
Hertz-Mindlin with bonding	Norfolk sandy loam soil [52] Clay soil in South China [31] Loam clay soil [32]	[53,54]	[53]	Steel [53] Steel [54]	Steel [54]	Between soil particles [55] Between soil particle and contact material [55]	Shear modulus [53] Poisson's ratio [53] Bond normal stiffness [52] Bond shear stiffness [52] Critical stress of the bond [32]
PBCM *	Sandy loam soil [22,56] Sandy soil [56] Sandy soil [27,57]	[56]	[56]				Normal and shear particle stiffness [22,56] Cohesion strength [58] Young modulus [27] Elasticity modulus of particle; elasticity Modulus of bond; damping coefficient [59]
HSCM *	Sandy loam soil [28]	[28]	[28]	Steel [28]	Steel [28]		
HSCM with LCM *	Sandy loam soil in North-west area [29] Grapevine antifreezing soil in Ningxia, China [30] Light loam in North China Plain [23]	[23,29]	[23,29]	Steel [23]	Q235 steel rubber [30] Steel [23]	Between soil particle and Q235steel/rubber [30]	Shear strength, yield strength [29] Surface energy [23]
Hertz-Mindlin with JKR * Cohesion	Clayey black soil [34] Lateritic soil particles in hot areas [36]	[34]	[34,36]	65 Mn; PTFE * [34]	65 Mn; PTFE [34] 28 Mn; B5 board [36]	Between soil particles and 65 Mn/PTFE [34] Between soil particles [36]	Surface energy between soil particles and 65 Mn/PTFE [34] Surface energy between soil particles [36] Poisson's ratio [34]

* Note: Define of the abbreviations. DEM (discrete element method), HMCM (Hertz–Mindlin contact model), PBCM (parallel bond contact model), HSCM (hysteretic spring contact model), LCM (liner cohesion model), JKR (Johnson–Kendall–Roberts), PTEE (polytetrafluoroethylene).

2.2. DEM Model Establishment of Crop Seed and Parameter Calibration

The profile of crop seeds is usually irregular, the main process of model establishment is: determine the profile and establish a 3D outline model; establish a DEM seed model with particles; calibrate the key parameters affecting seed dynamic and mechanistic behaviors and verify the results with experiments. Common crop seeds have no cohesion between each other in the operation of agricultural machinery; a Hertz–Mindlin (no slip) model, which has no cohesion, has been widely adopted [4].

2.2.1. 3D Model Establishment of Crop Seeds

There were three common approaches to determining crop seed profile: direct determining, slice modeling, and 3D-scanning. The direct determining method means determine the length, width, and thickness of crop seeds with a micrometer or vernier caliper directly, and calculating the equivalent diameter and degree of sphericity [61–63]. Based on the difference of seed profile and size, a model can be established by classification [64–66]. The slice modeling method slices crop seeds into a certain thickness and collects an outline profile of each section, then imports them into 3D modeling software to establish a 3D model [67,68]. The 3D-scanning method uses 3D laser scanner or minitype CT scanner to scan the seed profile and obtain the point cloud data, then imports the data into 3D modeling software. This was more accurate than the other two approaches [69,70]. By these methods, crop seed models of wheat, maize, rice, potato, flax, oilseed rape, soybean, and black pepper have been established in recent studies, as listed in Table 2 [64,65,67,71–79].

A crop seed model is usually established by a particle agglomeration method, and most studies found that the simulation accuracy increases with the increase of particle quantity of the model, but the time spent increased simultaneously [69,80]; however, some researchers found that specific crop seed models were more accurate when the particle number was less [71,81].

2.2.2. The Contact and Constitutive Parameters of a Crop Seed Model

Apart from outline profile, the DEM parameters of a crop seed model also include contact parameters such as friction coefficient, elasticity modulus, and so on. The repose angle test was widely used to determine the static and rolling friction coefficient between crop seeds, and its calibration process was similar with the soil parameters. As listed in Table 1, the static and rolling friction coefficients of many kinds of crop seed have been calibrated [78,82–92]. The static and rolling friction coefficients between crop seeds and contact materials were usually determined with a slope sliding method or self-made coefficient determining apparatus [64]; collision coefficients of restitution were calibrated with seed dropping tests and collision tests [74,75]. Using a static compression test, stiffness coefficient, elasticity modulus, Poisson's ratio, and shear modulus were calibrated [62,72,93]. Some other studies calibrated the damping coefficient with a uniaxial loading test [94–96]. Common crop seed models and the key DEM parameters are listed in Table 2.

In real operation processes, air resistance and magnetic forces may also exist. Binelo et al. [77] established an aerodynamic engine, through which the seed suspending velocity, drag coefficient, orientation, and projection area were added in the model. In this way, the air friction of the total seed could be calculated. In order to model the magnetic force of seeds coated with magnetic powder, Wang [97] and Hu et al. [98] created a force plug-in with C programming language using numerical fitting, then loaded the magnetic power through an API interface of EDEM software (DEM Solutions Ltd., Edinburgh, UK) to model the magnetic attraction behavior. Yu et al. [99–101] established a maize-ear model with Agridem software, which they developed. A corncob was modeled by particles with connections, and special balls were set between the niblet and corncob particle to connect them together, the connections could be broken off, so that the breakage of the corncob and maize threshing process could be simulated.

Table 2. DEM model and key parameters of common crop seeds.

Seed Type	3D Model	Key DEM Parameters							
		Coefficient between Soil Particles			Coefficient between Soil Particles and Contact Material			Elasticity Modulus	Damping Coefficient
		Static Friction	Rolling Friction	Collision Restitution	Static Friction	Rolling Friction	Collision Restitution		
Wheat	 [71,72]	[71,82]	[71]	[82]	Steel [82] Low-carbon steel [71]	Low-carbon steel [71]	[82]		Between wheat seed; wheat seed and steel [82]
Maize	 [64–67]	[64,83]	[64,71, 83]		Crylic acid [83] Low-carbon steel [71]	Crylic acid [83] Aluminium; Organic glass [90] Low-carbon steel [71]			
Rice	 [73]  [74]	[74]	[74]				[74]		
Potato seed	 Sphere; big ellipsoid particle; small ellipsoid [65]	[84]		[65]	Steel; plastic [65]	Steel; plastic [84]	Steel [65]		
Flax	 [75]	[75]	[75]	[75]	Organic glass; Aluminum cylinder [75]	Organic glass; Aluminum cylinder [75]	Organic glass; steel [75]	[75]	
Rapeseed	 [76]	[85,86]	[85,86]	[85]	Abs; aluminium alloy [85] Organic glass [87]	Abs; aluminium alloy [85] Organic glass [76,87]	Abs; aluminium alloy [85]	[95]	[95]

Table 2. *Cont.*

Seed Type	3D Model	Key DEM Parameters							
		Coefficient between Soil Particles			Coefficient between Soil Particles and Contact Material			Elasticity Modulus	Damping Coefficient
		Static Friction	Rolling Friction	Collision Restitution	Static Friction	Rolling Friction	Collision Restitution		
Soybean	 [77,78]	[78]	[78]				[78]		
Black pepper	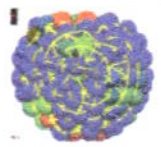 [79]	[79]	[79]		steel [79]	steel [79]	[79]		

2.3. Other Agricultural Materials

The establishment of a straw model has been the latest development in DEM research. Straws physical structure is complicated; in the preliminary stage, a rigid straw model was mostly studied, which could be used to analyze the straw movement, burying, and distribution in the operation of soil tillage [102–106]. By adding contact points or a stick key between two rigid straw models, moment and force could be transferred, and a flexible straw model with elasticity was established [107–109]. Furthermore, Schramm et al. [110], Wang et al. [111], and Liu et al. [112] calibrated the Young's modulus, cohesion damping coefficient, elasticity modulus, and cohesion parameters through a cantilever beam test and three point bending test. For the straw cutting behavior, Guo [113] and Zhang [114] modeled the cutting process of rattan straw and maize straw scarfskin, respectively. Liao et al. [115,116] established fodder rape straw models in the bolting stage and early pod stage, respectively, and calibrated the friction coefficient and cohesion parameters through a repose angle test and straw cutting test. Using a straw compression test and cutting test, Zhang et al. [117] calibrated the key mechanic parameters of Young's modulus, bending strength, and elasticity modulus of a BPM (bonded particle model) straw model; the model could simulate the four straw forms of short, standard, long, and unbroken straw after being crushed [118–120]. Table 3 summarizes the established straw models and their key DEM parameters in the literature.

Fertilizer models were mostly represented by single spherical particles: the static and rolling friction coefficient between urea particles and the static and rolling friction coefficient between urea particles and ABS (acrylonitrile-butadiene-styrene) and PVC (polyvinyl chloride) material were calibrated through repose angle test and slope sliding test [121,122]; and the terminal velocity of large particle urea, DAP (diammonium phosphate), and potassium sulfate was determined through CFD-DEM coupling simulation [123]. Based on the mix uniformity of three fertilizers after fertilizing, Yuan et al. [124] calibrated the rolling friction coefficient between nitrogen fertilizer, phosphate fertilizer, and potassic fertilizer, and calibrated the static and rolling friction coefficient between the fertilizer and a conveyor.

There have also been studies that established a DEM model of fodder, vermicomposting nursery substrate, and pig manure organic fertilizer treated with Hermetia illucen, and calibrated the contact and constitutive DEM parameters [125–127]. Coetzee and Lombard [128] established a grape model to predict the removal of berries from the stems (Figure 3); grape berries were modeled by single DEM particles, and the stem was modeled by a branch of particles that stick together and which could be broken off, and the model could simulate the grape picking and stem breaking process.

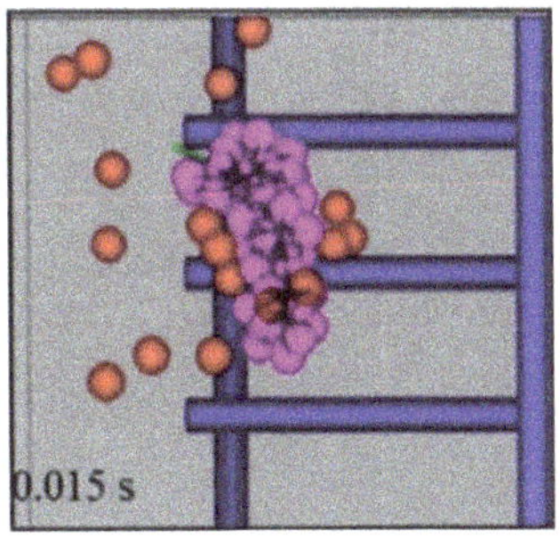

Figure 3. Grape picking process modeling [128].

Table 3. DEM models and key parameters of straw.

Straw Type	Straw Model	Specialty	Contact Parameters	Constitutive Parameters
Oat straw	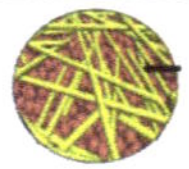 [104]	Rigid	Friction coefficient between straw and blade [104]	Straw stiffness [104]
Maize straw	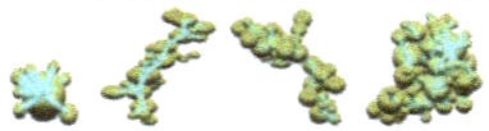 [117]		Normal and shear contact stiffness, in normal and shear critical stress [117]	
Fodder rape straw in bolting stage	 Simulation Experiment [115]	Cuttable	Collision coefficient of restitution between straw and between straw and steel; static and rolling coefficient between straw and between straw and steel [115]	Normal and shear stiffness; normal and shear critical stress; bond radius [115]
Fodder rape straw in early pod stage	 [116]		Collision coefficient of restitution between straw and between straw and steel; static and rolling coefficient between straw and between straw and steel [116]	Elasticity modulus, shearing modulus and Poisson's ratio [116]
	 [110]		Viscous damping coefficient [110]	Young's modulus [110]
Wheat straw	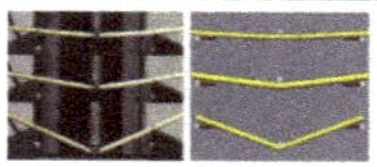 Experiment Simulation [109]	Flexible	Friction coefficient between straw, Friction coefficient between straw and steel [118]	Young's modulus [118,120] Bending strength [108] Tension modulus [120]
	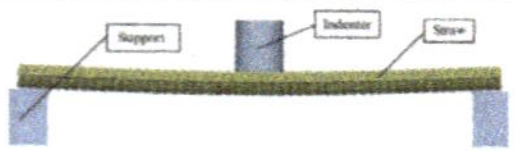 [112]		Static friction coefficient between straw and steel, restitution coefficient [112]	Shearing modulus; elasticity modulus Bond radius, shear and normal cohesion stiffness [112]

3. DEM Simulation of Agricultural Machinery Operation Processes

The interaction relationship between agricultural material and agricultural machinery during its operational process has been the focus of DEM modeling research [14]. The DEM simulations of typical agricultural machinery operation processes are as follows:

3.1. Simulation of Tillage and Soil Preparation

3.1.1. Simulation of Rotary Tillage

DEM could record the resistance and torque of a rotary blade in a small time interval, the variation trends of the simulated working resistance curve in all directions were drawn, and they were found to be similar to the theoretical resistance curve and the soil bin experimental results [129–131]. For torque requirements, the change of key stages can be obtained and typical variation curves can be illustrated with DEM (Figure 4) [130,132–134]. The experiment curve had a second peak value, while the simulation one did not, indicating the simulation remains to be improved [135].

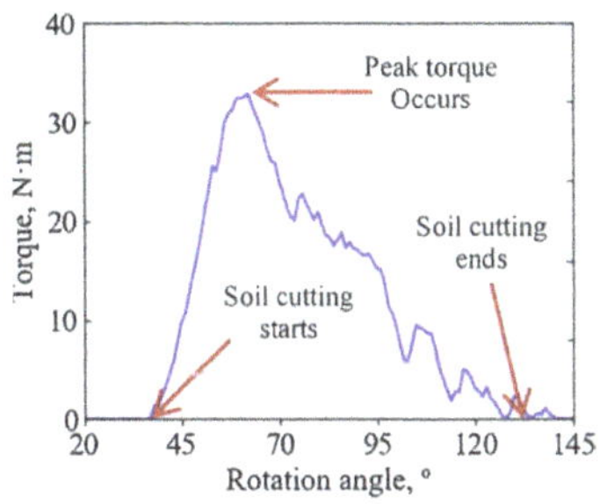

Figure 4. Typical torque variation curve during rotary tillage soil cutting process [132].

For the aspect of soil and straw disturbance, DEM software can monitor the path, contact force, displacement, velocity, and accelerated velocity of single particles, and analyze their dynamic characteristics under different working depths and rotating speeds microscopically [31,102] (Figure 5a). It can also record the soil and straw disturbance produced by a rotary tillage device quantitatively in a macroscopic view [60,103], as shown in Figure 5c,d; Zheng et al. [136] marked soil blocks in different colors in a DEM model and found that a rotary blade could shear, tear, overturn, and throw the soil block so as to make the tillage layer flat, comminute, and loosen. Zhao [137] compared velocity vector maps of different types of rotary blade (Figure 5d) and found that when the blade was tilted with the travel direction, straw could be thrown aside so that the seedbed could be cleaned up.

3.1.2. Simulation of Subsoiling

The purpose of subsoiling is to break up the plow-pan and loosen the soil. Zeng et al. [3] used a built-in measuring sphere in DEM to monitor the internal stress and porosity, with which the break up performance of a plow-pan was evaluated; while Ding et al. [32] evaluated the soil breakage performance by calculating the number of broken bonding keys of a DEM cohesive soil model (Figure 6a). Huang et al. [138] analyzed the microcosmic movement and macroscopic disturbance of soil particles in different positions, and clarified the soil dynamic principles during a subsoiling process (Figure 6b). By combining a DEM simulation with an orthogonal test method, the effect of the structural parameters of the subsoiler, such as if it has a wing, the install height of the wing, the distance between two subsoilers, and the operational parameters (subsoiling depth, travel speed on working resistance, and soil disturbance) have been studied [139–142]. Based on the above-mentioned results, various subsoilers, such as a polyline soil-breaking blade subsoiler and convex blade subsoiler (Figure 6c), have been developed [23,143–146].

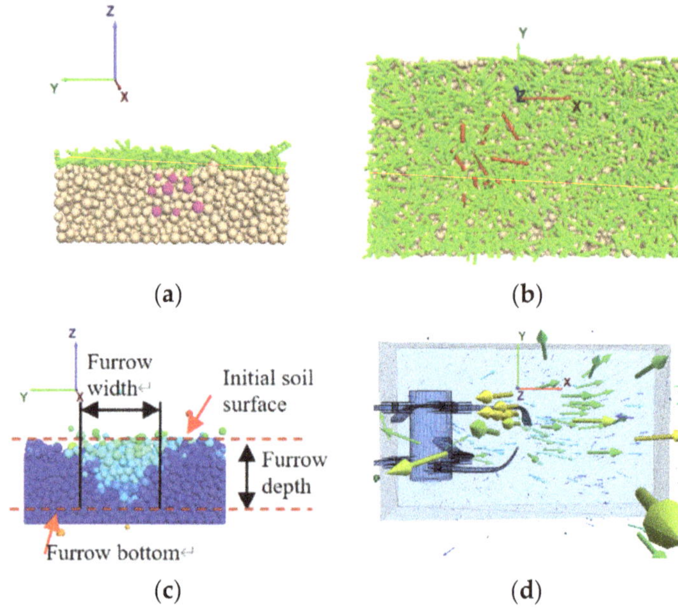

Figure 5. Monitoring soil and straw disturbance in the rotary tillage process (x,y,z reperesents the axes of the 3D DEM model). (**a**) Soil particles [60]; (**b**) Straw [60]; (**c**) Soil disturbance in DEM [60]; (**d**) Straw velocity vector map in DEM [137].

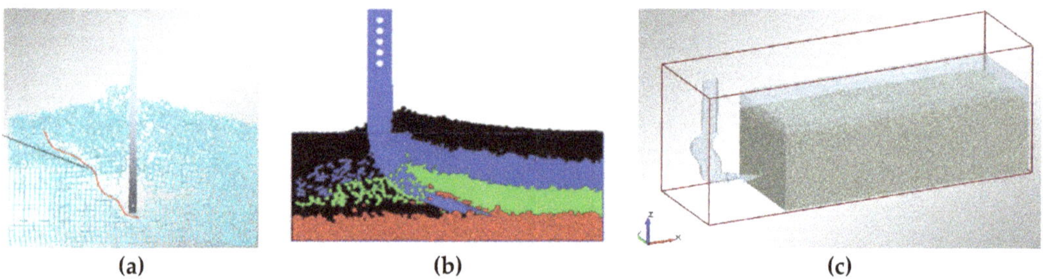

Figure 6. Simulation of a typical subsoiling process [32]. (**a**) Simulation of bonding breakage. (**b**) Soil disturbance in vertical direction [138]. (**c**) Simulation of soil particle velocity [144].

3.1.3. Simulation of Plowing

The aim of plowing is to bury the surface soil and straw into the deeper soil layer. With the established compound soil model with different layers, Ucgul et al. [45,47] recorded soil particle movement at different depths. It was found that when the plowing depth was 300mm, only 10% of the surface soil remained at a 100 mm depth, 53% soil particles were buried at 200–300mm depth, and the simulation results were verified with field experiments. Saunders et al. [147] studied the relationship between the soil burying performance of a moldboard skimmer and the drag force, and found that DEM could predict working resistance more accurately than the other existing theoretical models.

3.1.4. Simulation of Interrow Tillage

In the simulation research of interrow tillage, by monitoring soil particle movement and the breakage of bonding keys, Cheng [148] analyzed the influence of rotary speed, travel speed, bending angle of the hiller blade, number of hiller blades, angle of the shell and soil breakage, number of soil particles being thrown upward, and distribution of the thrown soil. Liu [149] studied the influence of tillage depth, working speed, and angle of sweep wing on soil breakage rate, hilling height, and hilling width. This research was helpful for understanding the interrow tillage mechanism and provided reference for the design and optimization of structures and operation parameters of its key components.

3.1.5. Simulation of Soil Compaction and Traction

Soil is compacted during the operation of agricultural machinery. DEM could model the compaction of tires on soil, the sinkage of the tire, and the resistance and torque requirement of the tire [150,151]. Difficulties lie in the modeling of the pressure delivered from the tire to the soil surface, and three approaches are commonly adopted at present. The first was to model the pressure delivery with a built-in Sohne model (Figure 7a), with which the stress distribution could be simulated [152–154]. The second established a tire model with a particle agglomeration method, so that the tire has gravity and its sinkage could be simulated (Figure 7b) [155]. The third was with the help of a CFD (computational fluid dynamics)–DEM coupling method, where a tire was modeled with CFD and the soil was modeled with DEM (Figure 7c), and the dynamic characteristics of the tire on a sandy soil road could be simulated [156,157].

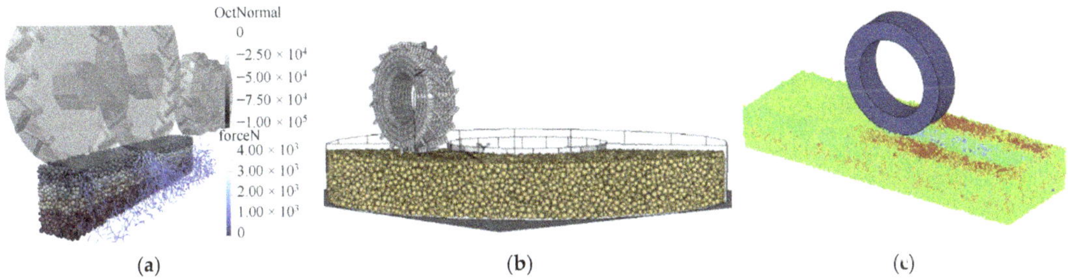

Figure 7. Three typical soil compression modeling methods. (**a**) Preinstall stress in the model [154]; (**b**) Particle agglomerate method [155]; (**c**) CFD–DEM coupling method [156].

Traction performance can be simulated and evaluated by DEM. Nakanishi et al. [158] investigated the effect of lug height, lug thickness, number of lugs, and wheel diameter on the net traction of a wheel. Zeng et al. [159] predicted the tire rutting, vertical force, tractive force, and sinkage of a tire–sand interaction with DEM–FEM, and it was in good agreement with soil bin experiment. Nishiyama et al. [157] proposed a FEM–DEM coupling method for tire traction analysis, in which the soil model was transferred from a FEM to DEM model only when the tire was approaching, once the tire left, the model became a FEM model again. With this approach, the simulation time could be reduced significantly.

3.2. *Simulation of Seeding*

3.2.1. Simulation of Furrow Opening

With the help of DEM, recent studies have conducted single parameter tests to analyze the effect of working depth, rake angle, and travel speed on furrow opening performance and resistance [160–162]. By establishing opener models with different structures: Tamas et al. [53] compared the operation performance of different furrow openers; Tekeste et al. [40] studied the effect of worn openers with different surface profiles on working resistance; and Ucgul et al. [28,163] studied the effect of soil characteristics on furrow opening performance by changing the moisture content and compaction level of the soil model. Combined with an orthogonal test, Zhang et al. [164] and Liu et al. [165] optimized the key structure parameters of an opener and the moldboard shape of a furrow opening device, respectively.

3.2.2. Simulation of Seed Metering

In DEM simulation, the motion state of seed flow can be monitored in real time, and the effect of structure parameters and operation parameters of metering device on the metering performance can be evaluated [166] (Figure 8); for example, the effect of structure and operation parameters of a metering wheel on the beginning angle and ending angle of seed cleaning performance [167], and the motion states of maize seed groups being affected by vibration in field operations [168]. The force transfer of seed flow can also be monitored. Tian [169] studied the variation of the maximum force of rice seed in the seed taking stage of an ejection ear spoon device, which was difficult to obtain through an experiment method, and analyzed the reason for seed miss-filling and re-filling. Taking the metering percentage of a pass as the response target, a lot of structure and operation parameters have been designed and optimized with the help of DEM, such as depth, length, and section size of the model-holes of a hill-seeding centralized metering device [170]; and the arc radius, center angle, and lateral spacing of the middle plate of the seed taking claw of a garlic seed-picking device [171].

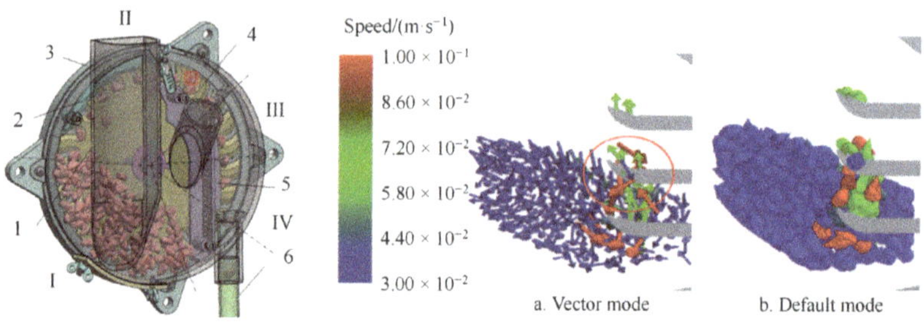

Figure 8. Typical seed metering process simulation. (**a**) Seeding filling process of a maize seed metering device [166]; (**b**) Seed taking process of a garlic metering device [171]. 1.Upper shell. 2. Seed cleaning device. 3. Lower shell. 4. Metering disc. 5. Seed pushing board. 6. Seed delivery tube. I. Seed filling zone. II. Seed cleaning zone. III. Seed guidance zone. IV. Seed pushing zone.

The CFD–DEM coupling method, which models airflow with CFD and models crop seeds with DEM, could monitor not only the airflow velocity and distribution in the metering device and metering tube, but also the movement of seeds in the airflow, so as to analyze: the effect of the structure of a pressurized tube on seed motion, and airflow in the tube of an air-assisted centralized metering device for rapeseed and wheat (Figure 9) [72,172]; the effect of the structure of three different metering discs of inside-filling air-blowing seed metering device on the drag force of the seed in a large circular shape [173]; the effect of the diameter and length of metering tube on airflow and motion of seed flow [174]; the seed filling performance in the adsorption, removing, and separation stages of a pneumatic seed metering service, with guided assistant filling [175], providing references for the design and optimization of air-assisted metering devices.

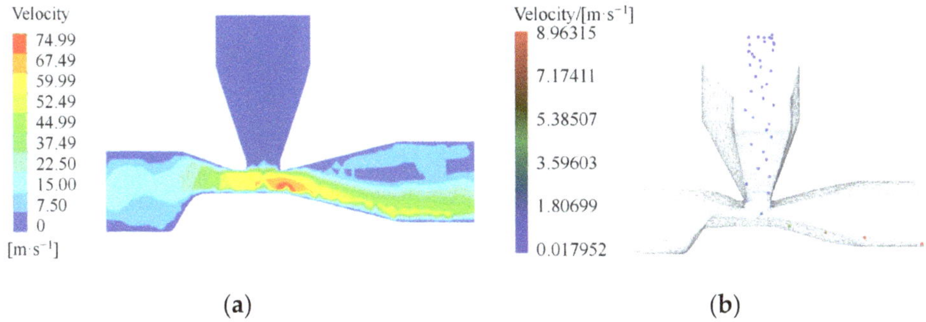

Figure 9. The coupling simulation of the metering process of a metering device for wheat and rapeseed [72] (**a**) CFD–DEM coupling simulation of airflow; (**b**) CFD–DEM coupling simulation of seed flow.

3.2.3. Simulation of Fertilizer Metering

In DEM, it is convenient to record the trajectory of fertilizer particles in the metering process, determine the spatial distribution position, distribution evenness index, etc. With the obtained data, a set of factor influencing the movement of fertilizer particles and fertilizer metering performance can be analyzed, such as the feeding rate, feeding angle and feeding position angle's effect on the distribution in lateral distance of a fertilizer spreader with centrifugal swing disk [176] (Figure 10a); the effect of the inner diameter of the screw blade, screw pitch, outlet distance, number of screw heads, and the blocking wheel opening width on the coefficient of variation of fertilization stability [177].

With CFD–DEM coupling simulation, Gao [178] studied the effect of the rotary speed of a metering shaft on the metering quantity of a metering device, and analyzed the effect of position of the maximum airflow velocity and airflow velocity in the entrance on the movement of fertilizer in the tube; Liu et al. [179] analyzed airflow distribution and movement of a fertilizer group under different inlet velocities by monitoring the fertilizer accumulation performance through DEM, and evaluated the injection performance (Figure 10b).

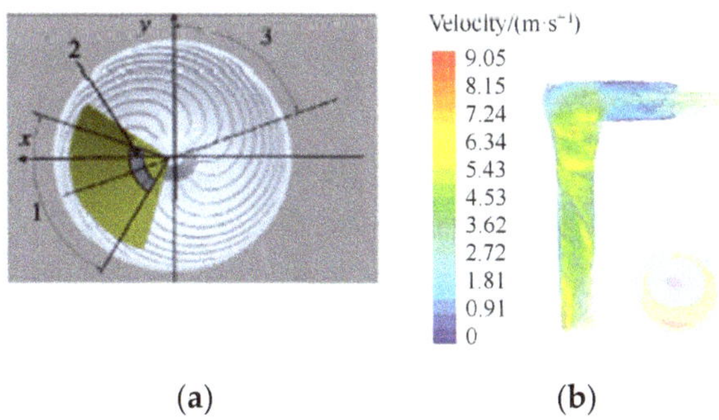

Figure 10. Simulation of fertilizer metering processes. (**a**) Simulation of fertilizing evenness [176]; (**b**) Simulation of fertilizer particle movement in the airflow [179]. 1. Feeding angle 2. Feeding zone 3. Feeding position angle.

Zeng et al. [5] monitored the dynamic characteristics between cotyledon and soil particles in soybean seedling emergence in DEM. Zhou et al. [180] analyzed the contact between seed and soil in a quantitative way, by monitoring their contact numbers, and providing new ideas for the understanding of seedbed preparation.

3.3. Simulation of Crop Harvesting

3.3.1. Simulation of Material Transfer

DEM modeling of the crop harvesting process includes material transfer, threshing, and cleaning. In order to analyze the wheat harvesting process, Wang et al. [181] adopted a coupling simulation method with EDEM–Recurdyn (EDEM, DEM Solutions Ltd., Edinburgh, UK; Recurdyn, FunctionBay Inc., Seoul, Korea), in which the wheat was simulated by DEM and the harvest combine was simulated by FEM, and the translate characters, velocity in an axial direction, and regional flow rate of material quantity in the continuous delivery process of the wheat were analyzed (Figure 11). Wang et al. [182] modeled rice plants with connected balls and simulated the transfer process from the conveyer to the threshing units of a combine harvester.

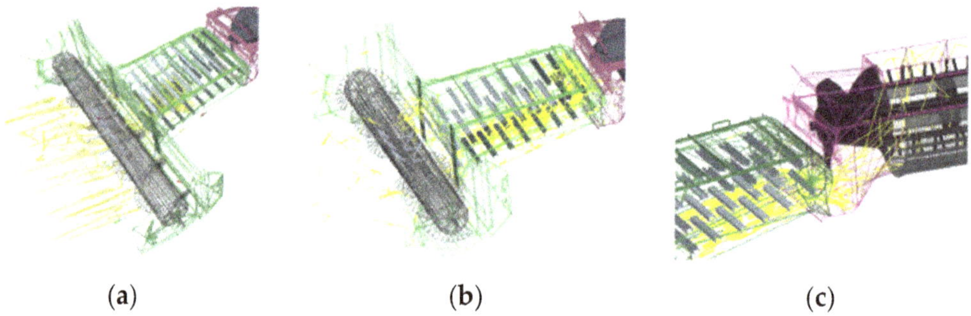

Figure 11. Simulation of wheat movement in a conveyor system [181]. (**a**) Wheat in the spiral conveyor; (**b**) wheat in the incline conveyor; (**c**) wheat in the feeding head of a spiral roller.

3.3.2. Simulation of Threshing

With the established corncob DEM model, Yu [100] modeled the maize threshing process of a drum-type corn threshing device (Figure 12), and analyzed the effect of rotating speed and feeding rate on the threshed rate and force of the corncob. Mou et al. [183] modeled the breaking process of maize kernels in a silage harvesting process, and obtained the optimal parameter combination of teeth number, blade depth and crushing gap, and the rotary speed of the knife roll of the threshing device. With the threshing simulation of rice, the large deformation and fragmentation of the rice plants was modeled by Wang et al. [182].

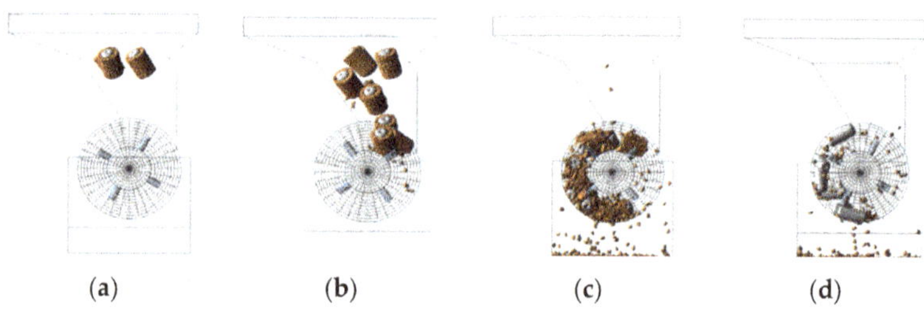

Figure 12. Simulation of a maize threshing process [100]. (**a**) Generation of corncob; (**b**) start of threshing; (**c**) during threshing; (**d**) end of threshing.

3.3.3. Simulation of Cleaning

To study the separation of straw and grain in the screening process, Lenaerts et al. [107] analyzed the properties of straw and grains on screening speed (Figure 13a). Li et al. [184] and Wang et al. [185] effectively analyzed the influence of the operation parameters of vibrational amplitude, frequency, and direction angle on screening time and efficiency. Han [186] and Ma et al. [187] optimized the structure of separation sieves (Figure 13b).

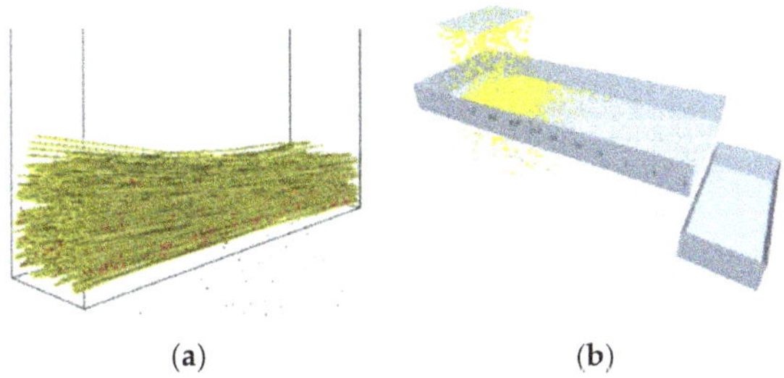

Figure 13. Simulation of grain screening process. (a) Simulation of grain–straw separation [107]; (b) simulation of sieving process of maize grain [187].

Using a CFD–DEM coupling method, the movement of particles in the air-screen cleaning was monitored, and the effect of inlet airflow velocity on the longitudinal velocity and vertical height of grains and short straws was investigated [188]. Xu et al. [189] studied the centroid velocities of grains, stems, and light impurities in the air-and-screen cleaning process, and monitored their degree of dispersion. With these data, the cleaning performance was analyzed.

Wei et al. [33] modeled the separation of soil and potatoes during the potato harvesting process and clarified the effect of structure and operation parameters of a wavy separating sieve on the collision between potatoes and soil clod breakage.

3.4. Simulation of Post-Harvest Processes

3.4.1. Simulation of Grain Conveying

With a DEM simulation, it was observed that the rotation angles of paddy grains increased with the increase of feeding direction. Adjusting the grain vertical direction into the hulling region by controlling the feeding angle could increase the one-time hulling ratio and reduce the breakage ratio (Figure 14) [190]. Chen et al. [191] also found that paddy rice, not only has a translational motion, but also rotates due to the shear of grain flow, which leads to a change in the orientation angle, which was determined by the depth of the grain layer. This study could provide guidance for the design and optimization of feeding systems.

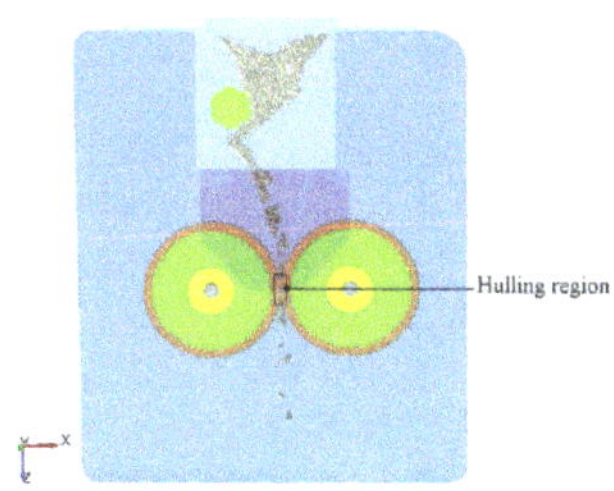

Figure 14. Simulation of rice grain conveying and hulling [190] (x,y,z reperesents the axes of the 3D DEM model).

Wang et al. [192] studied the effect of the inclination angle of returning plate and belt travel speed on particle return rate of wheat granules in a returning device for a fully enclosed belt conveyor; the returning mechanism was clarified and the optimal combination parameters of the two factors was obtained.

3.4.2. Simulation of Storage and Discharge of Seed in Silos

With DEM, the discharge process of a hopper was studied, it was found that the shape of the particle had a significant influence on discharge rate [193]. Using the sum of vertical forces exerted by rapeseed particles on the walls and floor of a receiving container, a difference in the mass flow rates was observed [87]. By recording the position, velocity, and force of wheat particles in a silo with DEM, the location of the front of the rarefaction wave was inferred [194]. Then the formation and propagation of the wave in the discharging process was analyzed. Zaki et al. [195] studied the influence of orifice shape on the discharging process of a flat-bottomed cylindrical silo. Horabik et al. [196] studied the effect of particle shape and filling method on the radial profile of the normal pressure on the bottom of a shallow silo.

3.4.3. Simulation of Grain Separation

Separators were used to separate grains from stones or other impurities. Kannan et al. [197] simulated the separation process of the discarding of a heavy product (stones) and accumulation of a light product (grains) on the deck of a destoner. Then, the effect of deck inclination vibration speed and fluidizing air velocities at the surface on the segregation performance was investigated and their optimal combination value was obtained [198].

Meng et al. [199] conducted simulation research on the separation of whole and broken rice in an indented cylinder separator; the motion trajectory of whole and broken rice with different rotational speed ratios was recorded, and the effect of indent number on the escape angles was investigated. Then the optimal angle of the trough was decided.

4. Research Prospects

DEM could conveniently establish simulation models of agricultural machines and their components, and it was possible to quickly adjust structure and operation parameters within DEM to conduct simulations and collect movement and dynamic data of agricultural materials. Therefore, DEM has become an important approach to help with the design and optimization of the structure and working parameters of agricultural machines and their components. However, the application of research on DEM still remains in its infancy, its application breadth and depth should be developed further.

The DEM modeling accuracy of agricultural materials should be promoted to a higher degree. Most of the existing DEM models of agricultural materials show isotropy and were constituted by spherical particles with the same mechanical properties. However, the construction of agricultural material is usually complicated, and differences exist between different parts of the seed, maize straw cortex and flesh, grape fruit and stem, etc. These differences make agricultural materials isotropic, due to which the physical and mechanical characters remain to be studied further and the models need to be improved accordingly to promote accuracy and reliability, such as the establishment of compound root–soil models. The studies of deformation, failure, and breakage of crop grains and straws are still in a primary stage, and the contact model and model parameters need to be studied further with simulation research of movement and dynamic behaviors, and considering the bending, twisting, cutting, smashing, and damaging behaviors of agricultural materials.

There are also challenges in improving the basic theories of DEM. The contact models in DEM are hypothetical models which approximate the real materials. Certain differences of mechanical relationships exist from the real situation, for example, the adherence between soil particles and tillage equipment still lacks a reliable contact model. To this end, the DEM needs to be improved in its fundamental theories, to make the nonlinear mechanical and dynamic behavior produced by simulations better match the real states. One of the main advantages of DEM is that it is less time consuming, but in the data obtaining process, the mechanical and dynamic information of each DEM particle needs to be calculated. When the DEM particle number is increased, the time spent may increase in a many-fold manner; therefore, how to carry out the calculations more effectively, and improve simulation efficiency is one of the most important research issues.

There is also a need to conduct coupling research along with other methods. DEM is mainly suitable for the research of discrete materials, however, in the interaction process between agricultural materials and agricultural machinery, both discrete materials and continuous mediums should be used. By solid–fluid coupling, complex field situations could be modeled with the consideration of moisture content, air resistance, magnetic force, and so on, with the help of FEM and many-body dynamics. Through solid–solid coupling, the monitoring of the specific resistance of different positions of agricultural machinery components may be realized, and the complex motion of agricultural machinery parts could be modeled; and thus, the agricultural machinery structures could be better designed and optimized.

Author Contributions: Conceptualization: H.Z., Y.H. and Z.Z. H.Z. reviewed the literature and wrote the initial draft of the paper with assistance from Y.H. and Z.Z., Z.L. and W.L. contributed to revising the manuscript. All authors have read and agreed to the published version of the manuscript.

Funding: This work was supported by the grant of China Postdoctoral Science Foundation (No. 2020M683578), the grant of the Fundamental Research Funds for the Central Universities (No. 2452019204) and the innovation chain funds of major industry in Shanxi Province (2018ZDCXL-NY-03-06).

Institutional Review Board Statement: Not applicable.

Informed Consent Statement: Not applicable.

Data Availability Statement: Not applicable.

Conflicts of Interest: The authors declare no conflict of interest.

References

1. Gilandeh, Y.A.; Fazeli, M.; Roshanianfard, A.; Hernández, J.L.H.; Penna, A.F.; Miranda, I.H. Effect of different working and tool parameters on performance of several types of cultivators. *Agriculture* **2020**, *10*, 145. [CrossRef]
2. Yezekyan, T.; Benetti, M.; Armentano, G.; Trestini, S.; Sartori, L.; Marinello, F. Definition of reference models for power, mass, working width, and price for tillage implements. *Agriculture* **2021**, *11*, 197. [CrossRef]
3. Zeng, Z.W.; Chen, Y.; Zhang, X. Modelling the interaction of a deep tillage tool with heterogeneous soil. *Comput. Electron. Agric.* **2017**, *143*, 130–138. [CrossRef]
4. Horabik, J.; Molenda, M. Parameters and contact models for DEM simulations of agricultural granular materials: A review. *Biosyst. Eng.* **2016**, *147*, 206–225. [CrossRef]
5. Zeng, Z.W.; Chen, Y.; Long, Q. Simulation of cotyledon-soil dynamics using the discrete element method (DEM). *Comput. Electron. Agric.* **2020**, *174*, 1–9. [CrossRef]
6. Karmakar, S.; Ashrafizadeh, S.R.; Kushwaha, R.L. Experimental validation of computational fluid dynamics modeling for narrow tillage tool draft. *J. Terramech.* **2009**, *46*, 277–283. [CrossRef]
7. Nematollahi, E.; Zare, S.; Moghaddam, M.M.; Ghasemi, A.; Ghorbani, F.; Banisi, S. DEM-based design of feed chute to improve performance of cone crushers. *Miner. Eng.* **2021**, *168*, 106927. [CrossRef]
8. Cundall, P.A. A computer model for simulation progressive large scale movement in blocky rock system. In Proceedings of the Symposium of International Society of Rock Mechanics, Nancy, France, 4–6 October 1971.
9. Cundall, P.A.; Strack, O.D.L. A discrete numerical model for granular assemblies. *Geotechnique* **1979**, *29*, 47–65. [CrossRef]
10. Fleissner, F.P. *Object Oriented Simulation with Lagrangian Particle Methods*; Shaker Verlag: Aachen, Germany, 2010.
11. Ergenzinger, C.; Seifried, R.; Eberhard, P. A discrete element model to describe failure of strong rock in uniaxial compression. *Granul. Matter* **2010**, *13*, 341–364. [CrossRef]
12. Smilauer, V.; Catalano, E.; Chareyre, B.; Dorofeenko, S.; Duriez, J.; Dyck, N.; Elias, J.; Er, B.; Eulitz, A.; Gladky, A.; et al. Yade Documentation, 2nd ed. The Yade Project. 2015. Available online: http://yade-dem.org/doc/ (accessed on 20 November 2015). [CrossRef]
13. Bravo, E.L.; Tijskens, E.; Suárez, M.H.; Cueto, O.G.; Ramon, H. Prediction model for non-inversion soil tillage implemented on discrete element method. *Comput. Electron. Agric.* **2014**, *106*, 120–127. [CrossRef]
14. Shmulevich, I. State of the art modeling of soil–tillage interaction using discrete element method. *Soil Tillage Res.* **2010**, *111*, 41–53. [CrossRef]
15. Coetzee, C.J.; Els, D.N.J. Calibration of discrete element parameters and the modelling of silo discharge and bucket filling. *Comput. Electron. Agric.* **2009**, *65*, 198–212. [CrossRef]

16. Tsuji, T.; Nakagawa, Y.; Matsumoto, N.; Kadono, Y.; Takayama, T.; Tanaka, T. 3-D DEM simulation of cohesive soil-pushing behavior by bulldozer blade. *J. Terramech.* **2012**, *49*, 37–47. [CrossRef]
17. Obermayr, M.; Dressler, K.; Vrettos, C.; Eberhard, P. Prediction of draft forces in cohesionless soil with the Discrete Element Method. *J. Terramech.* **2011**, *48*, 347–358. [CrossRef]
18. Asaf, Z.; Rubinstein, D.; Shmulevich, I. Determination of discrete element model parameters required for soil tillage. *Soil Tillage Res.* **2007**, *92*, 227–242. [CrossRef]
19. Zhang, R.; Han, D.L.; Ji, Q.L.; He, Y.; Li, J.Q. Calibration methods of sandy soil parameters in simulation of Discrete Element Method. *Trans. Chin. Soc. Agric. Mach.* **2017**, *48*, 49–56.
20. Ucgul, M.; Fielke, J.M.; Saunders, C. Three-dimensional discrete element modelling of tillage: Determination of a suitable contact model and parameters for a cohesionless soil. *Biosyst. Eng.* **2014**, *121*, 105–117. [CrossRef]
21. Ucgul, M.; Fielke, J.M.; Saunders, C. 3D DEM tillage simulation: Validation of a hysteretic spring (plastic) contact model for a sweep tool operating in a cohesionless soil. *Soil Tillage Res.* **2014**, *144*, 220–227. [CrossRef]
22. Mak, J.; Chen, Y.; Sadek, M.A. Determining parameters of a discrete element model for soil-tool interaction. *Soil Tillage Res.* **2012**, *118*, 117–122. [CrossRef]
23. Zheng, K.; He, J.; Li, H.W.; Diao, P.S.; Wang, Q.J.; Zhao, H.B. Research on polyline soil breaking blade subsoiler based on subsoiling soil model using Discrete Element Method. *Trans. Chin. Soc. Agric. Mach.* **2016**, *47*, 62–72.
24. Potyondy, D.O.; Cundall, P.A. A bond-particle model for rock. *Int. J. Rock Mech. Min. Sci.* **2004**, *41*, 1329–1364. [CrossRef]
25. Itasca. *User's Manual for PFC3D Version PFC5.0*; Itasca Consulting Group, Inc.: Minneapolis, MN, USA, 2017.
26. Zhang, R.; Li, J. Simulation on mechanical behavior of cohesive soil by Distinct Element Method. *J. Terramech.* **2006**, *43*, 303–316. [CrossRef]
27. Tamas, K.; Jori, I.J. The influence of the soil water content in the soil-tool DEM model. *Prog. Agric. Eng. Sci.* **2015**, *11*, 43–70. [CrossRef]
28. Ucgul, M.; Fielke, J.M.; Saunders, C. Three-dimensional discrete element modelling (DEM) of tillage: Accounting for soil cohesion and adhesion. *Biosyst. Eng.* **2015**, *129*, 298–306. [CrossRef]
29. Shi, L.R.; Zhao, W.Y.; Sun, W. Parameter calibration of soil particles contact model of farmland soil in northwest arid region based on discrete element method. *Trans. Chin. Soc. Agric. Eng.* **2017**, *33*, 181–187.
30. Ma, S.; Xu, L.M.; Yuan, Q.C.; Niu, C.; Zeng, J.; Chen, C.; Wang, S.S.; Yuan, T.X. Calibration of discrete element simulation parameters of grapevine anti freezing soil and its interaction with soil-cleaning components. *Trans. Chin. Soc. Agric. Eng.* **2020**, *36*, 40–49.
31. Fang, H.M.; Ji, C.Y.; Farman, A.C.; Guo, J.; Zhang, Q.Y.; Chaudhry, A. Analysis of soil dynamic behavior during rotary tillage based on Distinct Element Method. *Trans. Chin. Soc. Agric. Mach.* **2016**, *47*, 22–28.
32. Ding, Q.S.; Ren, J.; Belal, E.A.; Zhao, J.Q.; Ge, S.Y.; Li, Y. Analysis of subsoiling process in wet clayey paddy soil. *Trans. Chin. Soc. Agric. Mach.* **2017**, *48*, 38–48.
33. Wei, Z.C.; Su, G.L.; Li, X.Q.; Wang, F.M.; Sun, C.Z.; Meng, Q.Y. Parameter optimization and test of potato harvester wavy sieve based on EDEM. *Trans. Chin. Soc. Agric. Mach.* **2020**, *51*, 109–122.
34. Li, J.W.; Tong, J.; Hu, B.; Wang, H.B.; Mao, C.Y.; Ma, Y.H. Calibration of parameters of interaction between clayey black soil with different moisture content and soil-engaging component in northeast China. *Trans. Chin. Soc. Agric. Eng.* **2019**, *35*, 130–140.
35. Ajmal, M.; Roessler, T.; Richter, C.; Katterfeld, A. Calibration of cohesive DEM parameters under rapid flow conditions and low consolidation stresses. *Powder Technol.* **2020**, *374*, 22–32. [CrossRef]
36. Xing, J.J.; Zhang, R.; Wu, P.; Zhang, X.R.; Dong, X.H.; Chen, Y.; Ru, S.F. Parameter calibration of discrete element simulation model for lateritic soil particles in hot areas of Hainan Province. *Trans. Chin. Soc. Agric. Eng.* **2020**, *36*, 158–166.
37. Ucgul, M.; Saunders, C.; Fielke, J.M. A method of quantifying Discrete Element Method simulations of top soil burial from a mouldboard plough. In Proceedings of the ASABE International Meeting, Orlando, FL, USA, 17–20 July 2016.
38. Ucgul, M.; Saunders, C.; Fielke, J.M. Discrete element modelling of top soil burial using a full scale mouldboard plough under field conditions. *Biosyst. Eng.* **2017**, *160*, 140–153. [CrossRef]
39. Ucgul, M.; Saunders, C. Simulation of tillage forces and furrow profile during soil-mouldboard plough interaction using discrete element modelling. *Biosyst. Eng.* **2019**, *190*, 58–70. [CrossRef]
40. Liu, J.A. *Study on Subsoiler Parameters Optimization and Comprehensive Effect of Subsoiling Based on the Discrete Element Method*; China Agricultural University: Beijing, China, 2018.
41. Bourrier, F.; Kneib, F.; Chareyre, B.; Fourcaud, T. Discrete modeling of granular soils reinforcement by plant roots. *Ecol. Eng.* **2013**, *61*, 646–657. [CrossRef]
42. Tekeste, M.Z.; Way, T.R.; Syed, Z.; Schafer, R.L. Modeling soil-bulldozer blade interaction using the discrete element method (DEM). *J. Terramech.* **2020**, *88*, 41–52. [CrossRef]
43. Wang, X.Z.; Zhang, S.; Pan, H.B.; Zheng, Z.Q.; Huang, Y.X.; Zhu, R.X. Effect of soil particle size on soil-subsoiler interactions using the discrete element method simulations. *Biosyst. Eng.* **2019**, *182*, 138–150. [CrossRef]
44. Milkevych, V.; Munkholm, L.J.; Chen, Y.; Nyord, T. Modelling approach for soil displacement in tillage using discrete element method. *Soil Tillage Res.* **2018**, *183*, 60–71. [CrossRef]
45. Yang, Y.Y.; Li, M.; Tong, J.; Ma, Y.H. Study on the interaction between soil and the five-claw combination of a mole using the discrete element method. *Appl. Bionics Biomech.* **2018**, *08*, 1–10. [CrossRef] [PubMed]

46. Gursoy, S.; Chen, Y.; Li, B. Measurement and modelling of soil displacement from sweeps with different cutting widths. *Biosyst. Eng.* **2017**, *161*, 1–13. [CrossRef]
47. Obermayr, M.; Vrettos, C.; Eberhard, P.; Dauwel, T. A discrete element model and its experimental validation for the prediction of draft forces in cohesive soil. *J. Terramech.* **2014**, *53*, 93–104. [CrossRef]
48. Thakur, S.; Ahmadian, H.; Sun, J.; Ooi, J. Scaling of discrete element model parameters in uniaxial test simulation. In Proceedings of the 6th International Conference on Discrete Element Methods, Golden, CO, USA, 5–6 August 2013; pp. 457–462.
49. Marigo, M.; Stitt, E.H. Discrete Element Method (DEM) for industrial applications: Comments on calibration and validation for the modelling of cylindrical pellets. *KONA Powder Part J.* **2015**, *32*, 236–252. [CrossRef]
50. Tekeste, M.Z.; Tollner, E.W.; Raper, R.L.; Way, T.R.; Johnson, C.E. Non-linear finite element analysis of cone penetration in layered sandy loam soil-Considering precompression stress state. *J. Terramech.* **2009**, *46*, 229–239. [CrossRef]
51. Mckyes, E.; Ali, S. The cutting of soil by harrow blades. *Terramechanics* **1977**, *14*, 43–58. [CrossRef]
52. Tekeste, M.Z.; Balvanz, L.R.; Hatfield, J.L.; Ghorbani, S. Discrete element modeling of cultivator sweep-to-soil interaction: Worn and hardened edges effects on soil-tool forces and soil flow. *J. Terramech.* **2019**, *82*, 1–11. [CrossRef]
53. Chen, Y.; Munkholm, L.J.; Nyord, T. A discrete element model for soil-sweep interaction in three different soils. *Soil Tillage Res.* **2013**, *126*, 34–41. [CrossRef]
54. Tamas, K. The role of bond and damping in the discrete element model of soil-sweep interaction. *Biosyst. Eng.* **2018**, *169*, 57–70. [CrossRef]
55. Linde, V.J. Discrete Element Modeling of a Vibratory Subsoiler. Master's. Thesis, Department of Mechanical and Mechatronic Engineering, University of Stellenbosch, Matieland, South Africa, 2007.
56. Sadek, M.A.; Chen, Y. Feasibility of using PFC3D to simulate soil flow resulting from a simple soil-engaging tool. *Trans. ASABE* **2015**, *58*, 987–996.
57. Dai, F.; Song, X.F.; Zhao, W.Y.; Zhang, F.W.; Ma, H.J.; Ma, M.Y. Simulative calibration on contact parameters of Discrete Elements for covering soil on whole plastic film mulching on double ridges. *Trans. Chin. Soc. Agric. Mach.* **2019**, *50*, 49–56.
58. Pue, J.D.; Emidio, G.D.; Flores, R.D.; Bezuijen, A.; Cornelis, W.M. Calibration of DEM material parameters to simulate stress-strain behaviour of unsaturated soils during uniaxial compression. *Soil Tillage Res.* **2019**, *194*, 104303. [CrossRef]
59. Korncl, T.; István, J.J.; Mouazen, A.M. Modelling soil-sweep interaction with discrete element method. *Soil Tillage Res.* **2013**, *134*, 223–231.
60. Zhao, H.B.; He, J.; Li, H.W.; Mao, Y.J.; Hu, H.N.; Zhang, Z.Q.; Liu, P. Comparison on soil, straw disturbance and resistance of conventional and plain-straight blade for strip-tillage with Discrete Element Method. *Int. Agric. Eng. J.* **2018**, *27*, 229–240.
61. Guzman, L.J.; Chen, Y.; Landry, H. Discrete element modeling of seed metering as affected by roller speed and damping coefficient. *Trans. ASABE* **2019**, *63*, 189–198. [CrossRef]
62. Yan, H.; Yu, J.Q.; Kou, X.X. A study on boundary modeling of three-dimensional Discrete Element Method based on Pro/ENGINEER. In Proceedings of the ICRTMME, Shenzhen, China, 7–28 January2011.
63. Yan, H. *Anew Kind of Method for the Optimizated Design of Combination Inner-Cell Corn Precision Seed Metering Device*; Jilin University: Changchun, China, 2012.
64. Wang, Y.X.; Liang, Z.J.; Zhang, D.X.; Cui, T.; Shi, S.; Li, K.H.; Yang, L. Calibration method of contact characteristic parameters for corn seeds based on EDEM. *Trans. Chin. Soc. Agric. Eng.* **2016**, *32*, 36–42.
65. Liu, W.Z.; He, J.; Li, H.W.; Li, X.Q.; Zheng, K.; Wei, Z.C. Calibration of simulation parameters for potato minituber based on EDEM. *Trans. Chin. Soc. Agric. Mach.* **2018**, *49*, 125–135.
66. Liu, W.Z.; Li, H.W.; Ma, S.C.; Lu, C.Y.; He, J.; Li, X.Q.; Wei, Z.C.; Xu, Q.M.; Liu, P.; Su, G.L. Parameters analysis and calibration of potato minituber required in EDEM based on the angle of repose of particle stacking test. *IAEJ* **2018**, *27*, 1–14.
67. Shi, L.R.; Wu, J.M.; Sun, W.; Zhang, F.W.; Sun, B.G.; Liu, Q.W.; Zhao, W.Y. Simulation test for metering process of horizontal disc precision metering device based on discrete element method. *Trans. Chin. Soc. Agric. Eng.* **2014**, *30*, 40–48.
68. Shi, L.R.; Zhao, W.Y.; Wu, J.M.; Zhang, F.W.; Sun, W.; Dai, F.; Wang, L.J. Application of slice modeling technology in finite element analysis of agricultural products. *J. Chin. Agric. Mech.* **2013**, *34*, 95–98.
69. Liu, C.L.; Wang, Y.L.; Song, J.N.; Li, Y.N.; Ma, T. Experiment and discrete element model of rice seed based on 3D laser scanning. *Trans. Chin. Soc. Agric. Eng.* **2016**, *32*, 294–300.
70. Yan, H. 3D scanner-based corn seed modeling. *Appl. Eng. Agric.* **2015**, *32*, 181–188.
71. Chen, Z.P.; Carl, W.; Eric, V. Determination of material and interaction properties of maize and wheat kernels for DEM simulation. *Biosyst. Eng.* **2020**, *195*, 208–226. [CrossRef]
72. Lei, X.L.; Liao, Y.T.; Liao, Q.X. Simulation of seed motion in seed feeding device with DEM-CFD coupling approach for rapeseed and wheat. *Comput. Electron. Agric.* **2016**, *131*, 29–39. [CrossRef]
73. Li, H.; Zeng, S.; Luo, X.; Fang, L.; Liang, Z.; Yang, W. Design, DEM simulation, and field experiments of a novel precisionseeder for dry direct-seeded rice with film mulching. *Agriculture* **2021**, *11*, 378. [CrossRef]
74. Zhang, S.; Tekeste, M.Z.; Li, Y.; Gaul, A.; Zhu, D.Q.; Liao, J. Scaled-up rice grain modelling for DEM calibration and the validation of hopper flow. *Biosyst. Eng.* **2020**, *194*, 196–212. [CrossRef]
75. Shi, L.R.; Sun, W.; Zhao, W.Y.; Yang, X.P.; Feng, B. Parameter determination and validation of discrete element model of seed potato mechanical seeding. *Trans. Chin. Soc. Agric. Eng.* **2018**, *34*, 35–42.

76. Xu, L.Z.; Wei, C.; Liang, Z.; Chai, X.Y.; Li, Y.M.; Liu, Q. Development of rapeseed cleaning loss monitoring system and experiments in a combine harvester. *Biosyst. Eng.* **2019**, *178*, 118–130. [CrossRef]
77. Binelo, M.O.; Lima, R.F.; Khatchatourian, O.A.; Stransky, J. Modelling of the drag force of agricultural seeds applied to the discrete element method. *Biosyst. Eng.* **2019**, *178*, 168–175. [CrossRef]
78. Cunha, R.N.; Santos, K.G.; Lima, R.N.; Duarte, C.R.; Barrozo, M.A.S. Repose angle of monoparticles and binary mixture: An experimental and simulation study. *Powder Technol.* **2016**, *303*, 203–211. [CrossRef]
79. Ghodki, B.M.; Kumar, K.C.; Goswami, T.K. Modeling breakage and motion of black pepper seeds in cryogenic mill. *Adv. Powder Technol.* **2018**, *29*, 1055–1071. [CrossRef]
80. Markauskas, D.; Álvaro, R.G.; Kačianauskas, R.; Zdancevičius, E. Maize grain shape approaches for DEM modelling. *Comput. Electron. Agric.* **2015**, *118*, 247–258. [CrossRef]
81. Tekeste, M.Z.; Mousaviraad, M.; Rosentrater, K.A. Discrete Element Model calibration using multi-responses and simulation of corn flow in a commercial grain auger. *Trans. ASABE* **2018**, *61*, 1743–1755. [CrossRef]
82. Liu, F.Y. *Discrete Element modeling of the Wheat Particles and Short Straw in Cleaning Devices*; Northwest A&F University: Yangling, China, 2018.
83. Mousaviraad, M.; Tekeste, M.Z.; Rosentrater, K.A. Calibration and validation of a Discrete Element Model of corn using grain flow simulation in a commercial screw grain auger. *Trans. ASABE* **2017**, *60*, 1403–1415. [CrossRef]
84. Shi, L.R.; Ma, Z.T.; Zhao, W.Y.; Yang, X.P.; Sun, B.G.; Zhang, J.P. Calibration of simulation parameters of flaxed seeds using discrete element method and verification of seed-metering test. *Trans. Chin. Soc. Agric. Eng.* **2019**, *35*, 25–33.
85. Wang, D. *Design and Simulation of Centralized Centrifugal Precision Seed-Metering System for Rapeseed*; Huazhong Agricultural University: Wuhan, China, 2019.
86. Molenda, M.; Horabik, J. *Mechanical Properties of Granular Agro-Materials and Food Powders for Industrial Practice. Part I. Characterization of Mechanical Properties of Particulate Solids for Storage and Handling*; Institute of Agrophysics PAS: Lublin, Poland, 2005.
87. Parafiniuk, P.; Molenda, M.; Horabik, J. Discharge of rapeseeds from a model silo: Physical testing and discrete element method simulations. *Comput. Electron. Agric.* **2013**, *97*, 40–46. [CrossRef]
88. Barrios, G.K.P.; Carvalho, R.M.; Kwade, A.; Tavares, L.M. Contact parameter estimation for DEM simulation of iron ore pellet handling. *Powder Technol.* **2013**, *248*, 84–93. [CrossRef]
89. Elskamp, F.; Kruggel, E.H.; Hennig, M.; Teipel. A strategy to determine DEM parameters for spherical and nonspherical particles. *Granul. Matter* **2017**, *19*, 46. [CrossRef]
90. Shi, L.R.; Zhao, W.Y.; Sun, B.G.; Sun, W. Determination of the coefficient of rolling friction of irregularly shaped maize particles by using discrete element method. *Int. J. Agric. Biol. Eng.* **2020**, *13*, 15–25. [CrossRef]
91. Liu, F.Y.; Zhang, J.; Li, B.; Chen, J. Calibration of parameters of wheat required in discrete element method simulation based on repose angle of particle heap. *Trans. Chin. Soc. Agric. Eng.* **2016**, *32*, 247–253.
92. Hao, J.J.; Long, S.F.; Li, H.; Jia, Y.L.; Ma, Z.K.; Zhao, J.G. Development of Discrete Element Model and calibration of simulation parameters for mechanically-harvested yam. *Trans. Chin. Soc. Agric. Eng.* **2019**, *35*, 34–42.
93. Wu, M.C.; Cong, J.L.; Yan, Q.; Zhu, T.; Peng, X.Y.; Wang, Y.S. Calibration and experiments for discrete element simulation parameters of peanut seed particles. *Trans. Chin. Soc. Agric. Eng.* **2020**, *36*, 30–38.
94. Yu, Q.X.; Liu, Y.; Chen, X.B.; Sun, K.; Lai, Q.H. Calibration and experiment of simulation parameters for panax notoginseng seeds based on DEM. *Trans. Chin. Soc. Agric. Mach.* **2020**, *51*, 123–132.
95. Wojtkowski, M.; Pecen, J.; Horabik, J.; Molenda, M. Rapeseed impact against a flat surface: Physical testing and DEM simulation with two contact models. *Powder Technol.* **2010**, *198*, 61–68. [CrossRef]
96. Liu, F.Y.; Zhang, J.; Chen, J. Construction of visco-elasto plasticity contact model of vibratory screening and its parameters calibration for wheat. *Trans. Chin. Soc. Agric. Eng.* **2018**, *34*, 37–43.
97. Wang, J. *Dynamic Modeling and Simulation Analysis of Seed-Metering Process of a Permanent—Magnet Magnetic Plate Precision Seed-Metering Device*; Jiangsu University: Zhenjiang, China, 2012.
98. Hu, J.P.; Zhou, C.J.; Hou, C.; Wang, J. Simulation analysis of seed-filling performance of magnetic plate seed-metering device by discrete element method. *Trans. Chin. Soc. Agric. Mach.* **2014**, *45*, 94–98.
99. Yu, Y.J.; Yu, J.Y.; Li, Q.L.; Yu, J.Q.; Fu, C. Analysis on the contact interaction between thresher and corn ears based on the DEM. *Appl. Mech. Mater.* **2013**, *246–247*, 71–77. [CrossRef]
100. Yu, Y.J. *Research on Analysis Method of Corn Threshing Based on 3D DEM*; Jilin University: Changchun, China, 2013.
101. Yu, Y.J.; Yu, J.Q.; Chen, Z.; Fu, H. Design of 3-D DEM boundary modeling software. *Trans. Chin. Soc. Agric. Eng.* **2011**, *42*, 99–103.
102. Zeng, Z.W.; Ma, X.; Chen, Y.; Qi, L. Modelling residue incorporation of selected chisel ploughing tools using the discrete element method (DEM). *Soil Tillage Res.* **2020**, *197*, 104505. [CrossRef]
103. Shi, Y.Y.; Sun, X.R.; Wang, X.C.; Hu, Z.C.; Newman, D.; Ding, W.M. Numerical simulation and field tests of minimum-tillage planter with straw smashing and strip laying based on EDEM software. *Comput. Electron. Agric.* **2019**, *166*, 105021. [CrossRef]
104. Zeng, Z.W.; Chen, Y. Simulation of straw movement by discrete element modelling of straw-sweep-soil interaction. *Biosyst. Eng.* **2019**, *180*, 25–35. [CrossRef]
105. Huo, L.L.; Tian, Y.S.; Zhao, L.X.; Yao, Z.L.; Hou, S.L.; Meng, H.B. Research on physical property of crop straw and test methods. *Renew. Energy Resour.* **2011**, *29*, 86–92.

106. Zhang, T.; Liu, F.; Zhao, M.Q.; Ma, Q.; Wang, W.; Fan, Q.; Yan, P. Determination of corn stalk contact parameters and calibration of Discrete Element Method simulation. *J. China Agric. Univ.* **2018**, *23*, 120–127.
107. Lenaerts, B.; Aertsen, T.; Tijskens, E.; Ketelaere, B.; Ramon, H.; Baerdemaeker, J.; Saeys, W. Simulation of grain-straw separation by a discrete element approach with bendable straw particles. *Comput. Electron. Agric.* **2014**, *101*, 24–33. [CrossRef]
108. Leblicq, T.; Smeets, B.; Ramon, H.; Saeys, W. A discrete element approach for modelling the compression of crop stems. *Comput. Electron. Agric.* **2016**, *123*, 80–88. [CrossRef]
109. Leblicq, T.; Smeets, B.; Vanmaercke, S.; Ramon, H.; Saeys, W. A discrete element approach for modelling bendable crop stems. *Comput. Electron. Agric.* **2016**, *124*, 141–149. [CrossRef]
110. Schramm, M.; Tekeste, M.Z.; Plouffe, C.; Harby, D. Estimating bond damping and bond Young's modulus for a flexible wheat straw discrete element method model. *Biosyst. Eng.* **2019**, *186*, 349–355. [CrossRef]
111. Wang, Q.R.; Mao, H.P.; Li, Q.L. Simulation of vibration response of flexible crop stem based on discrete element method. *Trans. Chin. Soc. Agric. Mach.* **2020**, *51*, 131–137.
112. Liu, F.Y.; Zhang, J.; Chen, J. Modeling of flexible wheat straw by discrete element method and its parameters calibration. *Int. J. Agric. Biol. Eng.* **2018**, *11*, 42–46. [CrossRef]
113. Guo, Q. *Experimental Research on the Cutting Mechanism and Performance of Rattan Straw*; Jiangsu University: Zhengjiang, China, 2016.
114. Zhang, L.X. *Research on the Discrete Element Modeling Method of Corn Stalk's Mechanical Characteristics*; Northwest A&F University: Yangling, China, 2017.
115. Liao, Y.T.; Liao, Q.X.; Zhou, Y.; Wang, Z.T.; Jiang, Y.J.; Liang, F. Parameters calibration of Discrete Element Model of fodder rape crop harvest in bolting stage. *Trans. Chin. Soc. Agric. Mach.* **2020**, *51*, 73–82.
116. Liao, Y.T.; Wang, Z.T.; Liao, Q.X.; Wan, X.Y.; Zhou, Y.; Liang, F. Calibration of Discrete Element Model parameters of forage rape stalk at early pod stage. *Trans. Chin. Soc. Agric. Mach.* **2020**, *51*, 236–243.
117. Zhang, F.W.; Song, X.F.; Zhang, X.K.; Zhang, F.Q.; Wei, C.J.; Dai, F. Simulation and experiment on mechanical characteristics of kneading and crushing process of corn straw. *Trans. Chin. Soc. Agric. Eng.* **2019**, *35*, 58–65.
118. Sitkei, G. *Mechanics of Agricultural Materials*; Elsevier Science Pub. Co. Inc.: New York, NY, USA, 1986.
119. Annoussamy, M.; Richard, G.; Recous, S.; Guerif, J. Change in mechanical properties of wheat straw due to decomposition and moisture. *Appl. Eng. Agric.* **2000**, *16*, 657–664. [CrossRef]
120. Wright, C.; Pryfogle, P.; Stevens, N.; Hess, J.; Ulrich, T. Biomechanics of wheat/barley straw and corn stover. In *Twenty-Sixth Symposium on Biotechnology for Fuels and Chemicals*; Springer: Berlin/Heidelberg, Germany, 2005; Volume 121, pp. 5–19.
121. Liu, C.L.; Wei, D.; Song, J.N.; Li, Y.N.; Du, X.; Zhang, F.Y. Systematic study on boundary parameters of Discrete Element Simulation of granular fertilizer. *Trans. Chin. Soc. Agric. Mach.* **2018**, *49*, 82–89.
122. Wen, X.Y.; Jia, H.L.; Zhang, S.W.; Yuan, H.F.; Wang, G.; Chen, T.Y. Test of suspension velocity of granular fertilizer based on Edem-Fluent coupling. *Trans. Chin. Soc. Agric. Mach.* **2020**, *51*, 69–77.
123. Wen, X.Y.; Yuan, H.F.; Wang, G.; Jia, H.L. Calibration method of friction coefficient of granular fertilizer by discrete element simulation. *Trans. Chin. Soc. Agric. Mach.* **2020**, *51*, 115–122.
124. Yuan, J.; Xin, C.B.; Niu, Z.R.; Li, Y.; Liu, X.H.; Xin, S.; Wang, J.F.; Wang, L.H. Discrete element model simulation and verification of fertilizer blending uniformity of variable rate fertilization based on relevance vector machine. *Trans. Chin. Soc. Agric. Eng.* **2019**, *35*, 37–45.
125. Peng, F.; Wang, H.Y.; Fang, F.; Liu, Y.D. Calibration of Discrete Element Model parameters for pellet feed based on injected section method. *Trans. Chin. Soc. Agric. Mach.* **2018**, *49*, 140–147.
126. Luo, S.; Yuan, Q.X.; Gouda, S.; Yang, L.Y. Parameters calibration of vermicomposting nursery substrate with discrete element method based on jkr contact model. *Trans. Chin. Soc. Agric. Mach.* **2018**, *49*, 343–350.
127. Peng, C.W.; Xu, D.J.; He, X.; Tang, Y.H.; Sun, S.L. Parameter calibration of discrete element simulation model for pig manure organic fertilizer treated with Hermetia illucen. *Trans. Chin. Soc. Agric. Eng.* **2020**, *36*, 212–218.
128. Coetzee, C.J.; Lombard, S.G. The destemming of grapes: Experiments and discrete element modelling. *Biosyst. Eng.* **2013**, *114*, 232–248. [CrossRef]
129. Fang, H.M.; Ji, C.Y.; Zhang, Q.Y.; Guo, J. Force analysis of rotary blade based on distinct element method. *Trans. Chin. Soc. Agric. Eng.* **2016**, *32*, 54–59.
130. Xiong, P.Y.; Yang, Z.; Sun, Z.Q.; Zhang, Q.Q.; Huang, Y.Q.; Zhang, Z.W. Simulation analysis and experiment for three-axis working resistances of rotary blade based on discrete element method. *Trans. Chin. Soc. Agric. Eng.* **2018**, *34*, 113–121.
131. Zhu, X.M. The study on the power consumption of a rotary tiller in rototilling. *J. Anhui Inst. Technol.* **1986**, *5*, 121–133.
132. Zhao, H.B.; He, J.; Li, H.W.; Ma, S.C.; He, J.; Wang, Q.J.; Lu, C.Y.; Zheng, Z.Q.; Zhang, C. The effect of various edge-curve types of plain-straight blades for strip tillage seeding on torque and soil disturbance using DEM. *Soil Tillage Res.* **2020**, *202*, 104674. [CrossRef]
133. Lee, K.S.; Park, S.H.; Park, W.Y.; Lee, C. Strip tillage characteristics of rotary tiller blades for use in a dryland direct rice seeder. *Soil Tillage Res.* **2003**, *71*, 25–32. [CrossRef]
134. Marenya, M.O. Performance characteristics of a deep tilling rotavator. *Jpn. J. Appl. Phys. Suppl.* **2010**, *21*, 209–212.
135. Matin, M.A.; Fielke, J.M.; Desbiolles, J.M.A. Torque and energy characteristics for strip-tillage cultivation when cutting furrows using three designs of rotary blade. *Biosyst. Eng.* **2015**, *129*, 329–340. [CrossRef]

136. Zheng, K.; He, J.; Li, H.W.; Chen, L.Q.; Hu, H.N.; Liu, W.Z. Influence of working order on working quality and power consumption of subsoiling and rotary tillage combined machine. *Trans. Chin. Soc. Agric. Eng.* **2017**, *33*, 52–60.
137. Zhao, H.B. *Study on Driven Seedbed-Cleaning and Anti-Blocking Device of Residue Inter-Row Side-Throwing for Minimum till Wheat Seeding*; China Agricultural University: Beijing, China, 2019.
138. Huang, Y.X.; Hang, C.G.; Yuan, M.C.; Wang, B.T.; Zhu, R.X. Discrete Element Simulation and experiment on disturbance behavior of subsoiling. *Trans. Chin. Soc. Agric. Mach.* **2016**, *47*, 80–88.
139. Hang, C.G.; Gao, X.J.; Yuan, M.C.; Huang, Y.X.; Zhu, R.X. Discrete element simulations and experiments of soil disturbance as affected by the tine spacing of subsoiler. *Biosyst. Eng.* **2018**, *168*, 73–82. [CrossRef]
140. Hang, C.G.; Huang, Y.X.; Zhu, R.X. Analysis of the movement behaviour of soil between subsoilers based on the discrete element method. *J. Terramech.* **2017**, *74*, 35–43. [CrossRef]
141. Li, B.; Chen, Y.; Chen, J. Comparison of two subsoiler designs using the discrete element method (DEM). *Trans. ASABE* **2018**, *61*, 1529–1537. [CrossRef]
142. Wang, X.Z.; Yue, B.; Gao, X.J.; Zheng, Z.Q.; Zhu, R.Q.; Huang, Y.X. Discrete Element simulations and experiments of disturbance behavior as affected by mounting height of subsoiler's wing. *Trans. Chin. Soc. Agric. Mach.* **2018**, *49*, 124–136.
143. Sun, J.Y.; Wang, Y.M.; Ma, Y.H.; Tong, J.; Zhang, Z.J. DEM simulation of bionic subsoilers (tillage depth > 40 cm) with drag reduction and lower soil disturbance characteristics. *Adv. Eng. Softw.* **2018**, *119*, 30–37. [CrossRef]
144. Wang, J.Y. *Design and Experiment of Subsoiler-Stubble Chopper Device in No-Tillage*; Northeast Agricultural University: Haerbin, China, 2019.
145. Ma, Y.J.; Wang, A.; Zhao, J.G.; Hao, J.J.; Li, C.J.; Ma, L.P.; Zhao, F.W.; Wu, Y. Simulation analysis and experiment of drag reduction effect of convex blade subsoiler based on discrete element method. *Trans. Chin. Soc. Agric. Eng.* **2019**, *35*, 16–23.
146. Wang, Y.X.; Zhang, D.X.; Yang, L.; Cui, T.; Zhong, X. Modeling the interaction of soil and a vibrating subsoiler using the discrete element method. *Comput. Electron. Agric.* **2020**, *174*, 0168–1699. [CrossRef]
147. Saunders, C.; Ucgul, M.; Godwin, R.J. Discrete element method (DEM) simulation to improve performance of a mouldboard skimmer. *Soil Tillage Res.* **2021**, *205*, 104764. [CrossRef]
148. Cheng, Y.M. *The Ridging Performance Analysis & Optimization of Ridging Device of Potato Cultivator*; Xihua University: Chengdu, China, 2017.
149. Liu, X. *Design and Experiment of Cultivator Ridging Mechanism and Resistance Test Mount Device*; Jilin University: Changchun, China, 2020.
150. Khot, L.R.; Salokhe, V.M.; Jayasuriya, H.P.W.; Nakashima, H. Experimental validation of distinct element simulation for dynamic wheel-soil interaction. *J. Terramech.* **2007**, *44*, 429–437. [CrossRef]
151. Chen, J.Q. *Design and Experiment of Press Roller with Bionic Convex Structure*; Northeast Agricultural University: Haerbin, China, 2018.
152. Söhne, W. Druckverteilung im boden und bodenverformung unter schlepperreifen. *Grundl. Landtech. -Konstr.* **1953**, *5*, 49–63.
153. Lamandé, M.; Schjønning, P. Soil mechanical stresses in high wheel load agricultural field traffic: A case study. *Soil Res.* **2018**, *56*, 129–135. [CrossRef]
154. Pue, J.D.; Lamandé, M.; Schjønning, P.; Cornelis, W.M. DEM simulation of stress transmission under agricultural traffic Part 3: Evaluation with field experiment. *Soil Tillage Res.* **2020**, *200*, 104606. [CrossRef]
155. Du, Y.H.; Gao, J.W.; Jiang, L.; Jiang, L.H.; Zhang, Y.C. Numerical analysis on tractive performance of off-road wheel steering on sand using discrete element method. *J. Terramech.* **2017**, *71*, 25–43. [CrossRef]
156. Zhao, C.L.; Zang, M.Y. Application of the FEM/DEM and alternately moving road method to the simulation of tire-sand interactions. *J. Terramech.* **2017**, *72*, 27–38. [CrossRef]
157. Nishiyama, K.; Nakashima, H.; Yoshida, T.; Shimizu, H.; Miyasaka, J.; Ohdoi, K. FE-DEM with interchangeable modeling for off-road tire traction analysis. *J. Terramech.* **2018**, *78*, 15–25. [CrossRef]
158. Nakanishi, R.; Nakashima, H.; Miyasaka, J.; Ohdoi, K. Tractive performance analysis of a lugged wheel by open-source 3D DEM software. *J. Terramech.* **2020**, *92*, 51–65. [CrossRef]
159. Zeng, H.Y.; Xu, W.; Zang, M.Y.; Yang, P.; Guo, X.B. Calibration and validation of DEM-FEM model parameters using upscaled particles based on physical experiments and simulations. *Adv. Powder Technol.* **2020**, *31*, 3947–3959. [CrossRef]
160. Wang, Y.M.; Xue, W.L.; Ma, Y.H.; Tong, J.; Liu, X.P.; Sun, J.Y. DEM and soil bin study on a biomimetic disc furrow opener. *Comput. Electron. Agric.* **2019**, *156*, 209–216. [CrossRef]
161. Li, B.; Chen, Y.; Chen, J. Modeling of soil-claw interaction using the discrete element method (DEM). *Soil Tillage Res.* **2016**, *158*, 177–185. [CrossRef]
162. Barr, J.B.; Ucgul, M.; Desbiolles, J.M.A.; Fielke, J.M. Simulating the effect of rake angle on narrow opener performance with the discrete element method. *Biosyst. Eng.* **2018**, *171*, 1–15. [CrossRef]
163. Ucgul, M.; Fielke, J.M.; Saunders, C. Defining the effect of sweep tillage tool cutting edge geometry on tillage forces using 3D discrete element modelling. *Inf. Process. Agric.* **2015**, *2*, 130–141. [CrossRef]
164. Zhang, Q.S.; Liao, Q.X.; Ji, W.F.; Liu, H.B.; Zhou, Y.; Xiao, W.L. Surface optimization and experiment on ditch plow of direct rapeseed seeder. *Trans. Chin. Soc. Agric. Mach.* **2015**, *46*, 53–59.
165. Liu, X.P.; Zhang, Q.S.; Liu, L.C.; Wei, G.L.; Xiao, W.L.; Liao, Q.X. Surface optimization of ship type ditching system based on differential geometry and edem simulation. *Trans. Chin. Soc. Agric. Mach.* **2019**, *50*, 59–69.

166. Shi, S.; Zhang, D.X.; Yang, L.; Cui, T.; Li, K.H.; Yin, X.W. Simulation and verification of seed-filling performance of pneumatic-combined holes maize precision seed-metering device based on EDEM. *Trans. Chin. Soc. Agric. Eng.* **2015**, *31*, 62–69.
167. Yu, J.Q.; Shen, Y.F.; Niu, X.T.; Fu, H.; Ni, T.H. DEM simulation and analysis of the clearing process in precision metering device with combination inner-cell. *Trans. CSAE* **2008**, *24*, 105–109.
168. Zhang, T.; Liu, F.; Zhao, M.Q.; Liu, Y.Q.; Li, F.; Chen, C. Movement law of maize population in seed room of seed metering device based on discrete element method. *Trans. Chin. Soc. Agric. Eng.* **2016**, *32*, 27–35.
169. Tian, L.Q. *Mechanism Analysis and Experimental Study on the Rice Seed Sowing Device with Ejection Ear Spoon Type*; Northeast Agricultural University: Harbin, China, 2017.
170. Lei, X.L.; Yang, W.H.; Yang, L.J.; Liu, L.Y.; Liao, Q.X.; Ren, W.J. Design and experiment of seed hill-seeding centralized metering device for rapeseed. *Trans. Chin. Soc. Agric. Mach.* **2020**, *51*, 54–64.
171. Hou, J.L.; Wang, H.X.; Niu, Z.R.; Xi, R.; Li, T.H. Discrete element simulation and experiment of picking and clearing performance of garlic seed-picking device. *Trans. Chin. Soc. Agric. Eng.* **2019**, *35*, 48–57.
172. Lei, X.L.; Liao, Y.T.; Wang, L.; Wang, D.; Yao, L.; Liao, Q.X. Simulation of gas-solid two-phase flow and parameter optimization of pressurized tube of air-assisted centralized metering device for rapeseed and wheat. *Trans. Chin. Soc. Agric. Eng.* **2017**, *33*, 67–75.
173. Han, D.D.; Zhang, D.X.; Yang, L.; Cui, T.; Ding, Y.Q.; Dian, X.H. Optimization and experiment of inside-filling air-blowing seed metering device based on EDEM-CFD. *Trans. Chin. Soc. Agric. Mach.* **2017**, *48*, 43–51.
174. Liu, Y.Q.; Zhao, M.Q.; Liu, F.; Yang, T.J.; Zhang, T.; Li, F.L. Simulation and optimization of working parameters of air suction metering device based on discrete element. *Trans. Chin. Soc. Agric. Mach.* **2016**, *47*, 65–73.
175. Shi, S.; Liu, H.; Wei, G.J.; Zhou, J.L.; Jian, S.Q.; Zhang, R.F. Optimization and experiment of pneumatic seed metering device with guided assistant filling based on EDEM-CFD. *Trans. Chin. Soc. Agric. Mach.* **2020**, *51*, 54–66.
176. Liu, C.L.; Li, Y.N.; Song, J.N.; Ma, T.; Wang, M.M.; Wang, X.J.; Zhang, C. Performance analysis and experiment on fertilizer spreader with centrifugal swing disk based on EDEM. *Trans. Chin. Soc. Agric. Eng.* **2017**, *33*, 32–39.
177. Zha, X.; Zhang, G.; Han, Y.; Salem, A.E.; Fu, J.; Zhou, Y. Structural optimization and performance evaluation of blocking wheel-type screw fertilizer distributor. *Agriculture* **2021**, *11*, 248. [CrossRef]
178. Gao, G.B. *Design and Experiment of Key Parts of Side-Depth Fertilizer Device with Pneumatic Conveying for Paddy*; Northeast Agricultural University: Harbin, China, 2019.
179. Liu, Z.D.; Wang, Q.J.; Li, H.W.; He, J.; Lu, C.Y.; Yu, C.C. Fertilizer injecting route analysis and test for air-blowing seed-fertilizer hole-applicator via CFD-DEM coupling. *Trans. Chin. Soc. Agric. Eng.* **2019**, *35*, 18–25.
180. Zhou, H.B.; Chen, Y.; Sadek, M.A. Modelling of soil-seed contact using the Discrete Element Method (DEM). *Biosyst. Eng.* **2014**, *121*, 56–66. [CrossRef]
181. Wang, W.Z.; Liu, W.R.; Yuan, L.H.; Qu, Z.; He, X.; Lv, Y.L. Simulation and experiment of single longitudinal axial material movement and establishment of wheat plants model. *Trans. Chin. Soc. Agric. Mach.* **2020**, *51*, 170–180.
182. Wang, Q.R.; Mao, H.P.; Li, Q.L. Modelling and simulation of the grain threshing process based on the discrete element method. *Comput. Electron. Agric.* **2020**, *178*, 105790. [CrossRef]
183. Mou, X.D.; Jiang, H.X.; Sun, Y.C.; Xu, H.G.; Yao, Y.C.; Geng, D.Y. Simulation optimization and experiment of disc-type grain crushing device of silage corn harvester. *Trans. Chin. Soc. Agric. Mach.* **2020**, *51*, 218–226.
184. Li, H.C.; Li, Y.M.; Tang, Z.; Xu, L.Z.; Zhao, Z. Numerical simulation and analysis of vibration screening based on EDEM. *Trans. Chin. Soc. Agric. Eng.* **2011**, *27*, 117–121.
185. Wang, C.J.; Liu, Q.; Ma, L.Z.; Li, L. Cottonseed particle motion Law in 3-DOF hybrid vibration screen surface. *Trans. Chin. Soc. Agric. Eng.* **2015**, *31*, 49–56.
186. Han, M. *Study on the Maize Threshing and Cleaning Mechanism with Low Loss and Anti-Blocking and Its Linkage Control*; Jiangsu University: Zhengjiang, China, 2020.
187. Ma, Z.; Li, Y.M.; Xu, L.Z. Discrete-element method simulation of agricultural particles' motion in variable-amplitude screen box. *Comput. Electron. Agric.* **2015**, *118*, 92–99. [CrossRef]
188. Li, H.C.; Li, Y.M.; Fang, G.; Zhao, Z.; Xu, L.Z. CFD–DEM simulation of material motion in air-and-screen cleaning device. *Comput. Electron. Agric.* **2012**, *88*, 111–119. [CrossRef]
189. Xu, L.Z.; Li, Y.; Chai, X.Y.; Wang, G.M.; Liang, Z.W.; Li, Y.M.; Li, B.J. Numerical simulation of gas-solid two-phase flow to predict the cleaning performance of rice combine harvesters. *Biosyst. Eng.* **2020**, *190*, 11–24. [CrossRef]
190. Chen, P.Y.; Jia, F.G.; Liu, H.R.; Han, Y.L.; Zeng, Y.; Meng, X.Y. Effects of feeding direction on the hulling of paddy grain in a rubber roll huller. *Biosyst. Eng.* **2019**, *183*, 196–208. [CrossRef]
191. Chen, P.Y.; Han, Y.L.; Jia, F.G.; Meng, X.Y.; Xiao, Y.W.; Bai, S.G. DEM simulations and experiments investigating the influence of feeding plate angle in a rubber-roll paddy grain huller. *Biosyst. Eng.* **2021**, *201*, 23–41. [CrossRef]
192. Wang, Z.Y.; Yuan, F.L.; Wang, F.C.; Sun, H.N. Simulation test of returning performance of returning device for fully enclosed belt conveyor. *China Oils Fats* **2020**, *45*, 112–118.
193. Cleary, P.W.; Sawley, M.L. DEM modelling of industrial granular flows: 3D case studies and the effect of particle shape on hopper discharge. *Appl. Math. Model.* **2002**, *26*, 89–111. [CrossRef]
194. Kobylka, R.; Horabik, J.; Molend, M. Development of a rarefaction wave at discharge initiation in a storage silo: DEM simulations. *Particuology* **2018**, *36*, 37–49. [CrossRef]

195. Zaki, M.; Siraj, M.S. Study of a flat-bottomed cylindrical silo with different orifice shapes. *Powder Technol.* **2019**, *354*, 641–652. [CrossRef]
196. Horabik, J.; Parafiniuk, P.; Molenda, M. Stress profile in bulk of seeds in a shallow model silo as influenced by mobilisation of particle-particle and particle-wall friction: Experiments and DEM simulations. *Powder Technol.* **2018**, *327*, 320–334. [CrossRef]
197. Kannan, A.S.; Lassen, N.C.K.; Carstensen, J.M.; Lund, J.; Sasic, S. Segregation phenomena in gravity separators: A combined numerical and experimental study. *Powder Technol.* **2016**, *301*, 679–693. [CrossRef]
198. Kannan, A.S.; Jareteg, K.; Lassen, N.C.K.; Carstensen, J.M.; Hansen, M.A.E.; Dam, D.; Sasic, S. Design and performance optimization of gravity tables using a combined CFD-DEM framework. *Powder Technol.* **2017**, *318*, 423–440. [CrossRef]
199. Meng, X.Y.; Jia, F.G.; Qiu, H.L.; Han, Y.L.; Zeng, Y.; Xiao, Y.W.; Chen, P.Y. DEM study of white rice separation in an indented cylinder separator. *Powder Technol.* **2019**, *348*, 1–12. [CrossRef]

agriculture

MDPI

Article

Influence of Mechanical and Intelligent Robotic Weed Control Methods on Energy Efficiency and Environment in Organic Sugar Beet Production

Indrė Bručienė [1,*], Domantas Aleliūnas [1], Egidijus Šarauskis [1] and Kęstutis Romaneckas [2]

[1] Institute of Agricultural Engineering and Safety, Agriculture Academy, Vytautas Magnus University, LT-44248 Kaunas, Lithuania; domantas.aleliunas@vdu.lt (D.A.); egidijus.sarauskis@vdu.lt (E.Š.)
[2] Institute of Agroecosystems and Soil Sciences, Agriculture Academy, Vytautas Magnus University, LT 44248 Kaunas, Lithuania; kestutis.romaneckas@vdu.lt
* Correspondence: indre.bruciene@vdu.lt; Tel.: +370-614-26336

Abstract: Rapidly warming climate, tightening environmental requirements, an aging society, rising wages, and demand for organic products are forcing farming to be more efficient and sustainable. The main aim of this study was to perform an analytical analysis and to determine the energy use and GHG emissions of organic sugar beet production using different weed control methods. Seven different methods of non-chemical weed control were compared. Mechanical inter-row loosening, inter-row cutting and mulching with weeds, weed smothering with catch crops, and thermal inter-row steaming were performed in field experiments at the Experimental Station of Vytautas Magnus University (Lithuania, 2015–2017). The other three, namely, automated mechanical inter-row loosening with cameras for row-tracking, inter-row loosening with a diesel-powered robot, and inter-row loosening with an electric robot were calculated analytically. The results showed that the average total energy use of organic sugar beet production was 27,844 MJ ha^{-1}, of which manure costs accounted for 48–53% and diesel fuel for 29–35%. An average energy efficiency ratio was 7.18, while energy productivity was 1.83 kg MJ ha^{-1}. Analysis of GHG emissions showed that the total average GHG emissions to the environment from organic sugar beet production amounted to 4552 kg CO_{2eq} ha^{-1}, and the average GHG emissions ratio was 4.47. The most sustainable organic sugar beet production was achieved by using mechanical inter-row loosening with a diesel-powered robot for weed control.

Keywords: energy indicators; GHG emissions; organic farming; sugar beet; non-chemical weed control; robots

Citation: Bručienė, I.; Aleliūnas, D.; Šarauskis, E.; Romaneckas, K. Influence of Mechanical and Intelligent Robotic Weed Control Methods on Energy Efficiency and Environment in Organic Sugar Beet Production. *Agriculture* **2021**, *11*, 449. https://doi.org/10.3390/agriculture11050449

Academic Editor: José Pérez-Alonso

Received: 26 March 2021
Accepted: 13 May 2021
Published: 15 May 2021

1. Introduction

1.1. Weeds in Sugar Beet Production

Weeds are one of the main problems in crop production because they compete with major field crops for vital resources such as water, nutrients, space, and light [1–4]. As a result of this competition, the yield is significantly reduced, its quality deteriorates, and it is harder to harvest. Depending on the crops, weeds can reduce crop productivity by up to 40% [5–8].

The most common annual weed species in sugar beet crops in Lithuania are *Chenopodium album*, *Sinapis arvensis*, *Echinochloa crus galli*, *Stellaria media*, and perennial weed species dominates by *Taraxacum officinalis*, *Elytrigia repens*, and *Plantago major* [9].

Sugar beet is particularly sensitive to weed competition, which can reduce the yield of sugar beet roots by 26 to 100% [2,4,10–15], even at low weed density of five weeds per 1 m^2 [16,17]. Therefore, weed control is very important and can significantly increase the yield of sugar beet [11,18]. Although organic farms tend to require more labor and investment, organic farming is becoming increasingly more popular across the EU, with the main countries growing organic beets being Germany (52% of the European organic farmland) and Austria (23%) [19]. Sugar beet is also an important crop in local Lithuanian industry as the main raw material for white sugar.

The problem of weeds is particularly relevant in organic crops, where the yield is already 25–50% lower compared to traditional farming [1]. Detailed literature analysis by Seufert et al. [20], comparing the yield data from ecological and traditional practices, showed that the yield of organic plants is generally lower, ranging from 5 to 34% of traditional system yields. Therefore, the weed control method chosen, and good farm management practice, have a huge impact not only on crop productivity and its quality but also on the impact on the environment.

1.2. Ecological Weed Control Methods

The conventional method of weed control using chemical herbicides is not acceptable in organic farming practices. Therefore, mechanical weed control is an alternative to chemical weed control [2,4,21,22]. In addition, chemical weed control is not only expensive but also has a negative impact on the environment [11]. Weed control based on chemical treatment has increased costs and made some weeds resistant to herbicides [23]. Mechanical weed control removes weeds physically by eradicating them, i.e., by shredding all plants or by separating weed stems and leaves from their roots [2,15]. The scientific literature presents a variety of positive and negative results from studies related to mechanical weed control [24]. Kunz et al. [17] found that mechanical inter-row loosening alone or in combination with herbicides results in a similar sugar beet yield to conventional spraying.

Mulching is another method of weed control that can also be used in organic crops. Plant mulch residues on the soil surface alter the physical and chemical environment of weed seeds, thereby inhibiting weed germination and growth. As an alternative to mechanical weed control, thermal weed control with heat, electricity, flame, hot water, and foam can be used [4,25]. The work of Kunz [26] revealed that the use of a living mulch in sugar beet production can reduce the amount of herbicide by up to 65% and the number of weeds by up to 83%.

Often, a mechanical method of weed control requires large investments in energy, labor, and time, and its speed and accuracy are limited by a lack of staff skills and experience [27]. In some cases, a mechanical method of weed control alone may not be enough, as such a method is not always able to eradicate weeds growing very close to sugar beet. In such cases, a manual weed control method is often still used. Due to the potential risk of crop damage, especially when weed control is performed in crop row between the plants, it is necessary to use precise mechanical tools [28,29]. Accurate position control of implements or tools based on digital technologies such as various types of cameras (RGB, hyperspectral, multispectral, etc.), sensors, RTK-GPS, and machine learning [4,14,18,24,30–33] provide the ability to control weeds in crop rows close to plants without damaging the crop. These technologies allow not only the control of weeds in a specific field location in real time, but also the very accurate classification of crop plants and weeds. Automated precise weed control by self-propelled robots has been rapidly evolving and has become one of the key areas of agricultural robotics research over the past decade [18,32,34–39]. Kunz et al. [14] evaluated camera-steered mechanical weed control and found that such an approach improved the mechanical efficiency of weed control in sugar beet, maize, and soybean crops. Automated inter-row weeding controlled by cameras killed 78% of weeds, while manual control killed only 65%. Automated steering with the help of visual cameras allows the operator not only to direct the cultivator's working parts closer to the main crop rows, but also to increase the driving speed from 4 km h^{-1} in manual steering to 7–10 km h^{-1} with automated steering. Automated and robotic techniques facilitate the operator's work, reduce labor time costs, and increase the efficiency of mechanical weeding compared to conventional weed control methods [17]. In addition, the smaller dimensions and weights of robots compared to traditional heavy tractors result in lower soil compaction [38–40].

1.3. Energy Use and GHG Emissions

Sustainable development requires management of energy use and greenhouse gas (GHG) emissions in all production processes [41]. Haciseferogullari et al. [42] examined the energy balance of sugar beet production and found that the total energy input, total energy output, output/input ratio, and net energy consumed in chemical production were 19,760 MJ ha^{-1}, 378,491 MJ ha^{-1}, 19.15, and 18.15, respectively. Meanwhile, Yousefi et al. [43] showed that, for sugar beet growing in Western Iran, the total input and output was 49517 and 1,095,360 MJ ha^{-1}, respectively. The calculated energy use efficiency was 22.12 and the carbon efficiency ratio of sugar beet was 10.95. Dimitrijevic et al. [44] estimated the energy efficiency of sugar beet production and found that 0.93 MJ kg^{-1} of energy was needed to produce one kilogram of product. The results of Šarauskis et al. [45] showed that energy consumption could be 10 to 70% lower in organic farming compared to conventional farming practices.

Agriculture makes a significant contribution to GHG emissions, with as much as 24% of total net GHG emissions coming from Agriculture, Forestry, and Other Land Use (AFOLU) [46]. Tzilivakis et al. [47] estimated energy consumption and GHG emissions from sugar beet production in the UK and reported that the average global warming potential (GWP) per ton of sugar beet grown was 0.024 tons of CO_2 t^{-1}. Pishgar-Komleh et al. [41] estimated that the total GHG emissions from cucumber production were 82724 kg CO_{2eq} ha^{-1}, with the highest emissions from diesel (61%), electricity (19%), and manure (14%). Organic crop production is associated with lower average GHG emissions per hectare than conventional farming [19,47].

Substantial research can be found in the scientific literature, providing indicators of energy and GHG emissions from different crops. However, no research demonstrating an analysis of energy use and GHG emissions in organic sugar beet production, investigating different weed control methods involving robots, has been found. Therefore, the aim of this study is to analyze, evaluate, and compare the impact of conventional as well as robotic weed control methods regarding the energy and environmental performance in organic sugar beet production. Mechanical and thermal weed control methods were performed in field experiments, and automated and robotic weed control systems were calculated analytically.

2. Materials and Methods

2.1. Description of Weed Control Methods

This study examines seven different weed control methods in organic sugar beet crops. The baseline data needed for this study concerning four non-automated weed control methods—inter-row loosening (E1), inter-row cutting and mulching with weeds (E2), inter-row cutting and mulching with catch crop (E3), and inter-row steaming (E4)—were obtained after experimental research at the Experimental Station of Vytautas Magnus University in 2015–2017. To avoid repetition, the location and conditions of the research are described in detail in Romaneckas et al. [48]. The other three weed control methods for comparative analysis—automated inter-row loosening with cameras for row-tracking (E5), inter-row loosening with a diesel-powered robot (E6), and inter-row loosening with a solar-powered robot (E7) (Figure 1)—were based on analytical theoretical methods. All weed control treatments were performed three times per season.

The electric robot (E7) is designed for weed control and sowing of crops sown in wide rows. The robot operates fully automatically using highly accurate GPS technology and is independent of cameras or sensors to detect plants and weeds. Because the robot knows the position of each seed, it can start weeding at a very early stage—before the crop begins to germinate. Robots using camera systems that need to recognize crops and weeds do not have this capability. Three weeding wires in each row ensure weed removal between rows. Weed control in rows between crops is performed by a weeding arm. The electric motor pulls its weeding arm in and out of the row [49]. The diesel-powered robot (E6) controls weeds by recognizing them with the help of cameras. Such systems use weed detection algorithms and generate weed maps.

The technical data of the automated inter-row loosening (E5) machine were taken from the manufacturer's specification, and the working speed (7.2 km h^{-1}) was selected based on the results of tests performed by Kunz et al. [17]. One of the robots used to control weeds was powered by a diesel engine [50]. The inter-row loosening machine supplied with the robot was selected according to the power requirement specified by the manufacturer. The operating speed (5 km h^{-1}) of the diesel-powered robot (E6) was selected following the manufacturer's recommendations for the weed control operation. Another robot compared in the study was powered by electricity with solar modules and batteries (E7). This robot can work 8 h or more without interruption [49]. Its working speed (1.0 km h^{-1}) was the lowest of all the weed control machines.

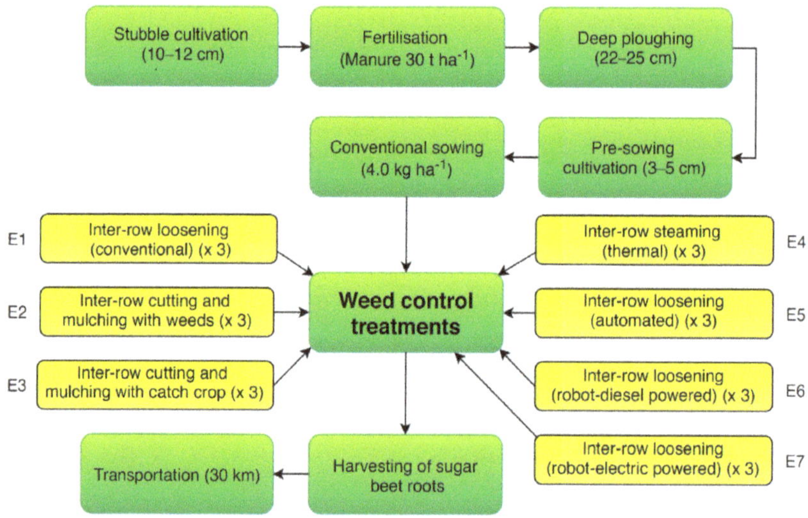

Figure 1. Scheme of weed control methods for comparative analysis. Note: (x3)—Technological operations were performed three times per season.

2.2. Calculation of the Energy Input/Output

This study examines the energy input for an area of 20 hectares of organic sugar beet production. In order to compare the energy efficiency of different weed control methods, energy input (agricultural machinery, fuel, working hours, etc.) and energy output were calculated. Energy input and technological indicators were assessed for each weed control method, including all organic sugar beet growing operations in accordance with the recommendations of the Lithuanian Institute of Agrarian Economics for agricultural companies, farmers, and other enterprises providing agricultural services [51]. Energy output was calculated using the results of sugar beet root yield, excluding leaf biomass left on the soil surface after harvest. In order to unify different units of measurement of agricultural input, a conversion method was applied. It uses input energy equivalents of fuel, human labor, agricultural machinery, seeds, etc., which are presented in Table 1.

Table 1. Energy equivalents used for the calculation of energy input and output.

Input/Output	Energy Equivalent	Unit	Reference(s)
Human labor	1.96	MJ h^{-1}	[52,53]
Diesel fuel	56.31	MJ L^{-1}	[53–55]
Propane	50.80	MJ kg^{-1}	[56,57]
Machinery	357.2	MJ h^{-1}	[58]
Electricity	11.93	MJ kWh^{-1}	[59,60]
Seed (sugar beet)	50.00	MJ kg^{-1}	[42,55,61,62]
Seed (catch crop, i.e., mustard)	34.20	MJ kg^{-1}	[62,63]
Manure	0.47	MJ kg^{-1} (FM)	[64,65]
Output—Sugar beet yield	17.28	MJ kg^{-1}	[62,66]

FM—fresh matter.

Energy input was calculated following the equation:

$$\text{Energy input}\ \left(\text{MJ ha}^{-1}\right) = \sum_{i=1}^{n} \text{Agricultural input}\ \left(\text{unit ha}^{-1}\right) * \text{Energy equivalent}\ \left(\text{MJ unit}^{-1}\right) \tag{1}$$

2.3. Energy Use Efficiency

Various energy efficiency parameters were calculated to evaluate and to compare different weed control methods. The energy efficiency ratio (Equation (2)) was determined by dividing the energy output of sugar beet (MJ ha^{-1}) by the total energy consumption (MJ ha^{-1}) [44,55,61,67]. Energy productivity (kg MJ^{-1}) is an indicator showing how many kilograms of sugar beet can be grown using one MJ of energy [44,55]. Net energy gain (MJ ha^{-1}) is expressed as the difference between the energy output and input [44,55,68,69]. Non-renewable energy includes diesel and machinery. Renewable energy consists of human labor, seeds, electricity, and manure.

$$\text{Energy efficiency ratio} = \frac{\text{Energy output}\ \left(\text{MJ ha}^{-1}\right)}{\text{Energy input}\ \left(\text{MJ ha}^{-1}\right)} \tag{2}$$

2.4. Fuel Consumption

One of very important direct energy inputs of sugar beet production is fuel. Diesel fuel consumption for the E1–E5 weed control methods was determined based on data published by LIAE [51]. For E6, fuel consumption was calculated according to the engine performance curve provided by the manufacturer [70], assuming that the robot was operating in economy mode and the engine crankshaft speed did not exceed 1600 min^{-1} (engine RPM did not exceed 1600 min^{-1}). Diesel was not used for the E7 method, but the energy use per hectare of weed control was calculated based on the hourly electricity consumption of the robot [49].

2.5. Assessment of GHG Emissions

Typically, GHG emissions into the atmosphere are expressed in carbon dioxide equivalent (CO_{2-eq}). The conversion factors presented in Table 2 were used to estimate GHG emissions from different organic sugar beet production systems. The GHG emissions from each sugar beet production process were calculated by estimating the input of diesel, propane gas, electricity, seeds, fertilizers, and machinery and multiplying them by the corresponding GHG factors—Equation (3):

$$\text{GHG emission}\ \left(\text{kg CO}_{2\text{eq}}\ \text{ha}^{-1}\right) = \sum_{i=1}^{n} \text{Agricultural input}\ \left(\text{unit ha}^{-1}\right) * \text{GHG coefficient}\ \left(\text{kg CO}_{2\text{eq}}\ \text{unit}^{-1}\right) \tag{3}$$

The sustainability of different weed control methods for organic sugar beet production was assessed using the ratio of the output of the obtained production to the total input of GHG emissions—Equation (4) [71]:

$$\text{GHG emission ratio} = \frac{\text{GHG emission output } (\text{kg CO}_{2\text{eq}}\ \text{ha}^{-1})}{\text{GHG emission input } (\text{kg CO}_{2\text{eq}}\ \text{ha}^{-1})} \quad (4)$$

Table 2. GHG coefficients (kg CO_{2-eq} unit^{-1}) of agricultural input and output.

Particulars	Description of Input and Output	Unit	GHG Coefficient	Reference(s)
A. Input				
Machinery	Tillage, sowing, crop care, harvesting, and transportation	MJ^{-1}	0.071	[41,60,72,73]
Diesel fuel	Tillage, sowing, crop care, harvesting, and transportation	L^{-1}	2.76	[41,60,71,73]
Electricity	Weed control with electric-powered robot	kWh^{-1}	0.78	[60,74]
Propane	Weed control with thermal inter-row steaming	kg^{-1}	0.63	[71]
Seeds (sugar beet)	Main sowing material	kg^{-1}	3.54	[75]
Seeds (catch crop, i.e., mustard)	Inter-row sowing material	kg^{-1}	2.63	[76]
Manure	Farmyard manure as the main organic fertilizer	kg^{-1}	0.126	[41,77]
B. Output				
Sugar beet	Yield of sugar beet roots	kg^{-1}	0.40	[78]

2.6. Sugar Beet Yield

The yield of organic sugar beet roots in experiments E1–E4 was determined by the weighing method, expressed in tons per hectare (t ha^{-1}). Sugar beet yield for automated-robotic weed control methods was assumed based on the yield obtained using traditional mechanical inter-row loosening (E1). It was assumed that the sugar beet yield in automated mechanical inter-row loosening (E5) was 11% higher due to automation [17]. Kunz et al. reported that precision hoeing technologies increased the yield of sugar beet by 37%, while, after conventional hoeing, it increased by only 26% [17]. For robotic loosening methods (E6, E7), due to lower soil compaction, the sugar beet yield was assumed to have an additional increase of 7% [79] compared to the sugar beet yield obtained with traditional mechanical loosening (E1). This percentage increase was used based on the presentation by Shockley and Dillon [79] that, due to the light configuration of the autonomous vehicles, soil compaction could be reduced, leading to a 7% increase in crop yields.

2.7. Statistical Analysis

Data from the research in organic sugar beet production was statistically processed using the Tukey HSD test [80]. Significant differences in sugar beet energy and GHG emissions, as well as the differences in yield between the weed control methods analyzed, were determined by calculating the least significant difference for a 95% level of confidence ($p < 0.05$). Different letters in the figures (a, b, c, etc.) indicate a statistically significant difference.

3. Results

3.1. Fuel Consumption for Sugar Beet Production

For all weed control methods, except for inter-row cutting and mulching, the same working width of the machines was chosen—3.0 m. Other technological indicators of the machines differed because the implementation of different inter-row farming practices requires different power and working time and unequal field productivity. These factors also led to differences in fuel consumption. The main technological indicators of the weed control technological operations are presented in Table 3.

Table 3. Technological indicators of weed control operations (calculated by the authors, according to [51]).

Inter-Row Farming Practice	Machinery Power (kW)	Working Width (m)	Field Capacity (ha h^{-1})	Working Time (h ha^{-1})	Fuel Consumption (L ha^{-1})
Inter-row					
Sowing	45	3.00	1.48	0.68	6.60
Cutting and mulching	67	4.00	2.12	0.47	4.21
Steaming	45	3.00	0.52	1.92	11.5
Loosening (conventional)	54	3.00	0.80	1.25	5.04
Loosening (automated)	83	3.00	1.92	0.52	3.71
Loosening (robot, diesel-powered)	54	3.00	1.50	0.67	5.19
Loosening (robot, electric-powered)	1	3.00	0.27	3.75	3.00 *

* kWh ha^{-1}.

Sugar beet production requires high fuel input due to the intensive use of machinery for soil preparation, weed control, harvesting, and transportation. The total fuel consumption for sugar beet cultivation ranged from 150.7 to 185.3 L ha^{-1}, depending on the weed control method applied (Figure 2).

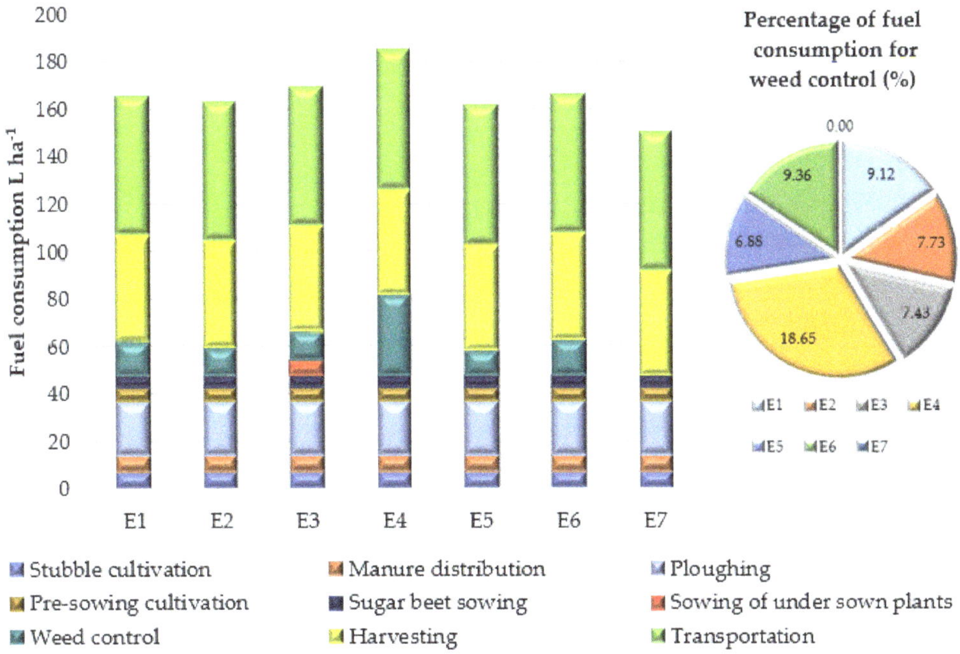

Figure 2. Total fuel consumption of organic sugar beet production using different weed control methods. E1—mechanical weed control (inter-row loosening), E2—mechanical weed control (inter-row cutting and mulching with weeds), E3—smothering weed control (inter-row cutting and mulching with catch crop), E4—thermal weed control (steaming), E5—inter-row loosening (automated), E6—inter-row loosening (robot—diesel powered), and E7—inter-row loosening (robot—electric powered).

3.2. Sugar Beet Yield

Crop care and the chosen method of weed control affect the yield of sugar beet. Analytical calculations showed that the highest average sugar beet yields (65.2 t ha^{-1}) were predicted in the variants using robotic weed control methods (E6, E7), when mechanical inter-row loosening is performed with diesel and electric robots (Figure 3). The lowest average sugar beet yield (33.5 t ha^{-1}) in 2015 and 2017 was obtained by applying inter-row cutting and mulching with weeds (E2), and in 2016 it was obtained by applying inter-row cutting and mulching with catch crop (E3) (36.4 t ha^{-1}). Such large differences in yields between the years were caused by meteorological conditions. Significant differences in the yield of organic sugar beet ($p < 0.05$) were found only by comparing mechanical weed control methods with other methods—E2, E3, and E4, except for E1 and E4. No significant difference was found in the assessment of sugar beet yields between years.

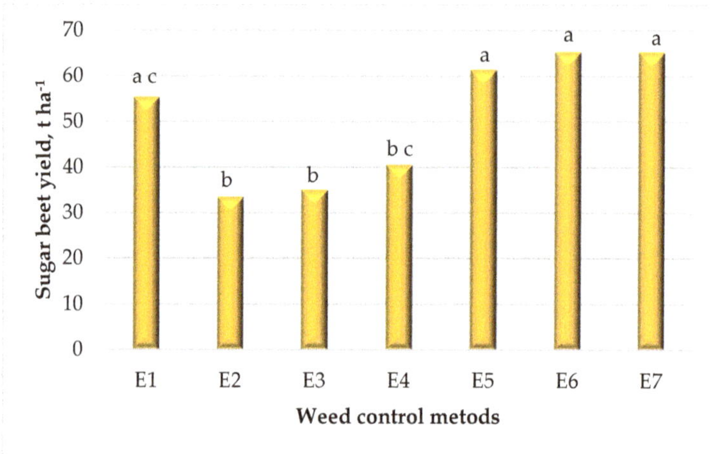

Figure 3. Sugar beet yield in average under different weed control methods.

3.3. Energy Indicators

The total average energy consumption for sugar beet production over the three-year period, regardless of the weed control method, ranged from 26671 MJ ha^{-1} to 29615 MJ ha^{-1} (Figure 4). The largest share of energy consumption was accounted for by 48–53% for manure and 29–35% for diesel fuel. The energy consumption of agricultural machinery averaged 14.7% of the total energy consumption and depended on total working time per hectare.

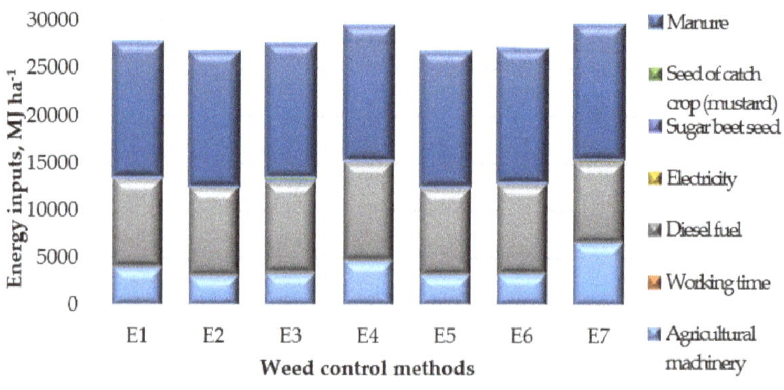

Figure 4. The impact of weed control methods on the energy consumption of organic sugar beet production.

The energy consumption of working time, electricity, sugar beet, and catch crops accounted for an insignificant share of the total energy consumption—2.4%. In our study, the average net energy gain was 171879 MJ ha^{-1} and the average energy productivity of sugar beet production was 1.83 kg MJ^{-1}. Application of the E6 weed control method resulted in the highest energy productivity—2.41 kg MJ^{-1}, achieved when the average three-year sugar beet yield was 65.2 t ha^{-1}. The ratio of renewable energy sources to total energy consumption was the highest (53.7%) when applying automated inter-row loosening with cameras (E5). Renewable energy sources can be reused and are non-consumable. However, they are limited and have a probability of exhaustion [81]. The ratio of non-renewable energy sources was the highest (51.4%) when applying thermal inter-row loosening (E4).

Assessment of the energy efficiency ratio indicates whether a plant production technology gains energy (>1) or loses energy (<1) [44]. When applying different weed control methods for organic sugar beet production, the highest energy efficiency ratio (Figure 5) was achieved by mechanical inter-row loosening with a diesel-powered robot E6 (9.16) and the lowest—by E2 (5.19).

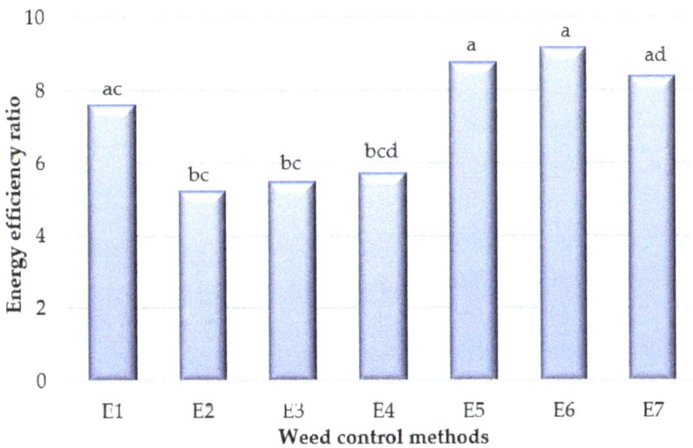

Figure 5. The impact of weed control methods on energy efficiency ratio.

3.4. Greenhouse Gas Emissions

The results of calculated GHG emissions are presented in Figure 6. An evaluation of different weed control methods in organic sugar beet production revealed that, due to high nitrogen concentration, organic fertilizers were the most important factor. Total GHG emissions from organic sugar beet production ranged from 4471 to 4693 kg CO_{2eq} ha^{-1}, depending on the weed control method applied.

The best GHG emissions ratios, showing the sustainability of organic sugar beet production technologies, were achieved when applying mechanical inter-row loosening with a diesel-powered robot (E6), an electric robot (E7), and automated inter-row loosening (E5), 5.80, 5.56, and 5.49, respectively (Figure 7).

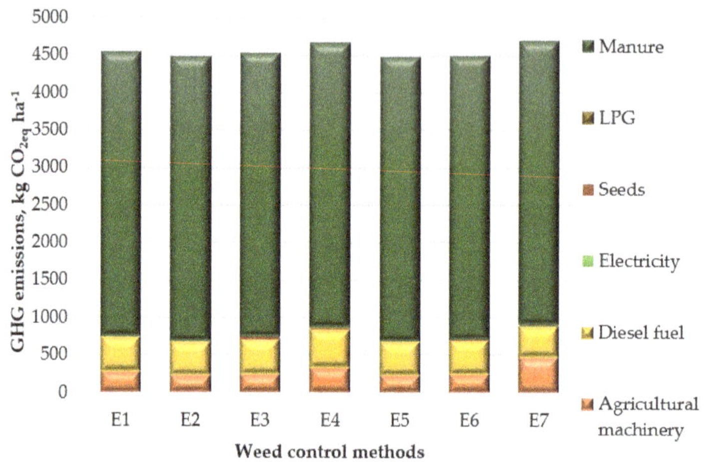

Figure 6. The impact of weed control methods on GHG emissions into the environment.

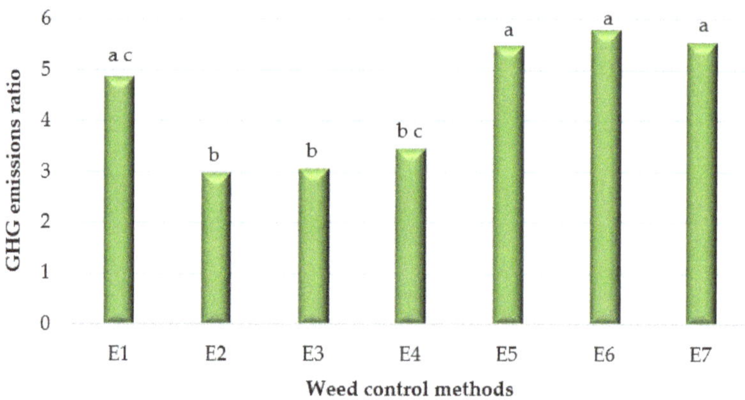

Figure 7. The impact of weed control methods on GHG emissions ratio.

4. Discussion

This section discusses the impact of seven different mechanical, thermal, and intelligent robotic weed control methods on fuel consumption, energy consumption, GHG emissions, and yield in organic sugar beet production. The main limitation of this study was that it was not possible to perform research on all methods under the same conditions. Four weed control methods E1–E4 were investigated in an experimental field and the results were obtained under real soil and climatic conditions. Meanwhile, for comparison, for the automated and robotic weed control methods, E5–E7, due to lack of equipment in the country, sugar beet yield and the other results required for this study were calculated using previous results obtained by other researchers.

4.1. Fuel Consumption

The most fuel-intensive technological operation was transportation, accounting for about 35.5% of total input, followed by harvesting (26.9%) and ploughing (13.6%). The remaining fuel input for land preparation, manure spreading, sowing, and weed control accounted for about 24% of the total input. The E7 weed control method had the lowest fuel consumption due to the electric robot. Automated inter-row loosening had the lowest fuel consumption compared with the other diesel-powered methods, as using visual cameras for row-tracking allows operating at higher speeds. The results of a study by Gerhards et al. [82] demonstrated that automatic camera-controlled weeding was 7 to 15% more efficient at 8 km h^{-1} compared to 4 km h^{-1}. The most polluting method of weed control from an environmental point of view was the thermal inter-row steaming (E4) due to low working speed. Erdal et al. [61] obtained very similar results (170.3 L ha^{-1}) when determining the fuel consumption for sugar beet production.

Analyzing the percentage of diesel fuel consumption for weed control, it was found that the highest fuel consumption of 18.65% was in the E4 treatment with inter-row steaming. Others behind it were the diesel-powered E6 and conventional E1 treatments, which consumed 9.36 and 9.12% of the total fuel used in sugar beet production, respectively.

4.2. Sugar Beet Yield

The average yield of organic sugar beet in the European Union is about 50 t ha^{-1} [19]. In our study, it was found that the sugar beet yield depended on the weed control method and varied from 65.2 t ha^{-1} using robots E6 and E7 to 33.5 t ha^{-1} using the mechanical weed control method E2 with inter-row cutting and mulching with weeds. Researchers from other countries have obtained similar results for the yields of sugar beet. Majidi et al. [12] showed that the yield of sugar beet roots ranged from 27.2 to 73.7 t ha^{-1} when using chemical weed control methods, and the lowest root yield was obtained in the control (20.7 t ha^{-1}). The study of Jursik et al. [10] of sugar beet crop in the Czech Republic when applying manual weed control revealed that the yield of sweet roots can vary on average from 38.9 to 63.2 t ha^{-1}. Gerhards et al. [13] found that the losses of crop yield might vary depending on the sugar beet growing region. Field tests in Germany and the Russian Federation showed that the sugar beet yields were higher in Germany (82.9–95.0 t ha^{-1}) compared to those obtained in Russia (37.9–54.6 t ha^{-1}).

The lowest sugar beet yields in cutting and mulching treatments E2 and E3 can be explained by the fact that the rate of weed growth compared to crops is very fast, and if weeds are not destroyed and managed, they can begin to dominate throughout the field [83]. Although weeds in a sugar beet farm reduce crop yield and quality, they can also lead to higher labor and material losses [18]. Analysis of previous research data has shown that inter-row crops such as weeds and catch crops compete with sugar beet and negatively affect its root yield and sucrose content [84].

An important factor in the application of different weed control methods is soil compaction, which has a significant impact on crop growth. The low weight of the robots compared to traditional large and heavy machines makes it possible to reduce soil compaction. Sugar beet is sensitive to soil pressures due to reduced leaf area, final yield, and root quality parameters in compacted soils. A visual indicator of soil compaction is the depth of the roots, which is influenced by the pressures of the wheels of agricultural tractors and machines on the soil. Control of the frequency and position of agricultural machinery through the field on well-defined tracks can help reduce the impact of compaction on the soil [85]. Soil compaction has negative consequences such as decreased soil porosity, decreased aeration [86], decreased saturated hydraulic conductivity, and increased soil resistance to root penetration [87]. Typically, soil compaction results in negative crop growth conditions due to high mechanical root resistance, decreased soil aeration, and reduced water accumulation [88].

4.3. Energy Assessment

Assessment of energy consumption for different weed control methods applied revealed that the highest energy consumption was in the E7 (29,615 MJ ha^{-1}) and E4 (29,499 MJ ha^{-1}) variants. It was caused by the fact that working speed and work productivity were the lowest when working with a solar-powered robot and applying thermal weed control methods (E7—0.27 ha h^{-1}; E4—0.52 ha h^{-1}). For comparison, when applying conventional inter-row loosening (E1), work productivity was 0.80 ha h^{-1}.

Both Dimitrijevic et al. [44] and Erdal et al. [61], 28.9 and 24.2%, respectively, identified diesel fuel as the second most intensive energy user in sugar beet production. Other authors confirm that the thermal weed control method reduced productivity by 10% due to the additional time required to replenish propane gas tanks [89].

The estimated total average energy consumption of 27844 MJ ha^{-1} falls within the range of energy consumption values determined by other researchers: 11,100–35,800 MJ ha^{-1} [62,67]. Dimitrijevic et al. [44], after a three-year study in Serbia, obtained almost identical average energy consumption for sugar beet production (27849 MJ ha^{-1}). Other scientists' research showed that, in Turkey, at a sugar beet yield of 68 t ha^{-1}, the energy productivity was 1.98 kg MJ^{-1} [90]. Serbian researchers, studying the energy and economic indicators of sugar beet production, determined the energy productivity of sugar beet to be 1.57 kg MJ^{-1} when the yield was 48.2 t ha^{-1} [44].

The energy efficiency ratio indicator was mainly influenced by the sugar beet yield. A significant difference in energy efficiency ($p < 0.05$) was found between the automated-robotic inter-row loosening methods E5–E7 and E2 and E3. A significant difference was also found between the E4 and E5–E6 weed control methods. Šarauskis et al. [91] compared conventional and organic sugar beet cultivation methods and found that the energy efficiency ratio in organic crops was significantly lower, ranging from 4.02 to 8.21. In this study, the overall energy efficiency ratio of all sugar beet growing technologies was 7.18. Tzilivakis et al. [47] studied sugar beet production in the UK and found that sugar beet, on average, produces 9.7 times more energy than it consumes. Asgharipour et al. [55] found the energy efficiency ratio for chemical sugar beet production to be 13.4. In the studies of other researchers, significantly higher input-output ratios were obtained and ranged from 15.4 [62] to 26.45 [44]. This was mainly due to the use of more intensive chemicalized sugar beet growing technologies, which allowed the reduction of energy inputs and increased the yields. Organic sugar beet production uses more energy for weed control and organic fertilizers.

In some cases, a mechanical weed control method alone may not be sufficient. This method is not always able to eradicate weeds growing very close to sugar beets. In such cases, a manual weed control method is often still used.

4.4. GHG Emissions

Manure accounted for the largest share of GHG emissions, averaging about 83% of total emissions. On a lesser scale, but still significant, the environment was polluted by diesel, machinery, and seeds; 10%, 6.4%, and 0.4%, respectively. Electricity and LPG accounted for an insignificant share of GHG emissions. Other authors obtained similar research results. Šarauskis et al. [45] found that the largest source of GHG emissions when growing winter wheat and spring barley in organic and sustainable farming systems was usage of manure. It ranged from 74% of total emissions in small farms (2 ha) to 82% in large farms (80 ha). Yousefi et al. [43] found that, in sugar beet production, systems using irrigation electricity accounted for the largest share of GHG emissions (73%), followed by nitrogen fertilizers (15%) and diesel (7%).

Assessment of the environmental impact of weed control methods revealed that the largest impact on GHG emissions was related to machinery and diesel fuels. In terms of GHG emissions from fuel, the cleanest weed control methods were using an electric robot (E7) and automated mechanical inter-row loosening (E5) with GHG emissions from fuel being 416 and 447 kg CO_{2eq} ha^{-1}, respectively. Assessment of the environmental impact of the machinery alone showed that the lowest pollution was observed with the E2 and E5 weed control methods, 226 and 230 kg CO_{2eq} ha^{-1}, respectively, and the highest was with E7 (476 kg CO_{2eq} ha^{-1}). This was mainly caused by the very low productivity of an electric robot (0.27 ha h^{-1}).

The GHG emissions ratio results are due to higher sugar beet yields associated with the automation of the process and lower GHG costs. A slightly lower GHG emissions ratio (4.87) was obtained using conventional mechanical inter-row loosening (E1). There were no significant differences between these weed control methods. A significant difference was found only when comparing weed control methods E1, E5, E6, and E7 with the other methods—E2, E3, and E4, except for E1 and E4. The total average GHG emissions ratio was 4.47. The studies of Yousefi et al. [43], evaluating the efficiency of sugar beet production for environmental pollution, indicated that the GHG emissions ratio was 10.95.

5. Conclusions

In this study, an analytical analysis was performed to determine the energy use and GHG emissions in organic sugar beet production. It was estimated that, on average, organic sugar beet production consumes 27844 MJ ha^{-1} of energy, while energy output was 199724 MJ ha^{-1}. Manure (50.6%) and diesel (33.6%) accounted for the largest share of energy consumption. After evaluation of the efficient use of energy resources by applying different weed control methods, it was found that sugar beets could be grown in the most energy efficient way by using a diesel-powered robot. Energy efficiency, energy productivity, and net energy gain were equal to 7.18, 1.83 kg MJ^{-1}, and 171879 MJ ha^{-1}, respectively. The share of renewable and non-renewable energy input in total energy consumption was 51.7% and 48.3%, respectively.

The results showed that the lowest total energy consumption (26671 MJ ha^{-1}) was obtained when using an automated method of mechanical inter-row loosening with cameras. Meanwhile, the highest energy efficiency ratio (9.16) was achieved when using a diesel-powered robot.

The highest yield of organic sugar beet roots (65.2 t ha^{-1}) was found in those variants where sugar beet was grown when applying weed control with diesel and electric robots, and the lowest (33.5 t ha^{-1}), when applying inter-row mulching with weeds.

Analysis of GHG emissions revealed that the total greenhouse gas emissions ranged from 4471 to 4693 kg CO_{2eq} ha^{-1}, depending on the weed control method applied. The main source of GHG emissions was manure, which accounted for over 80% of all pollution. The highest sustainability of organic sugar beet production was achieved when applying inter-row loosening with a diesel powered robot.

The study found that intelligent robotic weed control methods, compared to conventional ones, can reduce energy costs, increase crop yields, and reduce GHG emissions into the environment, while improving the quality of the environment and the supply of organic food. Summarizing the results of the analysis, we can draw the important conclusion that reduction of the consumption of diesel fuel and organic fertilizers, such as manure, is the most important factor in controlling GHG emissions in organic sugar beet production. Due to the high precision and low weight of the machines, the use of automated and robotic methods in weed control has a promising future. However, the cost-effectiveness, technological reliability, and ease of use of robotic weed control methods in farms are important aspects that still need to be investigated in future research. New experimental studies are needed to compare changes in soil properties and compaction processes using robotic and conventional crop care machines. It is also important to assess and compare the effects of robotic weed control methods in real soil and meteorological conditions on differences in weed control efficiency, working time, energy consumption, and GHG emissions.

Author Contributions: Conceptualization, I.B. and E.Š.; methodology, I.B., D.A., and E.Š.; validation, I.B., E.Š., and K.R.; formal analysis, D.A. and K.R.; investigation, I.B. and D.A.; resources, I.B., E.Š., and K.R.; data curation, I.B. and D.A.; writing—original draft preparation, I.B., D.A., E.Š., and K.R.; writing—review and editing, E.Š. and K.R.; visualization, I.B. and D.A.; supervision, E.Š. All authors have read and agreed to the published version of the manuscript.

Funding: This project has received funding from European Social Fund (project No. 09.3.3-LMT-K-712-22-0315) under grant agreement with the Research Council of Lithuania (LMTLT).

Institutional Review Board Statement: Not applicable.

Informed Consent Statement: Not applicable.

Data Availability Statement: The data presented in this study are available on request from the corresponding author.

Conflicts of Interest: The authors declare no conflict of interest.

References

1. Kirchmann, H.; Bergström, L.; Kätterer, T.; Andrén, O.; Andersson, R. Can Organic Crop Production Feed the World? In *Organic Crop Production—Ambitions and Limitations*; Kirchmann, H., Bergström, L., Eds.; Springer: Dordrecht, The Netherlands, 2008; pp. 39–72. [CrossRef]
2. Cioni, F.; Maines, G. Weed control in sugarbeet. *Sugar Tech.* **2010**, *12*, 243–255. [CrossRef]
3. Jabran, K.; Chauhan, B.S. *Non-Chemical Weed Control*; Academic Press: Cambridge, MA, USA, 2018; p. 172. ISBN 978-0-12-809881-3.
4. Cook, S.K.; Davies, L.R.; Pickering, F.; Tatnell, L.V.; Huckle, A.; Newman, S.; Whiteside, C.; White, C.; Talbot, D.; Holmes, H.; et al. *Weed Control Options and Future Opportunities for UK Crops*; Research Review No. CP 182/1807258; Agriculture and Horticulture Development Board: Kenilworth, UK, 2019.
5. Oerke, E.C.; Dehne, H.W. Safeguarding production-losses in major crops and the role of crop protection. *Crop. Prot.* **2004**, *23*, 275–285. [CrossRef]
6. Oerke, E.C. Crop losses to pests. *J. Agric. Sci.* **2006**, *144*, 31–43. [CrossRef]
7. Pimentel, D. Pest control in world agriculture. *Agric. Sci.* **2009**, *2*, 272–293.
8. Sanchez-Bayo, F. Impacts of agricultural pesticides on terrestrial ecosystems. In *Ecological Impacts of Toxic Chemicals*; Bentham Science Publishers Ltd.: Potomac, MD, USA, 2011; pp. 63–87. [CrossRef]
9. Romaneckas, K.; Romaneckienė, R.; Šarauskis, E.; Pilipavičius, V.; Sakalauskas, A. The effect of conservation primary and zero tillage on soil bulk density, water content, sugar beet growth and weed infestation. *Agron. Res.* **2009**, *7*, 73–86.
10. Jursik, M.; Holec, J.; Soukup, J.; Venclova, V. Competitive relationships between sugar beet and weeds in dependence on time of weed control. *Plant Soil Environ.* **2018**, *5*, 108–116. [CrossRef]
11. Jalali, A.H.; Salehi, F. Sugar beet yield as affected by seed priming and weed control. *Arch. Agron. Soil. Sci.* **2013**, *59*, 281–288. [CrossRef]
12. Majidi, M.; Heidari, G.; Emam, Y. Qualitative characteristics of sugar beet as affected by different broadleaf herbicides combinations. *Iran Agric. Res.* **2017**, *36*, 1–6. [CrossRef]
13. Gerhards, R.; Bezhin, K.; Santel, H.J. Sugar beet yield loss predicted by relative weed cover, weed biomass and weed density. *Plant Prot. Sci.* **2017**, *53*, 118–125. [CrossRef]
14. Kunz, C.; Weber, J.F.; Peteinatos, G.G.; Sokefeld, M.; Gerhards, R. Camera steered mechanical weed control in sugar beet, maize and soybean. *Precis. Agric.* **2018**, *19*, 708–720. [CrossRef]
15. Bhadra, T.; Mahapatra, C.K.; Paul, S.K. Weed management in sugar beet: A review. *Fundam. Appl. Agric.* **2020**, *5*, 147–156. [CrossRef]
16. Kropff, M.J.; Spitters, C.J.T. A simple model of crop loss by weed competition from early observations on relative leaf area of the weeds. *Weed Res.* **1991**, *31*, 97–105. [CrossRef]
17. Kunz, C.; Weber, J.F.; Gerhards, R. Benefits of Precision Farming Technologies for Mechanical Weed Control in Soybean and Sugar Beet—Comparison of Precision Hoeing with Conventional Mechanical Weed Control. *Agronomy* **2015**, *5*, 130–142. [CrossRef]
18. Sabanci, K.; Aydin, C. Smart robotic weed control system for sugar beet. *J. Agric. Sci. Technol.* **2017**, *19*, 73–83.
19. European Commission. *EU Agricultural Outlook for Markets and Income, 2018–2030*; European Commission, DG Agriculture and Rural Development: Brussels, Belgium, 2018. Available online: https://ec.europa.eu/info/files/report-eu-agricultural-outlook-2018-30_en (accessed on 3 October 2020).
20. Seufert, V.; Ramankutty, N.; Foley, J.A. Comparing the yields of organic and conventional agriculture. *Nature* **2012**, *485*, 229–232. [CrossRef]
21. Kunz, C.; Weber, J.F.; Gerhards, R. Comparison of different mechanical weed control strategies in sugar beets. In Proceedings of the Conference 27th Deutsche Arbeitsbesprechung über Fragen der Unkrautbiologie und -Bekämpfung, Braunschweig, Germany, 23–25 February 2016.
22. Peruzzi, A.; Martelloni, L.; Frasconi, C.; Fontanelli, M.; Pirchio, M.; Raffaelli, M. Machines for non-chemical intra-row weed control in narrow and wide-row crops: A review. *J. Agric. Eng.* **2017**, *48*, 57–70. [CrossRef]
23. Stevanato, P.; Chiodi, C.; Broccanello, C.; Concheri, G.; Biancardi, E.; Pavli, O.; Skaracis, G. Sustainability of the sugar beet crop. *Sugar Tech.* **2019**, *21*, 703–716. [CrossRef]
24. Machleb, J.; Peteinatos, G.G.; Kollenda, B.L.; Andújar, D.; Gerhards, R. Sensor-based mechanical weed control: Present state and prospects. *Comput. Electron. Agric.* **2020**, *176*. [CrossRef]
25. Slaughter, D.C.; Giles, D.K.; Downey, D. Autonomous robotic weed control systems: A review. *Comput. Electron. Agric.* **2008**, *61*, 63–78. [CrossRef]
26. Kunz, C. Integrated Weed Control in Sugar Beet (Beta vulgaris), Using Precision Farming Technologies and Cover Cropping. Ph.D. Thesis, University of Hohenheim, Stuttgart, Germany, 2017.
27. Perez-Ruiz, M.; Brenes, R.; Urbano, J.M.; Slaughter, D.C.; Forcella, F.; Rodríguez-Lizana, A. Agricultural residues are efficient abrasive tools for weed control. *Agron. Sustain. Dev.* **2018**, *38*, 1–8. [CrossRef]

28. Pedersen, S.M.; Fountas, S.; Blackmore, S. Agricultural robots—Applications and economic perspectives. In *Service Robot Applications*; Takahashi, Y., Ed.; InTech: Rijeka, Croatia, 2008; pp. 369–382.
29. Van Evert, F.K.; Samson, J.; Polder, G.; Vijn, M.; van Dooren, H.; Lamaker, A.; van der Heijden, G.; Kempenaar, C.; van der Zalm, T.; Lotz, B. A robot to detect and control broad-leaved dock (*Rumex obtusifolius* L.) in grassland. *J. Field Robot.* **2011**, *28*, 264–277. [CrossRef]
30. Norremark, M.; Sorenson, C.G.; Jorgenson, R.N. HortiBot: Comparison of present and future phytotechnologies for weed control—Part III. In Proceedings of the 2006 ASAE Annual Meeting American Society of Agricultural and Biological Engineers, Portland, OR, USA, 9–12 July 2006; p. 1.
31. Piron, A.; van der Heijden, F.; Destain, M.F. Weed detection in 3D images. *Precis. Agric.* **2011**, *12*, 607–622. [CrossRef]
32. Perez-Ruiz, M.; Slaughter, D.C.; Gliever, C.; Upadhyaya, S.K. Tractor-based Real-time Kinematic-Global Positioning System (RTK-GPS) guidance system for geospatial mapping of row crop transplant. *Biosyst. Eng.* **2012**, *111*, 64–71. [CrossRef]
33. Heno, S.; Viou, L.; Khan, M.F.R. Sugar beet production in France. *Sugar Tech.* **2018**, *20*, 392–395. [CrossRef]
34. Perez-Ruiz, M.; Slaughter, D.C.; Fathallah, F.A.; Gliever, C.J.; Miller, B.J. Co-robotic intra-row weed control system. *Biosyst. Eng.* **2014**, *126*, 45–55. [CrossRef]
35. Bawden, O.; Ball, D.; Kulk, J.; Perez, T.; Russell, R. A lightweight, modular robotic vehicle for the sustainable intensification of agriculture. In *Proceedings of the 16th Australasian Conference on Robotics and Automation 2014*; Australian Robotics and Automation Association Inc.: Brisbane, Australia, 2014; pp. 1–9. Available online: https://eprints.qut.edu.au/82219 (accessed on 5 October 2020).
36. Jensen, K.; Larsen, M.; Nielsen, S.H.; Larsen, L.B.; Olsen, K.S.; Jorgensen, R.N. Towards an Open Software Platform for Field Robots in Precision Agriculture. *Robotics* **2014**, *3*, 207–234. [CrossRef]
37. Choi, K.H.; Han, S.K.; Han, S.H.; Park, K.; Kim, K.; Kim, S. Morphology-based guidance line extraction for an autonomous weeding robot in paddy fields. *Comput. Electron. Agric.* **2015**, *113*, 266–274. [CrossRef]
38. Grimstad, L.; Johan, P. The thorvald II agricultural robotic system. *Robotics* **2017**, *6*, 24. [CrossRef]
39. Utstumo, T.; Urdal, F.; Brevik, A.; Dorum, J.; Netland, J.; Overskeid, O.; Berge, T.W.; Gravdahl, J.T. Robotic in-row weed control in vegetables. *Comput. Electron. Agric.* **2018**, *154*, 36–45. [CrossRef]
40. Perez, T.; McCool, C.; Bawden, O.; Kulk, J. Robotic weeding–From concept to trials. In Proceedings of the 7th Asian-Australasian Conference on Precision Agriculture, Hamilton, New Zealand, 16–18 October 2017.
41. Pishgar-Komleh, S.H.; Omid, M.; Heidari, M.D. On the study of energy use and GHG (greenhouse gas) emissions in greenhouse cucumber production in Yazd province. *Energy* **2013**, *59*, 63–71. [CrossRef]
42. Haciseferogullari, H.; Acaroglu, M.; Gezer, I. Determination of the energy balance of the sugar beet plant. *Energy Sources* **2003**, *25*, 15–22. [CrossRef]
43. Yousefi, M.; Khoramivafa, M.; Mondani, F. Integrated evaluation of energy use, greenhouse gas emissions and global warming potential for sugar beet (*Beta vulgaris*) agroecosystems in Iran. *Atmos. Environ.* **2014**, *92*, 501–505. [CrossRef]
44. Dimitrijević, A.; Gavrilović, M.; Ivanović, S.; Mileusnić, Z.; Miodragović, R.; Todorović, S. Energy use and economic analysis of fertilizer use in wheat and sugar beet production in Serbia. *Energies* **2020**, *13*, 2361. [CrossRef]
45. Šarauskis, E.; Masilionytė, L.; Juknevičius, D.; Buragienė, S.; Kriaučiūnienė, Z. Energy use efficiency, GHG emissions, and cost-effectiveness of organic and sustainable fertilisation. *Energy* **2019**, *172*, 1151–1160. [CrossRef]
46. IPCC. Climate Change 2014: Mitigation of Climate Change. Working Group III Contribution to the Fifth Assessment Report of the Intergovernmental Panel on Climate Change. 2014. Available online: https://www.ipcc.ch/report/ar5/wg3/ (accessed on 2 October 2020).
47. Tzilivakis, J.; Warner, D.J.; May, M.; Lewis, K.A.; Jaggard, K. An assessment of the energy inputs and greenhouse gas emissions in sugar beet (Beta vulgaris) production in the UK. *Agric. Syst.* **2005**, *85*, 101–119. [CrossRef]
48. Romaneckas, K.; Adamavičienė, A.; Šarauskis, E.; Balandaitė, J. The impact of intercropping on soil fertility and sugar beet productivity. *Agronomy* **2020**, *10*, 1406. [CrossRef]
49. FarmDroid. The Product Sheet FD20. 2020. Available online: https://farmdroid.dk/en/welcome/ (accessed on 7 October 2020).
50. Agrointelli. Robotti, Our Technology. 2020. Available online: https://www.agrointelli.com/robotti/ (accessed on 1 October 2020).
51. LIAE. Prices of Mechanized Agricultural Services (Electronic Resource). Part 1. Main Tillage, Vilnius. 2018. Available online: https://www.laei.lt/?mt=leidiniai&straipsnis=1275&metai=2018 (accessed on 17 October 2020).
52. Tabar, I.B.; Keyhani, A.; Rafiee, S. Energy balance in Iran's agronomy (1990–2006). *Renew. Sustain. Energy Rev.* **2010**, *14*, 849–855. [CrossRef]
53. Yadav, G.S.; Das, A.; Lal, R.; Babu, S.; Meena, R.S.; Saha, P.; Singh, R.; Datta, M. Energy budget and carbon footprint in a no-till and mulch based rice–mustard cropping system. *J. Clean. Prod.* **2018**, *191*, 144–157. [CrossRef]
54. Taylor, A.E.B.; O'Callaghan, P.W.; Probert, S.D. Energy audit of an English farm. *Appl. Energy* **1993**, *44*, 315–335. [CrossRef]
55. Asgharipour, M.R.; Mondani, F.; Riahinia, S. Energy use efficiency and economic analysis of sugar beet production system in Iran: A case study in Khorasan Razavi province. *Energy* **2012**, *44*, 1078–1084. [CrossRef]
56. Mombarg, H.F.M.; Kool, A.; Corré, W.J.; Langeveld, J.W.A.; Sukkel, W. Energy and Climate Yardstick, Methodology and Equations, Wageningen. 2003, p. 26. Available online: https://library.wur.nl/WebQuery/wurpubs/fulltext/20355 (accessed on 6 October 2020). (In Dutch).

57. Bos, J.F.F.P.; De Haan, J.; Sukkel, W.; Schils, R.L.M. Energy use and greenhouse gas emissions in organic and conventional farming systems in The Netherlands. *NJAS Wagen. J. Life. Sci.* **2014**, *68*, 61–70. [CrossRef]
58. Tabatabaeefar, A.; Emamzadeh, H.; Varnamkhasti, M.G.; Rahimizadeh, R.; Karimi, M. Comparison of energy of tillage systems in wheat production. *Energy* **2009**, *34*, 41–45. [CrossRef]
59. Chauhan, N.S.; Mohapatra, P.K.; Pandey, K.P. Improving energy productivity in paddy production through benchmarking—An application of data envelopment analysis. *Energy Convers. Manag.* **2006**, *47*, 1063–1085. [CrossRef]
60. Ilahi, S.; Wu, Y.; Raza, M.A.A.; Wei, W.; Imran, M.; Bayasgalankhuu, L. Optimization approach for improving energy efficiency and evaluation of greenhouse gas emission of wheat crop using data envelopment analysis. *Sustainability* **2019**, *11*, 3409. [CrossRef]
61. Erdal, G.; Esengün, K.; Erdal, H.; Gündüz, O. Energy use and economical analysis of sugar beet production in Tokat province of Turkey. *Energy* **2007**, *32*, 35–41. [CrossRef]
62. Reineke, H.; Stockfisch, N.; Märländer, B. Analysing the energy balances of sugar beet cultivation in commercial farms in Germany. *Eur. J. Agron.* **2013**, *45*, 27–38. [CrossRef]
63. Wegener, U. Dauerhafte Bodenbearbeitungsverfahren in Zuckerrübenfruchtfolgen—Ertragsbildung, Rentabilität, Energiebilanz und Bodenerosion im Vergleich. Ph.D. Thesis, Georg-August-Universität Göttingen, Cuvillier, Göttingen, Germany, 2001.
64. Stout, B.A. *Handbook of Energy for World Agriculture*; Elsevier: Amsterdam, The Netherlands, 2012.
65. Soni, P.; Taewichit, C.; Salokhe, V.M. Energy consumption and CO_2 emissions in rainfed agricultural production systems of Northeast Thailand. *Agric. Syst.* **2013**, *116*, 25–36. [CrossRef]
66. DLG. DLG-Futterwerttabellen. In *Wiederkäuer*, 7th ed.; DLG: Frankfurt/Main, Germany, 1997.
67. Jacobs, A.; Brauer-Siebrecht, W.; Christen, O.; Götze, P.; Koch, H.J.; Rücknagel, J.; Märländer, B. Silage maize and sugar beet for biogas production in crop rotations and continuous cultivation–energy efficiency and land demand. *Field Crops Res.* **2016**, *196*, 75–84. [CrossRef]
68. Martin-Gorriz, B.; Soto-García, M.; Martínez-Alvarez, V. Energy and greenhouse-gas emissions in irrigated agriculture of SE (Southeast) Spain. Effects of alternative water supply scenarios. *Energy* **2014**, *77*, 478–488. [CrossRef]
69. Eskandari, H.; Attar, S. Energy comparison of two rice cultivation systems. *Renew. Sustain. Energy Rev.* **2015**, *42*, 666–671. [CrossRef]
70. Kubota Corporation. Products and Solutions–Engines. 2020. Available online: https://www.kubota.com/products/engines/index.html (accessed on 30 November 2020).
71. Lal, R. Carbon emission from farm operations. *Environ. Int.* **2004**, *30*, 981–990. [CrossRef]
72. Dyer, J.A.; Desjardins, R.L. Carbon dioxide emissions associated with the manufacturing of tractors and farm machinery in Canada. *Biosyst. Eng.* **2006**, *93*, 107–118. [CrossRef]
73. Nabavi-Pelesaraei, A.; Abdi, R.; Rafiee, S. Applying artificial neural networks and multi-objective genetic algorithm to modelling and optimization of energy inputs and greenhouse gas emissions for peanut production. *Int. J. Biosci.* **2014**, *4*, 170–183. [CrossRef]
74. Casey, J.W.; Holden, N.M. Analysis of greenhouse gas emissions from the average Irish milk production system. *Agric. Syst.* **2005**, *86*, 97–114. [CrossRef]
75. Trimpler, K.; Stockfisch, N.; Marlander, B. The relevance of N fertilization for the amount of total greenhouse gas emissions in sugar beet cultivation. *Eur. J. Agron.* **2016**, *81*, 64–71. [CrossRef]
76. West, T.O.; Marland, G. A synthesis of carbon sequestration, carbon emissions, and net carbon flux in agriculture: Comparing tillage practices in the United States. *Agric. Ecosyst. Environ.* **2002**, *91*, 217–232. [CrossRef]
77. Li, X.M.; Kotelko, M. An integrated manure utilization system (imus): Its social and environmental benefits. In Proceedings of the 3rd International Methane and Nitrous Oxide Mitigation Conference, Beijing, China, 17–21 November 2003.
78. BioGrace. BioGrace-i Calculation Rules. Version 4d. Harmonised Calculations of Bioenergy Greenhouse Gas Emissions in Europe. 2015. Available online: https://www.biograce.net/home (accessed on 9 October 2020).
79. Shockley, J.M.; Dillon, C. An Economic Feasibility Assessment for Adoption of Autonomous Field Machinery in Row Crop Production. In Proceedings of the 14th International Conference on Precision Agriculture, Montreal, QC, Canada, 24–27 June 2018.
80. Tukey, J.W. Methodology, and the statistician's responsibility for both accuracy and relevance. *J. Am. Stat. Assoc.* **1979**, *74*, 786–793. [CrossRef]
81. Unakıtan, G.; Aydın, B. A comparison of energy use efficiency and economic analysis of wheat and sunflower production in Turkey: A case study in Thrace Region. *Energy* **2018**, *149*, 279–285. [CrossRef]
82. Gerhards, R.; Kollenda, B.; Machleb, J.; Möller, K.; Butz, A.; Reiser, D.; Griegentrog, H.W. Camera-guided Weed Hoeing in Winter Cereals with Narrow Row Distance. *Gesunde Pflanzen.* **2020**, *72*, 403–411. [CrossRef]
83. Ahmad, M.T.; Tang, L.; Steward, B.L. Automated mechanical weeding. In *Automation: The Future of Weed Control in Cropping Systems*; Springer: Dordrecht, The Netherlands, 2014; pp. 125–137.
84. Romaneckas, K.; Adamaviciene, A.; Eimutyte, E.; Pupaliene, R.; Sarauskis, E.; Marks, M.; Kimbirauskienė, R.; Cekanauskas, S. Impact of non-chemical weed control methods on the soil and sugar beet root chemical composition. *J. Elem.* **2018**, *23*, 1215–1227. [CrossRef]
85. Marinello, F.; Pezzuolo, A.; Cillis, D.; Chiumenti, A.; Sartori, L. Traffic effects on soil compaction and sugar beet (*Beta vulgaris* L.) taproot quality parameters. *Span. J. Res.* **2017**, *15*, 1–8. [CrossRef]

86. McHugh, A.D.; Tullberg, J.N.; Freebairn, D.M. Controlled traffic farming restores soil structure. *Soil Till. Res.* **2009**, *104*, 164–172. [CrossRef]
87. Valdes-Abellan, J.; Jiménez-Martínez, J.; Candela, L.; Tamoh, K. Comparison among monitoring strategies to assess water flow dynamic and soil hydraulic properties in agricultural soils. *Span. J. Agric. Res.* **2015**, *13*, e12-001. [CrossRef]
88. Da Silva, A.P.; Kay, B.D. The sensitivity of shoot growth of corn to the least limiting water range of soils. *Plant Soil* **1996**, *184*, 323–329. [CrossRef]
89. Khelifi, M. Energy use and time requirements for different weeding strategies in grain corn. *Can. Biosystem. Eng.* **2001**, *43*, 2.13–2.21.
90. Baran, M.F.; Gokdogan, O. Determination of energy balance of sugar beet production in Turkey: A case study of Kırklareli Province. *Energy Effic.* **2016**, *9*, 487–494. [CrossRef]
91. Šarauskis, E.; Romaneckas, K.; Kumhála, F.; Kriaučiūnienė, Z. Energy use and carbon emission of conventional and organic sugar beet farming. *J. Clean. Prod.* **2018**, *201*, 428–438. [CrossRef]

agriculture

Review

Research Progress and Development of Mechanized Potato Planters: A Review

Zhiqi Zheng [1], Hongbo Zhao [1], Zhengdao Liu [1], Jin He [2] and Wenzheng Liu [3,*]

[1] College of Mechanical and Electronic Engineering, Northwest A&F University, Yangling 712100, China; zhiqizheng@nwsuaf.edu.cn (Z.Z.); zhaohb@nwafu.edu.cn (H.Z.); liuzd@nwafu.edu.cn (Z.L.)
[2] College of Engineering, China Agricultural University, Beijing 100083, China; hejin@cau.edu.cn
[3] College of Enology, Northwest A&F University, Yangling 712100, China
* Correspondence: lwzheng@nwafu.edu.cn; Tel.: +86-188-1066-2985

Abstract: Potato is one of the most important crops in the world to ensure food security for the world's fast-growing population. Mechanized planting is crucial for improving production and per unit area yield of potato. To meet the agronomic requirements of potato planting, various types of potato planters have been developed according to the existing differences in geographical environment, cropping system, farm scale, and economic status among different countries and regions. This paper summarizes the research progress and application status of potato planters all over the world. Specifically, the seed-metering devices of potato planters are classified firstly. Then, potato planters are classified into three types: potato planters for manual/animal, two-wheel tractor, and four-wheel/crawler tractor traction. Furthermore, the characteristics of some typical commercial potato planters are discussed. Finally, some suggestions are proposed in order to promote the development of potato planters.

Keywords: potato planter; seed metering device; mechanization; research progress

Citation: Zheng, Z.; Zhao, H.; Liu, Z.; He, J.; Liu, W. Research Progress and Development of Mechanized Potato Planters: A Review. *Agriculture* **2021**, *11*, 521. https://doi.org/10.3390/agriculture11060521

Academic Editor: José Pérez-Alonso

Received: 20 April 2021
Accepted: 29 May 2021
Published: 3 June 2021

1. Introduction

Potato (*Solanum tuberosum* L.), the world's third most important crop and first most important non-cereal food crop, has been regarded as a key source to ensure food security for the world's fast-growing population [1–3]. According to data from FAO, the potato industry is undergoing major changes. Before the early 1990s, potatoes were mostly grown and consumed in Europe, North America, and the former Soviet Union. Since then, there has been a dramatic increase in potato demand and production in Asia, Africa, and the Americas, where the output rose from less than 30 million tons in the early 1960s to more than 261 million tons in 2019. Furthermore, FAO data show that since 2005, production in developing countries has exceeded that from developed countries, and it is estimated that potato has provide nutrition for over a billion people [4]. Nowadays, potato has been the vital economic crop in many countries. For example, potato contributes 2.9% to agricultural gross domestic product (GDP) in India. In Egypt, the potato is the second most important strategic crop, accounting for about 25% of total Egyptian agricultural exports, and has been ranked as one of the most important vegetable crops [5].

At present, global potato production and food security are being greatly affected by many factors, such as arable land, climate, population, and resources [6–8]. Therefore, it is important to achieve a high quality, high yield, improved efficiency, and sustainability of potato production and to accelerate potato industrialization to ensure long-term food security [9–11]. As can be seen in Figure 1, the world has witnessed a continuous increase of potato production since 2000, and potato production showed a 14.7% increase from 2000 to 2009, from 322.8 million tons (Mt) to 370.4 Mt. However, the overall potato harvested area showed a 38.2% decrease from 2000 to 2019, from 19.9 million ha (Mha) to 17.3 Mha.

Therefore, it can be confirmed that the per unit area yield of potato has been increasing steadily [12].

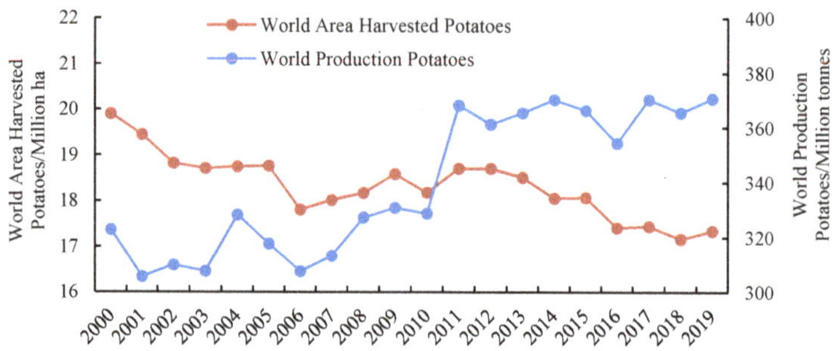

Figure 1. Production/yield quantities of potatoes in the world, 2000–2019. Date from reference [12]. Copyright 2021 FAO.

From 2000 to 2019, the top five highest average potato production countries were China, India, the Russian Federation, Ukraine, and the United States of America. Moreover, as can be seen from Figure 2, during this 20-year period, production in China and India has increased dramatically, while potato production has kept a dynamic balance in the Russian Federation, Ukraine, and the USA. In the 2019 rankings of the top ten potato producing countries, China and India occupy the first and second places, while the Russian Federation occupies the third.

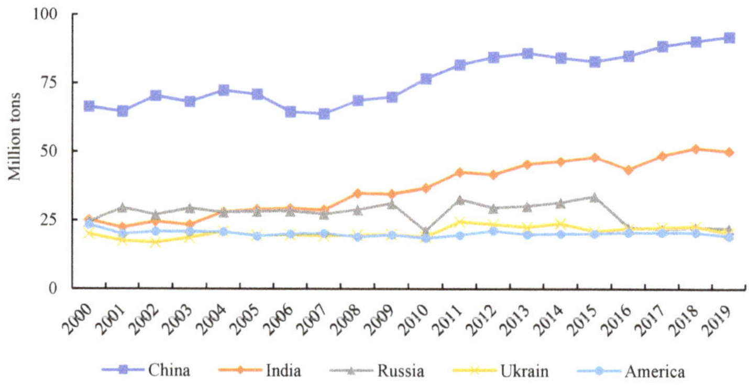

Figure 2. Potato production variation of the top five potato producing countries during the period 2000–2019. Date from reference [12]. Copyright 2021 FAO.

As can be seen from Figure 3, during 2000 to 2019, Asia has been the major producer of potatoes, with 45.13% of the global potato production during that period, and Europe is ranked a close second with a portion of 35.49%. The Americas and Africa rank third and fourth with 12.55% and 6.31%, respectively, while Oceania only has 0.51% of the global production share of potatoes, which ranks as last. Across the world, more than 17 Mha of farmland are devoted to potato production, with a yield of about 21.36 t/ha in 2019. According to Table 1, by analyzing the data from 2000 to 2019, it can be confirmed that the growing area has been decreasing while potato quantity has increased in the regions of the Americas, Europe, and Oceania, and the yield has increased simultaneously. In addition, potato quantity has increased along with the growing area in the regions of Africa and Asia, which also saw yield growth at the same time. Consequently, the world's potato per unit area yield achieves steady growth without considering the decline in the growing area of potato, which is in good concordance with the above analysis and results.

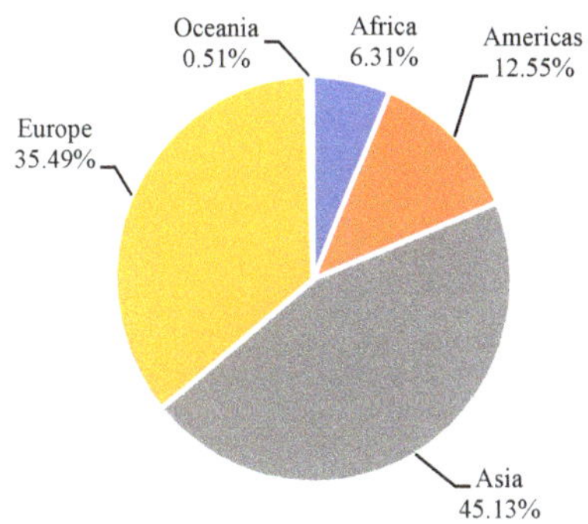

Figure 3. Global production share of potatoes in different continents during 2000–2019. Date from reference [12]. Copyright 2021 FAO.

Table 1. Yields of potato in different continents. Date from ref. [12]. Copyright 2021 FAO.

	Harvested Area 2000 (Mha)	Harvested Area 2019 (Mha)	Quantity 2000 (Mt)	Quantity 2019 (Mt)	Yeild 2000 (t/ha)	Yeild 2019 (t/ha)
World	19.89	17.34	322.77	370.44	16.23	21.36
Africa	1.25	1.76	13.18	26.53	10.54	15.07
Americas	1.92	1.54	42.19	45.08	21.97	29.27
Asia	7.97	9.30	121.49	189.81	15.24	20.41
Europe	8.70	4.70	144.20	107.26	16.57	22.82
Oceania	0.54	0.43	1.70	1.74	3.15	4.05

There are several factors that affect the global potato production, among which mechanization is crucial for improving production and per unit area yield of potato. Planting, cultivating, and harvesting are the three main components of potato production. Among potato production components, planting is one of the most important steps, which is complex, labor-intensive, and difficult to mechanize, influencing the production quality and final output of potato greatly [13,14]. The entire planting operation can be divided into tillage, ditching, spraying, seeding, fertilizing, ridging, etc. Depending on the method employed for planting, these functions are performed by different machines, or even by one machine in a single pass over the field. Reducing labor intensity, improving working efficiency, and ensuring seeding quality are advantages of mechanized potato planting [15,16]. Therefore, potato planters are especially significant for the development of the potato industry. This paper aims to analyze the research progress and application status of potato planters in the world. Firstly, the seed metering device of potato planters was clarified. Then, potato planters were classified into three types according to the drive mode: potato planters for manual/animal, two-wheel tractor, and four-wheel/crawler tractor traction (Figure 4).

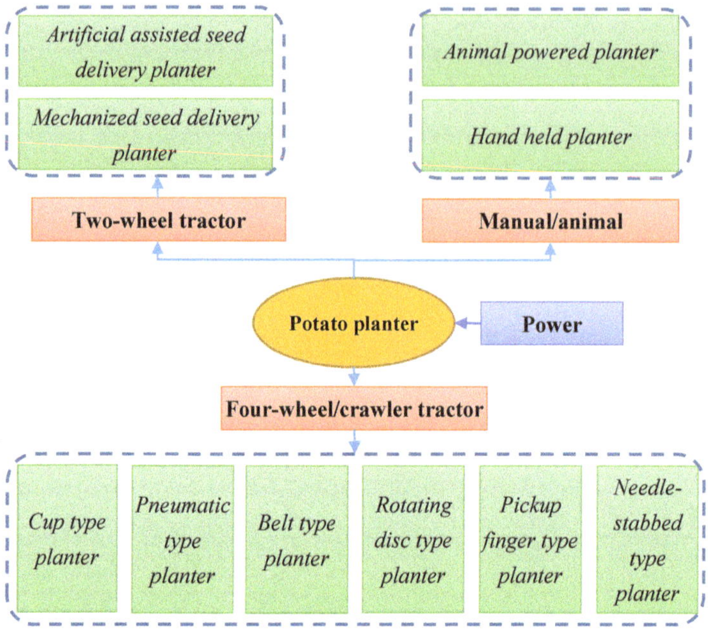

Figure 4. Schematic diagram of categories of potato planter.

2. Review of Seed Metering Devices of Potato Planters

The seed metering device is the key part of a potato planter [17,18]. According to the planting method, this paper summarizes the structural characteristics of the potato seed metering devices and classifies them as follows: cup, pneumatic, belt, rotating disc, pickup finger, needle-stabbed, etc.

2.1. Cup-Type Seed Metering Device

The cup-type seed metering device (Figure 5) is the most widely used seeding device for potato planters around the world [13,16,19]. The metering process of the cup-type potato metering device can be divided into five phases: seed filling, seed transporting, seed cleaning, seed translocation, and seed dropping. Under gravity, the potato seeds flow to the bottom of the seed box, where seed filling occurs. The driving pulley, which is powered by the potato planter's land wheel, drives the seed metering device. The seed metering belt/chain close to the seed filling area moves upward, and each metering cup installed on the seed metering belt/chain scoops 1–2 potato seeds in turn. When the cups move to the seed cleaning area, the surplus potato seeds that lay in the cups will be cleaned up by the seed cleaning device and fall back into the seed box. Once the cup reaches the highest point and crosses the driving pulley, the potato seeds fall on the back of the next cup. The two spoons adjacent to each other have a separate space, which guarantees there is only one potato seed on the back of each cup. Potato seeds are carried by the cups to the seed release point, then drop into the bottom of the seed ditch, and finally the whole seeding process is completed [20,21].

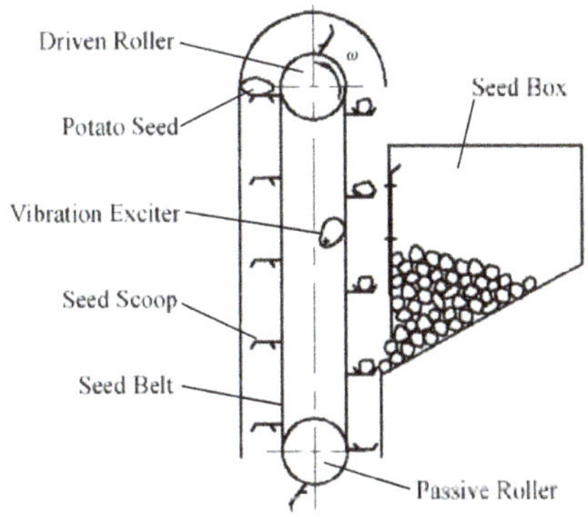

Figure 5. Cup-type seed metering device. Reprinted with permission from ref. [19]. Copyright 2016 Chinese Heilongjiang Agricultural Commission.

According to the transporting mode, the cup-type seed metering device can be divided into belt-cup and chain-cup types. Compared with the chain-cup-type metering device, the belt-cup type has a simple structure, which is not expensive and its working speed can be adjusted easily, but it is more likely to slip during the operation process, leading to poor reliability of transportation and making it difficult to guarantee seeding space uniformity. For the chain-cup-type metering device, the transportation quality is good, but it has a high damage rate of potato seeds, leading to yield loss [16,19]. Several studies were conducted on improving the working performance of the cup-type seed metering device. For example, Buitenwerf et al. [13] built a mathematical model to predict the planting accuracy and planting capacity of cup-belt seed metering devices. Lü et al. [22] designed a belt-cup-type seed metering device, which was suitable for seeding whole potato seeds or sliced potato seeds. In order to solve the problem of sowing loss in the chain cup-type seed metering device, Sun et al. [23] and Niu et al. [24] designed a sowing loss monitoring and compensation system, respectively. Laboratory tests showed good performance of the system. To improve the efficiency and seeding precision of potatoes, Zhang et al. [25] designed a novel vacuum and spoon belt combination metering device.

2.2. Pneumatic-Type Seed Metering Device

The pneumatic-type seed metering device (Figure 6) operates with high negative pressure to attract and hold potato seeds when removing them from the seed box, and the redundant seeds will be cleared away by a related seed cleaning device and returned to the seed box again for later use. The remaining seeds are released in a suitable area (with no negative pressure) and drop into the seedbed. There are several advantages of pneumatic-type seed metering devices, such as low seed damage rate, high metering quality, high adaptation to seeds of different sizes and shapes, and high operation speed [26,27]. Due to these advantages, a series of studies were conducted on the design and optimization of the pneumatic-type seed metering devices.

McLeod et al. [6] designed a prototype pneumatic seeding device to select and deliver a single true potato seed to a specific location. Lü et al. [15,20] proposed a novel design scheme for potato air-suction metering. Field test results confirmed that the working performance met the requirements of precision seeding. Lai et al. [28,29] designed a vibrating feeding pneumatic disc metering device for mini-tuber through theoretical calculation based on the physical properties of mini-tuber. Mao et al. (2013b) [30] designed and optimized a precision pneumatic metering device with declined disc for virus-free mini-potato seed. This type of device not only improved the working efficiency but also reduced the damage on seed potato.

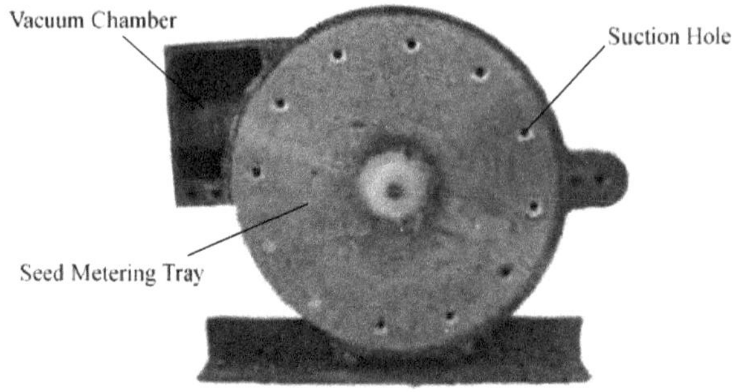

Figure 6. Pneumatic-type seed metering device. Reprinted with permission from ref. [31]. Copyright 2019 Chinese Society for Agricultural Machinery.

2.3. Belt-Type Seed Metering Device

The belt-type seed metering device uses a belt to capture, deliver, and sow potato seeds. This kind of seed metering device can be divided into two forms: moving-belt type and cup-belt type.

2.3.1. Moving-Belt-Type Seed Metering Device

The main components of moving-belt-type seed metering devices are the seed box, seed transportation device, side belt, mid-belt, etc. Potatoes are dropped from the seed box to the mid-belt through the seed transportation device. Both sides of the mid-belt are arranged with a side belt, and the direction of the linear speed between the mid-belt and side belt is converse. Potatoes on the mid-belt are sorted into a single row, then transported to the seed-throwing area, and finally drop into the seedbed. The redundant potatoes are removed from mid-belt to side-belts. The side belts transport the potatoes to the end sides and prepare them for moving to the mid-belt again for single-row sorting and seeding. This kind of seed metering device has a simple structure, small size, low rate of seed injury, and stronger adaptability to various shapes of potato seeds [32].

There are several studies about moving-belt-type seed metering devices. For example, He. [32] designed a belt type of potato seed sowing device; theory, analysis, and experiment were conducted and confirmed that the design satisfied the requirements of precision seeding. Liu et al. [33] developed a mechanical planting technology and designed a related seeding device based on the principle of forced vibration, which could promote the mechanization level of potato micro-seed planting. Nowadays, some commercial planters have applied this kind of seed metering device (Figure 7), such as Standen Engineering Ltd., Grimme UK Ltd., Trinkel Ltd., etc.

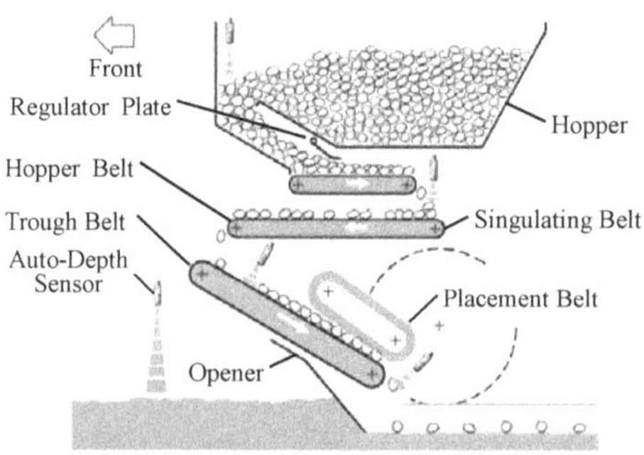

Figure 7. Moving-belt-type seed metering device used in commercial planter.

2.3.2. Cup-Belt-Type Seed Metering Device

The seed conveyor belt (Figure 8) is designed with grooves, the grooves are arranged in equal uniform spacing, which guarantees seeding space uniformity. Potatoes are filled into the grooves, relying on the flowability, by hand. The grooves move with the movement of the seed conveyor, and potato seeds drop from the grooves into the seedbed at the seed-dropping mouth. This type of seed metering device has a simple structure and is easy to operate. Isher Engineering Works Ltd. (Meerut, India) and Vasilios Kelidis & Co (Kato Nevrokopi, Greece). use this kind of seed metering device to plant whole or sliced potato seeds. However, this type of seed metering device needs to employ humans to put the potato seeds into the grooves, which leads to high labor strength and cost.

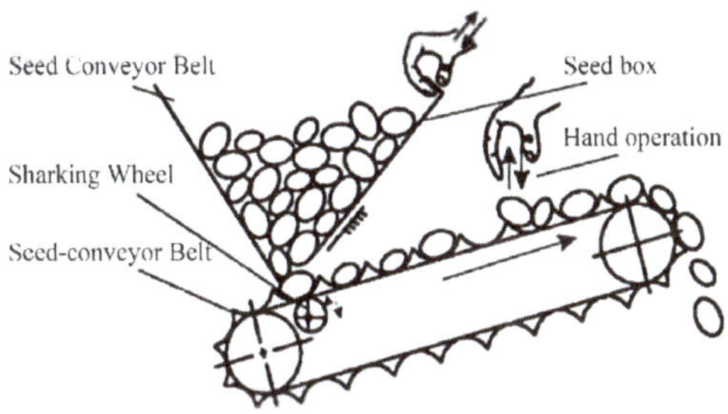

Figure 8. Cup-belt-type seed metering device. Reprinted with permission from ref. [31]. Copyright 2019 Chinese Society for Agricultural Machinery.

2.4. Rotating Disc Seed Metering Device

For the rotating disc seed metering device (Figure 9), the main components are a rotating disc, transaction system, seed box, cover plate, and seed tube. Certain numbers of round holes or grids with the same diamcter/dimensions are arranged uniformly around the center of the rotating disc. The rotating disc is nested within the cover plate, and the cover plate is fixed upon the frame. Moreover, the seed tube is set underneath the rotating disc and penetrates through the cover plate, and a seed dropping mouth is generated at the point of the joint parts of the seed tube and cover plate. Seeds are filled into the round holes/grids successively by hand from the seed box. When the round holes/grids turn to the position of the seed dropping mouth, potato seeds located in the round holes/grids will fall into the seed tube under gravity. Finally, potato seeds drop into the seedbed through the seed tube. The characteristics of this device are as follows: (1) high adaptability (it is suitable for irregularly shaped seeds); (2) no seed damage and blocking; (3) low rate of miss/multiple seeding index; (4) high intensity of labor force, resulting in high operation costs.

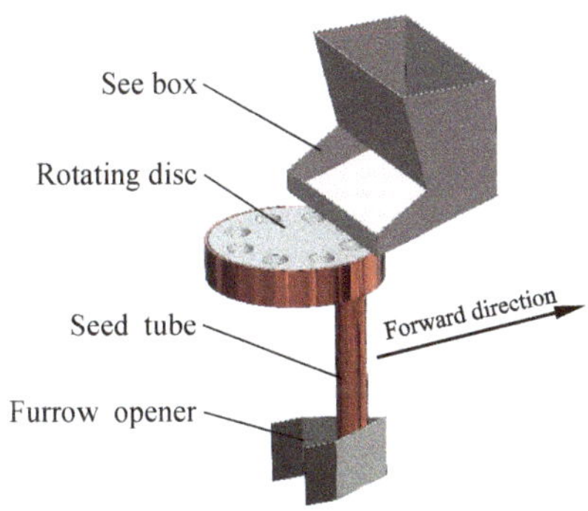

Figure 9. Revolving disc seed metering device.

There were several applications of patents related to the rotating disc seed metering device in the 1990s, such as the patents of US S736981 A [34], US 744984 A [35], US 997674 A [36], US 2642909 A [37], etc. As this type of seed metering device is suitable for seeding irregularly shaped seeds, Lü et al. [38] designed a rotating disc seed metering device which could plant the sliced and whole potatoes with the same seed metering device. Nowadays, with its applicability in planting whole or cut potato seeds, the rotating disc seed metering device has been widely used by many international agricultural machinery enterprises, such as F.lli Spedo, Ferrari Costruzioni Meccaniche, US Small Farm Equipment, Garmach, and Ganesh Raj Industries.

2.5. Pickup Finger Seed Metering Device

For the pickup finger seed metering device (Figure 10), a vertical disk equipped with a certain number of spring-loaded fingers are rotating in a seed box. The fingers touch the vertical disk under elastic force and open in sequence when traveling to the bottom of the seed box, and each finger picks up a single potato seed. Then, the potato seeds are carried by the fingers with the rotation of the vertical disk and open in sequence again when traveling to the seed dropping area, and release seeds above the seed furrow. This kind of metering device is suitable for seeding whole or cut potato seeds. In order to improve the seeding performance of the pickup finger seeding device, Chen S.Y. [39] designed a novel device, in which the finger closes and opens along with peripheral direction. Field tests indicated that it has a good performance.

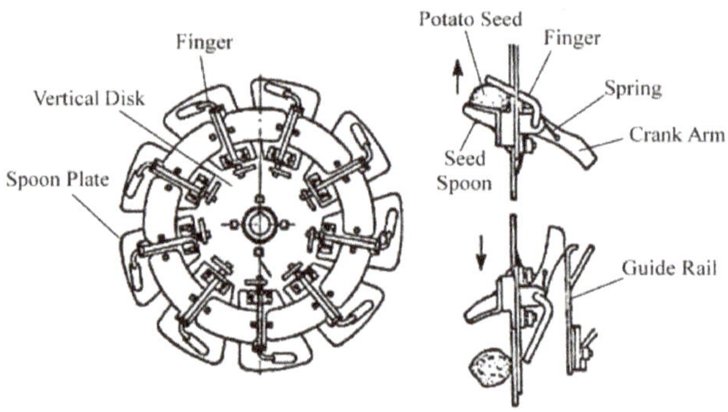

Figure 10. Pickup finger seed metering device. Reprinted with permission from ref. [31]. Copyright 2019 Chinese Society for Agricultural Machinery.

2.6. Needle-Stabbed Seed Metering Device

For the needle-stabbed seed metering device (Figure 11), elastic plastic needles are mounted on the edge of a vertical disk and rotate with the vertical disk in a seed box. Each needle stabs and grabs one seed in sequences in the seed feeding area. Potato seeds are carried by needles with the rotation of the vertical disk, and then the needles are pulled out from the potato seed as they travel to the seed dropping area and seeds are dropped into the seedbed. Due to the mechanical structure of the needle-stabbed seeding device, the shape and size of potato seeds have little impact on its planting spacing uniformity.

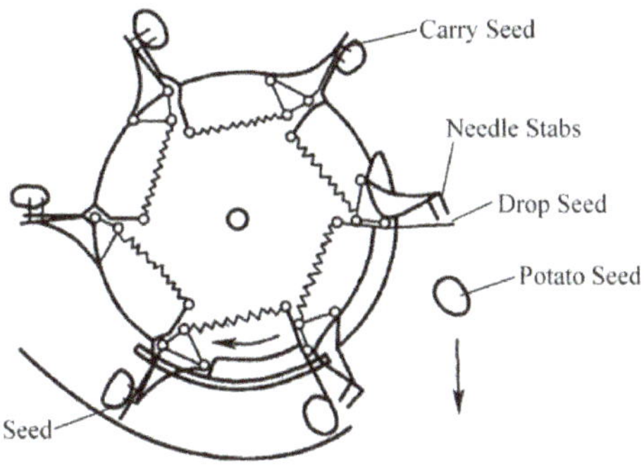

Figure 11. Needle-stabbed seed metering device. Reprinted with permission from ref. [31]. Copyright 2019 Chinese Society for Agricultural Machinery.

Seeding performance is an important factor for the seed metering device. Misener G.C. [40] evaluated the seed piece distribution patterns achieved by cup- and needle-stabbed-type potato planters through a large number of trials. Results indicated that the needle-stabbed seed metering device operated slightly more effectively than the cup type. Furthermore, based on the premier research, Misener G.C. [41] studied the effect of seed size and type on seeding performance of potato planters. It was concluded that the size of the seed had little effect on the miss/multiple seeding index, and whole potato seeds had a smaller miss-seeding index than the cut potato seed.

3. Application Status of Potato Planter

This section summarizes currently available potato planters around the world and classifies these machines into three categories: potato planters for manual/animal, two-wheel tractor, and four-wheel/crawler tractor traction.

3.1. Manual/Animal-Powered Potato Planters

Potato planters for human/animal power represent the most basic form of mechanization. In order to enhance productivity regarding energy efficiency and ergonomics, it is necessary to improve the design quality of the equipment. In addition, a demand-driven development and innovation system is essential for the adoption of improved human/animal-powered equipment. Manual/animal traction potato planters usually have a small body and simple structure, are light weight, and are easy to operate. These planters are mainly designed for small plots of land, terracing mountain, or hilly areas [42]. However, this type of planter lacks appeal nowadays because of its low working efficiency and high labor intensity. Some typical planters reviewed are as follows.

3.1.1. Hand-Held Potato Planter

Jab planter (Figure 12a): The jab planter is a typical manual potato planter used for a long period. The planter has a hinged "beak", which closes and opens under the control of a handbrake (Figure 12a). Potato seed drops into the "beak" when it drives into the soil. The "beak" opens when the foot on the long handle contacts the land surface or clenches the handbrake, then the seed drops into the soil at a certain depth.

(a) (b) (c)

Figure 12. Hand-held potato planters. (**a**) Jab planter. Reprinted with permission from ref. [43]. Copyright 2020 Stand 'n Plant; (**b**) double-handle bulb planter. Reprinted with permission from ref. [44]. Copyright 2020 Groves Nurseries & Garden Centre; (**c**) manual-drawn potato planter. Reprinted with permission from ref. [45]. Copyright 2008 Central Potato Research Institute.

Double-handle bulb planter (Figure 12b): This planter mainly contains two blades (which can be closed and opened upon a hinge), two handles, and a hinge; the mechanism enables one to make a hole and deposit the potato seed in one step under foot-driven power. In looser soils, the planter places the potato seed into the cone-shaped reservoir created by the blades, then drives the blades into the ground and opens them to separate the soil and release the potato seed in the soil. In compacted soils, the planter is used much like a post-hole digger, which removes a plug of soil before planting.

Sneeboer potato planter: The Sneeboer potato planter is a potato manual seeding tool commonly used in small areas, such as gardens and small-scale farms. The planter includes a T-handle, a planting tube, and a long handle. The planting tube is a cone shape designed to prevent clogging when planting seed potatoes. Moreover, there is an edged swing fixed on the planting tube, which is used for foot-driving the tube into soil conveniently.

Hand-pushed potato planter: This planter is designed for seeding potato micro-bulbs (e.g., potato mini-tuber) and other large grains of crop seeds, such as onion, garlic, tulip, etc. It has a container for seeds, an opener to place the seeds, a driving wheel to provide power for the seeding device, and a depth wheel to control the seeding depth. In addition, the type of seeding device is a disk brush. The farmer walks behind the seeder and controls it by handlebars. The total weight of the seeder is 30 kg, the working efficiency is 0.11 ha/h, and the seeding inaccuracy is less than 9.0%.

Manual-drawn potato planter (Figure 12c): The Central Potato Research Institute of India developed a two-row manually drawn potato planter of cup type [45], which is suitable for small and marginal farmers. The seed metering device consists of two rubberized belts with 16 MS sheet cups riveted on each belt which have been made spring-loaded in order to counter jerks. The planter works more efficiently with graded potato seeds. The working efficiency is 0.5 ha/d, the miss-seeding rate is less than 0.5%.

3.1.2. Animal-Powered Potato Planter

As the popularity of potatoes grows around the world, inventions to improve potato planting began flooding the market. In the 18th century, the first animal-powered potato planter was developed to promote working efficiency. Since then, the animal-powered potato planter developed rapidly because of its high efficiency, low manual power consumption, and good seeding quality. The animal-powered potato planter is usually drawn by horse or ox, and completes the operation of furrow opening, seeding, covering, and ridging in one pass. However, the application of the animal-powered potato planter became less and less after the tractor came into being. Some once-popular animal-powered potato planters are shown as follows (Figure 13). The John Deere One-Row Planter is suitable for planting whole or cut, treated or un-treated potato seeds. Ridges of high-wide, low-narrow, or with any desired variation can be obtained by adjusting the characteristic parameters of the soil covering disc. In addition, there is a fertilizing device attached on the planter for fertilizing. For the animal-drawn potato planter developed by Central Potato Research Institute of India [45], potato seeds are dropped and covered while making a ridge simultaneously. The seed metering device consists of a hollow rotor with cups fixed on it. A manually operated clutch and a lifting arrangement have been provided. Seeding space can be adjusted in the range of 16–30 cm. The working deficiency and seed damage rate are 0.9 ha/day and less than 0.5%, respectively, while with small seed missing rate.

(a) (b)

Figure 13. Animal-powered potato planters. (**a**) John Deere. Reprinted with permission from ref. [45]. Copyright 2008 Central Potato Research Institute; (**b**) animal drawn. Reprinted with permission from ref. [45]. Copyright 2008 Central Potato Research Institute.

3.2. *Two-Wheel Tractor-Driven Potato Planter*

There are greater interests in mechanization for crop planting in smallholder agriculture. However, mechanized planters mounted on four-wheel tractors are not suitable for small-size farmland. Therefore, two-wheel tractors are widely used by small farmers to provide power for mechanized potato planting in small-size farmland because of their small size and easy maintenance, etc. [46,47]. According to the way of clutching potato seeds from hopper to the seed drop tube, the two-wheel tractor-driven potato planters can be divided into artificial assisted and mechanized seed delivery types. The cup-type seed metering device is usually used for seed delivery. Some typical artificial assisted and mechanized seed delivery planters for two-wheel tractor are described as follows.

3.2.1. Artificial Assisted Seed Delivery Planters

The main components of this planter (Figure 14) are a seed drop tube, furrow opener, soil covering plates, etc. Furthermore, there is a rotary tiller in front of the furrow opener to conduct rotary tillage operation. The seed box filled with potato seeds is fixed on the tractor and not far from the operator. In this way, the operator can control the tractor and put the seeds into the seed drop tube at the same time during the operation of potato seeding, which enables single-person operation. As can be seen from the planter's working principle, the seeding space is determined by the operator. Furthermore, the planter can be operated by tow operators, one for tractor control and the other for seed dropping.

Figure 14. Artificial assisted seed delivery planter. Reprinted with permission from ref. [48]. Copyright 2020 Google.

3.2.2. Mechanized Seed Delivery Planters

The mechanized seed delivery planter avoids dropping seeds by artificial assistance. Compared with the artificial assisted seed delivery planter, it obtains a higher level of mechanization. There are various models of this kind of planter produced in China, Italy, and Ukraine that are light in weight, matching with small power, and seeding in a single row. These machines are powered by two-wheel tractors with a power range of 8 to 12 hp, net weight of about 100 kg, and working speed of below 3 km/h in general. Although these planters are made by different manufacturers, their differences are not obvious, and their seed metering device is usually the cup type. Some typical commercial potato planters are shown in Table 2.

Table 2. Comparison of mechanized seed delivery planters.

Country	Company	Model	Picture	Technical Characteristics
Ukraine	ROSTA Scientific Production Company	KSM-1 [49]		Cup-chain seed metering device; matched type, mounted; single one seeding row; plate heaves are designed on the edge of ground wheels to prevent slip; equipped with a fertilizer distributor (optional); seeding space, 25/33 cm; working speed, 1.7–2.2 km/h; working efficiency, 0.2–0.3 ha/h; suitable seed size, 4–6 cm; net weight, 50 kg.
Italy	F.LLI Spedo	SPA/M [50]		Cup-chain seed metering device; matched type, mounted; single one seeding row; two adjustable self-cleaning bedder discs allow variable coverage; seeding depth adjustable; seeding space, 25/30/35/40 cm; net weight, 84 kg.
China	Weifang Luke Machinery Co., Ltd.	2CM-1B [51]		Cup-chain seed metering device; matched type, mounted; single one seeding row; attached with a fertilizing device; the fertilizer quantity can be adjusted easily and conveniently; different options for ridge shape can be obtained by changing the angle between two bedder discs and fixed position; net weight, 110 kg, working depth, 8–15 cm, seeding space, 12 cm; matched power, 8–12 hp.

Table 2. *Cont.*

Country	Company	Model	Picture	Technical Characteristics
China	Yucheng Qichen Machinery Co., Ltd.	Hand tractor belt [52]		Cup-chain seed metering device; matched type, mounted; single one seeding row; net weight, 30 kg; sowing depth, 8–15 cm; working width, 12 cm; matched power, 8–12 hp.

3.3. *Four-Wheel/Crawler Tractor-Driven Potato Planter*

Potato planters powered by four-wheel/crawler tractors are the most widely used potato seeding machines all over the world, ranging from large-scale farms to small plot lands. There are several kinds of four-wheel/crawler tractor-driven potato planters. As can be seen from the above discussion, the seed metering device is the key part of a potato planter. Therefore, according to the type of seed metering device, the four-wheel/crawler tractor-driven potato planters can be divided into cup, pneumatic, belt, rotating disc, pickup finger, and needle-stabbed types.

3.3.1. Cup-Type Potato Planter

The cup-type potato planter is the most commonly used seeding machine because of its easy operation, steady working performance, low cost, and convenience in adjusting seeding space. Referring to its seeding ability, there are many different specifications from large machinery to small machinery. For the small-scale planter, the matched power is generally below 20 hp and with a single seeding row. For the large/medium-scale planter, the matched power is usually larger than 20 hp, and with two or more seeding rows. The typical cup-type potato planters are shown as follows.

(1) Cup-type potato planter in small scale

This planter is designed for small plot lands or terracing mountain due to its small size and light weight; the matched power generally is less than 20 hp, which matches a tractor with a three-point suspension. Several researchers such as Hamad et al. [53], Al-Gaadi et al. [54], Ebrahem et al. [5], Wakchaure et al. [55], Gavrilov et al. [56], and Wang et al. [57,58] have made efforts to develop and modify the cup-type potato planter in small scale, to promote the development of this kind of potato planter.

Nowadays, the whole plastic mulching and furrow planting pattern of potato has been widely applied in the arid areas on the Loess Plateau of Northwest China. Several researchers have developed combined potato planters to adapt to this cultivation pattern. For instance, Shi et al. [59] developed a combine potato seeder which could throw and cover soil on a film edge. Dai et al. [60,61] invented a combined potato planting machine for covering soil on top of a full film surface. Key components were designed and operation parameters were confirmed. Sun et al. [62,63] designed a potato planter in 2017 and developed a double-rank multi-rod hill-drop potato planter on plastic film in 2018, respectively. Results indicated that both of the planters work well. In order to realize the hill-drop under plastic film in potato seeding, Sun et al. [64] designed a double-crank multi-rod mechanism and built a prototype.

Furthermore, there are several commercial planters around the world. Some typical machines are shown in Table 3.

Table 3. Comparison of the cup type potato planter in small scale.

Country	Company	Model	Picture	Technical Characteristics
Poland	Bomet Sp. z o.o. Sp. K.	BOMET [65]		Cup-chain seed metering device; matched type, mounted; single one seeding row; seeding space, 29/32/35 cm; planting depth, 10–15 cm; net weight, 130 kg; matched power, 20 hp.
Italy	F.LLI Spedo	SPA/T [50]		Cup-chain seed metering device; matched type, mounted; single one seeding row; an adjustable coulter forms a furrow while a chain with cups settles the seeds to the desired depth; two adjustable bedder discs allow variable coverage; seeding space, 25/30/35/40 cm; loading capacity, 50 kg; matched power, 15 hp.
USA	US Small Farm Equipment Co.	P-10M [66]		Cup-chain seed metering device; matched type, mounted; single one seeding row; sprocket set for six seeding space changes from 20.3 to 38.1 cm; traction tires for less drive wheel slippage results in better seeding space; working efficiency is 0.4 ha/h at forward speed of 3.2 to 6.4 km/h.
China	Changzhou LEFA Industry and Trade Co., Ltd.	LF-PT32 [67]		Cup-chain seed metering device; matched type, mounted; single one seeding row; net weight, 138 kg; matched power, 20–50 hp; seeding space, 32 cm; row space, 51 cm.

(2) Cup-type potato planter in large/medium scale

This kind of planter is suitable for large/medium plots of land; it has a complicated structure and multiple function with high efficiency. Due to its high efficiency, the required power is larger than 20 hp in general. In addition, this kind of potato planter can seed two rows or more in a single pass, and it could be trailed or suspended with the tractor.

Several studies were conducted in design and optimization of the cup-type potato planter in large/medium scale. For example, Mari et al. [68] carried out research to evaluate the performance of the four-row potato planter. Conclusions indicated that this planter is labor-saving. In order to enhance the seeding quality, Lü et al. [69] analyzed the bouncing and casting process of the cup-type potato seed metering device and the performance of the flow deflector, and put forward the optimization methods. Aiming at reducing the costs of potato seeds and labor of workers, a fully automatic in-situ seeding machine for cutting seed potatoes was developed by Cho et al. [70]. Aiming to reduce the inaccuracy of fertilizer in cup-type potato planters, Yang et al. [71] proposed a separated layer fertilization technology based on surface drainage and V-type anti-blocking structure.

The cup-type potato planter in large/medium scale is the most commonly used potato planter all over the world. There are lots of enterprises and various brands of this type equipment. Some typical machines are listed in Table 4.

Table 4. Comparison of cup type potato planter in large/medium scale.

Country	Company	Model	Picture	Main Structures and Properties	Technical Characteristics
INDIA	SWAN AGRO	NSE PPR-2 [72]		Matched type, mounted; two seeding rows; two furrow openers; two seed metering devices; row spacing, 60–66 cm; working depth, 13–15 cm; working efficiency, 0.51 ha/h; required power, 40–60 hp.	Fully automatic; agitator available to avoid seeds block problem; double seeding prevention through conveyor vibration mechanism; fertilizer tank is attached with the machine.
Italy	F.LLI Spedo	SPA-2/D [50]		Matched type, mounted; two seeding rows; two disc openers; four discs with 46 cm; two seed metering devices; pure weight, 480 kg; loading capacity, 500 kg; seeding capacity, 0.4–0.6 ha/h; row distance, 70–90 cm; required power, 50 hp.	The disc openers are attached with a compression spring, allowing work on strong soil; the inclined belt permits work on hilly ground; the hopper is equipped with a rubberized canvas to prevent massive loading of tubers on the belt.
England	Standen Engineering Ltd.	SP330 [73]		Matched type, mounted; three seeding rows; stainless steel openers and floating side shields; stainless forming hood; three seed metering devices; seeding space, 13–45 cm; required power, 120 hp.	Infrared proximity sensors control the hydraulic floor drive, maintain a regulated level of seeds in the cup area; large upper cup belt rollers lessen the associated 'flicking' action and possible chit breakage.
China	Shandong Transce Agricultural Machinery Technology Co., Ltd.	2CMX-4B [74]		Matched type, mounted; four seeding rows; four furrow openers; four seed metering devices; manually operated lane marker; row spacing, 90 cm; working width, 360 cm; required power, 125.1–179.9 hp.	Adopting advanced technologies of vibration, lifting seeding, and planting depth control; a disc-type rubber expands the application spacing to 70–90 cm.
Germany	GRIMME Landmaschinenfabrik GmbH & Co. KG	GL 420 [75]		Matched type, mounted; two seeding rows; infinitely adjustable mechanical shaker; manually operated lane marker; two control valves; four seed metering devices; distance between rows, 75–90 cm; loading capacity is 1200 kg; minimum required power, 122.4 hp.	Potato cultivation with improved protection against erosion; a shaping board for building ridges with specific characteristics from loose to firm soil with a smooth surface; optionally equipped with the systems of Clever Planting, Section Control.
Belgium	AVR	AVR Ceres 450 [76]		Matched type, trailed; four seeding rows; four hydraulic drive seed metering devices, 40 cups in each planting element; four wheels; hydraulic hopper, capacity is 3500–4000 kg; row spacing, 70–90 cm.	Each individual seeding row assembly is disconnected manually; Culti Control Option maintains constant seeding depth; the ridging hood is operated as standard by an automatic pressure control.

Table 4. *Cont.*

Country	Company	Model	Picture	Main Structures and Properties	Technical Characteristics
Holland	Dewulf	CP 42 [77]		Matched type, trailed; four seeding rows; four V-shaped furrow openers; four seed metering devices; distance between rows, 75–90 cm; mechanical or hydraulic drive type; working efficiency, 1.5 ha/h.	Eagle-Eye system monitors the planting process; cups are positioned at a forward angle of 8° in the upwards section to ensure excellent pick-up; GPS and Precision-Planting system ensure seeding quality.
USA	Double L	9560 [78]		Matched type, trailed; six seeding rows; six seed metering devices; seeding space, 0–64.77 cm; row spacing, 71.12–101.60 cm; maximum planting speed, 9.66 km/h; minimum required power, 170 hp; hopper capacity, 6600 kg.	Seed flow components prevent seed bridging; equipped with fertilizer function; attached with controllers and receive GPS, radar, and ground speed data from the tractor.
USA	SPUDNIK Equipment CO. LLC	8312 [79]		Matched type, trailed; twelve seeding rows; front toolbars; folding type machine; twelve seed metering devices; row spacing, 86.4–96.5 cm; adjustable to plant 3, 6, 9, or 12 rows; seed hopper capacity, 400 cwt.	It is foldable and can be transported from field to field within road restrictions; seed bowl sensors keep the seed bowl at optimum levels; sensors measure the depth on both sides of the front toolbar.

3.3.2. Pneumatic-Type Potato Planter

The pneumatic-type potato planter has the advantages of a low seed damage rate, high adaptation to seeds size and shape, high sowing speed, etc. Therefore, the development of this kind of planter has been a hot spot of current research. Mao. [80] designed a precision pneumatic planter for virus-free mini-potato, in which the seed metering device has a declined disc. Lü et al. [81] developed an air-powered precision potato planter. The seed metering device applied the novel form with the distributed multi suction arms, and suction nozzles are installed at the end of each arm. In order to meet the conditions of small-plot operation in hilly and mountainous areas and enhance the mechanization level for mini-tuber, Wang et al. [82] designed a precision single-row air-suction planter.

Though the development of the pneumatic-type potato planter has a bright future, few producers manufacture and sell this kind of potato planter around the world nowadays. The successful examples available were from the enterprises of Crary Industries from the US and ERME from France (Figure 15).

Specifically, Crary Industries produces the Lockwood 600 series of pneumatic potato planters, which include three models: 604, 606 and 608. Figure 15a presents the Lockwood 606 pneumatic potato planter. The seed metering device in this planter series adopted a vacuum wheel with stainless steel arms, and the flexible urethane cups allow for higher planting speed while maintaining seed placement accuracy. In addition, this type of planter is suitable for planting whole or sliced potato seeds. The specifications of this series of planter are shown as follows. (1) The hitch type between planter and tractor is pull or semi-mount, and the minimum required power is 140, 180, and 225 hp, respectively. (2) Seeding rows are 4, 6, and 8, respectively. (3) A new larger hopper design eliminates the use of extensions. (4) A radar-controlled hydraulic drive delivers more accurate seeding space, and seeding space can be adjusted in the range of 15.2–45.7 cm. (5) Row spacing is 81.3–101.6 cm, and the maximum planting speed is as fast as 10.5 km/h.

(**a**) (**b**)

Figure 15. Pneumatic-type potato planter. (**a**) Lockwood 606. Reprinted with permission from ref. [83]. Copyright 2021 Crary Industries; (**b**) PLMS 4 rows. Reprinted with permission from ref. [84]. Copyright 2020 Engineering & Manufacturing Aircraft Tools.

Furthermore, ERME produces the PLMS series pneumatic potato planter. This type of planter is equipped with the traditional vacuum seed metering device and includes seeding rows of 3, 4, 5, 6, 8, and 12. The four-seeding-rows planter is popular (Figure 15b). Take the four-seeding-rows planter as an example; its specifications are represented as follows. There are four seed metering devices, and the minimum row spacing is 35 cm. A vibrator is applied in each seed metering device to ease the flow of seeds. Each row is independent and mounted on a parallelogram system. Planting depth can be adjusted by the rear depth wheels, and the planting density can be adjusted by a set of sprockets. The normal working speed is 2.2 km/h. The pure weight is just 560 kg, which is suitable for the small-area and light-weight tractor.

3.3.3. Belt-Type Potato Planter

The belt planter has a strong adaptability for planting non-calibrated seed potatoes, various sizes, or longish varieties. Several studies were conducted on the design and optimization of this planter. For example, Meijer et al. [85] developed an automatic belt-type planter for pre-sprouted seed. On a laboratory model the shape of the bed, the belt material, and belt speeds were analyzed. Results showed that the planter performed well and the seed damage was low. Misener et al. [86] modified a two-row, horizontal-cup belt planter. Performance tests were conducted and the results showed that with modifications the planter performed well. Arsenault et al. [87] constructed a two-row planter with good performance. Momin et al. [88] carried out a study to evaluate the performance of a two-row planter, and obtained the suitable combination of operation parameters. Aiming at enhancing the mechanization level of potato micro-seed planting, Liu et al. [18] designed a vibration-based seeder.

Due to its strong adaptability, low seed damage, and simple structure, some commercial belt-type potato planters are used in mechanical potato planting. Table 5 shows varieties of this type of machine.

Table 5. Comparison of belt-type potato planters.

Country	Company	Model	Picture	Technical Characteristics
Netherlands	Dewulf	Miedema Structural 30 [77]		Matched type, trailed; three seeding rows; minimum required power, 80.2 hp; distance between rows, 4–100 cm; seed hopper capacity, 3500 kg; maximum working speed, 11 km/h; the unique planting system limits friction between seeds and prevents damage to seeds; the drive and agitators are powered by hydraulics; small turning circle; possible to switch between planting two and three rows easily.
Germany	GRIMME Land-maschinenfabrik GmbH & Co. KG	GB 430 [75]		Matched type, trailed; four seeding rows; minimum required power, 122.4 hp; row spacing, 70–91.4 cm; seed hopper capacity, 3500 kg; maximum working speed, 10 km/h; chassis equipped portal pendulum axle, high maneuverability with 35° steering angle; combined with the integrated furrow openers, the Flow-Board XL ensures constant soil cover on seed potatoes.
England	Standen Engineering Ltd.	ZENO 21 [73]		Matched type, mounted; two seeding rows; matched power, 102 hp; row space adjustable, 76–102 cm; open-center, closed-center, or load-sensing hydraulic systems; depth control can be either manual or hydraulic; a sensor fitted to the RH wheel axle monitors the forward speed.
Netherlands	WIFO-Anema BV	MP20 [89]		Matched type, mounted; two seeding rows; row space, 75/90 cm; seed box capacity, 1100 L; seeding object, potato mini-tuber or garlic; pressure on the depth wheel can be controlled by a spring; position of the discs can be easily adjusted to change the ridge shape; spraying device can be equipped for applying chemicals during planting.

3.3.4. Rotating-Disc-Type Potato Planter

The rotating-disc-type potato planter has strong adaptability, low seed damage, high seeding quality, and low multiple/miss-seeding rate; but manually assisted seed feeding results in high labor intensity and low working efficiency. Furthermore, this type of planter is generally used on small scales. Experts put forward studies that mainly focus on the design of this type of planter. For example, Kohser F. [34], Springer H.J. [35], Heikes A.T. [36], and Osborne J.E. [37] et al. make efforts in structural innovation in terms of the seed metering device and machine, respectively. The scientific research team of Northeast Agricultural University, using this technology, developed a 2CMB2-type two-row planter [31,38]. The development of this planter is mainly targeted at the mixed cropping area with one crop one year and two crops one year systems in Southwest China and the wintering area in South China where the plot is small and the slope is large. Some typical rotating-disc-type potato planters are shown in Table 6.

Table 6. Comparison of rotating-disc-type potato planters.

Country	Company	Model	Picture	Technical Characteristics
Italy	F.LLI Spedo	TPA-PP2 [50]		Matched type, mounted; two seeding rows; matched power, 29.9 hp; row spacing, 17–45 cm; each element is equipped with a large alveolar plate; seeds are covered by the discs with adjustable inclination or by the ridgers with adjustable wings.
Italy	IMAC Co.	PPS-2F [90]		Matched type, mounted; two seeding rows; required power, 27.2 hp; row space, 70–90 cm; seeding space, 14–28 cm; seed hopper capacity, 150 kg; pure weight, 165 kg; working efficiency, 1–2 ha/h.
USA	US Small Farm Equipment Co.	RP-20M [66]		Matched type, mounted; two seeding rows; row spacing can be adjusted according to requirements; maximum working speed, 3.2 km/h; planting depth, 3.3–13.3 cm; row spacing, 107–120 cm; loading capacity, 90 kg; simple and easy to maintain; quick-change sprocket set.
Ukraine	ROSTA Scientific Production Company	SLM-1-2 [49]		Matched type, mounted; two seeding rows; working efficiency, 0.11 ha/h; seeding accuracy, 92% at least; seeding space, 3.3–6.7 cm; whole weight, 32 kg; suitable for seeding onions-bulbs, garlic, bulbs of a tulip, a gladiolus, potato mini-tubers, etc.
Slovenia	TEHNOS, d.o.o.	SK2 [91]		Matched type, mounted; two seeding rows; required power, 35.4–65.3 hp; adjustable row space, depth and width of cultivation; working efficiency, 0.1–0.3 ha/h; with a fertilizer dispensing device and interrow cultivator; machine weight, 320 kg.

3.3.5. Pickup-Finger-Type Potato Planter

The pickup-finger-type potato planter has its own characteristics, such as complex structure, suitability for seeding a certain size of seeds, low working speed, and poor seeding quality. In order to promote the development of this kind of planter, some studies were conducted. For example, Frase et al. [92] invented a type of pickup-finger-type potato planter for planting seeds of different sizes and shapes, which avoids trouble for changing seed plates and seed selecting. Plant spacing uniformity is the main parameter of the seed metering device. Boydaş et al. [93,94] investigated the performance of a pickup finger potato planter in laboratory and optimized the combined operation parameters. However, the seeding performance is significantly affected by seed shape as well as size. Moreover, the operation speed is limited to 7 to 9 km/h because of its mechanical structure. Meanwhile, seeds damage is likely to occur during the process of seeds pickup. In the context of high working performance and efficient modern agricultural development, the pickup-finger-type potato planter will be replaced in the longer term [95]. Due to its poor stability and seeding quality, this type of planter has narrow application in agricultural production. Only a few enterprises such as Harriston manufacture and produce the commercial pickup-finger-type potato planter. Figure 16 presents the Harriston Clamp potato planter; it has six seeding rows, and the hitch type between planter and tractor is pull. Seeding space and bowl height can be easily adjusted by the tractor, which has a cab equipped with GPS velocity sensors for precise seed planting.

Figure 16. Harriston Clamp potato planter. Reprinted with permission from ref. [96]. Copyright 2020 Harriston Industries.

3.3.6. Needle-Stabbed-Type Potato Planter

The research on needle-stabbed-type potato planters is mainly concentrated in the United States. During the 19th and 20th centuries, some experts applied a number of technical invention patents around the development and innovation of this type of planter. However, this type of planter has the vital disadvantages of promoting seeds damage and virus infection. Nowadays, while few commercial needle-stabbed-type potato planters are available in potato planting, the typical machines are shown as follows. There is the Lockwood 500 series needle-stabbed-type potato planter developed by Crary Industries, which includes three models: 504, 506, and 508. The hitch type between planter and tractor is semi-trailed. The number of seeding rows are 4, 6, and 8, respectively. Both the row space and seeding space are adjustable, and in the range of 81 to 102 cm and 17.8 to 40.6 cm, respectively. The working speed is 0–10.4 km/h, and the matched power is 100.1, 135 and 175 hp, respectively. Figure 17a presents the Lockwood 504 potato planter series of needle-stabbed-type potato planter produced by Harriston with seeding rows of 4, 6, and 8. The planter employs the novel seeding sensor technology, independent seeding element, and hydraulic driving system. Figure 17b shows the Harriston six-row potato planter.

(a) (b)

Figure 17. Needle-stabbed potato planters. (**a**) Lockwood 504. Reprinted with permission from ref. [83]. Copyright 2021 Crary Industries; (**b**) PLMS 4 rows. Reprinted with permission from ref. [96]. Copyright 2020 Harriston Industries.

3.4. *Comparision of the Above Potato Planters in Different Types*

By exploring the development of the technology and equipment of potato planting, it can be confirmed that the cup-type planter is the most widely used planting equipment in the world currently and will continue its dominant position in the future. Due to its simple structure, easy operation, and steady working performance, various specifications of the cup-type potato have been developed from large machinery to small machinery. Moreover, the cup-type potato planter not only can be powered by two/four-wheel tractors in large/small scale, but also can be powered by animal in certain conditions. Compared with the cup-type potato planter, the pneumatic- and belt-type planters have strong adaptability for planting non-calibrated seed potatoes, with a low seeds damage rate, among which, the pneumatic-type planter has no requirement for graded potato seeds and does little damage to seeds. In addition, it can be operated at high speed (more than 10 km/h) with good seeding performance. In addition, the working efficiency and quality can both be guaranteed, which means the pneumatic-type planter has a bright future. While for the belt-type planter, seeding performance is affected by the size uniformity of potato seeds. High seed size uniformity results in good seeding quality; poor seed size uniformity results in poor seeding quality. Therefore, in order to obtain good seeding results, potato seeds need to be graded by size before planting with the belt-type planter. It can be inferred that use of the belt-type planter might increase slightly and at a slow rate in the future.

For the rotating-disc-type potato planter, though it has the advantages of simple machine structure, easy operation, feasibility for seeding various standard potato seeds (cut or whole), and good seeding quality, the manual assistant feeding seeds for dropping makes it highly labor intensive with low working efficiency. This type of planter is generally used on small-scale farmland, terracing mountain, or hilly areas. Though the rotating-disc-type planter is not widespread around the world, it has prospects in developing countries and some developed countries where farmland scale is limited because of the terracing mountain and hilly areas. For the pickup-finger-type potato planter, the irregular shape and size of potato seeds easily leads to a high rate of miss-seeding and multiple seeding. Furthermore, under the pressure of pickup fingers, potato seeds are likely to be damaged during the grasping process. Thus, the pickup-finger-type potato planter has a limited development potential in the future. Similar to the pickup-finger-type planter, the needle-stabbed-type potato planter also faces its development problems as a result of the vital issue of seeds damage and virus infection. Despite the fact that the needle-stabbed-type potato planter has the advantages of high adaptability to different sizes and shapes of seeds with good seeding performance, the vital issue of seeds damage hinders its sustainable development. Furthermore, manual/animal-powered potato planters are usually small, light, simple, and easily manufactured, utilized, and maintained. These planters are invariably used on small farms or hilly regions due to the complicated landform, wavy terrain, and poor traffic conditions. Therefore, the manual/animal-powered potato planter has its necessity of existence.

In summary, potato planters ranging from small- to large-scale with different types are available around the world. Farm size and characteristics (landform, slope, etc.) determine the application of potato planter types. Generally, large-sized farmlands are suitable for employing large-scale potato planters, while small-sized farmlands or hilly regions are suitable for choosing small-scale potato planters. It can be concluded that the agricultural land per capita in developed and developing countries may have some potential relationship with the selection of appropriate mechanical planting technology. As can be seen from Figure 18a, Africa goes through an obvious decline in agricultural land per capita, while the Americas, Asia, and Europe show the inconspicuous decline trend. The agricultural land per capita of Asia and Europe is 0.37 ha and 0.62 ha, lower than the world average level of 0.63 ha. As can be seen in Figure 18b, China and India are the world's top two countries for potato production, while their agricultural land per capita rates are just 0.37 ha and 0.13 ha, respectively. Considering China and India are developing countries, the size of the majority of farmlands is smaller than 1 ha. It can be confirmed that small- or medium-scale potato planters are suitable for potato production in these two countries [97]. Specifically, cup-type potato planters obtain the dominant position for potato production in China and India currently and their position will be steady in future. Most other countries in Asia are in a similar situation. For Europe, the agricultural land per capita is a little lower than that of the world level. However, most counties in Europe are economically advanced with fewer people engaged in farming. The percentage of large-sized farms in Europe is higher than in Asia, which is about 30%. Nevertheless, two-thirds of farms are still smaller than 5 ha in size [98]. For instance, there are 111,740 farms smaller than 2 ha in size operating a total of 212,000 ha of agricultural land [99,100]. There are numerous farms in Europe, with different geologies, topographies, and climates. Various kinds of potato planters are available for potato production in Europe, such as large-scale planters (cup type, belt type, etc.) for large-sized farm, small-scale planters (cup type, rotating disc type, manual-powered type, etc.) for small-sized farmland and fields in hilly regions. In North America, the United States and Canada are the main potato production countries. Take the United States as an example, where there are about 2,023,400 farms, the total farming areas are about 363.2 Mha, and the average farm size is estimated at 179.7 ha [101]. These farms are particularly concentrated in the Great Plains, a vast expanse of flat land that is beneficial to applying large-scale agricultural machines. Consequently, potato planting machines are large-scale, integrated, and automatic in the United States, and the types of potato planter can be cup, pneumatic, belt, pickup-finger or needle-stabbed types. In Latin America, most crop production is based on subsistence agriculture implemented by resource-poor smallholder farmers [102]. A smallholder farm is generally less than 3 ha and dispersed around a village. Despite the high possession of agricultural land per capita (Figure 18a), farms are of small size due to the fragmentation of agricultural land in Latin America, which has a negative impact on the lower field efficiency of machines in irregular plots. Cup-type potato planters with medium or small scales are spread widely in this area. Oceania possesses the largest agricultural land per capita. Australia and New Zealand are the main potato production countries in this region. As developed countries, farms larger than 20 ha in Australia and New Zealand take up 70% of the land [97], which is beneficial to applying large- or medium-scale potato planting machines such as cup- or pneumatic-type planters. For Africa, it is predominantly rural, with 54% of the population engaged in agriculture. Most farmers cultivate small, fragmented parcels of farmland, yet are responsible for food production, making the smallholder farm sector a key player in the continent's rural economy [103]. Due to poverty, the ecological intensification of agriculture has seldom been addressed in the context of the smallholder farming systems that characterize rural Africa. Therefore, whether deliberately or not, much of African agriculture has remained rather "ecological" [102]. Considering the conditions of the economy and agricultural production, potato planters of manual/animal-powered, two-wheel tractor-driven and four-wheel tractor-driven types with small scale are applicable to service potato production in Africa.

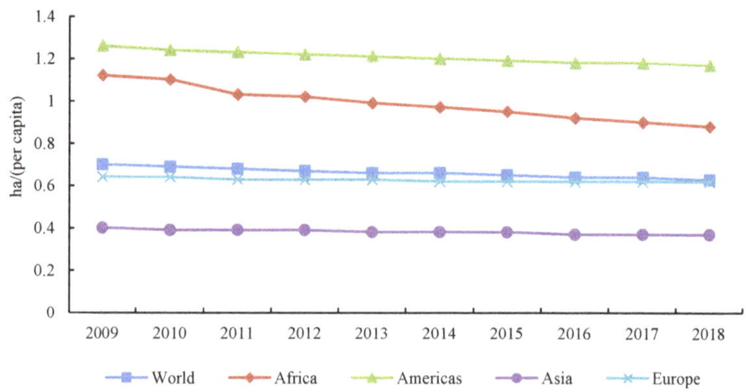

(**a**)

Figure 18. *Cont.*

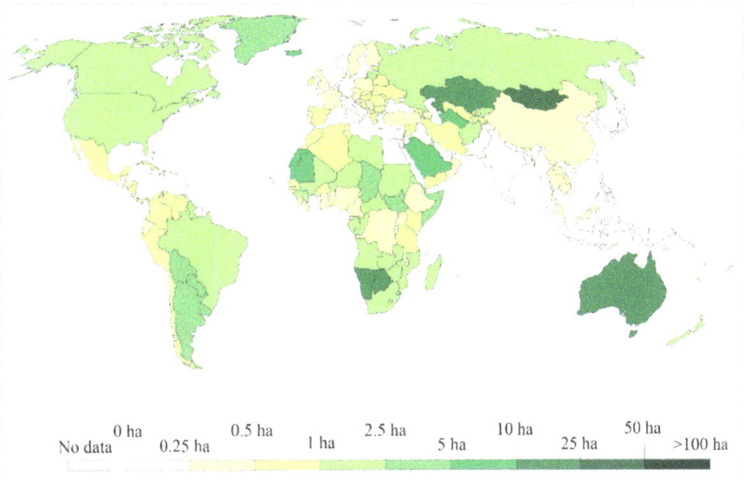

(b)

Figure 18. (**a**) Agricultural land per capita by continent, 2009–2018 (Note: Oceania is not shown in Figure 18a due to its large agricultural area combined with its tiny population). Date from ref. [104,105]. Copyright 2021 Land Use and FAO; (**b**) global agricultural land per capita in 2018. Date from references [104,105]. Copyright 2021 Land Use and FAO.

4. Conclusions and Recommendations

Currently, there are various types of potato planters in small, medium, and large scales around the world. However, due to the differences in economy, cropping systems, and geographical and environmental conditions, the level of the development of mechanical potato planting is unbalanced. In order to promote the development of high-performance potato planters, some recommendations are presented as follows.

(1) It is necessary to make efforts in studying the working mechanism of potato planters. Nowadays, problems of poor seeding quality and unstable operation performance exist. To solve them, it is essential to do further research on key technologies in ditching, seeding, fertilizing, and ridging. Moreover, exploring new techniques and using new materials are helpful for promoting a planter's quality and stability.

(2) Improving the level of automation and intelligentization of potato planters will be the key point of research in the future, among which, digital technologies (i.e., sensors, GPS, machine vision, seeding quality monitoring, etc.) would become the essential components to enhance working quality and efficiency. Thus, the further exploration of potato planting machines should be taken into consideration.

(3) Potato planters suitable for the local economic and geographic conditions in different areas should be promoted. For example, the manual/animal-powered potato planters or the small-scale potato planters powered by tractor are matched with small plot lands, terracing mountain, or hilly areas due to their stability and safety; large/medium potato planters powered by tractor are matched with large/medium-scale farms.

(4) The rapid development of potato planters needs policy support, which includes providing adequate research projects and funds, and encouraging researchers to study innovations in potato planting technologies. Furthermore, a specific targeted subsidy system needs to be established to ensure these planters are affordable for local farmers.

(5) Farmers in different areas are faced with various kinds of problems when using the present commercial potato planter. It is important to solve these problems, which need

universities and research institutions cooperating closely with companies. Researchers from universities and research institutions should propose solutions or technologies, and companies should apply these solutions and technologies in time to improve the quality and applicability of potato planters.

Author Contributions: Conceptualization: Z.Z. and W.L.; Z.Z. reviewed the literature and wrote the initial draft of the paper with assistance from H.Z., Z.L., J.H. and W.L.; W.L. contributed to revising the manuscript. All authors have read and agreed to the published version of the manuscript.

Funding: This work was supported by the Key Industry Chain Innovation Project of Shaanxi Province (Grant No. 2018ZDCXL-NY-03-06).

Institutional Review Board Statement: Not applicable.

Informed Consent Statement: Not applicable.

Data Availability Statement: Not applicable.

Conflicts of Interest: The authors declare no conflict of interest.

References

1. Birch, P.; Bryan, G.; Fenton, B.; Gilroy, E.; Hein, I.; Jones, J.; Prashar, A.; Taylor, M.; Torrance, L.; Toth, I. Crops that feed the world 8: Potato: Are the trends of increased global production sustainable? *Food Secur.* **2012**, *4*, 477–508. [CrossRef]
2. Gaiero, P.; Speranza, P.; de Jong, H. Introgressive hybridization in potato revealed by novel cytogenetic and genomic technologies. *Am. J. Potato Res.* **2018**, *95*, 607–621. [CrossRef]
3. Monte, M.; Rey, F.; Carboni, M.; Castellote, M.; Sucar, S.; Norero, N.; Colman, S.; Massa, G.; Colavita, M.; Feingold, S. Genetic diversity in argentine andean potatoes by means of functional markers. *Am. J. Potato Res.* **2018**, *95*, 286–300. [CrossRef]
4. Tanios, S.; Eyles, A.; Tegg, R.; Wilson, C. Potato tuber greening: A review of predisposing factors, management and future challenges. *Am. J. Potato Res.* **2018**, *95*, 248–257. [CrossRef]
5. Ebrahem, I.Z.; Ayman, A.E.; Guidetti, R. A new small potato planter for Egyptian agriculture. *J. Agric. Eng.* **2011**, *42*, 7–13. [CrossRef]
6. Mcleod, C.D.; Misener, G.C.; Tai, G.; Caissie, R. A precision seeding device for true potato seed. *Am. Potato J.* **1992**, *69*, 255–264. [CrossRef]
7. Raigond, B.; Verma, A.; Pathania, S.; Sridhar, J.; Kochhar, T.; Chakrabarti, S.K. Development of a reverse transcription loop-mediated isothermal amplification for detection of potato virus a in potato and in insect vector aphids. *Crop Prot.* **2020**, *137*, 105296. [CrossRef]
8. Parajuli, R.; Matlock, M.D.; Thoma, G. Cradle to grave environmental impact evaluation of the consumption of potato and tomato products. *Sci. Total Environ.* **2021**, *758*, 143662. [CrossRef]
9. Jong, D.H. Impact of the potato on society. *Am. J. Potato Res.* **2016**, *93*, 415–429. [CrossRef]
10. Zhang, H.; Xu, F.; Wu, Y.; Hu, H.-H.; Dai, X.-F. Progress of potato staple food research and industry development in China. *J. Integr. Agric.* **2017**, *16*, 2924–2932. [CrossRef]
11. Su, W.; Wang, J. Potato and food security in China. *Am. J. Potato Res.* **2019**, *96*, 100–101. [CrossRef]
12. FAO. FAOSTAT on Production/Yield Quantities of Potatoes in World, 2000–2019. Available online: http://www.fao.org/faostat/en/ (accessed on 2 February 2021).
13. Buitenwerf, H.; Hoogmoed, W.B.; Lerink, P.; Muller, J. Assessment of the behaviour of potatoes in a cup-belt planter. *Biosyst. Eng.* **2006**, *95*, 35–41. [CrossRef]
14. Liu, W.; He, J.; Li, H.; Li, X.; Zheng, K.; Wei, Z. Calibration of simulation parameters for potato minituber based on EDEM. *Trans. Chin. Soc. Agric. Mach.* **2018**, *49*, 125–135, 142. [CrossRef]
15. Lü, J.; Yang, Y.; Li, Z.; Shang, Q.; Li, J.; Liu, Z. Design and experiment of an air-suction potato seed metering device. *Int. J. Agric. Biol. Eng.* **2016**, *9*, 33–42. [CrossRef]
16. Liu, W.; Jin, H.; Li, X.; Wei, Z.; Zhang, Z.; Kan, Z.; Zhao, H.; Zhang, Y. Research progress of potato planter. *J. Agric. Mech. Res.* **2018**, *40*, 7–13. [CrossRef]
17. Du, H.; Shang, S.; Yang, R.; Wang, D. Research and analysis on mechanized potato seed sowing techniques. *J. Agric. Mech. Res.* **2011**, *33*, 214–217, 221. [CrossRef]
18. Liu, W.; He, J.; Li, H.; Li, X.; Wei, Z.; Liu, P. Design and experiment of vibration-arranging based seeder for potato micro-seed. *Trans. Chin. Soc. Agric. Mach.* **2019**, *50*, 70–80, 116. [CrossRef]
19. Zheng, D.; Wei, C.; Du, W.; Fan, M.; Liu, G. The study and analysis of cup seed sowing technology and uniformity. *J. Agric. Mech. Res.* **2016**, *38*, 106–109. [CrossRef]
20. Lu, J.; Yang, Y.; Shang, Q.; Li, Z.; Li, J.; Liu, Z.; Wang, Y. Performance optimization test on air-suction potato seed metering device with positive pressure airflow and zero-speed seeding. *Trans. Chin. Soc. Agric. Eng.* **2016**, *32*, 40–48. [CrossRef]

21. Wang, X.; Tang, H.; Wang, J.; Lü, J.; Li, Z. Optimized design and experiment on double-row cross spoon-belt potato precision seed metering device. *Trans. Chin. Soc. Agric. Mach.* **2016**, *47*, 82–90. [CrossRef]
22. Lü, J.; Yang, Y.; Li, Z.; Tian, Z.; Shang, Q.; Wu, J.E. Design and experiment of cup-belt type potato seed-metering device. *Trans. Chin. Soc. Agric. Eng.* **2016**, *32*, 17–25. [CrossRef]
23. Sun, W.; Wang, G.; Wu, J. Design and experiment on loss sowing testing and compensation system of spoon-chain potato metering device. *Trans. Chin. Soc. Agric. Eng.* **2016**, *32*, 8–15. [CrossRef]
24. Niu, K.; Fang, X.; Liu, Y.; Lu, C.; Yuan, Y. Optimized design and performance evaluation of an electric cup-chain potato metering device. *Int. J. Agric. Biol. Eng.* **2017**, *10*, 36–43. [CrossRef]
25. Wanzhi, Z.; Chenglong, L.; Zhaoqin, L.; Xieteng, Q.; Haoyu, L.; Jialin, H. Optimized Design and Experiment on Novel Combination Vacuum and Spoon Belt Metering Device for Potato Planters. *Math. Probl. Eng.* **2020**, *2020*, 1–12. [CrossRef]
26. Zhai, G.; Bao, D.; Wang, Z.; Yang, L.; Li, F. Design of metering device key parts of pneumatic grass seeder. *J. Earth Sci. Eng.* **2014**, *45*, 47–51. [CrossRef]
27. Singh, R.C.; Singh, G.; Saraswat, D.C. Optimisation of design and operational parameters of a pneumatic seed metering device for planting cottonseeds. *Biosyst. Eng.* **2005**, *92*, 429–438. [CrossRef]
28. Lai, Q.; Ma, W.; Su, W.; Hu, Z.; Xing, J. Design and experiment of pneumatic disc seed-metering device for mini-tuber. *Trans. Chin. Soc. Agric. Mach.* **2016**, *47*, 30–37. [CrossRef]
29. Lai, Q.; Ma, W.; Liu, S.; Su, W.; Zhang, Z. Simulation and experiment on seed-filling performance of pneumatic disc seed-metering device for mini-tuber. *Trans. Chin. Soc. Agric. Mach.* **2017**, *48*, 44–53. [CrossRef]
30. Mao, Q.; Huang, M.; Huang, J.; Duan, H. Experiment on pneumatic metering device with declined disc for virus-free seed potato. *Chin. Potato J.* **2013**, *27*, 112–118.
31. Li, Z.; Wen, X.; Lu, J.; Li, J.; Yi, S.; Qiao, D. Analysis and prospect of research progress on key technologies and equipments of mechanization of potato planting. *Trans. Chin. Soc. Agric. Mach.* **2019**, *50*, 1–16. [CrossRef]
32. He, Y.J. The Research of the New Potato Seed Sowing Device for Planter. Master's Thesis, China Agricultural University, Beijing, China, 2006.
33. Liu, W.; He, J.; Li, H.; Li, X.; Lu, C.; Wei, Z.; Su, G.; Zhao, H.; Liu, P.; Wang, C. Seeding performance optimization on vibration-arranging type seeding device for potato micro-seed. *Trans. Chin. Soc. Agric. Eng.* **2019**, *35*, 1–11. [CrossRef]
34. Kohser, F. Potato Planter. US0736981A, 25 August 1903.
35. Springer, H.J. Potato Planter. US0744984A, 24 November 1903.
36. Heikes, A.T. Potato Planter. US0997674A, 11 July 1911.
37. Edgar, O.J. Potato Planter. US2642909A, 23 June 1953.
38. Lü, J.; Tian, Z.; Ying, Y.; Shang, Q.; Wu, J.; Li, Z.; Wang, X. The development situation, existing problems and development trend of potato machinery. *J. Agric. Mech. Res.* **2015**, *37*, 258–263. [CrossRef]
39. Chen, S. Study on the device for discharging and supplying potatoes of potato planter. *Cereals Oils Process.* **1982**, 17–24.
40. Misener, G.C. Relative performance of cup and pick type potato planters. *Can. Agric. Eng.* **1979**, *21*, 131–134.
41. Misener, G.C. Potato planters—Uniformity of spacing. *Trans. ASAE* **1982**, *25*, 1504–1505. [CrossRef]
42. Gulati, S.; Singh, M. Design and development of a manually drawn cup type potato planter. *Potato J.* **2003**, *30*, 61–62.
43. Stand 'n Plant. The Hand-Held Jab Planter. Available online: https://www.standnplant.com/index.php/planter (accessed on 3 March 2020).
44. Groves Nurseries & Garden Centre. Potato Planter Tool. Available online: https://grovesnurseries.co.uk/Potato-Planter-Tool (accessed on 15 October 2020).
45. Singh, M.; Gulati, S.; Singh, S. *Farm. Equipment and Machines for Potato Production*; Central Potato Research Institute: Himachal Pradesh, India, 2008.
46. Bell, R.W.; Haque, M.E.; Johansen, C.; Vance, W.; Kabir, M.E.; Musa, M.A.; Mia, M.N.N.; Neogi, M.G.; Islam, M.A. Mechanised minimum soil disturbance establishment and yield of diverse crops in paddy fields using a two-wheel tractor-mounted planter suitable for smallholder cropping. *Exp. Agric.* **2018**, *54*, 755–773. [CrossRef]
47. Pushpitha, N.P.G.; Weerasinghe, K.D.N.; Maier, D. Modification of a two-wheel tractor as a versatile power machine for post disaster recovery programs. *Procedia Eng.* **2018**, *212*, 614–621. [CrossRef]
48. Kim, P. Nibbi AE 11 with Homemade Potato Planter Powered by Two Wheel Tractor. Available online: https://www.youtube.com/watch?v=GJKj8qKH-YA (accessed on 15 October 2020).
49. ROSTA Scien-tific Production Company. KSM-1 Potato Planter for Motor Cultivator. Available online: http://www.rostaltd.com/en/equipment-for-motor-cultivators/seeder-ksm-1 (accessed on 18 October 2020).
50. F.LLI Spedo. Automatic Potato Planter Baby. Available online: http://www.spedo.it/en/products/potato-planter/automatic-potato-planter-baby/ (accessed on 15 October 2020).
51. Weifang Luke Machinery Co., Ltd. Potato Planter 1 Row. Available online: https://www.alibaba.com/product-detail/potato-planter-1-row_60161270110.html?spm=a2700.details.maylikeexp.1.4a804065hwyjKR (accessed on 15 October 2020).
52. Yucheng Qichen Machinery Co., Ltd. Hand Tractor Belt Potato Planter Seeder. Available online: https://qcnj.en.alibaba.com/product/60724247362-805831511/Hand_tractor_belt_Potato_planter_seeder.html?spm=a2700.icbuShop.41413.17.50be47c972UUZt (accessed on 15 October 2020).

53. Hamad, S.A.; Ismail, Z.E.; Hemeda, M.A. Development of a potato planter to plant tuber pieces with previously grown buds. *J. Agric. Sci. Mansoura Univ.* **1994**, *19*, 2321–2343.
54. Al-Gaadi, K.A.; Marey, S.A. Effect of forward speed and tuber characteristics on tuber spacing uniformity for a cup-belt potato planter. *Middle East. J. Sci. Res.* **2011**, *8*, 753–758.
55. Prashant, W.; Shriram, A.; Vikas, R.; Swapnil, D.; Onkar, B.; Atish, P. Design and fabrication of potato planting machine. *Int. J. Eng. Technol. Res.* **2017**, *4*, 1057–1060.
56. Gavrilov, V.; Semenov, A.; Novikov, A.; Ivanov, V.; Terent'Ev, A. Uniform planting of sprouted tubers semi-automatic potato planter. *Vestn. Kazan State Agrar. Univ.* **2020**, *15*, 64–70. [CrossRef]
57. Wang, G.; Sun, W.; Chen, L.-D.; Zhang, H.; Liu, X.; Li, H.; Yang, X.; Yang, H. Realization of an integrated seeding and compensating potato planter based on one-way clutch. *Int. J. Agric. Biol. Eng.* **2020**, *13*, 79–87. [CrossRef]
58. Wang, G.; Sun, W.; Zhang, H.; Liu, X.; Li, H.; Yang, X.; Zhu, L. Research on a kind of seeding-monitoring and compensating control system for potato planter without additional seed-metering channel. *Comput. Electron. Agric.* **2020**, *177*, 105681. [CrossRef]
59. Shi, L.; Yang, X.; Zhao, W.; Sun, W.; Li, R.; Sun, B. Design and test of potato combine seeder with throwing and covering soil on film edge. *Trans. Chin. Soc. Agric. Mach.* **2018**, *49*, 129–137. [CrossRef]
60. Dai, F.; Xin, S.; Zhao, W.; Liu, F.; Xin, B.; Ma, M. Design and experiment of combined potato planting machine for covering soil on top of full film surface. *Trans. Chin. Soc. Agric. Mach.* **2017**, *48*, 56, 76–83. [CrossRef]
61. Dai, F.; Zhao, W.; Song, X.; Xin, S.; Liu, F.; Xin, B. Operating parameter optimization and experiment of device with eevating and covering soil on plastic-film. *Trans. Chin. Soc. Agric. Mach.* **2017**, *48*, 88–96. [CrossRef]
62. Sun, W.; Liu, X.; Wang, H.; Zhang, H.; Wu, J.; Yang, X.; Wang, G. Design and test of double crank multi-rod hill-drop potato planter on plastic film. *Trans. Chin. Soc. Agric. Eng.* **2018**, *34*, 34–42. [CrossRef]
63. Sun, W.; Liu, X.; Zhang, H.; Wang, H.; Tian, B. Design of potato casingsoil planter in all-in-one machine combined with fertilizing, sowing, ridging, complete film mulching and planting line covering. *Trans. Chin. Soc. Agric. Eng.* **2017**, *33*, 14–22. [CrossRef]
64. Sun, W.; Simionescu, P. Parameter analysis and field tests of a double crank multi-rod under plastic-film hill-drop mechanism potato planter. *Am. J. Potato Res.* **2020**, *97*, 256–264. [CrossRef]
65. Bomet Sp. z o.o. Sp. K. Sadzarka do ziemniaków jednorzędowa Gemini. Available online: https://www.bomet.pl/Sadzarka-do-ziemniakow-jednorzedowa-Gemini.html (accessed on 20 October 2020).
66. US Small Farm Equipment Co. 1-Row Cup Type and Assisted-Feed Planter. Available online: https://ussmallfarm.com/product/planters/ (accessed on 10 September 2020).
67. Changzhou LEFA Industry and Trade Co. Ltd. 1 Row Potato Planter (20–50 HP Tractor). Available online: https://cnlefa.en.made-in-china.com/ (accessed on 18 May 2021).
68. Mari, G.R.; Memon, S.A.; Leghari, N.; Brohi, A.D. Evaluation of tractor operated potato planter. *J. Appl. Sci.* **2002**, *2*, 889–891. [CrossRef]
69. Lu, J.; Wang, Y.; Li, Z.; Dui, H.; Liu, Z.; Li, J.; Sun, H.; Peng, M. Performance analysis and experiment of cup-belt type patato seed-metering device with flow deflector. *Trans. Chin. Soc. Agric. Eng.* **2017**, *33*, 19–28. [CrossRef]
70. Cho, Y.; Choi, I.S.; Kim, J.D.; Oh, J.; Lee, D.H. Performance test of fully automatic potato seeding machine by iln-situ process of cutting seeds. *J. Biosyst. Eng.* **2017**, *42*, 147–154. [CrossRef]
71. Yang, R.; Yang, H.; Lian, Z.; Zhang, H.; Guo, D.; Zheng, Y. Design and experiment of separated Layer fertilization furrow opener for potato planter. *Trans. Chin. Soc. Agric. Mach.* **2018**, *49*, 104–113. [CrossRef]
72. SWAN AGRO. Cup Type Potato Planter. Available online: https://www.swanagro.in/en/product/potato-planter/29 (accessed on 30 October 2020).
73. Standen Engineering Ltd. Standen Potato Planter. Available online: https://standen.co.uk/products/planters/ (accessed on 15 October 2020).
74. Shan-Dong Transce Agricul-Tural Machinery Technology Co., Ltd. 2CMX-4B Potato Planter. Available online: http://www.xs-tc.com/malingshuzhongzhijixie/xinghao-2CMX-4Bsilongsixing_39.html (accessed on 7 August 2019).
75. GRIMME Landmaschinenfabrik GmbH & Co. KG. Potato Planting Technology. Available online: https://www.grimme.com/de/producttypes/legetechnik/ (accessed on 17 October 2020).
76. AVR. AVR Ceres 450. Available online: https://avr.be/en/node/47 (accessed on 15 May 2020).
77. Dewulf. Structural-Belt and Miedema Cup Planters. Available online: https://www.dewulfgroup.com/en/product/planting/planters/ (accessed on 15 October 2020).
78. Double L. Double L Potato Planter. Available online: https://www.doublelglobal.com/potato-planter.php (accessed on 15 October 2020).
79. SPUDNIK Equipment Company LLC. 12 Row Folding Conventional Planter. Available online: https://www.spudnik.com/equipment_planting/equipment_planting.php (accessed on 20 March 2021).
80. Mao, Q. Design and research on key components of the planter for virus-free mini-potato seeds. Master's Thesis, Huzhong Agricultural University, Wuhan, China, 2013.
81. Lü, J.; Yi, S.; Tao, G.; Mao, X. Design and experiment of precision air-suction type planter for potato. *Trans. Chin. Soc. Agric. Eng.* **2018**, *34*, 16–24. [CrossRef]
82. Wang, F.; Sun, K.; Lai, Q.; Dong, J.; Su, W.; Yu, Q. Design and experiment of minituber precision single-row air-suction planter. *Trans. Chin. Soc. Agric. Mach.* **2020**, *51*, 66–76. [CrossRef]

83. Crary Industries. Air Cup S-Series and Needle-Stabbed Planters. Available online: http://www.lockwoodmfg.com/Lockwood-Potato-Equipment/Planters/ (accessed on 10 March 2021).
84. Engineering & Manufacturing Aircraft Tools. PLMS Pneumatic Type Potato Planter. Available online: http://www.erme-france.com/ (accessed on 15 November 2020).
85. Meijer, E.; Frederiks, J. Development of an automatic planter for presprouted seed. *Potato Res.* **1975**, *18*, 451–454. [CrossRef]
86. Misener, G.C.; Mcleod, C.D. A plot planter for potatoes. *Am. Potato J.* **1988**, *65*, 289–293. [CrossRef]
87. Arsenault, W.J.; Platt, H.W.; Pippy, E.; Cannon, A. A small plot potato planter. *Can. Agric. Eng.* **1996**, *38*, 145–147.
88. Momin, M.A.; Sarker, M.R.I.; Hossain, M.M. Field performance of a tractor operated semi-automatic potato planter. *J. Bangladesh Agril. Univ.* **2006**, *4*, 391–399.
89. WIFO-Anema BV. Minituber Planter. Available online: https://www.wifo.nl/en/products/planting-machines/planting-machines/minituber-planter (accessed on 15 October 2020).
90. IMAC Co. Semi-Automatic Potato Planter. Available online: https://www.imac-rondelli.it/eng/planters.html (accessed on 15 October 2020).
91. TEHNOS, d.o.o. Potato Planter with a Fertilizer Dispensing Device and Interrow Cultivator. Available online: http://en.tehnos.si/potato-planter-with-a-fertilizer-dispensing-device-and-interrow-cultivator (accessed on 16 October 2020).
92. Frase, R.J.; O'neil, G.R. Finger-Type Planter. US3638829A, 1 February 1972.
93. Boydas, M.G. Effect of cup size, seed characteristics and angular speed on the performance of an automatic potato planter under laboratory conditions. *J. Agric. Sci.* **2017**, *23*, 317–327. [CrossRef]
94. Boydas, M.G.; Uygan, F. Influence of seed physical properties and speed on the external mechanical damage Index and In-row spacing uniformity in an automatic potato planter. *J. Agric. Sci.* **2012**, *18*, 126–136. [CrossRef]
95. Murray, J.R.; Tullberg, J.N.; Basnet, B.B. *Planters and Their Components: Types, Attributes, Functional Requirements, Classification and Description*; Ustralian Centre for International Agricultural Research: Canberra, Australia, 2006; p. 140.
96. Harriston Industries. Harriston Clamp and Needle-Stabbed Potato Planter. Available online: http://www.harriston-mayo.com/harriston/ (accessed on 15 November 2020).
97. Adamopoulos, T.; Restuccia, D. The size distribution of farms and international productivity differences. *Am. Econ. Rev.* **2014**, *104*, 1667–1697. [CrossRef]
98. Eurostat. Farms and Farmland in the European Union-Statistics. Available online: https://ec.europa.eu/eurostat/statistics-explained/index.php?title=Farms_and_farmland_in_the_European_Union_-_statistics (accessed on 18 May 2021).
99. Lowder, S.K.; Skoet, J.; Raney, T. The Number, Size, and Distribution of Farms, Smallholder Farms, and Family Farms Worldwide. *World Dev.* **2016**, 16–29. [CrossRef]
100. Laure, L.; Latruffe, P. Does land fragmentation affect farm performance? A case study from Brittany, France. *Agric. Syst.* **2014**, *129*, 68–80. [CrossRef]
101. United States Department of Agriculture. *Farms and Land in Farms, 2019 Summary*; National Agricultural Statistics Services: Washington, DC, USA, 2020; pp. 5–6.
102. Tittonell, P.; Giller, K.E. When yield gaps are poverty traps: The paradigm of ecological intensification in African smallholder agriculture. *Field Crop. Res.* **2013**, *143*, 76–90. [CrossRef]
103. Kuivanen, K.S.; Alvarez, S.; Michalscheck, M.; Adjei-Nsiah, S.; Descheemaeker, K.; Mellon-Bedi, S.; Groot, J.C.J. Characterising the diversity of smallholder farming systems and their constraints and opportunities for innovation: A case study from the northern region, Ghana. *NJAS-Wagening. J. Life Sci.* **2016**, *78*, 153–166. [CrossRef]
104. Hannah, R.; Max, R. Land Use. Available online: https://ourworldindata.org/land-use (accessed on 18 May 2021).
105. FAO. FAOSTAT on Agricultural Land Per Capita. Available online: http://www.fao.org/faostat/en/ (accessed on 18 May 2021).

agriculture

MDPI

Article

Design and Experiment of Row-Controlled Fertilizing–Weeding Machine for Rice Cultivation

Yangjie Shi [1], Xiaobo Xi [1,*], Hao Gan [2], Xiang Shan [1], Yifu Zhang [1], Hui Shen [1] and Ruihong Zhang [1]

[1] School of Mechanical Engineering, Yangzhou University, Yangzhou 225000, China; DX120180059@yzu.edu.cn (Y.S.); shanxiang@yzu.edu.cn (X.S.); zyfu@yzu.edu.cn (Y.Z.); hshen@yzu.edu.cn (H.S.); zhangrh@yzu.edu.cn (R.Z.)

[2] Department of Biosystems Engineering and Soil Science, University of Tennessee, Knoxville, TN 37996, USA; hgan1@utk.edu

* Correspondence: xbxi@yzu.edu.cn

Abstract: Current approaches to topdressing and weeding operations for rice cultivation present several disadvantages, such as poor precision, low efficiency, serious environmental pollution, and so on. This paper presents a row-controlled fertilizing–weeding machine to improve the precision of fertilizing and weeding operations and to reduce the heavy pollution associated with rice cultivation. A proportional–integral–derivative algorithm was adopted to realize accurate fertilization control, and an automatic driving system for agricultural machinery based on BeiDou navigation was used for accurate row-controlled operation. Accuracy testing and field experiments were carried out. The results show that the fertilization control system can stabilize the speed to within 0.55 s of the desired speed with a standard deviation of around 0.32 r·min^{-1}. The row-controlled operation ensures the lateral deviation is within ± 5 cm at operating speeds below 5 km·h^{-1}. The high uniformity and accuracy of fertilization meet agronomic requirements and rice cultivation standards, and the weeding performance is acceptable at working speeds below 5 km·h^{-1}.

Keywords: rice; fertilization; weeding; precision agriculture

Citation: Shi, Y.; Xi, X.; Gan, H.; Shan, X.; Zhang, Y.; Shen, H.; Zhang, R. Design and Experiment of Row-Controlled Fertilizing–Weeding Machine for Rice Cultivation. *Agriculture* **2021**, *11*, 527. https://doi.org/10.3390/agriculture11060527

Academic Editor: José Pérez-Alonso

Received: 22 April 2021
Accepted: 4 June 2021
Published: 6 June 2021

1. Introduction

Topdressing and weeding operations are repeatedly carried out throughout the rice cultivation process and have an important impact on rice yields. In the middle and lower reaches of the Yangtze River in China, the main topdressing method is mechanical sprinkling, which is highly efficient. However, it is difficult to achieve a uniform distribution of fertilizer, resulting in low fertilizer utilization rates [1,2]. The most widely used weeding methods are chemical-based. Plant protection machinery can achieve remarkable control of spraying operations. However, overuse of herbicides can lead to serious soil pollution and ecological damage [3,4].

Side deep fertilization and mechanical weeding technologies are used to ensure high fertilizer utilization rates and reduce chemical herbicide use [5]. By applying fertilizer to soil at a certain depth, it can be more easily absorbed by the root system, which can effectively increase fertilizer utilization rates by 15–20%; at the same time, yields can be increased by up to 20% [6,7]. Existing side deep fertilization devices exhibit poor performance, which is mainly reflected in low fertilization uniformity, unstable operation, and frequent clogging of fertilizer delivery pipelines [8,9]. In response to these problems, Wang et al. [10] adopted a spiral fertilizer discharge wheel to the improve fertilization uniformity of a side deep fertilizing device. The coefficient of variation of fertilization stability was less than 7.8%. Shan Zeng et al. [11] and Kim et al. [12] designed pneumatic conveying systems suitable for granular fertilizer that improve the uniformity of fertilizer distribution and prevent blockage of the conveying pipe. In particular, Kim et al. showed that the application uniformity (coefficient of variation) in both the transverse and the longitudinal directions were in the range of 11.2–13.1 and 2.9–15.3%, respectively. Moreover, the application accuracy ranged from 81.9–97.4% for the working speed range of the device. Yu et al. [13] designed a controller for variable-rate application of granular fertilizer, which adjusts the amount of fertilizer based on working speed and improves stability. The response time of the controller (time to reach a steady application rate when the application rate is suddenly changed) ranged from 1.5 to 3.0 s, which was considered acceptable.

Mechanical weeding is an environmentally friendly weeding technology and can be implemented using various approaches, including ploughing, harrowing, and intertillaging. These methods can eliminate the use of chemical herbicides and therefore, protect the ecological environment [14,15]. However, ensuring effective weeding while avoiding mechanical damage to rice seedlings remains a challenge [16]. Perez-Ruiz et al. [17] proposed a fully automatic intra-row mechanical weed knife path control system for transplanted row crops. The system uses a real-time kinematics (RTK) global positioning system (GPS) to automatically detect crop planting geolocations and to control the path of pairs of intra-row weed knives travelling between crop plants along the row centerline. Field trial results show that, on average, the size of the close-to-crop zone can be maintained within $\pm$ 0.5 cm of the target size with a standard deviation of 0.94 cm at 0.8 $\text{km}\cdot\text{h}^{-1}$ and 1.39 cm at 1.6 $\text{km}\cdot\text{h}^{-1}$. Liu et al. [18] proposed a deep learning-based method for bending diagnosis of seedling line and guidance line extraction, which was shown to effectively prevent damage to seedlings caused by weeding machine components. Chen et al. [19] designed a hydraulic control system based on linear active disturbance rejection to prevent damage to seedlings caused by weeding components in paddy fields. The average seedling injury rate with the alignment control system was 3.9%. Wang et al. [20] designed a 3SCJ-1 type weeding machine that uses new weeding components for weed management in transplanted organic rice. The average weeding rate was 78.02%.

Although the above-mentioned approaches solve several common problems associated with fertilizing and weeding operations, they are only optimized for separately carrying out fertilizing or weeding operations. However, often only certain aspects of fertilizing or weeding operation are optimized, and from the agronomic point of view, combining these two operations to improve the efficiency has not been considered. This paper presents a row-controlled fertilizing–weeding (RFW) machine designed to meet the agronomic requirements of rice cultivation. The machine is capable of simultaneously performing fertilization and weeding operations and realizing row-controlled operation. An automatic driving system for agricultural machinery developed by our research group and based on China's BeiDou navigation system is adopted to improve the efficiency and quality of both the fertilizing and weeding operations. An experimental study was carried out to assess the performance of the prototype. The results provide a theoretical basis, as well as a reference, for further research on combined fertilization, and weeding technologies for rice cultivation.

2. Materials and Methods

2.1. The Design of the RFW Machine

2.1.1. Overall Structure

The RFW machine consists of a fertilizer box, fan, motor, fertilizer distribution apparatus, air–fertilizer mixing chamber, pipe, height adjustment device, furrower, gearbox, weeding wheel, spring, fertilizer tank, and depth-limiting plate. The machine is connected to the tractor body via the suspension frame. The overall structure of the RFW machine is shown in Figure 1. Technical details are listed in Table 1.

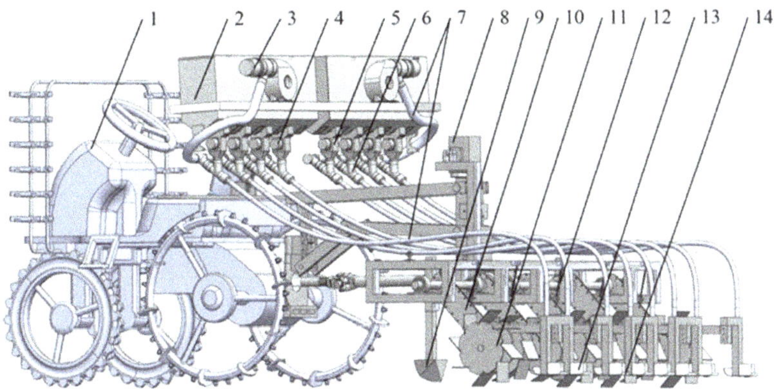

Figure 1. Overall structure of RFW machine. 1. Tractor; 2. Fertilizer box; 3. Fan; 4. Motor; 5. Fertilizer apparatus; 6. Air–fertilizer mixing chamber; 7. Pipe; 8. Height adjustment device; 9. Furrower; 10. Chain transmission box; 11. Weeding wheel; 12. Limiter spring; 13. Fertilizer tank; 14. Depth-limiting plate.

Table 1. Technical specifications of RFW machine.

Project	Parameter
Working width	2.5 m
Number of operation lines	8
Fertilizer box volume	350 L
Working speed	0–6 km h^{-1}
Fan model	SEAFLO SFBB1-320-02
Type of fertilizer apparatus	Grooved wheel fertilizer apparatus
Motor model	GPG-05SC

A pneumatic conveying method was adopted. Power is provided by the fan to convey the fertilizer. The rotation speed of the fertilizer apparatus is controlled by the motor speed, which therefore controls the amount of fertilizer dispensed. The height adjustment device adjusts the height of the machine relative to the ground, allowing the RFW machine to move easily through the field. The gearbox transfers the output power of the tractor to the weeding wheel, which then rotates. The furrower is used to trim ditches and plays a role in removing weeds between ditches. The depth-limiting plate and spring control the depth of the weeding wheel in the soil.

Fertilizer is discharged from the fertilizer tank through a square hole in the center of the depth-limiting plate. The RFW machine is equipped with an automatic driving system for agricultural machinery based on China's BeiDou navigation system, which improves the accuracy of the row-controlled operation and ensures the weeding wheel stays between crop rows [21].

2.1.2. Working Principle

As shown in Figure 2, during the weeding operation, the weeding wheels plough shallow soil to dig out or bury weeds under the soil, destroying the root growth environment of the weeds. During the fertilizing operation, fertilizer particles are fed into the air–fertilizer mixing chamber under the action of the motor, then mixed with the high-speed air flow produced by the fan. The fertilizer particles enter the fertilizer tank through the pipe due to the combined action of drag forces generated by the air flow and gravity, and are finally discharged through the square hole in the center of the depth-limiting plate. The fertilizer particles fall into the soil immediately after it has been ploughed by the weeding wheel and are concentrated in the middle of the seedling row, thus realizing deep side fertilization. The angle of the steering wheel is constantly adjusted during the row-controlled fertilizing and weeding operations, which effectively improves the efficiency and quality of both operations.

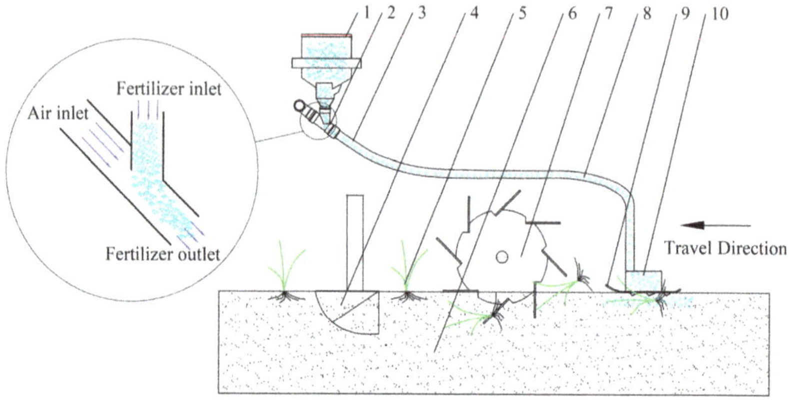

Figure 2. Working principle of RFW machine. 1. Fertilizer box; 2. Air–fertilizer mixing chamber; 3. Pipe; 4. Furrower; 5. Weeds; 6. Soil; 7. Weeding wheel; 8. Fertilizer; 9. Depth-limiting plate; 10. Fertilizer tank.

2.1.3. Control System

The fertilization control system was designed to improve the uniformity and accuracy of fertilization. The system uses a proportional–integral–derivative (PID) algorithm to achieve closed-loop control of the motor speed. The working principle of the system is shown in Figure 3. First, an RTK-based global navigation satellite system (GNSS) receiver monitors the working speed of the machine in real time and transmits the speed signal to an onboard computer. The desired motor speed is calculated by the computer based on the desired amount of fertilizer and machine parameters, then the desired speed is transmitted to the fertilization controller. The controller generates a pulse-width-modulated (PWM) signal, which is regulated by the PID controller, and the signal is sent to the motor drive module. The motor speed is controlled by changing the output voltage. At the same time, the speed measurement module feeds the actual working speed back to the controller, which alters the duty cycle of the PWM signal based on the above feedback, thereby realizing real-time control of the motor speed and precision fertilization.

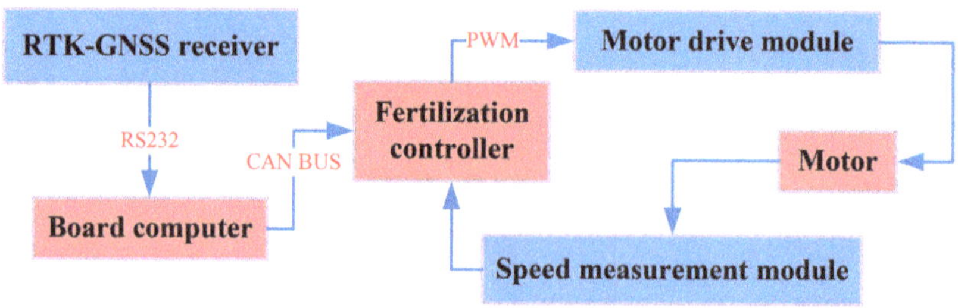

Figure 3. Principle of fertilization control system.

Based on the desired amount of fertilizer, the relationship between the desired motor speed and working speed should meet the following requirements:

$$\begin{cases} dS(t) = \frac{w}{600} V_G(t)dt \\ dM_0(t) = M_0 dS(t) \\ dM_0(t) = nqN_0(t)dt \end{cases} \quad (1)$$

where $S(t)$ is the corresponding operation area at time t, ha; $M_0(t)$ is the desired amount of fertilizer per area corresponding to time t, kg·ha^{-1}; $N_0(t)$ is the desired motor speed at time t, r·min^{-1}; $V_G(t)$ is the actual working speed at time t, km·h^{-1}; w is the working width, m; M_0 is the desired amount of fertilizer per unit area, kg·ha^{-1}; n is the number of fertilizer apparatuses; and q is the amount of fertilizer applied by the fertilizer apparatus during a single turn, kg.

The machine parameters can be used to obtain the control function for the desired motor speed as follows:

$$N_0(t) = \frac{5V_G(t)M_0}{192} \quad (2)$$

To achieve optimal control, it is necessary to adjust the PID controller parameters [22]. After repeated tests and tuning, the PID control law was obtained as:

$$U(t) = 278.9err(t) + 159 \int err(t)dt + 0.03\frac{derr(t)}{dt} \quad (3)$$

where $U(t)$ is the output of the PID controller and $err(t)$ is the deviation between the desired speed and actual speed.

Proper function of the row-controlled system relies on the automatic driving system for agricultural machinery based on BeiDou navigation. The operating principle of the row-controlled system is shown in Figure 4. First, the system reads the original path data for rice seeding and planting, which is used as a baseline for planning the desired path. The lateral and heading deviation are obtained by comparing the real-time position and heading information of the RFW machine with the desired path. Then, the theoretical rotation angle is calculated according to the deviations of the two desired values, within a predetermined error value limit. The actual controlled rotation angle is calculated based on the feedback signal from the rotation angle measuring device [23]. The motor drives the electronically controlled steering wheel, and the rotation angle is controlled to realize accurate row-controlled operation.

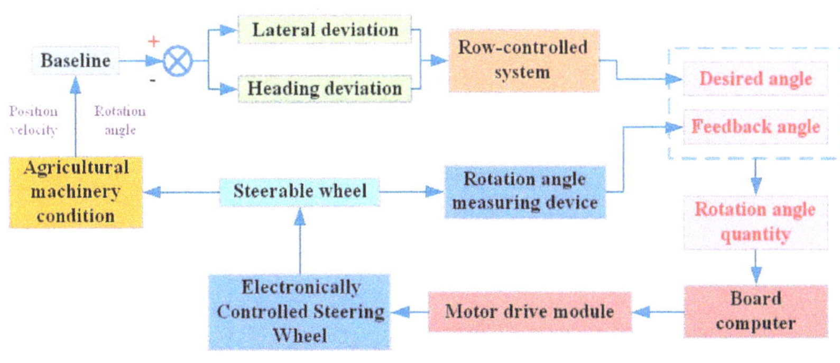

Figure 4. Principle of row-controlled system.

2.2. Accuracy Testing of Control System

2.2.1. Analysis of Response Characteristics of Fertilization Control System

To ensure fertilization uniformity in the longitudinal direction, the fertilization control system should quickly respond to changes in the working speed to ensure the motor speed stabilizes around the desired speed within a relatively short period of time. The motor speed was tested and analyzed under load conditions. In the experiment, the motor speed was programmed to change every 15 s in the range 10–30 r·min^{-1}, and the motor speed data were recorded every 100 ms by the speed measurement module. Finally, the variation of motor speed with load was obtained.

2.2.2. Analysis of Row-Controlled System Accuracy

The purpose of row-controlled operation is to reduce mechanical damage to rice seedlings caused by weeding components during the weeding operation. Deviation of the row-controlled operation must be controlled within a certain range. Deviation is mainly due to system error, vibration of the machine, and suspension deflection. Moreover, the operating speed will also influence the amount of deviation caused by these sources. Thus, the influence of the operating speed of the machine on the deviation of the row-controlled operation must be analyzed to determine the appropriate working speed range.

A field experiment was carried out on 28 June 2020 in an experimental field of Jinyun Agricultural Science and Technology Development Co., Ltd., Jiangdu District, Yangzhou City, Jiangsu Province, China (119.512° E, 32.562° N). During the test, the working speed of the machine was set to 3, 4, 5, and 6 km·h^{-1} and the machine travelled 150 m along the desired path. The actual working path was monitored by a high-precision satellite receiver installed on the longitudinal center line of the machine. Thirty monitoring points were selected along the path. Afterwards, the lateral deviation was calculated. The lateral deviation was taken as the offset distance between the actual working path and the desired path. As the absolute lateral deviation increases, the accuracy of the row-controlled operation decreases.

2.3. Field Experiment

The experiment was carried out according to NY/T1003-2006, Technical Specification of Quality Evaluation for Fertilizing Machinery [24]. Rice cultivation operations were performed by machines equipped with an automatic driving system for agricultural machinery based on BeiDou navigation. The row spacing was 0.25 m, the rice seedling height was in the range of 0.2 to 0.3 m, and the weed height was in the range of 0.1 to 0.2 m. Urea with a particle size range of 0.85 to 2.8 mm was selected as the fertilizer.

2.3.1. Fertilization Uniformity

The fertilization uniformity obtained using the proposed fertilization control system was measured under static conditions. Consistency in the amount of fertilizer applied to each row was analyzed using the coefficient of variation (*CV*). A value of less than 13% meets the technical requirements (NY/T 1003-2006). The eight fertilizer outlets were labelled as 1–4 on the left side and 5–8 on the right side of the machine. The *t*-tests were performed to verify the fertilization uniformity on each side based on data obtained from fertilizer outlets on the left and right sides of the machine. The coefficient of variation of the amount of fertilizer in each row was calculated as:

$$CV = \frac{\sqrt{\sum_{i=1}^{n} \frac{(x_i - \overline{x})^2}{n-1}}}{\overline{x}} \cdot 100\% \quad (4)$$

where x_i is the amount of fertilizer in each row, g; $\overline{x}$ is the average amount of fertilizer in each row, g; and n is the number of test rows (i.e., number of fertilizer outlets).

In the experiment, the motor speed was set to 10, 15, 20, 25, or 30 r·min^{-1}. The number of test rows was 8 and the test time was 5 min. Fertilizer dispensed from each outlet was collected in 8 separate buckets. At the end of each test, the fertilizer in each bucket was weighed and recorded. The test was repeated three times at each speed and the average values were calculated. The coefficient of variation of the amount of fertilizer in each row was determined using Equation (4). The total amount of fertilizer dispensed from each fertilizer outlet was measured and independent sample *t*-tests were performed using the average values. The significance level was set to $\alpha = 0.05$.

2.3.2. Fertilization Accuracy

The deviation of fertilization rate (γ_s) was used to evaluate fertilization accuracy. Deviation of the total amount of fertilizer refers to the ratio of the absolute value of the difference between the actual amount and the desired amount, expressed as:

$$\gamma_s = \frac{\left|\frac{10,000(W_b - W_a)}{S} - F\right|}{F} \cdot 100\% \tag{5}$$

where W_b is the mass of fertilizer added to the fertilizer box at the start of the experiment, kg; W_a is the mass of residual fertilizer in the fertilizer box at the end of the experiment, kg; S is the operation area, m^2; and F is the desired amount of fertilizer, kg·ha^{-1}.

The test plot was divided into ten areas (2.5 m × 100 m), the working speed was set to 3 km·h^{-1}, and the desired amount of fertilizer was set as 150, 250, or 350 kg·ha^{-1}, respectively. The total mass of fertilizer in the fertilizer box was weighed before and after the experiment, then the deviation of the total amount of fertilizer in the test area was calculated using Equation (5). The deviation of the total amount of fertilizer should be below 15%.

2.3.3. Weeding Performance

Weeding performance is mainly reflected in the removal effect of weeds per unit area and the degree of mechanical damage to rice seedlings. Here, weeding rate (λ) and seedling injury rate per unit area (η) were used as evaluation indexes [25]. The test plot was divided into four 2.5 m × 100 m areas, and five 2.5 m × 1 m areas within these four areas were selected for testing. The working speeds were set as 3, 4, 5, and 6 km·h^{-1}, respectively. The weeding rate and seedling injury rate were calculated using the following formulas:

$$\lambda = \frac{Q_b - Q_a}{Q_b} \cdot 100\% \tag{6}$$

where λ is the weeding rate, %; Q_b is the total number of weeds in the test area before weeding; and Q_a is the total number of weeds in the test area after weeding.

$$\eta = \frac{U_b}{U_a} \cdot 100\% \tag{7}$$

where η is the seedling injury rate, %; U_b is the total number of damaged rice seedlings in the test area (stem broken or bent and epidermis damaged); and U_a is the total number of rice seedlings in the test area.

Each test was repeated three times at each speed and the average values were calculated. The evaluation index was estimated for the overall area based on those of the smaller test areas. According to the rice weeding requirements, the weeding rate should be above 75% and the injury rate should be below 5%. The weeding rate and injured seedling rate were statistically analyzed using analysis of variance (ANOVA), with a significance level of $\alpha = 0.05$.

3. Results and Discussion

3.1. Analysis of Control System Accuracy

3.1.1. Response Characteristics of Fertilization Control System

Figure 5 shows the response of the actual motor speed to changes in the target speed. The response times of the system to achieve a stable speed close to the target speed were 0.55, 0.31, 0.32, 0.41, and 0.35 s. Thereafter, the actual motor speed tended to fluctuate up and down around the desired speed within an acceptable range. Standard deviations of the actual speed after stabilization were 0.26, 0.32, 0.21, 0.14, and 0.21 r·min^{-1}. The response time of the system was 0.31–0.55 s, which was shorter than that of the controller designed by Yu et al. [13], which ranged from 1.5 to 3.0 s, while the standard deviations of the actual speed after stabilization were very close. Thus, the control accuracy of the system is high and stable control can be achieved.

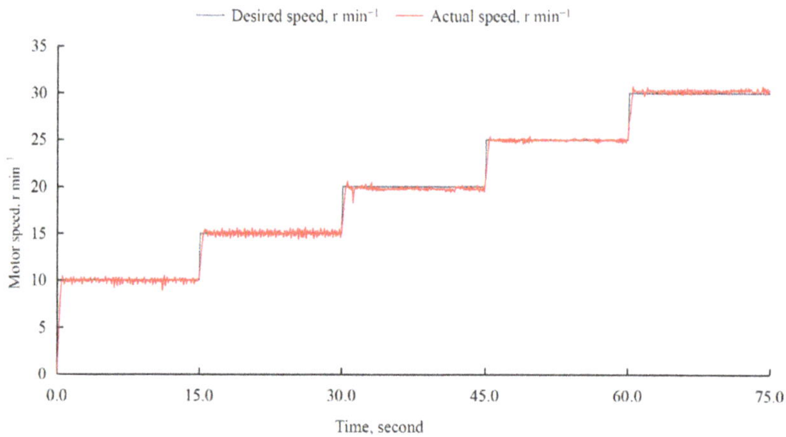

Figure 5. Desired speed versus actual motor speed.

3.1.2. Row-Controlled System Accuracy

Figure 6 shows the lateral deviation of the row-controlled operation at different speeds. The lateral deviation increased with increasing motor speed. The ranges of lateral deviation were −3.15–3.18, −3.82–4.13, −5.32–5.33, and −6.43–5.76 cm at working speeds of 3, 4, 5, and 6 km·h^{-1}, respectively. According to the rice cultivation requirements for the middle and lower reaches of the Yangtze River in China, the row spacing should be 25 cm. Based on the above results and technical specifications of the machine, a lateral deviation of ± 5 cm is considered acceptable. When the lateral deviation is too large, unstable movement of the lateral offset of the weeding wheel may damage seedling roots. To reduce the damage caused by the weeding wheel, the working speed should be controlled to below 5 km h^{-1}.

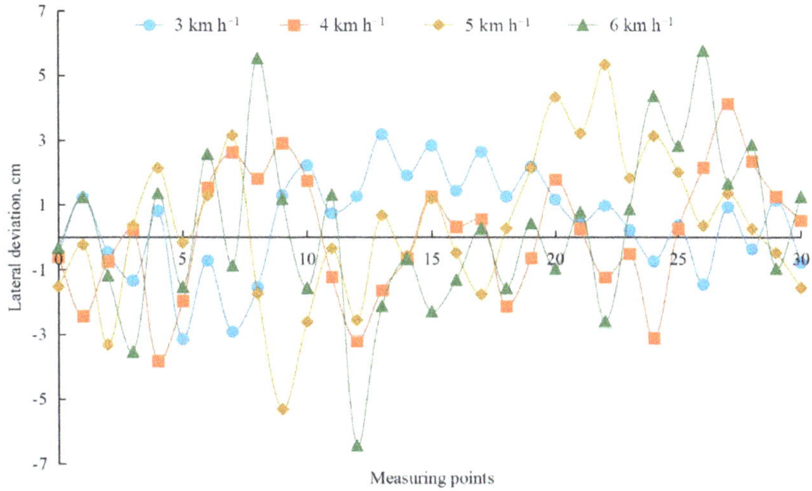

Figure 6. Lateral deviation of row-controlled operation at different speeds.

3.2. *Results of Field Experiment*

3.2.1. Fertilizer Application Uniformity

Table 2 shows the amount of fertilizer applied in each row at different motor speeds. The *CV* of the amount of fertilizer in each row ranged from 1.14 to 2.50%, which meets the requirement of *CV* less than 13%. As the motor speed increased, the amount of fertilizer in each row increased linearly. At the same time, the *CV* of the amount of fertilizer in each row decreased, indicating the application uniformity among rows increased.

Table 2. Statistical analysis of consistency in the amount of fertilizer in each row.

Speed (r·min^{-1})	Amount of Fertilizer (g)								*CV* (%)
	1	2	3	4	5	6	7	8	
10	1026.55	1008.25	1051.75	1041.25	1014.25	1072.61	1000.35	1057.75	2.50
15	1523.48	1500.38	1562.03	1537.73	1500.15	1580.93	1488.08	1573.80	2.35
20	2010.81	1989.42	2051.21	2034.63	1978.32	2082.51	1963.42	2053.23	2.06
25	2473.38	2462.25	2528.88	2523.13	2465.13	2541.13	2426.63	2525.13	1.66
30	2926.83	2917.95	2981.72	2990.41	2923.23	2988.75	2881.65	2980.95	1.41

Table 3 shows the results of an independent sample *t*-test for the average amount of fertilizer dispensed from outlets on the left and right sides of the machine. The corresponding *p* values for each speed show there is no significant difference in the amount of fertilizer dispensed from the left and right fertilizer outlets.

Table 3. Results of independent sample *t*-test for the average amount of fertilizer dispensed from the left and right fertilizer outlets.

Speed (r min^{-1})	t Stat	*p*
10	−0.22	0.83
15	−0.18	0.87
20	0.07	0.95
25	0.23	0.82
30	0.34	0.74

Figure 7 shows the amount of fertilizer applied during a single turn of the fertilizer apparatus at different motor speeds. At a constant motor speed, the amount of fertilizer fluctuates slightly. Overall, the amount of fertilizer was consistent, with a standard deviation of less than 0.52 g. The fluctuation is likely related to differences in the characteristics of different fertilizer distribution apparatuses. Nonetheless, as the speed increased, the amount of fertilizer tended to decrease.

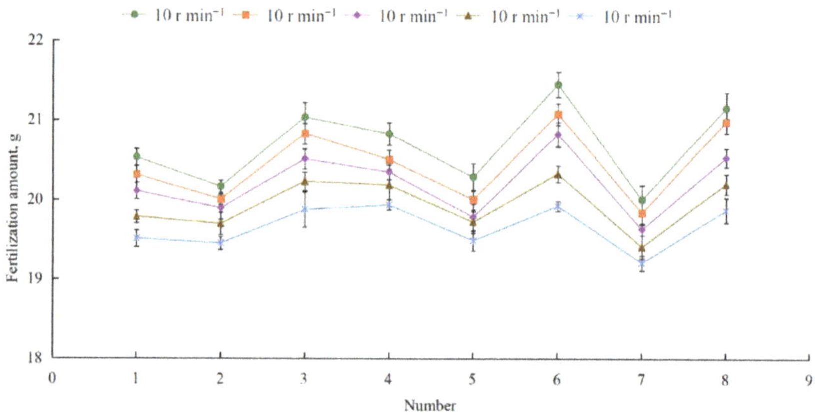

Figure 7. Amount of fertilizer applied during a single turn of the fertilizer distribution apparatus at different working speeds.

3.2.2. Fertilizer Application Accuracy

Figure 8 compares the desired and measured fertilization rates for three different target values. When the desired fertilization rate was 150 kg·ha^{-1}, the measured fertilization rate was generally higher than the desired fertilization rate; whereas, when the desired fertilization rate was 350 kg·ha^{-1}, the measured fertilization rate was generally lower. The reason for this may be that at a constant working speed, the motor speed must increase to increase the amount of fertilizer dispensed. However, as the motor speed increases, the amount of fertilizer applied by the fertilizer distribution apparatus during a single turn will decrease, leading to a decrease in fertilization rate, which is consistent with the results presented in Figure 7.

Table 4 shows the deviation between the desired and measured fertilization rates. When the desired fertilization rates were 150, 250, and 350 kg·ha^{-1}, the deviation ranges of the actual fertilization rates were 1.27–6.80, 0.04–5.00, and 1.00–4.66%, respectively. The deviation of less than 15% meets the fertilization requirements and the application accuracy of the machine meets the design requirements.

Compared with the *CV* and deviation obtained by Wang et al. [10], Shan Zeng et al. [11] and Kim et al. [12], the change in the *CV* and deviation shows the same trend: the faster the motor speed is, the smaller the *CV* is. The larger the desired fertilization rate is, the smaller the deviation is. Under the same motor speed, the values of *CV* and deviation of this paper are smaller, e.g., the *CV* obtained by Kim et al. [12] were in the range of 11.2 to 13.1%, and the deviation ranged from 2.6 to 18.1%. It proves that the uniformity and accuracy of fertilizer application of the prototype are worthy of affirmation.

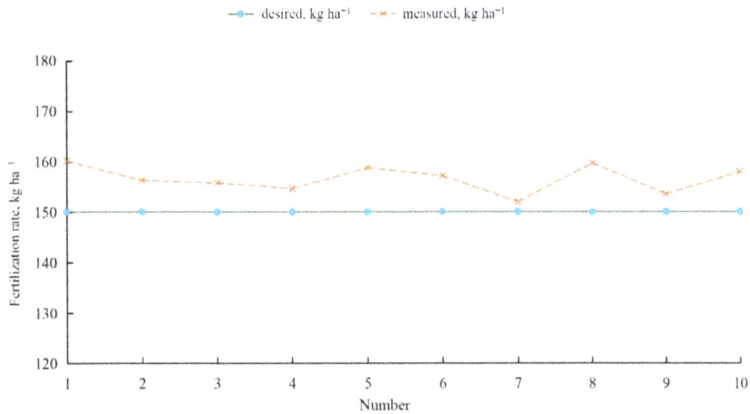

(**a**)

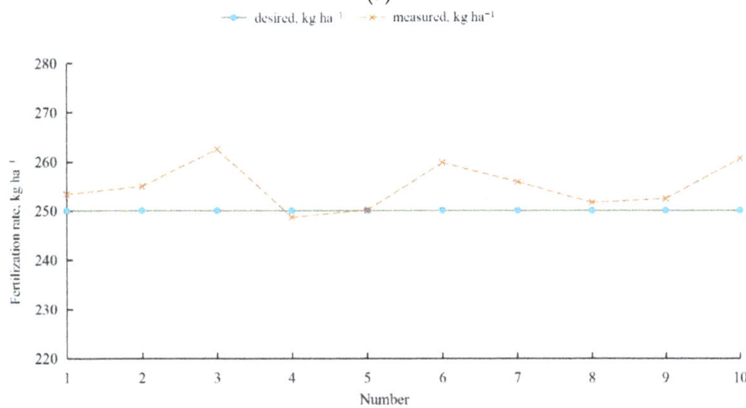

(**b**)

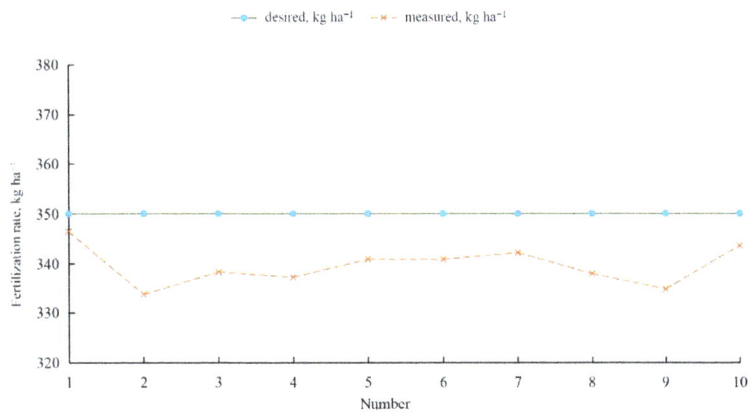

(**c**)

Figure 8. Desired versus measured fertilization rates for target values of (**a**) 150, (**b**) 250, and (**c**) 350 kg·ha^{-1}.

Table 4. Deviation of measured fertilization rate from desired values.

Desired (kg·ha^{-1})	γ_s (%)									
	1	2	3	4	5	6	7	8	9	10
150	6.80	4.20	3.87	3.07	5.87	4.80	1.27	6.47	2.33	5.27
250	1.32	1.96	5.00	0.56	0.04	3.88	2.32	0.64	0.96	4.20
350	1.00	4.66	3.37	3.69	2.63	2.63	2.26	3.49	4.37	1.86

3.2.3. Weeding Performance

Table 5 shows statistical data for weeding rate and seedling injury rate at different working speeds. In a working speed range of 3 to 6 km·h^{-1}, the weeding rate was 78.3–83.1%, which meets the requirements of the rice weeding operation. Table 6 shows the ANOVA results for the variation in average weeding rate with working speed. The *p*-value was far less than 0.05, indicating that the working speed has a significant impact on the weeding rate. The weeding rate decreased with increasing working speed, suggesting that the weeding wheel is less effective at higher speeds. As the working speed of the machine increases, the slip rate of the weeding wheel also increases and the ploughing effect is reduced, which leads to a decrease in weeding rate.

Table 5. Weeding rate and seedling injury rate at different working speeds of the machine.

Speed (km·h^{-1})	3	4	5	6
Weeding rate (%)	83.1 ± 3.5	81.7 ± 1.7	79.8 ± 2.6	78.3 ± 2.1
Seedling injury rate (%)	3.2 ± 0.7	4.1 ± 0.5	5.6 ± 0.4	7.3 ± 0.9

Table 6. Analysis of variance results for influence of working speed on weeding rate.

Source	SS	df	MS	F	*p*-Value	F Crit
Between groups	39.31583	3	13.10528	103.4627	9.71E-07	4.066181
Within groups	1.013333	8	0.126667			
Total	40.32917	11				

In the working speed range of 3 to 6 km·h^{-1}, the seedling injury rate was 3.2–7.3%. Table 7 shows the ANOVA results for the variation of seedling injury rate with working speed. The *p*-value is far less than 0.05, indicating that the working speed has a significant impact on seedling injury rate. The seedling injury rate increased with increasing working speed, which suggests that increasing the speed will reduce the accuracy of the row-controlled system, which is consistent with the results presented in Figure 6. Moreover, a seedling injury rate above 5% does not meet rice weeding operation requirements. To meet the requirements of the rice weeding operation in terms of weeding injury rate, the working speed must be below 5 km·h^{-1}.

Table 7. Analysis of variance results for variation of seedling injury rate with working speed.

Source	SS	df	MS	F	*p*-Value	F Crit
Between groups	29.07	3	9.69	33.70435	6.9E-05	4.066181
Within groups	2.3	8	0.2875			
Total	31.37	11				

Regarding the existing technical solutions of mechanical weeding proposed by Chen et al. [19] and Wang et al. [20], both solutions can achieve a good weeding effect. Under the speed of 3 km·h^{-1}, the weeding rate and seedling injury rate of the existing mechanical weeding equipment can reach 78.02% and 3.9%, respectively. At this speed, the weeding rate and seedling injury rate of the prototype were 83.1% and 3.2%, respectively. The performance of the prototype is slightly better than the above two kinds of equipment, but the gap is not big.

Figure 9 shows the weeding effect of the machine. From a comparison of the number of weeds and the damage to rice seedlings in the operation area and non-operation area, no obvious weed residues were observed in the operation area. The damage to rice seedlings was relatively low, which meets the requirements of the rice weeding operation and demonstrates good performance of the machine.

Figure 9. Results of weeding and fertilizing operations using an RFW machine.

3.3. Limitations

The findings of this study have to be seen in light of some limitations. First, constrained by the work principle of the row-controlled system, the machine is only applicable to paddy fields which were planted by using the automatic driving system for agricultural machinery based on BeiDou navigation. Of course, with the increasing popularity of the automatic driving system for agricultural machinery, the limitation will be greatly reduced. Secondly, due to the fact that there is no similar machine on the market, this paper only evaluates the performance of the machine according to the existing standards and existing technical solutions, and does not carry out comparative tests. At last, the scope of the field experiments are not substantial enough to understand the adaptability of the machine to different regions and soil environments. The adaptability of the device should be verified in the future.

4. Conclusions

To improve the efficiency and quality of rice fertilization and weeding operations and to reduce agricultural non-point source pollution, a row-controlled fertilizing–weeding machine was developed. Accurate fertilization control and row-controlled operation were realized using a PID algorithm and automatic driving system for agricultural machinery based on BeiDou navigation. The fertilization control system can stabilize the speed of the machine to within 0.55 s of the desired speed with a standard deviation of around 0.32 $r \cdot min^{-1}$. The row-controlled operation ensures a lateral deviation of $\pm$5 cm at operating speeds below 5 $km \cdot h^{-1}$, which meets the operational accuracy requirements of rice cultivation. The field test results show that the uniformity and accuracy of fertilization of the RFW machine are high. The coefficient of variation of the amount of fertilizer in each row ranged from 1.14 to 2.50%, with corresponding deviation in corresponding fertilization rates below 6.80%. The proposed system meets all requirements of the Chinese Technical Specification of Quality Evaluation for Fertilizing Machinery (NY/T1003-2006). In the working speed range of 3 to 6 $km \cdot h^{-1}$, the weeding rate was 78.3–83.1% and the seedling injury rate was 3.2–7.3%. Therefore, working speeds below 5 $km \cdot h^{-1}$ are acceptable for rice weeding operations.

In terms of operation efficiency, the machine integrates the fertilization and weeding operations to eliminate single-step operations, and greatly improves the operating efficiency. In terms of operation quality, precision fertilization and row-controlled operation are achieved using a PID controller and BeiDou navigation system, which were shown to improve the uniformity and accuracy of fertilization and effectively reduce the seedling injury rate while ensuring the desired weeding effect. In terms of environmental protection, side deep fertilization effectively increases fertilizer utilization rates and reduces runoff losses from the fertilizer. In addition, mechanical weeding avoids the use of herbicides, which can significantly reduce the ecological impact of rice cultivation on the farmland environment. Further research on the RFW machine will be of great significance in the field of rice cultivation.

Author Contributions: Conceptualization, Y.S. and X.X.; methodology, Y.S., X.X., and H.G.; software, Y.S., H.G., and Y.Z.; validation, Y.S., X.X., H.G., and Y.Z.; formal analysis, Y.S., H.G., X.S., and Y.Z.; investigation, Y.S., H.G., X.S., and Y.Z.; data curation, Y.S. and X.S.; writing—original draft preparation, Y.S.; writing—review and editing, Y.S. and X.X.; visualization, X.X., H.S., and R.Z.; supervision, X.X., X.S., H.S. and R.Z.; project administration, H.S. and R.Z.; funding acquisition, H.S. and R.Z. All authors have read and agreed to the published version of the manuscript.

Funding: This project was supported by the Interdisciplinary Project of Yangzhou University Crop Science Special Zone (yzuxk202007), Jiangsu Modern Agricultural Machinery Equipment and Technology Demonstration and Promotion Project (NJ2020-17), and the Science and Technology Plan Projects in Jiangsu Province of China (BE2018302).

Institutional Review Board Statement: Not applicable.

Informed Consent Statement: Not applicable.

Data Availability Statement: Not applicable.

Acknowledgments: The authors would like to thank the technical support of their teacher and supervisor. We also appreciate the assistance provided by team members during the experiments. Moreover, we would like to thank Yangzhou Huilong Machinery Manufacturing Co., Ltd. for manufacturing the device. Additionally, we sincerely appreciate the work of the editor and the reviewers of the present paper.

Conflicts of Interest: The authors declare no conflict of interest.

References

1. Qinpu, L.; Jingrong, S.; Lijie, P. Comparative study on fertilization intensity and integrated efficiency in China and Euro-American major countries. *Trans. Chin. Soc. Agric. Eng.* **2020**, *36*, 9–16. [CrossRef]
2. Sun, C.; Chen, L.; Zhai, L.; Liu, H.; Wang, K.; Jiao, C.; Shen, Z. National assessment of nitrogen fertilizers fate and related environmental impacts of multiple pathways in China. *J. Clean. Prod.* **2020**, *277*, 123519. [CrossRef]
3. Khan, M.A.; Costa, F.B.; Fenton, O.; Jordan, P.; Fennell, C.; Mellander, P.-E. Using a multi-dimensional approach for catchment scale herbicide pollution assessments. *Sci. Total Environ.* **2020**, *747*, 141232. [CrossRef]
4. Guan, X.; Chen, X.; Qiu, C.; Qian, Y.; Chen, J.; Shao, C.; Xie, J.; Deng, G.; Peng, C. Effects of long-term herbicide application on the crops in soybean-peanut rotations in the red soil upland of Southern China. *Field Crop. Res.* **2020**, *248*, 107723. [CrossRef]

5. Conghua, Z.; Xuyi, L.; Huizhe, C.; Hui, W.; Yuyuan, O.; Junqi, Y.; Baoming, H.; Xi, L. Effects of side deep placement of nitrogen fertilizer on yield and nitrogen utilization efficiency of mechanized direct-seeded rice. *J. Nucl. Agric. Sci.* **2020**, *34*, 2051–2058. [CrossRef]
6. Liu, T.; Fan, D.; Zhang, X.; Chen, J.; Li, C.; Cao, C. Deep placement of nitrogen fertilizers reduces ammonia volatilization and increases nitrogen utilization efficiency in no-tillage paddy fields in central China. *Field Crop. Res.* **2015**, *184*, 80–90. [CrossRef]
7. Zhu, C.; Xiang, J.; Zhang, Y.; Zhang, Y.; Zhu, D.; Chen, H. Mechanized transplanting with side deep fertilization increases yield and nitrogen use efficiency of rice in Eastern China. *Sci. Rep.* **2019**, *9*, 5653. [CrossRef] [PubMed]
8. Lintao, Y.; Xiaolin, C.; Siming, S.; Xiang, L.; Chenming, D.; Dongrong, W. Design of synchronous precision fertilizer machine for rice transplanting. *J. Agric. Mech. Res.* **2020**, *42*, 75–80. [CrossRef]
9. Chen, X.; Luo, X.; Wang, Z.; Zhang, M.; Hu, L.; Zeng, S.; Mo, Z. Experiment of synchronous side deep fertilizing technique with rice hill-drop drilling. *Trans. Chin. Soc. Agric. Eng.* **2014**, *30*, 1–7. [CrossRef]
10. Jinfeng, W.; Guanbao, G.; Wuxiong, W.; Jinwu, W.; Dongwei, Y.; Bowen, C. Design and experiment of key components of side deep fertilization device for paddy field. *Trans. Chin. Soc. Agric. Mach.* **2018**, *49*, 92–104. [CrossRef]
11. Shan, Z.; Zhenhui, Z.; Zhou, Y.; Xiwen, L.; Yipeng, T.; Zewen, M. Design and test of airflow layered fertilizer system for rice direct seeder. *Trans. Chin. Soc. Agric. Eng.* **2020**, *36*, 1–9. [CrossRef]
12. Kim, Y.; Kim, H.; Ryu, K.; Rhee, J. Fertiliser application performance of a variable-rate pneumatic granular applicator for rice production. *Biosyst. Eng.* **2008**, *100*, 498–510. [CrossRef]
13. Yu, J.; Kim, Y.; Ryu, K.H. Development of a Controller for Variable-rate Application of Granular Fertilizer. *J. Biosyst. Eng.* **2006**, *31*, 108–114. [CrossRef]
14. Mohanty, D.K.; Bhuyan, J. Evaluation of Different Mechanical Weed Control Methods in Rice Field. *J. Krishi Vigyan* **2020**, *8*, 115. [CrossRef]
15. Brandsæter, L.O.; Mangerud, K.; Andersson, L.; Børresen, T.; Brodal, G.; Melander, B. Influence of mechanical weeding and fertilisation on perennial weeds, fungal diseases, soil structure and crop yield in organic spring cereals. *Acta Agric. Scand. Sect. B Plant Soil Sci.* **2020**, *70*, 318–332. [CrossRef]
16. Guixiang, T.; Jinwu, W.; Wenqi, Z.; Chunliang, N.; Jiale, Z. Herbicidal mechanism and key components design for paddy weeding device. *Trans. Chin. Soc. Agric. Mach.* **2015**, *46*, 57–63. [CrossRef]
17. Pérez-Ruiz, M.; Slaughter, D.; Gliever, C.; Upadhyaya, S. Automatic GPS-based intra-row weed knife control system for transplanted row crops. *Comput. Electron. Agric.* **2012**, *80*, 41–49. [CrossRef]
18. Liu, F.; Yang, Y.; Zeng, Y.; Liu, Z. Bending diagnosis of rice seedling lines and guidance line extraction of automatic weeding equipment in paddy field. *Mech. Syst. Signal Process.* **2020**, *142*, 106791. [CrossRef]
19. Xueshen, C.; Guijin, F.; Xu, M.; Yu, J.; Long, Q.; Zhujian, H. Design and experiment of control system for weeding alignment in rice field based on linear active disturbance rejection control. *Trans. Chin. Soc. Agric. Eng.* **2020**, *36*, 19–27. [CrossRef]
20. Jinwu, W.; Chao, L.; Xin, L.; Xiang, L.; Jinfeng, W. Design and experiment of 3SCJ-1 type weeding machine for paddy field applied to film mulching and transplanting. *Trans. Chin. Soc. Agric. Mach.* **2018**, *49*, 102–109. [CrossRef]
21. Yangjie, S.; Xiaobo, X.; Fei, W.; Ruihong, Z.; Yifu, J.; Junjic, G. Design and experimental study on automatic driving system of motor type agricultural machinery based on Beidou navigation. *J. Northeast. Agric. Univ.* **2019**, *50*, 88–96. [CrossRef]
22. Liping, H.; Chunying, L.; Zhenmin, W.; Tianqi, Y.; Yunxia, Q. PID parameter optimization of electric variable fertilizer control system. *J. Agric. Mech. Res.* **2020**, *42*, 32–36. [CrossRef]
23. Xiaobo, X.; Yangjie, S.; Xiang, S.; Qi, Z.; Yifu, J.; Junjie, G.; Jianfeng, Z.; Ruihong, Z. Obstacle avoidance path control method for agricultural machinery automatic driving based on optimized Bezier. *Trans. Chin. Soc. Agric. Eng.* **2019**, *35*, 82–88. [CrossRef]
24. The Ministry of Agriculture of the People's Republic of China. *Technical Specifications of Quality Evaluation for Fertilizing Machinery: NY/T 1003-2003*; The Ministry of Agriculture of the People's Republic of China: Beijing, China, 2006.
25. Jinqing, L.; Zhifeng, L.; Pengrong, W.; Zihui, L.; Jicheng, L.; Zhongyuan, L.; Deqiu, Y. Design and experiment of driving-type crushing-weeding multi-functional potato cultivator. *Trans. Chin. Soc. Agric. Eng.* **2019**, *35*, 1–8. [CrossRef]

agriculture

Article

Evaluation of the Effect of Different Hand-Held Sprayer Types on a Greenhouse Pepper Crop

Julián Sánchez-Hermosilla [1,*], Victor J. Rincón [1], Francisco C. Páez [2], José Pérez-Alonso [1] and Ángel-Jesús Callejón-Ferre [1]

1 Department of Agricultural Engineering, Agrifood Campus of International Excellence (CeiA3), Research Center CIMEDES, University of Almería, Ctra. Sacramento s/n, 04120 Almería, Spain; vrc294@ual.es (V.J.R.); jpalonso@ual.es (J.P.-A.); acallejo@ual.es (Á.-J.C.-F.)

2 Research Centre "IFAPA-Cabra", Andalusian Government, 14940 Cabra, Spain; franciscoc.paez@juntadeandalucia.es

* Correspondence: jusanche@ual.es; Tel.: +34-950015107; Fax: +34-950015491

Abstract: The cultivation of vegetables in greenhouses is characterised by high planting density and environmental conditions that favour the development of pests and diseases. These are mainly controlled using plant protection products applied with a hand-held sprayer. This is inefficient low-tech equipment that is difficult to calibrate. The study evaluates one hand-held spray gun and two hand-held spray lances that are widely used in greenhouse vegetable crops. The tests were carried out on a pepper crop at two different developmental stages. Plant canopy deposition and losses to the ground were quantified using a colorimetric method based on applying a tartrazine solution. The results show that the flat-fan spray lance obtains a more uniform spray distribution in the plant canopy and results in losses to the ground that are between 2 and 2.75 times less than when using the other hand-held sprayers tested.

Keywords: greenhouse; hand-held sprayer; plant protection products; crop protection

Citation: Sánchez-Hermosilla, J.; Rincón, V.J.; Páez, F.C.; Pérez-Alonso, J.; Callejón-Ferre, Á.-J. Evaluation of the Effect of Different Hand-Held Sprayer Types on a Greenhouse Pepper Crop. *Agriculture* **2021**, *11*, 532. https://doi.org/10.3390/agriculture11060532

Academic Editor: Dimitrios S. Paraforos

Received: 30 April 2021
Accepted: 7 June 2021
Published: 9 June 2021

1. Introduction

In south-eastern Spain, greenhouse vegetable cultivation has increased substantially over recent years, with the cultivated surface area now amounting to around 32,234 ha [1]. This has made Almería province one of the leading vegetable cultivation areas in Europe, with a total production of 3.62 million tonnes in the year 2019–2020, representing approximately 25% of the national vegetable production. Pepper is the most important crop, accounting for 26% of the production, followed by tomato and cucumber, at 22% and 16%, respectively [1]. This sector has been very competitive in recent years compared to other production areas due to the relatively low production costs.

Pest and disease control in this type of crop is mainly carried out through the use of plant protection products, although in recent years the implementation of integrated pest management programs (IPMs) has contributed to their reduction [2–6]. However, IPM practices are still only implemented on a small scale and given the high concentration of greenhouses in the area, the risk of environmental pollution due to pesticide use has increased. There are studies that have detected the presence of pesticides in the soil [7,8] as well as in surface and groundwater.

The negative effects of plant protection products on the environment and human health has led to the development of important legislative initiatives. At the European level, Directive 2009/128/EC [9] on the sustainable use of pesticides was launched in 2009, with the aim of establishing a framework to reduce the risks to human health and the environment and to encourage the use of pest control techniques that reduce or eliminate the use of pesticides. More recently, in 2019, the European Commission presented its Green Deal communication [10], which strives to achieve a sustainable economy in the European

Union, supported by various strategies such as the Farm to Fork Strategy, which sets out a 50% reduction in the use and risk of pesticides by 2030 [11].

In general, the foliar application of plant protection products is considered an inefficient process because only a fraction of the sprayed liquid is retained on the vegetal canopy whilst the rest is lost to the ground or to drift [12,13]. Likewise, an application is considered adequate when the spray deposition on the vegetal canopy is close to the control threshold for the infestation or disease and is homogenously distributed, yet with minimal losses to the ground or to drift [13,14]. Therefore, to study spraying equipment in the field, it is necessary to quantify the quantity and spatial distribution of the spray deposition on the vegetal canopy and the losses produced, mainly to the ground and to drift. In the case of greenhouse crops, the major losses are to the ground, since drift losses do not occur given the closed nature of the facilities.

Numerous studies have shown that the spraying equipment is one of the most influential factors regarding the rational use of plant protection products on different crop types, including greenhouse horticultural crops [14–21]. The equipment most used on greenhouse horticultural crops are hand-held spray guns and spray lances [22–25] as these are low-cost, easily maintained devices that adapt well to different crop configurations and greenhouse types. Agricultural workers usually work with this type of equipment at high pressures (>20 bar) and high application rates [16,26,27]. For these reasons, and because this equipment is difficult to calibrate and its handling is influenced by the operator's expertise, inefficient spraying occurs, which poses an important risk to the environment. Generally, the distribution of plant protection products on the vegetal canopy using hand-held sprayers is not very uniform and significant losses to the ground occur [14,21,27]. Furthermore, using this equipment results in a high risk of operator exposure [19,28–32], which can lead to adverse health effects.

The abovementioned drawbacks have prompted the use of other greenhouse application equipment over recent years, based on vertical spray booms mounted on manual trolleys [14,33], on self-propelled vehicles [21,34] or on autonomous vehicles [18,24,35]. Employing vertical spray booms allows one to optimise the application of plant protection products by reducing the application rate and ground losses [21,36]. However, this equipment has still only been used on a few greenhouse horticultural crops.

The objective of this work is to analyse and compare the distribution of plant protection products from the hand-held sprayers present in a pepper crop plant canopy along with the losses to the ground caused by each of them. The tests were carried out with the most common hand-held sprayers used on greenhouse crops in south-eastern Spain. Specifically, we have evaluated the deposition in plant canopy and ground losses originating from one hand-held spray gun and two hand-held spray lances, each of which are widely used in the area.

2. Materials and Methods

2.1. Spraying Equipment Used

In the tests, one hand-held spray gun and two hand-held spray lances that are widely used in the area have been evaluated. Their characteristics are described below:

1. Trigger spray gun (Braglia Srl, Reggio Emilia, Italy): this spray gun (Figure 1a) is equipped with a nozzle consisting of a hexagonal brass body with helical grooves and an outlet hole (1.5 mm in diameter). It has a side lever that moves an internal rod, allowing the liquid to be driven through the central part of the nozzle or through the helical grooves. This produces an adjustable spray pattern, from a straight concentrated stream to a hollow cone. A trigger opens or closes the liquid outlet.
2. Rotating-handle spray lance (Sirfran S.L., Alicante, Spain): this spray lance (Figure 1b) is equipped with a cone nozzle that has a variable turbulence chamber and an outlet hole (1.5 mm in diameter). It has a rotating handle to switch the spray lance valve from the closed to the maximum flow position. As the handle is turned, the depth of

the turbulence chamber is modified, changing the spray pattern from a hollow-coned spray to a straight stream.

3. Flat-fan spray lance (Novi Fan S.L., Almería, Spain): this spray lance (Figure 2a) is equipped with two steel flat-fan nozzles, which form angles of 40° and 70° from the outlet tube axis (Figure 2b).

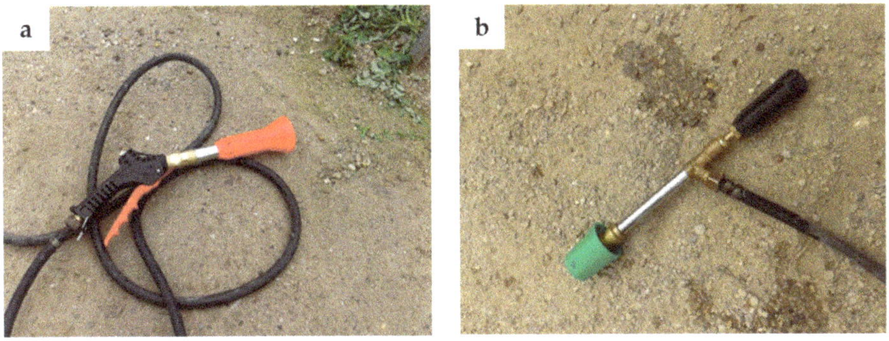

Figure 1. Trigger spray gun (**a**). Rotating-handle spray gun (**b**).

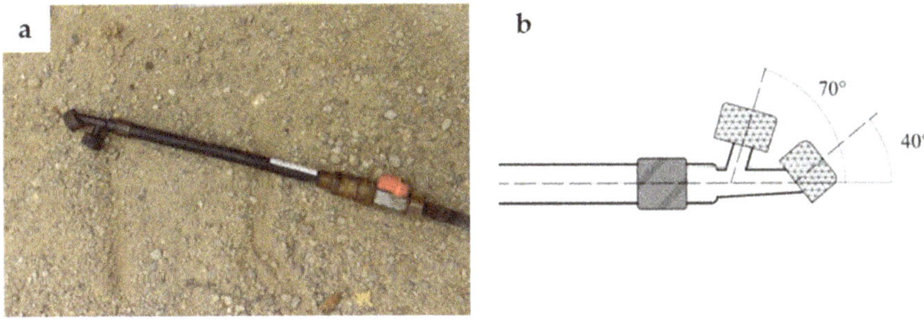

Figure 2. Flat-fan spray lance (**a**). Position of the twin flat-fan nozzles (**b**).

The hand-held sprayers were connected to a forklift carrying a 100 L tank and a membrane pump (M-30, Imovilli Pompe S.R.L., Reggio Emilia, Italy) by means of a 25 m long hose (17mm in diameter). The pressure and flow data were recorded at the pump outlet using a datalogger (DataChart 1250, Monarch Instrument, Amherst, NH, USA) equipped with a pressure sensor (ARAG s.r.l., Reggio Emilia, Italy) and a flow sensor (ORION Visual Flow, ARAG s.r.l., Reggio Emilia, Italy).

2.2. Experimental Design

The tests were conducted in the southern half (45 × 20 m^2) of an 1800 m^2 greenhouse located at the Foundation ANECOOP-UAL experimental facilities at the University of Almería (36°52′ N, 2°17′ W).

A pepper crop was used in the tests (*Capsicum annuum*, cv. Melchor), planted in a twin-row system, with a 2 m separation between the lines and 0.5 m between each pair of plants (2 plants/m^{-2}). The spray equipment tests were carried out at two different crop stages. The crop characteristics in the 2 tests carried out are listed in Table 1. The leaf area index (LAI) was measured from 6 plants taken at random in the greenhouse. The plants were completely stripped of their leaves and the surface area of each leaf was measured with an electronic planimeter (WinDias, Delta-T Devices Ltd., Cambridge, UK).

Table 1. Crop Characteristics.

Test	LAI	Crop Height (m)	Crop Depth (m)
1	3.20	1.44	1.13
2	4.75	1.75	1.08

The test area consisted of 25 crop lines, each of which was 18 m long with a north–south orientation, divided into 3 blocks. Each block contained 6 crop lines, 3 of which were used for Test 1 and the other 3 for Test 2, located alternately to avoid possible contamination between applications in the same test (Figure 3).

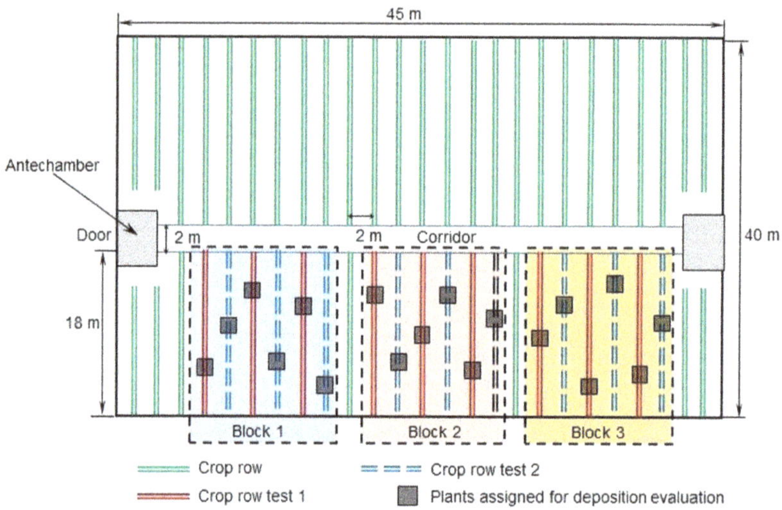

Figure 3. Floor plan of the greenhouse indicating the sampling blocks for the experimental plants.

In each block, and for each test, 3 applications were carried out, one for each hand-held sprayer on a different randomly assigned crop line. In each test line, a pair of plants was randomly selected for sampling; this was conducted in 12 different zones (at 4 depths and 3 heights; Figure 4). On the ground, coincident with the external layer (P1 and P4) and at the centre of each plant pair (Figure 4), artificial collectors were placed; these consisted of 30 × 80 mm filter paper strips (Filter-Lab Ref 1238, Filtros Anoia S.A., Barcelona, Spain). A total of 135 samples were taken in each trial ((12 zones + 3 soil samples) × 9 pairs of plants).

The working conditions (Table 2) were established in accordance with the common farming practices in the area; these consist of spraying at high pressures (approximately 2500 kPa) using high water volumes—for Test 1 and Test 2, these amounted to approximately 1100 L ha^{-1} and 1500 L ha^{-1}, respectively. The flow rates for each of the hand-held sprayers under these conditions are shown in Table 2—in Test 1, it was 4.53 L min^{-1} for the trigger spray gun, 4.38 L min^{-1} for the rotating-handle spray lance and 4.19 L min^{-1} for the flat-fan spray lance; in Test 2, the flow rates were 4.67, 4.35 and 4.21 L min^{-1} for the trigger spray gun, the rotating-handle spray lance and the flat-fan spray lance, respectively. The hand-held sprayers were tested on a horizontal test bench complying with the ISO 5682-1 standard [37] to determine the actual spray angle (Figure 5). The test bench was equipped with 36 grooves, each 50 mm wide and 1500 mm long.

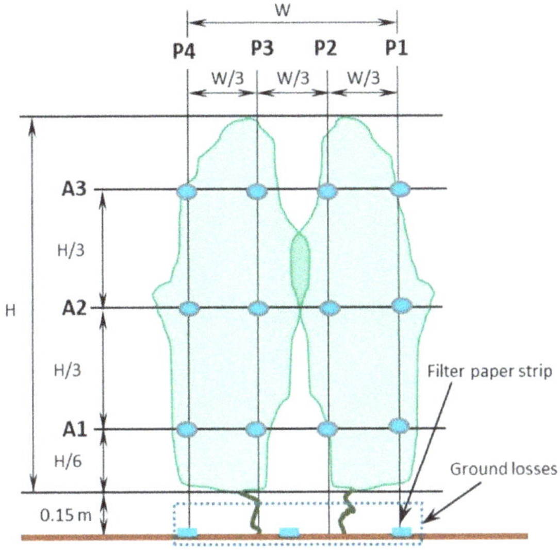

Figure 4. Sampling position (depth and height) for assessing the spray deposition on the canopy and the losses to the ground.

Table 2. Spray application parameters for the tests.

Applications		Nozzle No.	Spray Pressure (kPa)	Actual Spray Angle (Degrees)	Flow (L min^{-1})	Travel Speed (m s^{-1})	Application Rate (L ha^{-1})
Test 1	TSG	1	2440	60	4.53	0.63	1188.25
	RHSL	1	2565	46	4.38	0.68	1075.35
	FFSL	2	2559	90	4.19	0.62	1128.08
Test 2	TSG	1	2492	60	4.67	0.51	1517.12
	RHSL	1	2548	46	4.35	0.48	1498.45
	FFSL	2	2576	90	4.21	0.46	1511.16

TSG: trigger spray gun, RHSL: rotating-handle spray lance, FFSL: flat-fan spray lance.

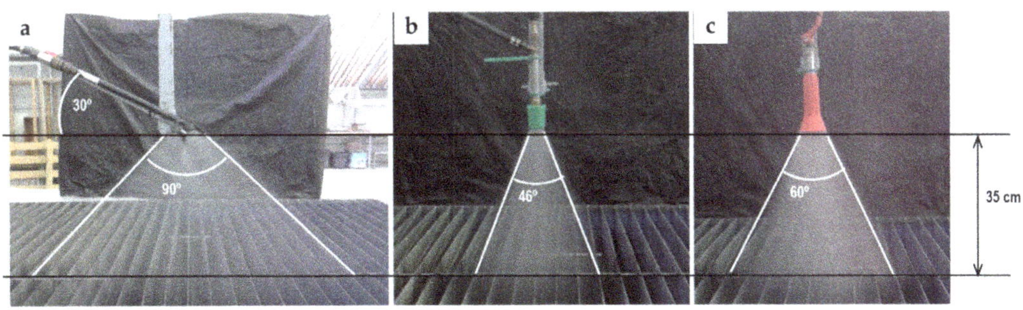

Figure 5. Testing the hand-held sprayers on a horizontal test bench (ISO 5682-1-compliant) to determine the actual spray angle: the flat-fan spray lance (**a**), rotating-handle spray lance (**b**) and trigger spray gun (**c**).

2.3. Measuring the Spray Deposition and Losses to the Ground

Tartrazine (Roha Europe, S.L.U., Torrent, Spain) was used as a tracer to quantify the deposition, using a tank concentration of approximately 10 g L^{-1}. After the tests were carried out in the greenhouse, 3 or 4 leaves were taken from each of the sampling areas listed above (Figure 4), depending on their size, to ensure that the total surface area sampled in each zone was approximately the same. These leaves were placed in 200 × 320 mm self-sealing bags. Likewise, the filter paper strips were placed in 120 × 180 mm self-sealing bags.

In the laboratory, the leaves were washed with 50 mL of distilled water in the same self-sealing bags, and after stirring for 1 min, they were stored in the dark for 30 min to prevent degradation by ambient light. Once the washing process was complete, the sheets were carefully removed from the bags, dried slightly by placing them on filter paper and then the surface was measured using an electronic planimeter (WinDias, Delta-T Devices Ltd., Cambridge, UK).

To extract the dye from the filter paper strips, a procedure similar to that used for the leaves was followed, but using 25 mL of distilled water to wash the sample in the same self-sealing bags.

The amount of tartrazine in the washing solution was determined at a wavelength of 425.5 nm using a dual-beam, UV-visible spectrophotometer (Helios Zeta, Thermospectronic, Cambridge, UK). As a baseline, the solution from washing the blank samples was used following the procedure described above.

To ascertain the actual tracer concentration in the tank, samples of the applied solution were taken at the outlet of the hand-held sprayers, recording a concentration of 8.21 g L^{-1} in Test 1 and 9.81g L^{-1} in Test 2. To compare the applications made with the different hand-held sprayers, the measured depositions were normalised, taking a reference application volume rate of 1100 L ha^{-1} in Test 1 and 1500 L ha^{-1} in Test 2, according to Equation (1).

$$d_n = \frac{d \cdot V_n}{C_T \cdot V} \quad (1)$$

where d_n is the normalised deposit (μL cm^{-2}), d is the tracer concentration per unit sample surface (μg cm^{-2}), V_n is the reference volume rate application (L ha^{-1}), V is the volume rate application (L ha^{-1}) and C_T is the tracer concentration of the tank for each treatment (g L^{-1}).

2.4. Statistical Analysis

The statistical analysis of the data was performed using SPSS v26.0 software (SPSS Inc., an IBM Company, Chicago, IL, USA). An analysis of variance (Anova) was performed, and the significant differences were evaluated using Duncan's test. Before the statistical analysis, the normality of the data was checked using the Kolmogorov–Smirnov test ($p < 0.05$), and the equality of variances by means of the Levene test.

3. Results and Discussion

3.1. Crop Deposition

There were no significant differences in canopy depositions made by the different hand-held sprayers used in the tests (Table 3), although the flat-fan spray lance deposited the most in both tests, followed by the trigger spray gun and the rotating-handle spray lance. Cerruto et al. [22] compared two spray lances similar to those used in this study on a tomato crop, obtaining comparable results. Likewise, Rincón et al. [27] obtained similar deposition results for the flat-fan spray lance on a greenhouse pepper crop.

Table 3. Normalised deposition on the canopy and losses to the ground (mean ± SD) in μL cm^{-2}.

Hand-Held Sprayer	Test 1		Test 2	
	Canopy	Ground	Canopy	Ground
TSG	0.97 ± 0.53a	7.39 ± 3.03a	0.97 ± 0.59a	8.69 ± 3.38a
RHSL	0.90 ± 0.54a	9.70 ± 5.460a	0.84 ± 0.68a	11.77 ± 6.61a
FFSL	1.06 ± 0.51a	3.76 ± 1.36b	1.09 ± 0.65a	4.28 ± 1.50b

TSG: trigger spray gun, RHSL: rotating-handle spray lance, FFSL: flat-fan spray lance. Means in the same column with the same letter do not differ significantly ($p < 0.05$; Duncan's test).

Furthermore, the flat-fan spray lance resulted in more uniform applications in the two tests, with coefficient variation (CV) values of 48.75% and 59.63% for Tests 1 and 2, respectively. The least uniform applications occurred with the rotating-handle spray lance, with variation coefficients of 60.27% and 79.76% in Tests 1 and 2, respectively. Similar results were obtained by Sánchez-Hermosilla et al. [21] in a greenhouse tomato crop. This may be due to the lower spray angle that this spray lance has, concentrating the droplets within a smaller spray width, which forces the operator to have good coordination between the arm movement and the feed to achieve adequate overlap and good distribution uniformity. This aspect is less limiting in the other devices tested, especially in the flat-fan spray lance, which, with a 90° spray angle, distributes the droplets in a wider arc (approximately 80 cm for applications made 35 cm from the vegetation), hence the higher application uniformity.

For the rotating-handle spray lance, both the closing and the adjustment of the output flow is performed using the rotating handle, which makes it difficult to keep the same handle position once the spray lance has been closed. Normally, the operator shuts off the spray lance when reaching the end of the crop line and opens it when starting a new line. All of this can also cause lower application uniformity for the rotating-handle spray lance than for the other equipment.

Another aspect contributing to the lack of application uniformity is the difficulty in reaching the internal areas of the plant canopy. As indicated by Braekman et al. [16], this is because only the finest droplets reach the inner part of the canopy. For all the hand-held sprayers tested, more than 69% of the quantified deposition occurred in the outer planes of the plant canopy (planes P1 and P4, Figure 4). The deposition in the interior planes (planes P2 and P3, Figure 4) was very low for all the devices, although in both tests it was somewhat higher for the flat-fan spray lance, with significant differences in the test on the more developed crop (Test 2) compared to the rotating-handle spray lance (Figure 6).

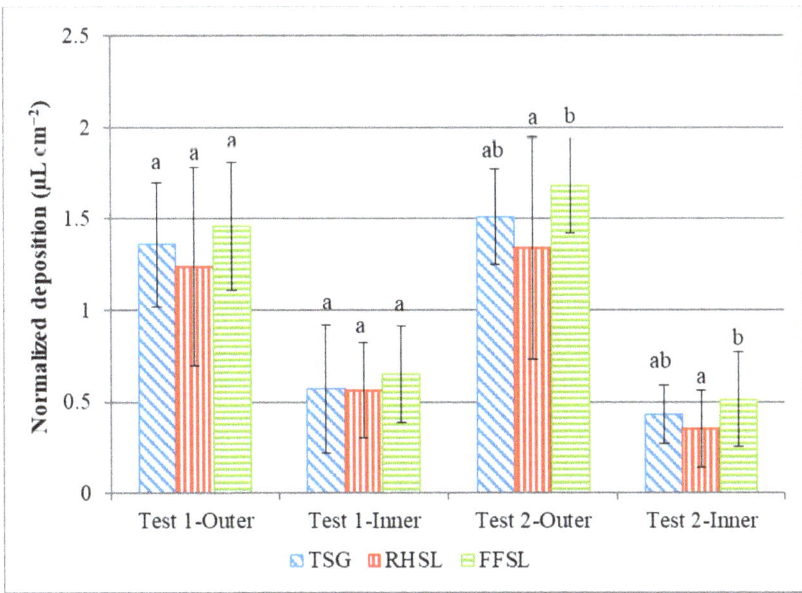

Figure 6. Normalised deposition (µL cm^{-2}) in the outer and inner zones of the canopy. The same letter means no significant differences (Duncan's test, $p < 0.05$). Bars signify the means ± SD of the data. TSG: trigger spray gun, RHSL: rotating-handle spray lance, FFSL: flat-fan spray lance.

3.2. Losses to the Ground

Ground loss values are shown in Table 3. One can observe that lower losses occurred with the flat-fan spray lance, while significant differences existed with regard to the losses caused by the other hand-held sprayers in the two tests. Cerruto et al. [22] also found lower ground losses when using a flat-fan spray gun on a greenhouse tomato crop.

The flat-fan spray lance results in ground losses 2 to 2.75 times less than for the other devices. This may be due to the orientation of the nozzles towards the vegetation. When the operator works with the flat-fan spray lance, the nozzles generate a flat jet which is approximately perpendicular to the vegetable canopy, moving from top to bottom; thus, the direct projection of droplets to the ground is very low, so most losses are due to the lower number of droplets that do not reach the vegetation or are due to runoff. This fact is confirmed when looking at the quantified losses in the internal and external zone, which are similar for this equipment in each of the tests (Table 4).

Table 4. Normalised deposition (in μL cm^{-2}) in the outer and inner zones of the ground (mean ± SD).

Application	Test 1		Test 2	
	Outer	Inner	Outer	Inner
TSG	8.59 ± 2.99a	4.98 ± 1.13a	10.03 ± 3.34a	6.02 ± 1.31ab
RHSL	12.57 ± 4.24b	3.97 ± 0.63a	15.19 ± 5.25b	4.93 ± 0.54bc
FFSL	3.80 ± 1.71c	3.66 ± 0.27a	4.36 ± 1.87c	4.11 ± 0.35c

TSG: trigger spray gun, RHSL: rotating-handle spray lance, FFSL: flat-fan spray lance. Means in the same column with the same letter do not differ significantly ($p < 0.05$; Duncan's test).

In contrast, for the other hand-held sprayers, the nozzles are located at the end of the devices, generating a tapered jet, the shaft of which forms an angle of approximately 20–25° to the crop line. This results in a direct projection of droplets towards the ground during the downward movement of the operator's arm, leading to high ground losses, which were greater in the external area than in the internal area for the two tests carried out (Table 4). In the external area, as well as the droplets received by direct spraying, we also quantified the losses caused by droplets not reaching the vegetation and falling as runoff. However, most of the quantified losses in the internal area were due to the falling of droplets that did not reach the vegetation or to the runoff; thus, their values are very similar to those obtained with the flat-fan spray lance.

4. Conclusions

The tests have shown that the plant canopy depositions do not present statistically significant differences for the different hand-held sprayers used, although they were somewhat higher for the flat-fan spray lance.

In general, there is a uniform distribution of spray in the plant canopy, with coefficients of variation greater than 48% for all the hand-held sprayers, mainly due to the arrangement of the nozzles and the opening of the spray jet, as well as the difficulty in reaching the internal vegetation areas. However, the increased spray angle of the flat-fan spray lance's opening allows for more uniform distributions, with coefficients of variation of less than 60% in the tests carried out.

As far as ground losses are concerned, these were between 2 and 2.75 times lower for the flat-fan spray lance than for the other hand-held sprayers tested, amongst which there were no statistically significant differences. This is due to the flat-fan spray lance's nozzle arrangement, which generates a flat jet during spraying which is approximately perpendicular to the vegetation.

In general, one can conclude that the flat-fan spray lance results in more uniform application with less ground losses than the other hand-held sprayers.

Author Contributions: Conceptualization and methodology, J.S.-H., V.J.R., F.C.P., J.P.-A. and Á.-J.C.-F.; software, J.S.-H. and V.J.R.; validation, J.S.-H., J.P.-A. and Á.-J.C.-F.; formal analysis, F.C.P. and J.P.-A.; investigation, J.S.-H. and V.J.R.; resources, J.S.-H.; data curation, J.S.-H. and V.J.R.; writing—original draft preparation, J.S.-H., V.J.R. and F.C.P.; writing—review and editing J.S.-H., V.J.R. and F.C.P.; visualization, J.S.-H., V.J.R., F.C.P., J.P.-A. and Á.-J.C.-F.; supervision, J.S.-H.; project administration and funding acquisition, J.S.-H. All authors have read and agreed to the published version of the manuscript.

Funding: This research was funded by Consejería de Economía, Conocimiento, Empresas y Universidad, Junta de Andalucía, grant number P12-AGR773 (co-financed with FEDER funds from the European Union).

Institutional Review Board Statement: Not applicable.

Informed Consent Statement: Not applicable.

Data Availability Statement: Not applicable.

Conflicts of Interest: The authors declare no conflict of interest.

References

1. Consejería de Agricultura, Pesca y Desarrollo Rural (CAPDR)—Andalusian Regional Government, Regional Ministry of Agriculture and Fishery. Summary of the Greenhouse Horticulture Crop Season. Almería 2019/20. Available online: https://www.juntadeandalucia.es/agriculturaypesca/observatorio/servlet/FrontController?action=RecordContent&table=11031&element=3311405&subsector=& (accessed on 8 April 2021). (In Spanish).
2. Glass, R.; Egea González, F.J. Biological control in the greenhouses of Almería and challenges for a sustainable intensive production. *Outlooks Pest Manag.* **2012**, *23*, 276–279. [CrossRef]
3. Cotes, B.; González, M.; Benítez, E.; De Mas, E.; Clemente-Orta, G.; Campos, M.; Rodríguez, E. Spider communities and biological control in native habitats surrounding greenhouses. *Insects* **2018**, *9*, 33. [CrossRef]
4. Rodríguez, E.; González, M.; Paredes, D.; Campos, M.; Benítez, E. Selecting native perennial plants for ecological intensification in Mediterranean greenhouse horticulture. *Bull. Entomol. Res.* **2018**, *108*, 694–704. [CrossRef]
5. Dáder, B.; Colomer, I.; Adán, Á.; Medina, P.; Viñuela, E. Compatibility of early natural enemy introductions in commercial pepper and tomato greenhouses with repeated pesticide applications. *Insect Sci.* **2020**, *27*, 1111–1124. [CrossRef] [PubMed]
6. Velasco, L.; Ruiz, L.; Galipienso, L., Rubio, L.; Janssen, D. A Historical Account of Viruses in Intensive Horticultural Crops in the Spanish Mediterranean Arc: New Challenges for a Sustainable Agriculture. *Agronomy* **2020**, *10*, 860. [CrossRef]
7. Padilla-Sánchez, J.A.; Romero-González, R.; Plaza-Bolaños, P.; Garrido Frenich, A.; Martínez Vidal, J.L. Residues and Organic Contaminants in Agricultural Soils in Intensive Agricultural Areas of Spain: A Three Years Survey. *Clean Soil Air Water* **2015**, *43*, 746–753. [CrossRef]
8. Plaza-Bolaños, P.; Padilla-Sánchez, J.A.; Garrido-Frenich, A.; Romero-González, R.; Martínez-Vidal, J.L. Evaluation of soil contamination in intensive agricultural areas by pesticides and organic pollutants: South-eastern Spain as a case study. *J. Environ. Monit.* **2012**, *14*, 1182–1189. [CrossRef] [PubMed]
9. European Union (EU). Directive 2009/128/EC of the European Parliament and of the Council of 21 October 2009 Establishing a Framework for Community Action to Achieve the Sustainable Use of Pesticides. Available online: https://eur-lex.europa.eu/legal-content/EN/TXT/PDF/?uri=CELEX:32009L0128&from=EN (accessed on 8 April 2021).
10. European Commission (EC). Communication from the Commission to the European Parliament, the Council, the European Economic and Social Committee and the Committee of the Regions—The European Green Deal, COM (2019) 640 Final. Available online: https://eur-lex.europa.eu/resource.html?uri=cellar:b828d165-1c22-11ea-8c1f-01aa75ed71a1.0002.02/DOC_1&format=PDF (accessed on 8 April 2021).
11. European Commission (EC). Communication from the Commission to the European Parliament, the Council, the European Economic and Social Committee and the Committee of the Regions—A Farm to Fork Strategy for a Fair, Healthy and Environmentally-Friendly Food System, COM/2020/381 Final. Available online: https://eur-lex.europa.eu/legal-content/EN/TXT/?uri=CELEX:52020DC0381 (accessed on 8 April 2021).
12. Braekman, P.; Foque, D.; Messens, W.; van Labeke, M.C.; Pieters, J.G.; Nuyttens, D. Effect of spray application technique on spray deposition in greenhouse strawberries and tomatoes. *Pest Manag. Sci.* **2010**, *66*, 203–212. [CrossRef]
13. Failla, S.; Romano, E. Effect of Spray Application Technique on Spray Deposition and Losses in a Greenhouse Vegetable Nursery. *Sustainability* **2020**, *12*, 7052. [CrossRef]
14. Sánchez-Hermosilla, J.; Rincón, V.J.; Páez, F.; Fernández, M. Comparative spray deposits by manually pulled trolley sprayer and a spray gun in greenhouse tomato crops. *Crop Prot.* **2012**, *31*, 119–124. [CrossRef]
15. Balsari, P.; Oggero, G.; Bozzer, C.; Marucco, P. An autonomous self-propelled sprayer for safer pesticide application in glasshouse. *Asp. Appl. Biol.* **2012**, *114*, 197–204.

16. Braekman, P.; Foqué, D.; van Labeke, M.-C.; Pieters, J.G.; Nuyttens, D. Influence of spray application technique on spray deposition in Greenhouse Ivy pot plants grown on hanging shelves. *HortScience* **2009**, *44*, 1921–1927. [CrossRef]
17. Farooq, M.; Salyani, M. Spray penetration into the citrus tree canopy from two air-carrier sprayers. *Trans. Am. Soc. Agric. Eng.* **2002**, *45*, 1287–1293. [CrossRef]
18. Rincón, V.J.; Grella, M.; Marucco, P.; Alcatrão, L.E.; Sanchez-Hermosilla, J.; Balsari, P. Spray performance assessment of a remote-controlled vehicle prototype for pesticide application in greenhouse tomato crops. *Sci. Total Environ.* **2020**, *726*, 138509. [CrossRef]
19. Rincón, V.J.; Páez, F.C.; Sánchez-Hermosilla, J. Potential dermal exposure to operators applying pesticide on greenhouse crops using low-cost equipment. *Sci. Total Environ.* **2018**, *630*, 1181–1187. [CrossRef]
20. Salyani, M.; Sweeb, R.D.; Farooq, M. Comparison of string and ribbon samplers in orchard spray applications. *Trans. ASABE* **2006**, *49*, 1705–1710. [CrossRef]
21. Sánchez-Hermosilla, J.; Rincón, V.J.; Páez, F.; Agüera, F.; Carvajal, F. Field evaluation of a self-propelled sprayer and effects of the application rate on spray deposition and losses to the ground in greenhouse tomato crops. *Pest Manag. Sci.* **2011**, *67*, 942–947. [CrossRef]
22. Cerruto, E.; Emma, G.; Manetto, G. Spray applications to tomato plants in greenhouses. Part 1: Effect of walking direction. *J. Agric. Eng.* **2009**, *3*, 41–48. [CrossRef]
23. Goossens, E.; Windey, S.; Sonck, B. Information service and voluntary testing of spray guns and other types of sprayer in horticulture. *Asp. Appl. Biol.* **2004**, *71*, 41–48.
24. Li, Y.J.; Li, Y.F.; Chen, R.H.; Li, X.S.; Pan, C.P.; Song, J.L. Evaluation of self-propelled high-energy ultrasonic atomizer on azoxystrobin and tebuconazole application in sunlit greenhouse tomatoes. *Int. J. Environ. Res. Public Health* **2018**, *15*, 1088. [CrossRef] [PubMed]
25. Valera-Martínez, D.L.; Belmonte-Ureña, L.J.; Molina-Aiz, F.D.; López-Martinez, A. *Greenhouse Agriculture in Almeria: A Comprehensive Techno-Economic Analysis*; Cajamar Caja Rural: Almería, Spain, 2016; p. 408.
26. Cerruto, E.; Emma, G.; Manetto, G. Spray applications to tomato plants in greenhouses. Part. 2: Effect of spray lance type. *J. Agric. Eng.* **2009**, *40*, 49. [CrossRef]
27. Rincón, V.J.; Sanchez-Hermosilla, J.; Páez, F.; Pérez-Alonso, J.; Callejón, A.J. Assessment of the influence of working pressure and application rate on pesticide spray application with a hand-held spray gun on greenhouse pepper crops. *Crop Prot.* **2017**, *96*, 7–13. [CrossRef]
28. Cerruto, E.; Emma, G.; Manetto, G. Operator contamination during pesticide application in tomato greenhouse. *Acta Hortic.* **2008**, *801*, 1531–1538. [CrossRef]
29. Nuyttens, D.; Braekman, P.; Windey, S.; Sonck, B. Potential dermal pesticide exposure affected by greenhouse spray application technique. *Pest Manag. Sci.* **2009**, *65*, 781–790. [CrossRef] [PubMed]
30. Nuyttens, D.; Windey, S.; Sonck, B.; Paper, C.; Rese, F.; Rese, F. Comparison of operator exposure for five different greenhouse spraying applications. *J. Agric. Saf. Health* **2004**, *10*, 187–195. [CrossRef]
31. Yarpuz-Bozdogan, N.; Bozdogan, A.M.; Daglioglu, N.; Erdem, T. Determination of dermal exposure of operator in greenhouse spraying. *Agric. Mech. Asia Africa Lat. Am.* **2017**, *48*, 33–38.
32. Cerruto, E.; Manetto, G.; Santoro, F.; Pascuzzi, S. Operator Dermal Exposure to Pesticides in Tomato and Strawberry Greenhouses from Hand-Held Sprayers. *Sustainability* **2018**, *10*, 2273. [CrossRef]
33. Llop, J.; Gil, E.; Llorens, J.; Gallart, M.; Balsari, P. Influence of air-assistance on spray application for tomato plants in greenhouses. *Crop Prot.* **2015**, *78*, 293–301. [CrossRef]
34. Balloni, S.; Caruso, L.; Cerruto, E.; Emma, G.; Schillaci, G. A prototype of self-propelled sprayer to reduce operator exposure in greenhouse treatment. In *Innovation Technology to Empower Safety, Health and Welfare in Agriculture and Agro-Food Systems, Proceedings of the International Conference, Ragusa, Italy, 15–17 September 2008*; Ragusa Safety Health Welfare: Ragusa, Italy, 2008.
35. González Sánchez, R.; Rodríguez Díaz, F.; Sánchez-Hermosilla López, J.; García Donaire, J. Navigation techniques for mobile robots in greenhouses. *Appl. Eng. Agric.* **2009**, *25*, 153–165. [CrossRef]
36. Sánchez-Hermosilla, J.; Páez, F.; Rincón, V.J.; Pérez-Alonso, J. Volume application rate adapted to the canopy size in greenhouse tomato crops. *Sci. Agric.* **2013**, *70*, 390–396. [CrossRef]
37. ISO. ISO 5682-1:2017–Equipment for Crop Protection—Spraying Equipment—Part 1: Test Methods for Sprayer Nozzles. Available online: https://www.iso.org/standard/60053.html (accessed on 8 April 2021).

agriculture

MDPI

Article

Equipment Performance, Costs and Constraints of Packaging and Transporting Rice Straw for Alternative Uses to Burning in the "Parc Natural l'Albufera de València" (Spain)

Antonio Torregrosa [1,*], Juan Miguel Giner [2] and Borja Velázquez-Martí [1]

1 Departamento Ingeniería Rural y Agroalimentaria, Universitat Politècnica de València, 46022 Valencia, Spain; borvemar@dmta.upv.es
2 Escuela Técnica Superior de Ingeniería Agronómica y del Medio Natural, Universitat Politècnica de València, 46022 Valencia, Spain; juagiva@etsiamn.upv.es
* Correspondence: torregro@dmta.upv.es; Tel.: +34-963877291

Abstract: Rice straw is a residue that causes significant environmental problems, as burning it causes CO_2 and ash emissions, while buried waste can cause issues associated with eutrophication. The extraction of straw from fields for alternative uses may contribute to solving these problems, but research into its economic viability is necessary. The straw can be used for crop mulching, biofuel, bedding for livestock, and so on. In this study, we analyse the work carried out by straw harvesting machines (rakes, balers, bundlers, and loaders) and calculate the costs of packing, road-siding, and transportation of the straw from the rice fields to stockage points in the producer area, as well as to locations outside of the rice production area, in order to assess the viability. The costs of all elemental operations were calculated. The costs of all the operations included between raking and unloading in the producer area stocking point ranged between 28.1 and 51 EUR t^{-1}. These costs were compared with the price of rain-fed cereal straw (wheat and barley), which is the most abundant, noting that the years in which rain-fed cereal straw reached high prices, rice straw could serve as a competitive product; however, in years when the former is cheap, it would be necessary to subsidise the harvesting of rice straw.

Keywords: straw; rice; baler; transport; economy; mechanisation

Citation: Torregrosa, A.; Giner, J.M.; Velázquez-Martí, B. Equipment Performance, Costs and Constraints of Packaging and Transporting Rice Straw for Alternative Uses to Burning in the "Parc Natural l'Albufera de València" (Spain). *Agriculture* **2021**, *11*, 570. https://doi.org/10.3390/agriculture11060570

Academic Editor: Mumtaz Cheema

Received: 28 April 2021
Accepted: 19 June 2021
Published: 21 June 2021

1. Introduction

The Parque Natural de L'Albufera de València (Spain) is one of the most important wetlands in Southern Europe. It is characterised by the lake and rice fields that surround it, the flooded lands of which serve as a refuge for a multitude of birds. Its management and preservation depend on various authorities: at the European level, it is part of the Natura 2000 network of European protected areas, as well as having been granted a number of LIFE projects. The legal jurisdiction affecting the conservation of l'Albufera de València is distributed among various authorities, meaning that it is sometimes difficult to know which one of them is in charge of each of the numerous aspects involved in the conservation of such an area. This fact provides an added difficulty when developing certain kinds of proposals [1].

Rice, as a crop, has a long history in the area, and is the most important in the Natural Park, with a production of 116,925 t and an area of 15,087 ha [2]. The traditional way of eliminating straw after harvesting was burning it, as this method has many advantages. It is a cheap system and has other positive effects, such as eliminating weed seeds and reducing the presence of other pathogens. However, the somewhat uncontrolled burning on days with prevailing winds meant that smoke was carried towards neighbouring cities, leading many citizens to complain about the presence of smoke and ash in the air. This confrontation, combined with the current heightened awareness of climate change, has led authorities to prohibit or limit this practice in Spain [3,4], as well as in other occidental countries such as the USA [5].

There are no easy alternatives to burning. Burying straw leads to problems with water eutrophication, the possible spread of some herbicide-resistant weeds, and pests and diseases becoming more difficult to control [3–6].

An alternative to burning or burying is the extraction of the straw from the fields, in order to put it to other uses, such as energy and biofuels, pulp and paper [5], construction materials [7], compost [8], livestock [9], mulch to prevent soil erosion [10], to reduce water losses [11], crop yields [11], weeds [12], and so on [13]. However, all these alternatives have a significant common cost, which is the packaging and subsequent transportation of the straw to the points of use, including biofuel industries, buildings, compost plants, animal farms, or agricultural fields for mulching [9–13].

In this study, a quantification of the magnitude of these costs was carried out, as well as monitoring the packing and loading of machines, analysing the possible routes based on the orography of the production area, and estimating the transport costs from the production site to destinations located within a 200 km radius [14,15].

The following phases are involved in the analysis of straw management:

(a) Regulation of the combine to produce long straw and windrows;
(b) Straw raking to improve the work of the balers;
(c) Baling;
(d) Road-siding;
(e) Loading and transportation to storage areas in the straw-producing area;
(f) Stockpiling in the production area;
(g) Loading in the production area and transportation to the destination;
(h) Stockpiling at the destination; and
(i) Final use of the straw.

Additionally, there are other, no less important factors which must be considered, such as the control of irrigation water, avoiding any crop burning that may spread to the areas where straw is being baled and, most importantly, how the collection and transportation is organised. The time period available is very short, as the fields must be flooded for environmental reasons and, moreover, keeping the fields dry implies energy costs due to the use of bilge pumps in the lower areas.

The handling of the combine harvester for straw baling is similar to that which would be employed for burning; that is, the cut should not be too low, in order to facilitate the subsequent drying of the straw. The machine does not chop the straw or spread it, thus facilitating its subsequent collection [5].

The straw remains can be collected directly with the balers but, if they are laid out with a windrowing machine, the subsequent work of the balers is made easier [5].

The baling can be done with small prismatic bales, which must later be grouped to facilitate mechanical handling. It can also be carried out with cylindrical bales or with large prismatic bales [16].

Road-siding is usually carried out by tractors equipped with fork loaders [5].

Transportation from the plot to the storage area at the origin is usually carried out by small trucks or tractors and trailers, due to the limited dimensions of the roads in the production area [5].

Unloading and stacking in the stockpile at the production area is usually carried out by dumping and/or manipulation with loaders coupled to a tractor, or with telehandlers. The stacking height is not usually excessive, as a long storage period is not foreseen. In addition, the straw is usually very humid, which can lead to problems with spontaneous combustion due to the high temperatures which can be reached inside the piles, if the stowage densities are too high [15].

The transportation of the straw from the production areas to the final destination is usually carried out with trucks. Tractors with loaders can be used, provided that the trucks are not excessively large or, whenever possible, they can be loaded from the sides. However, if they are large and have closed boxes, it is necessary to use telescopic handlers.

This study takes the measurements of the following aspects into consideration: the size of the straw and of the uncut remains left by the rice harvesters, the moisture of the straw, the working capacities of the rakes, prismatic balers of various sizes, the use of straw bale wrappers, the transportation of the straw, and its subsequent loading from the bed to the loading point at the foot of the plot. Transportation to the main stockpile in the production area and the working capacities of telescopic loaders in loading trucks for long-distance transport were also studied.

The transport of the straw to the final destination has a high influence on the total cost [17,18]; therefore, the main transport routes within the production area were evaluated and the costs of transportation by road from the production storage areas to possible destinations were estimated, taking the transport equipment used, the distance covered, and the bale density into consideration.

With all the above information, estimates were made for the cost of handling rice straw in L'Albufera de València, with a view to considering its economic viability for alternative uses, thus leading to an improvement in the environmental conditions of a place with such exceptional ecological value.

2. Materials and Methods

2.1. Straw Production Area and Transport Distances between the Fields and the Storage Point

The rice-producing area was divided into 13 zones. The plots are all accessed by paths which are parallel to the ditches, making transverse movements difficult. In addition, seven possible storage points were identified within the production area. These areas of harvesting and storage points were used as a reference for calculating the distances to be covered during transportation in the production area. The delimitation of these zones was carried out by direct observation of the transport routes and intermediate obstacles in Google Earth (Figure 1).

Figure 1. Image of rice crop area in the P.N. Albufera de València and the actuation zones into which it was divided (Image source: Google Earth. Modified by J.M. Giner).

Furthermore, the plot size was analysed using land register data. [19].

The centre of gravity (CG) of each zone was calculated and the shortest distance from that point to the closest road was measured (D_{CG}). The distance between the last point and the storage point was also calculated with Google Earth (Dpa). The full distance (Dp) between each production area and each storage point was calculated as the sum of the two previous values:

$$D = D_{CG} + Dp. \quad (1)$$

2.2. Straw Characterisation

Being the most abundant in the area, the biomass of the 'J. Sendra' rice variety was used.

To typify stubble, five 1 × 1 m squares were selected randomly in four fields, and the heights were measured. The stubble was cut at ground level and the samples were transferred to the laboratory, where they were weighed and dried.

To measure the amount of straw deposited by the combine harvester, four strips equivalent to the machine's cutting width (6 m) and 1 m in length were collected and weighed.

In addition, the straw baled in nine plots was estimated by counting the number of bales formed and, consequently, calculating the average weight of each bale.

Small bales (<25 kg) were weighed with an Alpha Tools scale with a capacity of 50 kg and a resolution of 10 g. The large prismatic bales were weighed with a PCE Ibérica dynamometer (http://www.pce-instrument.com, accessed on 10 October 2019) model PCE-PFG 10 K of 10 kN capacity and with a resolution of 0.5 N.

To measure the straw moisture, samples were dried in an oven at 60 °C for 48 h. The moisture, on a dry matter basis, was calculated as in Equation (2):

$$M_d = 100\,(W - DM)\,DM^{-1}, \quad (2)$$

where M_d is the percentage of moisture, W is the weight of a wet sample, and DM is the weight of the dry sample.

2.3. Equipment

The field capacity of the following equipment was measured using video records, GPS records, and posterior time analysis. The main specifications of the equipment analysed are detailed in Table 1.

Table 1. Equipment used for packaging and transporting in the production area.

Code	Description	Tractor	Implement	Activity
E1	Tractor 81 kW + Rake	New Holland T6 120	MUR, 6 m withd	Raking
E2	Tractor 70 kW + small baler	Landini REX 95 F	Welger AP 400	Baling small bales
E3	Tractor 75 kW + bale bundler	New Holland TN 95FA	Arcusin Multipack A14	Bundling small bales
E4	Tractor 125 kW + large baler	New Holland G170	NJ BB 940	Baling large bales
E5	Tractor 75 kW + Front loader	New Holland TN 95FA	Tenias S-100-2697	Road-siding
E6	Tractor 110 kW + front loader	Fendt 415 Vario	Fendt front loader	Small truck loading
E7	Small truck 28 big bales			Transport into the rice area
E8	Telehandler	JCB 535		Large truck loading
E9	Articulated large truck 64 big bales			Long distance transport

2.3.1. Raking

The straw was windrowed with a MUR rake. It has two vertical windlasses with a working width of 6 m. The rake was hitched to a New Holland T6 120, 88 kW tractor.

2.3.2. Baling and Bundling

The following balers were analysed:

- Small rectangular baler: Welger AP 400; and
- Big rectangular baler: NJ BB 940.

The small rectangular bales were packaged into groups of 12 bales using an Arcusin Multipack A14 bundler.

2.3.3. Road-Siding

The work of a New Holland TN 95FA tractor with a Tenias S-100-2697 front loader was measured when transporting the bales from the field to the road near the field.

2.3.4. Transport Equipment in the Production Zone

The most commonly used vehicles when transporting bales from the field to the storage area were small lorries. The field capacity of a truck able to transport seven rows of four large bales or bundles (28 in total) was measured using a GPS and video records.

2.3.5. Long-Distance Transport Equipment

Articulated trucks were used for long-distance transportation. The field capacity of the JCB 535 telehandler, used to load an articulated dump truck with a capacity of 68 large bales, was measured.

2.4. *Equipment Costs*

The estimation of the equipment cost was calculated according to the ASABE method [20,21].

2.5. *Long-Distance Transportation Costs*

The costs of transportation from the production area to the final destination were estimated using the data published by the 'Ministerio de Fomento' (Ministry of Development) [22].

2.6. *Total Costs*

To calculate the costs of straw handling and transport, the operations were divided into three steps (Figure 2). Step I included raking, baling, and road-siding. Step II included loading a small lorry or tractor trailer and transportation within the production area to a storage point and then being unloaded. Step III included loading to a large lorry for transportation to the final destination.

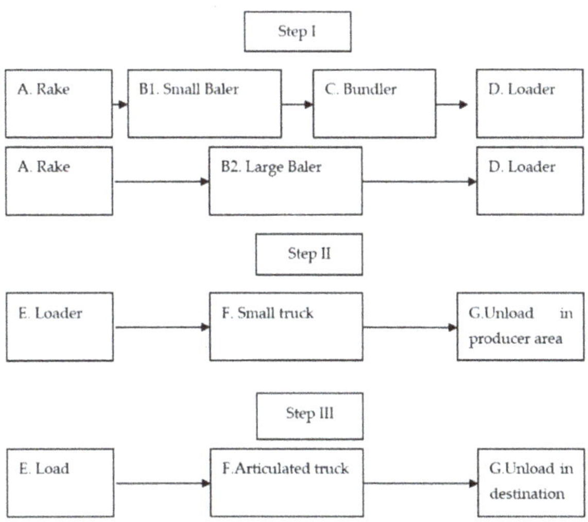

Figure 2. Diagram of operations.

The costs (EUR ha^{-1}) were calculated according to the following Equation (3):

$$C = C1 + C2 + C3, \tag{3}$$

where C is the total cost and C1, C2, and C3 are the costs of steps I, II, and III, respectively.

3. Results

3.1. Analysis of the Straw Production Area

According to the land registry data [23], in the P.N. there were 9354 registered plots, with a total area of 13,358 ha of rice crops, with an average surface of 1.43 ha/plot and a median of 0.94 ha plot^{-1}. Moreover, there are sometimes small channels and ridges between plots that make manoeuvring equipment difficult.

The vast majority of roads that the vehicles have to travel on are narrow, with single-lane traffic, making it necessary to use sidings when two vehicles need to cross on the same road. Fortunately, traffic is light, and these situations occur infrequently.

In Table 2, the surface, distances, and transport time from the harvesting zones to the storage points in the production zones are summarised. The average distance between the centre of gravity of the production area and the closest road (D_{CG}) was 1.19 km (sd ± 0.63 km), the average distance between zones to the storage point (Dpa) was 3.27 km (sd ± 1.42), and the average total distance from the road-siding to the storage point was 4.46 km (sd ± 1.36).

Table 2. Surfaces and transport distances into/within the straw production zone.

Harvesting Zone	Storage Point	Production Surface, ha	D_{CG}, km	Dpa, km	D, km
1	A1	370	1.53	4.60	6.13
2	A1	187	0.41	2.60	3.01
3	A1	720	1.79	3.10	4.89
4	A2	152	1.07	2.30	3.37
5	A2	897	1.87	2.10	3.97
6	A3	640	1.09	1.40	2.49
7	A5	1304	1.08	2.60	3.68
8	A5	2682	1.05	5.85	6.90
9	A4	666	2.65	1.90	4.55
10	A6	1476	0.73	5.90	6.63
11	A6	954	0.79	3.25	4.04
12	A6	1053	0.44	3.50	3.94
13	A7	1188	0.95	3.45	4.40
Average			1.19	3.27	4.46
Standard deviation			0.63	1.42	1.36

3.2. Straw and Stubble Characterisation

The straw and stubble were characterised (Table 3). The mean stubble height was 21 cm (sd ± 0.32), the mean wet weight was 373 g m^{-2} (sd ± 57), and the moisture on a dry basis was 89% (sd ± 49). Straw had a length of 30 cm (sd ± 4), a weight of 737 g m^{-2} (sd ± 130), and moisture on a dry basis of 78% (sd ± 37).

Table 3. Straw and stubble characteristics.

Place	Stubble			Straw				
	Height cm	Wet Mass g m^{-2}	Moisture on Dry Basis, %	Length, cm	Wet Mass, g m^{-2}	Moisture Dry Basis, %	kg ha^{-1} Wet	kg ha^{-1}, Dry
Catarroja	21.5	451	95	25	732	91	7320	3832
Albal	20.9	375	60	35	481	55	4810	3103
Romaní 1	21.6	320	45	30	459	43	4590	3210
Romaní II	21.2	348	155	28	475	125	4750	2111
Average	21.3	373	89	30	537	78	5370	3017
Standard deviation	0.32	57	49	4	130	37	1300	949

As can be seen, the straw has to be packed while it still has a high degree of moisture, a factor which can cause problems in its storage.

3.3. Straw Production

In the fields where the straw collected was measured, the results were highly variable, ranging from 1652 to 4456 kg ha^{-1} of wet straw. These values were lower than the total amount of straw produced by the fields, which have been estimated at around 6 t ha^{-1} of wet straw [3]. These differences were fundamentally due to the collection system. The balers only harvest a fraction of the deposited straw, as the farmers want a clean product without any mud. In fact, some samples of the straw that remained on the ground after packing were collected, and values to the order of 200 g m^{-2} (2 t ha^{-1}) of wet straw were found. No comprehensive measurements were made.

For subsequent machine cost calculations, an average production of 4.4 t ha^{-1} of harvested wet straw was considered.

3.4. Equipment Field Capacity

The field capacities were measured in relation to the machines used, as well as in comparison to the general capacities of similar equipment working with other winter cereals and in other countries. The terminology used was that recommended by ASABE [20].

3.4.1. Rake

The travel speed of the rake was 9.0 km h^{-1} and the working width was 6 m, giving a theoretical field capacity of 5.4 ha h^{-1}. However, as the plot dimensions are small, the field efficiency was low (estimated at 0.6), giving an effective working capacity of 3.2 ha h^{-1}.

3.4.2. Small Rectangular Baler

The baler produced 10.7 kg bales (sd $\pm$ 0.7) with dimensions 40 $\times$ 30 $\times$ 90 cm. The stubble was measured randomly to be 20 cm long. The straw was in ties. The amount of wet straw harvested was 1652 kg ha^{-1}; these values indicate low-density bales (99 kg m^{-3}), as well as a low amount of straw harvested.

The machine advanced at 2.3 km h^{-1} and the windrows were distanced at 6 m; therefore, the theoretical field capacity was 1.4 ha h^{-1}; however, frequent holdups due to problems with the tie rope led to an effective field capacity of 0.6 ha h^{-1}. Therefore, the field efficiency was as low as 42%.

The effective field capacity was similar to that reported by authors such as Márquez [23], but the field efficiency was much lower than that reported by other authors [24,25].

3.4.3. Small Bale Bundler

The small bales were grouped and packaged in the field with a bale bundler. The machine produced bundles of 12 bales, measuring 180 $\times$ 100 $\times$ 70 cm. The average speed was 3.3 km h^{-1}, the distance between bale lines was 6 m, and a field efficiency of 76% was measured; therefore, the effective field capacity was 1.5 ha h^{-1}.

3.4.4. Large Rectangular Baler

Speeds of 3.1 and 6.0 km h^{-1} were measured in 2019 and 2020, respectively. Straw lines were spaced at 7 m apart; therefore, considering an average speed of 4.5 km h^{-1}, the theoretical working capacity was 3.15 ha h^{-1}. Considering a field efficiency of 75%, the effective field capacity was 2.4 ha h^{-1}. The bales were 220 × 90 × 80 cm, with a volume of 1.58 m^3.

3.5. Road-Siding

The average distance for road-siding was 85 m (sd ± 21). In each trip, the loaders carried two large bales or bundles, and the duration of a full cycle was 3.0 min (sd ± 0.3); therefore, the theoretical field capacity was 40 bale h^{-1} (sd ± 4). If a field efficiency of 75% is assumed, the effective field capacity was 30 bale h^{-1}. At a weight of 300 kg bale^{-1}, this means 9000 kg h^{-1}, therefore, at 4.4 t ha^{-1}, the effective field efficiency will be 2.05 ha h^{-1}. Calculations were carried out with large bales.

3.6. Loading and Transporting Equipment in the Production Zone

The average transport speed observed for several rigid trucks and tractors with trailers in the production area, transporting rice between the fields and the factories, was 31 km h^{-1} (sd ± 4.3 km h^{-1}). The straw transport speed can be considered similar to that of these same vehicles transporting grain on the same roads.

3.7. Equipment Hourly Costs

In order to calculate the costs of each operation, the hourly costs of each machine were calculated beforehand, using the ASABE methodology [19]. The results are summarised in Table 4.

Table 4. Hourly costs (EUR h^{-1}) of the used equipment.

Code	Description	EUR h^{-1}
E1	Raking	39.2
E2	Small bale baling	43.9
E3	Bundling	79.0
E4	Large baler	116.6
E5	Road-siding	33.1
E6	Small truck loading	35.1
E7	Small truck, 7 × 4 big bales	45.0
E8	Telehandler	40.1
E9	Articulated truck, 16 × 4 big bales	EUR km^{-1}

3.8. Total Costs

3.8.1. Step I Costs

This step involved all the operations carried out, from raking to stacking at the roadside, including:

Raking (A);
Baling (B), field capacities of small (B1) and large (B2) rectangular balers;
Bundling small bales (C);
Road-siding with tractor equipped with front loader (D).

The registered effective field capacities are summarised in Table 5.

Table 5. Effective field capacity of each operation.

Task	Effective Field Capacity, ha h^{-1}	Equipment Hourly Costs, EUR h^{-1}	Equipment Costs, EUR ha^{-1}	Cords, EUR ha^{-1}	Total, EUR ha^{-1}
A. Raking	3.2	39.2	12.3	0	12.3
B1. Packing small bales	0.6	43.9	73.2	23.7	96.9
B2. Packing large bales	2.4	116.6	48.6	11.1	59.7
C. Grouping small bales	1.5	79.0	52.7	11.1	63.8
D. Road-siding	2.1	33.1	16.1	0	16.1

The cost (EUR ha^{-1}) of this step is

$$C1 - C_A + C_{Bi} + C_C + C_D. \quad (4)$$

Baling (B) was done with two types of machines. The costs for each type (B_1, small baler, and B_2, large baler), including the costs of the cords, are given in Table 6.

Table 6. Step I costs (EUR ha^{-1} and EUR t^{-1}).

Operations	A (EUR ha^{-1})	Bi (EUR ha^{-1})	C (EUR ha^{-1})	D (EUR ha^{-1})	C1 (EUR ha^{-1})	C1 (EUR t^{-1}) *
$A + B_1 + C + D$	12.3	96.9	63.8	16.1	189.1	43
$A + B_2 + D$	12.3	59.7	0	16.1	88.1	20

* Assuming 4.4 t ha^{-1}.

Managing small bales is considerably more expensive than managing large bales, due to the lower field capacity and the additional cost of making blocks of bales to facilitate mechanical managing.

3.8.2. Step II Costs

This step includes: (E) small truck loading, (F) transportation to storage in the production area and return, and (G) unloading.

Loading (E) was carried out with stacks four bales high. The average time to load one stack was 1.8 min (1.10–2.50 min stack^{-1}) but, as there were interruptions, the effective loading time can be estimated as 50% more (2.7 min stack^{-1}). Assuming the use of a two-axle truck or a trailer with a capacity of seven rows, the loading time would be 0.32 h.

If unloading (G) is done with a loader, it would take a similar amount of time to that of loading; therefore, we used the same value. Self-unloading trailers or lorries are not common in this area.

Transportation costs depend on the distance travelled. The distances from the geometric centres of the harvesting areas to the storage points in the production area, as well as the surfaces of the harvesting areas, were calculated, in order to estimate their straw production. Travel times were estimated as a function of distance. A travel speed of 25 km h^{-1} was assumed.

Considering a cost of 45 EUR h^{-1}—which is usual in similar equipment—the cost of this operation was calculated (Table 7). On average, for all areas, it was calculated at 44.9 EUR trip^{-1} (round trip). If the weight of the bales was 300 kg bale^{-1}, and 28 bales were carried per trip, the cost of this step would be 5.34 EUR t^{-1} or 1.6 EUR bale^{-1}.

The cost (EUR ha^{-1}) of this step was

$$C2 = C_E + C_{Ft} + C_G = 1.34 + 5.35 + 1.34 = 8.03\text{ €}\ t^{-1}, \quad (5)$$

where C2 is the cost of Step II, C_E is the cost of loading, C_{Ft} is the cost of transporting in the producer area, and C_G is the cost of unloading.

Table 7. Distances, surfaces, straw yield, and transport costs between the fields and the storage points in the production area.

Harvest Zone	Storage Point	Surface, ha	Straw Yield, t	D_{CG}, km	Dpa, km	D, km	Time, h Trip^{-1}	C_F EUR Trip^{-1}
1	A1	370	2246	1.53	4.60	6.13	1.13	50.9
2	A1	187	1137	0.41	2.60	3.01	0.88	39.6
3	A1	720	4371	1.79	3.10	4.89	1.03	46.4
4	A2	152	923	1.07	2.30	3.37	0.91	40.9
5	A2	897	5448	1.87	2.10	3.97	0.96	43.1
6	A3	640	3885	1.09	1.40	2.49	0.84	37.8
7	A5	1304	7915	1.08	2.60	3.68	0.93	42.0
8	A5	2682	16,283	1.05	5.85	6.90	1.19	53.6
9	A4	666	4041	2.65	1.90	4.55	1.00	45.2
10	A6	1476	8961	0.73	5.90	6.63	1.17	52.7
11	A6	954	5792	0.79	3.25	4.04	0.96	43.3
12	A6	1053	6393	0.44	3.50	3.94	0.96	43.0
13	A7	1188	7213	0.95	3.45	4.40	0.99	44.6
Average				1.19	3.27	4.46	1.00	44.9
Sd				0.63	1.42	1.36	0.11	4.9

3.8.3. Step III Costs

This step consisted of two parts: lorry loading and transportation to the final destination. The unloading at this point could be carried out by tilting or using the customer's own equipment; therefore, this cost was not included.

Loading can be carried out by tractors with loaders, if the truck is small, or (regardless of size, and if possible) from the sides. There are few trucks of this type in the area, the most common being articulated trucks with a closed-sided box; so, in the case of big trucks, it is usually necessary to use telescopic loaders.

The loading of articulated trucks was recorded on video and the times were analysed. An average time of 0.63 h truck^{-1} was measured. Hourly loader costs were estimated at 40 EUR h^{-1}, meaning that loading costs were 1.24 EUR t^{-1}, assuming 68 bales weighing 300 kg per truck. The average transport costs are 1.7 EUR km^{-1} [23].

Therefore, the cost of step III is the sum of loading (J) and transporting (K), the latter depending on the distance:

$$C3 = C_J + C_k = 1.24 + 1.7 \cdot 2\ d\ W^{-1}, \quad (6)$$

where C3 is the total cost [EUR (t km)$^{-1}$]; C_J is the loading cost (EUR t^{-1}); C_k is the transport cost (EUR km^{-1}), as a function of the one-way distance d (km); W is the truck load (t).

Table 8 shows an estimation of the transport costs for a range of one-way distances (25–150 km) from the stock in the production area to different destinations, calculated with a load of 20.4 t truck^{-1}.

Table 8. Transport to final destination cost, according to the one-way distance for a truck with a load of 20.4 t.

km	25	50	100	150
C3 (EUR t^{-1})	5	10	18	26

4. Discussion

The field capacities obtained in this study were poor, compared with related studies considering management of the straw of winter cereal crops such as barley or wheat and forage [23,26], included in the bibliography. The main reason for this was the difficulty of working in the small plots of L'Albufera de València, due to the use of traditional irrigation. In addition, the soft ground makes it even more challenging for machines to work well there, thus reducing overall efficiency. This agree with related studies by other authors from the USA [5,15–17].

When the individual operations were analysed, raking had an effective field capacity of 3.2 ha h^{-1}, which was on the same order as that reported by Jenkins et al. [15], ranging between 2.4 and 4.0 ha h^{-1}.

In the operation of baling with large balers, we measured 2.4 ha h^{-1}, similar to that reported by Jenkins et al. [15], ranging between 2.0 and 3.2 ha h^{-1}. When baling with a small baler, we achieved 0.6 ha h^{-1}, much lower that the values offered in previous studies (in the range of 1.2–2.0 ha h^{-1}). The reason probably relates to the concept of small bales as, in the USA, a small bale has a weight of approximately 40 kg [15] while, in our conditions, a small bale weighed between 15 and 20 kg. Small bale bundling achieved 1.5 ha h^{-1}.

In our conditions, road-siding had a capacity of 2.1 ha h^{-1}; much lower than the values given by Jenkins et al. [15] of 12–22 ha h^{-1}. The only explanation for this is that, in Valencia, small tractors are used and have to do more trips to harvest the same amount of straw.

Step I costs for large bales were 20 EUR t^{-1}, in contrast to 43 EUR t^{-1} for the small ones. The differences can be explained by the additional operation of bundling necessary to mechanically harvest the small bales. Bundling is expensive, as the machines have a high hourly cost, as well as an extra cord-related cost.

Step II costs were calculated for a bale density of 300 kg m^{-3}, resulting in 8.00 EUR t^{-1}. These costs are directly proportional to the bale density as, in any case, the maximum load capacity, in terms of weight of the transport equipment, was reached. These results agree with those of Kadam et al. [5] who, taking a density of 320 kg m^{-3} as reference, obtained a 40% increase in transportation costs when the bale density decreased to 270 kg m^{-3}, and a 20% decrease when the density increased to 400 kg m^{-3}.

Therefore, if the costs of steps I and II are combined, it is possible to compare them with the price of cereal straw at origin. The total costs of these two steps ranged between 28.1 and 51 EUR t^{-1}, which was in the range of half of the straw prices in 2019, which had an approximate value of 70 EUR t^{-1} [27]; however, in 2020, the price of cereal straw was about 42 EUR t^{-1} [28], which was less than the harvesting costs of small bales of rice straw.

Step III, the transport to destination points ranged from 5 EUR t^{-1} (for short distances of 25 km) to 26 EUR t^{-1} (for a 150 km one-way trip), using big trucks. Longer distances were not relevant, as the points of straw consumption were always to the West of Valencia, where we found the cereal growing areas, and it would not be possible to compete with either the price of rain-fed cereal or its transportation costs in years with normal straw yields.

5. Conclusions

The collection and packing processes of rice straw in the area of L'Albufera de València were found to not be very efficient, in comparison with the processes carried out for rain-fed cereal. This is due to the structural problems of the plots and the soft soil, which increased the associated costs.

Therefore, in years with low cereal straw prices, the process of collecting rice straw would run at a deficit, meaning that it would require some kind of official subsidy if farmers are to undertake the process; this also happens in other occidental rice producing countries, such as California in the USA, where straw-burning limitations have forced farmers to bale and transport straw, and the relevant authorities are subsiding these tasks [5].

However, in the years when straw reaches a high price, the costs of collecting rice straw from L'Albufera de València and transporting it to relatively close areas could be seen as a competitive activity. Furthermore, another advantage of rice straw is that, being an irrigated crop, production is regular and reliable, even in times of drought.

The packing and extraction of rice straw could, therefore, be viewed as a way to reduce the environmental impacts of burning or burying rice straw in the natural park.

Author Contributions: Conceptualization, A.T. and B.V.-M.; methodology, A.T. and J.M.G.; investigation, A.T. and J.M.G.; writing—original draft preparation, A.T.; writing—review and editing, A.T. and B.V.-M. All authors have read and agreed to the published version of the manuscript.

Funding: This research was funded by MINISTERIO DE CIENCIA, INNOVACIÓN Y UNIVERSIDADES. AGENCIA ESTATAL DE INVESTIGACIÓN, grant number RTC-2017-6249-2 titled "Desarrollo de un nuevo insumo para la agricultura sostenible: mulch de paja de arroz con incorporación de bacterias promotoras del crecimiento de las plantas (PGPB) y mecanización integral de procesos (SMART MULCH)".

Institutional Review Board Statement: Not applicable.

Informed Consent Statement: Not applicable.

Acknowledgments: We gratefully acknowledge Gestión Integral de Residuos Sólidos S.A. (GIRSA) and, in particular, the technician Pere López; as well as the farmer José Luis Pecino for facilitating the field works, and the technicians José V. Seguí and Cristina Zarzuela from the UPV.

Conflicts of Interest: The authors declare no conflict of interest.

References

1. L'albuferadevalència. Available online: http://albufera.valencia.es/en (accessed on 17 March 2021).
2. Food and Agriculture Organization of the United Nations. FAO Rice Market Monitor 2018, Volume XXI, Nº 1. Available online: http://www.fao.org/economic/est/publicaciones/publicaciones-sobre-el-arroz/seguimiento-del-mercado-del-arroz-sma/en/ (accessed on 21 January 2020).
3. Ribó, M.; Albiach, R.; Pomares, F.; Canet, R. *Alternativas de Gestión de la Paja de Arroz en la Albufera de Valencia*; Instituto Valenciano de Investigaciones Agrarias: Valencia, Spain, 2017; p. 9.
4. Sanchis, E. Emisiones de Gases en el Cultivo del Arroz: Efecto de la Gestión de la Paja. Master's Thesis, Universitat Politècnica de València, València, Spain, 2014.
5. Kadam, K.L.; Forrest, L.H.; Jacobson, W.A. Rice straw as a lignocellulosic resource: Collection, processing, transportation, and environmental aspects. *Biomass Bioenergy* **2000**, *18*, 369–389. [CrossRef]
6. Bockus, W.W.; Webster, R.K.; Wick, C.M.; Jackson, L.F. Rice residue disposal influences overwintering inoculum level of *Sclerotium oryzae* and stem rot severity. *Phitopatology* **1979**, *69*, 862–865. [CrossRef]
7. Chen, J.; Mohamed, E.; Yu, T.; Ren, Y.; Guo, Z.; Liu, S. Research progress of wheat straw and rice straw cement-based building materials in China. *Mag. Concr. Res.* **2017**, *70*, 84–95. [CrossRef]
8. Kaur, P.; Singh Kocher, G.; Sachdeva Taggar, M. Enhanced bio-composting of rice straw using agricultural residues: An alternate to burning. *Int. J. Recycl. Org. Waste Agricult.* **2019**, *8*, 479–483. [CrossRef]
9. Sarnklong, C.; Cone, J.W.; Pellikaan, W.; Hendriks, W.H. Utilization of Rice Straw and Different Treatments to Improve Its Feed Value for Ruminants: A Review. *Asian-Aust. J. Anim. Sci.* **2010**, *23*, 680–692. [CrossRef]
10. Prodoscimi, M.; Jordán, A.; Tarolli, P.; Keesstra, S.; Novara, A.; Cerdà, A. The immediate effectiveness of barley straw mulch in reducing soil erodibility and surface runoff generation in Mediterranean vineyards. *Sci. Total. Environ.* **2016**, *547*, 323–330. [CrossRef] [PubMed]
11. Cerdà, A.; González-Pelayo, Ó.; Giménez-Morera, A.; Jordán, A.; Pereira, P.; Novara, A.; Orenes, F.G. Use of barley Straw residues to avoid high erosion and runoff rates on persimmon plantations in Eastern Spain under low frequency–high magnitude simulated rainfall events. *Soil Res.* **2016**, *54*, 154–165. [CrossRef]
12. Bakker, M.M.; Govers, G.; Rounsevell, M.D. The crop productivity–erosion relationship: An analysis based on experimental work. *Catena* **2004**, *57*, 55–76. [CrossRef]
13. Ramakrishna, A.; Tam, H.M.; Wani, S.P.; Long, T.D. Effect of mulch on soil temperature, moisture, weed infestation and yield of groundnut in northern Vietnam. *Field Crop. Res.* **2006**, *95*, 115–125. [CrossRef]
14. Abril, D.; Navarro, E.; Abril, A. La paja de arroz. Consecuencias de su manejo y alternativas de aprovechamiento. *Rev. Fac. Agron.* **2009**, *17*, 69–79.
15. Jenkins, B.M.; Bakker-Dhaliwal; Summers, M.D.; Bernheim, L.G.; Lee, H.; Huisman, W.; Yan, L. Equipment performance, costs, and constraints in the commercial harvesting of rice straw for industrial applications. In Proceedings of the 2000 ASAE Annual International Meeting, Milwaukee, WI, USA, 9–12 July 2000; pp. 1–27.
16. Jenkins, B.M.; Toenjes, D.A.; Dobie, J.B.; Arthur, J.F. Performance of large balers for collecting rice straw. *Trans. ASAE* **1985**, *28*, 360–363. [CrossRef]
17. Summers, M.D.; Jenkins, B.M.; Hyde, P.R.; Williams, J.F.; Mutters, R.G.; Scardacci, S.C.; Hair, M.W. Biomass production and allocation in rice with implications for straw harvesting and utilization. *Biomass Bioenergy* **2003**, *24*, 163–173. [CrossRef]

18. Velázquez, B.; Torregrosa, A. Logistic models for distribution of straw in crops of fruit tree plots where mulch is applied. *Comput. Electron. Agric.* **2020**, *175*, 14. [CrossRef]
19. Sede electrónica del Catastro. Available online: https://www.sedecatastro.gob.es (accessed on 5 November 2019).
20. ASABE. *Agricultural Machinery Management Data*; ASAE D497.7, (R2015); ASABE: St. Joseph, MI, USA, 2011; p. 15.
21. ASABE. *Uniform Terminology for Agricultural Machinery Management*; ASAE S495 DEC01; ASAE Standards; ASAE: Washington, DC, USA, 2002; pp. 364–365.
22. Ministerio de Fomento. Observatorio de Costes del Transporte de Mercancías por Carretera. Ministerio de Transportes, Movilidad y Agenda Urbana 2017. Available online: https://www.fomento.es/MFOM.CP.Web/handlers/pdfhandler.ashx?idpub=TTW103 (accessed on 4 September 2020).
23. Márquez, L. *Maquinaria Agrícola Para la Recolección*; HB Ediciones: Madrid, Spain, 2014; p. 260.
24. Dobie, J.B.; Miller, G.E.; Mosley, R.H., III. Ground level harvest of rice straw. *Trans. ASAE* **1984**, *27*, 1263–1269. [CrossRef]
25. Pal, R.; Kumar, R.; Jalal, R.K.; Kumar, A. Economic and Environmental Performance of Straw Baler for Collection of Rice Residue Generated after Mechanical Harvesting by Combine Harvester. *Curr. J. Appl. Sci. Technol.* **2019**, *37*, 1–6.
26. Cédra, C. *Les Matériels de Récolte des Fourrages, Ensilage & Distribution*; Lavoiser Tec & Doc.: Paris, France, 1995; p. 395.
27. Rojo, S. La Tonelada de Paja se Está Pagando 33 Euros Más Cara que Hace un año. *El Norte de Castilla*; 30 August 2019. Available online: https://www.elnortedecastilla.es/castillayleon/tonelada-paja-pagando-20190830081243-nt.html (accessed on 18 December 2019).
28. Yébenes, J. La Paja de Cereales, sin Cotización por la Gran Oferta de Producción y la Menor Demanda de la Ganadería. *Lanza Diario de La Mancha*. Available online: https://www.lanzadigital.com/campo-2/la-paja-de-cereales-sin-cotizacion-por-la-gran-oferta-de-produccion-y-la-menor-demanda-de-la-ganaderia/ (accessed on 14 December 2020).

agriculture

MDPI

Review

Current Knowledge and Future Directions for Improving Subsoiling Quality and Reducing Energy Consumption in Conservation Fields

Shangyi Lou [1], Jin He [1,*], Hongwen Li [1], Qingjie Wang [1], Caiyun Lu [1], Wenzheng Liu [2], Peng Liu [1], Zhenguo Zhang [3] and Hui Li [4]

1 College of Engineering, China Agricultural University, Beijing 100083, China; xxshangy1@163.com (S.L.); lhwen@cau.edu.cn (H.L.); wangqingjie@cau.edu.cn (Q.W.); lucaiyun@cau.edu.cn (C.L.); liupeng_ctrc@cau.edu.cn (P.L.)
2 College of Enology, Northwest A&F University, Yangling 712100, China; lwzheng@nwafu.edu.cn
3 College of Mechanical and Electrical Engineering, Xinjiang Agricultural University, Xinjiang 830052, China; bigbearsun@163.com
4 Shandong Academy of Agricultural Machinery Sciences, Shandong 250100, China; Lihui7777777@163.com
* Correspondence: hejin@cau.edu.cn; Tel.: +86-10-62-737-300

Abstract: Subsoiling has been acknowledged worldwide to break compacted hardpan, improve soil permeability and water storage capacity, and promote topsoil deepening and root growth. However, there exist certain factors which limit the wide in-field application of subsoiling machines. Of these factors, the main two are poor subsoiling quality and high energy consumption, especially the undesired tillage depth obtained in the field with cover crops. Based on the analysis of global adoption and benefits of subsoiling technology, and application status of subsoiling machines, this article reviewed the research methods, technical characteristics, and developing trends in five key aspects, including subsoiling shovel design, anti-drag technologies, technologies of tillage depth detection and control, and research on soil mechanical interaction. Combined with the research progress and application requirements of subsoiling machines across the globe, current problems and technical difficulties were analyzed and summarized. Aiming to solve these problems, improve subsoiling quality, and reduce energy consumption, this article proposed future directions for the development of subsoiling machines, including optimizing the soil model in computer simulation, strengthening research on the subsoiling mechanism and comprehensive effect, developing new tillage depth monitoring and control systems, and improving wear-resisting properties of subsoiling shovels.

Keywords: subsoiling machine; subsoiling shovel design; anti-drag; tillage depth detection; tillage depth control; tillage depth stability; soil mechanical interaction

Citation: Lou, S.; He, J.; Li, H.; Wang, Q.; Lu, C.; Liu, W.; Liu, P.; Zhang, Z.; Li, H. Current Knowledge and Future Directions for Improving Subsoiling Quality and Reducing Energy Consumption in Conservation Fields. *Agriculture* **2021**, *11*, 575. https://doi.org/10.3390/agriculture11070575

Academic Editor: José Pérez-Alonso

Received: 21 May 2021
Accepted: 20 June 2021
Published: 23 June 2021

1. Introduction

As the important factor in crop growth, soil is a valuable agricultural resource, a significant production factor, and an indispensable foundation supporting the sustainable development of agriculture [1]. The ideal farmland soil is composed of 50% soil particles with organic matter and 50% pores; additionally, with respect to pores, the moisture and air respectively account for 25% [2]. However, due to long-term conventional ploughing practices, hardpan has been shaped, which is a typical form of soil compaction [3,4]. Consequently, the failure layer is formed, resulting in the rearrangement of soil particles under external forces [5,6], and giving rise to a reduction in soil porosity and an increase in bulk density [7–9]. The root-system penetrability is affected by soil compaction, and especially the high-level compaction inhibits the root growth [10]. The hardpan binds root proliferation, reduces root penetration, and decreases root length and dry matter [11]. In addition, the key enzymes required for plant respiration show a downward trend with increasing compaction stress, which leads to impaired root function and the loss of nutrient absorption [12–15].

Subsoiling can break the hardpan and alleviate the soil compaction without turning ploughing soil and disrupting the original topsoil structure [16]. Compared with conventional ploughing, the subsoiling tillage at 30 cm, 35 cm, and 40 cm decreased the mean bulk density by 4.59%, 7.13%, and 8.27% and reduced soil compactness by 17.62%, 23.63%, and 36.42%, respectively [17]. Meanwhile, subsoiling increased soil porosity, enhanced soil infiltration and water storage capacity, and decreased water inputs per growing day [18]. Additionally, in comparison with the rotary tillage, the subsoiling practice significantly increased the average root length density by 13% due to the better root penetration in both horizontal and vertical directions. Moreover, the average grain yield and dry matter weight of maize were respectively increased by 6.3% and 3.7% [19]. Furthermore, subsoiling improved the activity of plant-protective enzymes in maize root, reduced the degree of peroxide in cell membrane substances, and delayed the senescence of maize root. It is conducive to maintaining root system vitality in the later growth stage of maize [20]. Therefore, subsoiling is classified as a resource-saving and environmental-friendly technology, and is significant in promoting the sustainable development of agriculture [21–24].

In the early 1930s, some countries in North America and Europe began to use subsoiling technology to solve soil problems. Thanks to the significant effects of subsoiling on the repair and protection of arable land, the combination of tillage methods and comprehensive utilization of subsoiling technology has become increasingly close in these regions. The proportion of arable land that adopted subsoiling in Europe was 15.4% in 2005. Meanwhile, the proportion in North America exceeded 40% in 2006. Until 2015, the proportion in European and American areas had reached 60% [25]. In some Asian countries, subsoiling technology was adopted relatively late. To address the issue of the crop yield decline caused by soil compaction, China has promulgated a series of associated regulations to promote subsoiling and land preparation technologies since 2009 (Table 1). By the end of 2018, the subsoiling was practiced on 10.6 Mha, accounting for 7.41% of the total arable land [26]. At present, subsoiling technology is being widely used in scientific experiments and actual productions of various crops such as wheat, corn, cotton, sugar cane, tobacco, and soya beans.

Table 1. Relevant development regulations of subsoiling and land preparation technology in China.

Regulation	Related Content
Ministry of Agriculture and Rural Affairs: National agricultural machinery subsoiling and land preparation operation implementation plan (2016–2020) [27]	In 2016–2020, the nationwide annual operation area of subsoiling and land preparation via agricultural machinery exceeds 10 million ha
State Council: Government Work Report [a] (2015) [28]	Promote land renovation, add 13.33 million ha land under subsoiling and land preparation
State Council: Government Work Report (2014) [29]	Launch a pilot project to adopt subsoiling and land preparation technology covering 6.67 million ha
State Council: Central Document No.1 [b] (2014) [30]	Give great impetus to advance mechanization of subsoiling and land preparation
Ministry of Agriculture: National agricultural machinery subsoiling and land preparation operation implementation plan(2011–2015) [31]	A total of 71.3 million ha area in the nation will be subsoiled in 2011–2015
State Council: Opinions of the State Council on promoting sound and rapid development of agricultural mechanization and agricultural machinery industry (2010) [32]	Implement pilot projects to subsidize subsoiling and land preparation at suitable regions
State Council: Central Document No.1 (2010) [33]	Give great impetus to extend subsoiling and land preparation machinery
Ministry of Finance: Interim measures for the administration of the centralized use of the newly increased central funds for comprehensive agricultural subsidies for the building of basic grain capacity (2009) [34]	Bring conservation tillage methods such as subsoiling and land preparation into the key support scope of the newly increased agricultural funds for comprehensive subsidies
State Council: Central Document No.1 (2009) [35]	Implement pilot projects to subsidize the mechanized operation in the key link

[a] "Government Work Report" is a form of the official document of the Government of the People's Republic of China, which clearly points out the government's tasks for the current year; [b] "Central Document No.1" is the name originally given to the first document issued by the central authority every year, and focuses on the development of agriculture, rural areas, and farmers in 1982.

Subsoiling machines are indispensable to mechanized subsoiling, and their performance directly affects the hardpan's broken quality, topsoil structure, soil permeability, soil microbial quantity, root growth, and crop yield [36]. According to their working function, subsoiling machines can be divided into single-function subsoiling machines and multi-function subsoiling machines. The single subsoiling machines can be further divided into two categories: subsoiling ploughs and omni-directional subsoilers [37]. Some typically available subsoiling machines around the world are discussed in Table 2. Subsoiling machines in certain countries with large per-capita arable land areas (e.g., America, Canada, Australia, Russia) are mainly matched with high-power tractors which have several advantages, such as large tillage width, fast operation speed, high operation efficiency, and advanced machining technology. On the contrary, for those countries with small per-capita arable land areas, such as China and Japan, the per-capita arable land areas are respectively only about 0.08 ha and 0.03 ha, which are merely approximately one-sixth and one-twentieth of that in America, and about 0.038 and 0.014 times as large as Australia, respectively [38]. The subsoiling machines are mainly matched with small or medium-power tractors. The features of subsoiling machines in these countries include small size, light weight, flexible operation, and low cost. Nevertheless, current subsoiling machines have some problems with their performance, such as poor soil loosening quality, undesired soil disturbance, unstable tillage depth, high power consumption, and rapid wear of shovels.

Table 2. Comparison of different kinds of subsoiling machines.

Working Function	Type	Company	Country	Mechanism	Working Width (mm)	Matched Power (kW)	Features
Combined machine	DIABLO [39]	Maschio	Italy	Two-row winged shovels; Stubble breaking discs; Rear rollers with two rows of discs	5000–7000	300–400	Having ample clearance between shovels and rear discs
Combined machine	2730 [40]	John Deere	USA	Two rows of stubble breaking discs arranged symmetrically; Spring-tooth harrows; A suppress roller	4300–6800	250–460	Lower operation costs
Subsoiling plough	AP31 [41]	Agrowplow	Australia	Nine deep tillage ploughs; Two supporting wheels	2310–2970	73.5–132.3	Bolt-less for the quick changing of ploughs and points
Subsoiling plough	9200 [42]	Salford	Britain	Seven shovels with 0.99m spacing	5320	206	Equipped with automatic spring reset obstacle protection system
Omni-directional subsoiler	1S-300C [43]	Dahua	China	The chisel shovel with left and right winged shovels; Three rows beams; A suppress roller; Two gauge wheels	3000	147–191.1	With overload protection
Omni-directional subsoiler	1SQ-330 [44]	Aolong	China	Six side bended shovels; Two gauge wheels; A suppress roller	3300	99.2–154.4	Tips of two shovels arranged symmetrically forms an inverted trapezoid

In order to improve subsoiling quality and reduce energy consumption in the field with cover crops, several studies have been conducted on subsoiling shovel structure, anti-drag technologies, tillage depth detection and control, and soil-mechanical interaction [45–51]. Various subsoiling shovels are capable of meeting the needs of different regions in different countries. Reducing tillage resistance during subsoiling makes subsoiling machines more prominent with the advantages of resource saving and being environmentally friendly. Modern sensor detection technology and electronic-hydraulic control technology are used to accurately control the tillage depth of subsoiling machines, which benefits in terms of conserving moisture and preparing a good seedbed [19,52–55]. Additionally, soil-mechanical interaction mechanisms presented in micro and macro ways, mainly clarified via theoretical analysis and simulation, are an important means to realize the efficient development of the abovementioned studies. The integration of abovementioned technologies contributes greatly to the improvement of subsoiling quality and the reduction of energy consumption in the field with cover crops.

The goal of this study was to comprehensively review the existing literature related to technologies for improving subsoiling quality and reducing energy consumption currently being used in subsoiling machines. This article was based on the following aspects and organized as follows. Section 2 states the methods of design and optimization of subsoiling shovels, commenting on the different types of available shovels. Then, Section 3 explores five ways to reduce tillage resistance. Sections 4 and 5 provide a review of tillage depth detection and control, applying sensor detection technology and electro hydraulic control technology. Section 6 delves into the application of theoretical analysis and discrete element analysis in research on soil-mechanical interaction. Finally, in Section 7, recommendations are given for the future development of subsoiling machines.

2. Subsoiling Shovel Design

Subsoiling shovels are the main working parts of a subsoiling machine [37]. Their design has a significant influence on subsoiling quality and tillage resistance [56,57]. The common types of commercial shovels are chisel, winged, "V" shaped, and side bended shovels. Through optimizing the penetrating angle, the opening angle of winged shovels, and the shovel-handle structure, some subsoiling shovels with better working performance have been developed. The comparisons among these shovels are shown in Table 3.

Table 3. Subsoiling shovel type, model, country, working width, working depth, matched power, number, and feature.

Type	Model	Country	Working Width (cm)	Working Depth (cm)	Matched Power (KW)	Number	Feature
chisel	1S-225C [58]	Germany	200	50–65	55–129	3	Adjustable of chisels for any tractor track width
chisel	1S-150A (Figure 1a) [59]	China	150	25–30	36.8–51.5	3	Low tillage resistance
winged	PINOCCHIO200 [60]	Italy	130–300	45	50–160	3–7	Available in different widths and shovel numbers
winged	1S-200 (Figure 1b) [61]	China	200	300–400	70	5	Good flow of soil under the shovel
V shaped	1SQ-340 (Figure 1c) [62]	China	155	40–50	73.6–88.3	3	Forming mole drains at the bottom with smaller resistance
side bended	DALBO Ratoon [63]	Denmark	120–300	55	30–120	2–8	Low power requirement
side bended	1S-250 (Figure 1d) [64]	China	250	25–50	89–106.6	6	Special cambered surface

In order to improve quality, reduce resistance, and increase efficiency, a cambered subsoiler shank was designed by Chen [65]. The test showed that the subsoiling resistance was the smallest when the sweep angle was 50°. Based on the orthogonal experiment's results, i.e., the penetrating angle had a significant effect on subsoiling resistance, Liu et al. [47] optimized the chisel shank's structure and found that the resistance was minimal when the rack angle was 21°. Zhao et al. [66] designed a fitting curve-shaped shank and clarified the effect of the shovel tip on the soil particle's movement. The results of the comparison test between the fold-line-typed shank and the circular shank showed that the fitting curve-shaped shank effectively reduced soil disturbance and tillage resistance. By simulation analysis and digitized soil-bin experiments, Wang et al. [67] demonstrated that, on the circular shank, the optimum installing height of the winged shovel was 75 mm.

The shank type, penetrating angle, opening angle of the winged shovel, and shovel-handle structure are the main factors influencing the tillage resistance of subsoiling shovels. In order to improve subsoiling quality and reduce tillage resistance, these factors are used as the target parameters. Through subsoiling simulation and soil-bin and field trials, the effect of shovel structure on the tillage resistance as well as soil disturbance were analyzed and their impacts were clear. Then, these target parameters were optimized on the basis of the above results, and the shovel structure was accordingly improved to obtain better subsoiling quality and lower tillage resistance. Design and optimization of subsoiling shovels has always been one of the key technologies of subsoiling machines.

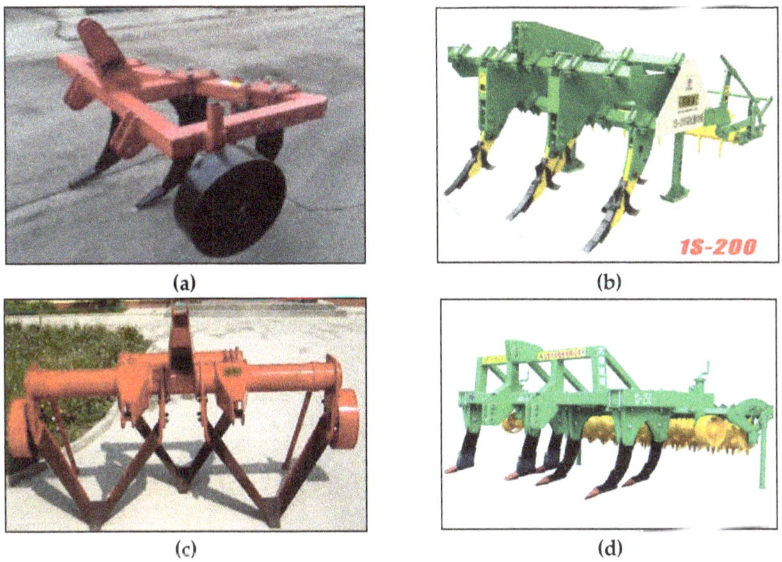

Figure 1. (**a**) Chisel shovels of the Hedongxiongfeng 1S-150A subsoiler (http://www.hedongxiongfeng.cn/pd.jsp?id=52#_pp=103_366 (accessed on 15 September 2020)); (**b**) Winged shovels of the Zhongnong-boyuan 1S-200 subsoiler (http://www.boyomac.com/product/17.html (accessed on 14 May 2021)); (**c**) V shaped shovels of the Woye 1SQ-340 subsoiler (https://www.nongjitong.com/product/3072.html (accessed on 15 September 2020)) (**d**) Side-bended shovels of the DaHua BaoLai 1S-250 subsoiler (https://www.dhbl.net/product/53.html (accessed on 15 September 2020)).

3. Anti-Drag

The energy consumption caused by subsoiling is three to five times as much as that of other planting procedures such as seeding, managing, and harvesting [68]. Decreasing tillage resistance during subsoiling is the main measure to reduce energy consumption. Several methods of decreasing resistance are available, such as vibration anti-drag technology, bionic anti-drag technology, layered subsoiling structure, shovel design with anti-drag structure, and surface coating technology of shovels.

3.1. Vibration Anti-Drag

Vibration anti-drag refers to adding an excitation source on the subsoiling machine to make subsoiling shovels constantly subject to the repeated positive and negative forces. The greatest advantage of the vibration subsoiling machine is its low traction resistance. According to the driving mode of excitation source, vibration machines can be divided into self-excited and forced-excited categories. Shovels of the forced-excited vibration subsoiler are capable of vibrating via connecting an eccentric mechanism. Sun et al. [69] adopted two eccentric discs to drive connecting rods and shovels reciprocating vibration in the vertical direction. The minimum resistance and energy consumption were obtained under the optimum combination of forward speed, vibration frequency, amplitude, angle, and speed ratio [70–73]. A typical forced-excited vibration subsoiler is shown in Figure 2. Self-excited vibrating can be obtained by applying elastic elements, such as pressure springs (Figure 3a), leaf springs (Figure 3b), and hydraulics [74–76] (Figure 3c). This anti-drag method not only has little impact on soil, but also protects shovels and other components [77].

Figure 2. DaHua BaoLai 1SZ-180 forced-excited vibration subsoiler (http://www.nongjitong.com/product/dahua_1sz-270_deep_loosening_machine.html (accessed on 12 November 2020)).

(**a**) (**b**) (**c**)

Figure 3. (**a**) Great Plains SS1300 subsoiler (http://www.greatplainsmfg.com.ua/en-gb/products/710/sub-soiler-narrow-tillage (accessed on 8 December 2020)); (**b**) Kverneland GLG-II subsoiler (https://www.kverneland.cn/node_80887/node_80897/node_80958# (accessed on 8 December 2020)); (**c**) AGRI-WELD subsoiler (https://www.plowmanbrothers.com/plowman-agricultural/ (accessed on 8 December 2020)).

3.2. Bionic Anti-Drag

The primary task of bionic anti-drag is to extract biological information, which includes the function, structure, process, or behavioral characteristics and mechanisms of the biological system. On this basis, a technology similar to the function of biological-systems was developed and applied to shovels to reduce resistance [78]. The contour line of the claw toe from house mice (Mus musculus) was extracted and found to feature an exponential function curve, and Zhang et al. [79] converted the upper contour line of the longitudinal profile of their (Mus musculus) claws into the technical parameters of the shovel's cutting edge curve. Then, the contour curve was enlarged properly and applied to the structural design of the cutting edge of the shovel handle. The comparative field experiment between the shovel with the feature of an exponential curve and the conventional shovel showed that the tillage resistance of the bionic shovel was reduced by 8.5–39.5% compared with the traditional shovel. Li et al. [80] obtained the accurate shape and dimension of a bear claw by a three-dimensional laser scanner and established a claw model. By the discrete element simulation, they found that the overall performance of the claw was the best when the rack angle was 30°. A bionic vibratory subsoiler was designed by Zhang et al. [81], which combined vibration anti-drag with bionic anti-drag. It was reported that the draft force of the bionic vibratory subsoiler was respectively reduced by 13–18% and 8.5–39.5% compared with the 1SZ-460 lever-type subsoiler and bionic anti-drag subsoiler.

In addition to the above two methods, the layered subsoiling structure is capable of reducing tillage resistance and mainly depends on two shovels. Along the forward direction, the shovel with a small depth is in front, while the shovel with a large depth is in the back [82,83]. The shovel design with an anti-drag structure optimized the handle or tip via experiment or simulation, thereby obtaining better subsoiling quality and less tillage resistance [84,85]. The surface-coating technology aims at obtaining better soil-removal performance of shovels by changing the properties of surface material, hence decreasing tillage resistance [86].

The above mentioned anti-drag technologies can effectively reduce resistance, whereas in studies of reducing shovel resistance, less consideration is given to the thickness of the three soil layers (tillage layer, hardpan, subsoil layer) and the soil's physical properties, as well as the soil smashing process. Apart from reducing tillage resistance, the power consumption of subsoilers and straw blocking of shovels should also be reduced. Although the forced-excited vibration subsoiler is capable of decreasing the traction, the required fuel consumption is much higher. In order to balance the traction and energy requirements and to improve the overall efficiency, better methods should be explored through further studies so as to drive the vibrator.

4. Tillage Depth Detection

Initially, the tillage depth could only be measured manually, which had great labor intensity and low detection accuracy. Additionally, these measured depth data were limited and discrete, and could therefore not reflect continuous changes including the variation of tillage depth with surface and time; consequently, it is unable to study the dynamic characteristics of depth. To this purpose, sensors, which have higher detection accuracy and provide a basis for depth adjustment of subsoiling machines, are used for measuring tillage depth. At present, there are two main kinds of sensors used in tillage depth detection: ultrasonic sensors and inclination sensors.

4.1. Tillage Depth Detection Based on Ultrasonic Sensor

The detection method of ultrasonic sensors utilizes a constant speed transmission of acoustic waves. The signal generator generates acoustic waves with a certain oscillation frequency and which travels through the air at a constant speed. The ultrasonic echo is generated when the acoustic wave encounters impurities and interfaces, and then it is received by the signal-receiving end. Therefore, the distance is obtained by calculating the formula of time interval and sound speed [87].

Tests were carried out to evaluate detection accuracy of the ultrasonic sensor on different fields by Mouazen et al. [88], and the ultrasonic sensor was installed on the bottom of the frame. Compared with the manual measurement values, it was found that the detection accuracy of the ultrasonic sensor was better in the soft sandy loam, while it was lower in the soil with cover crops. A tillage depth measurement device based on the ultrasonic sensor was designed by Li [89], and proved to behave better than the resistance-stain type measurement device in terms of measuring accuracy. Adamchuk et al. [90] developed a closed-loop automatic control system for tillage depth, and the ultrasonic sensor was applied to measure tillage depth. An online measuring device of subsoiling depth was proposed in the Chinese Academy of Agriculture Mechanization Sciences [91]. It used an ultrasonic sensor to detect the distance between the frame of the subsoiling machine and the surface, and thus the tillage depth was obtained. A tillage depth detection system was designed by Suomi et al. [50], which used an ultrasonic sensor installed in front of the frame with an inclination sensor mounted on the connection arm of the gauge wheel and the frame to measure the height and the angle; the depth was obtained by calculating the formula related to these two physical properties. The field experiment showed that the detection error of this system was within 10 mm.

4.2. Tillage Depth Detection Based on Inclination Sensor

Inclination sensors determine the position of the object according to the three-dimensional angle change, which should be attached to the objects' surface to measure the inclination of the object from the horizontal plane [92]. When the inclination sensor is used to detect tillage depth, it is often on the hitch device or profiling mechanism to measure the angle change. Then, the tillage depth is obtained by formula calculation.

A detector, consisting of a frame, an inclination sensor, and a sliding plate which undulated with the surface, was designed by Zhao et al. [93] and used to measure tillage depth. By detecting the relative angle between the frame and the tractor, the depth was acquired through geometric modelling of the depth and angle. Li et al. [94] developed a measurement device, and the inclination sensor was on the linkage of the frame and gauge wheel. By angle measurement and calculation, the tillage depth was obtained. Results showed that the error of this measurement method was within 6%. Xie et al. [95] installed inclination sensors on the lift arm of a tractor to detect angle change from the horizontal plane. The actual tillage depth was calculated according to the inclination of the lift arm and the geometrical dimension relation of the linkage mechanism. In addition to the abovementioned detection methods based on ultrasonic sensors and inclination sensors, potentiometers and encoders, of which the detection principle is similar to that of the inclination sensor, are also used to detect tillage depth [96,97].

In existing research on the technology of tillage depth detection, ultrasonic sensors and inclination sensors are widely used because of their low cost, simple working principles, and convenient installation methods. However, there still exist some problems. The detection accuracy of ultrasonic sensors would be affected by weeds, straws, and clods, which could lead to unstable detection results. In addition, inclination sensors can be used to measure the horizontal dip angle of the rear-suspension lifting arm. Although this method avoids the negative influence of surface unevenness and other abovementioned factors on detection accuracy, other problems may appear. As there are many connecting rods between the tractor boom and the suspension unit, drivers may adjust the length of the lift rod and the upper pull rod in practice, which leads to a change in the geometric parameter. Thus, it is inconvenient to use this method due to the need to recalibrate the sensor. Besides, the method, applying inclination sensors to measure the angle change of the profiling mechanism and calculate the mathematical model to get the tillage depth, results in the mismatch between the adjustment completion point and detection point of tillage depth. This is because the profiling mechanism is generally placed behind the tillage component, which leads to the delay of profiling.

5. Tillage Depth Control

Because of the complexity and variability of the field-work environment, the tillage depth often needs to be adjusted many times to achieve stable and consistent outcomes. It is essential to develop the control technology of tillage depth to obtain precise and efficient adjustments. Currently, there are two major control modes of tillage depth: the method of unified adjustment of each row and the method of independent adjustment of a single row. The adjusting mechanism belonging to the first adjustment mode is based mainly on the attachment position of the tillage machine and its tractor. This method commonly combines the hydraulic device of three-point suspension with mechanical and electrical technology to achieve automatic control of tillage depth [98]. With regard to the method of independent adjustment of a single row, separate adjustment mechanisms, hydraulic mechanisms, or electrical mechanisms need to be designed and installed on the machinery, and the number of mechanisms is equal to the number of subsoiling shovels. This method is capable of avoiding undesired tillage depth of each subsoiling shovel.

5.1. Method of Unified Adjustment of Each Row

The method of unified adjustment of each row aims at adjusting the tillage depth of the whole machine (adjusted on the three-point suspension) or a group of tillage components of the combined land-preparation equipment. Therefore, the hydraulic suspension system with electronic control technology (EHC) is an important support for this mode. The method of controlling depth using EHC mainly involved three control methods, including force control, position control, and integrated control of the two abovementioned modes. Long-term research on the automatic control of tillage depth based on these three methods has been conducted. In order to improve control accuracy, Lee et al. [48] developed a tillage depth control system. Some sensors were used to detect inclination of the three-point suspension, pitch angle of the tractor, and distance from the transducer to the surface. The drive circuit of the electromagnetic valve was controlled by a controller to change the state of the hydraulic circuit, and thus the position of the three-point suspension was adjusted. For the electro-hydraulic suspension system composed of a cartridge valve, a suspension control scheme based on the Controller Area Network (CAN) bus was proposed by Xie et al. [99] to improve the control accuracy, which mainly included the suspension subsystem ECU and two intelligent nodes. The experiment demonstrated that the developed nodes were capable of meeting the working requirements of the suspension system. Nie et al. [100] developed a tillage depth automatic control system, which was based on the original hydraulic suspension system and applied sensors, microcomputers, and stepping motors to realize automatic control. Moreover, the reliability and stability of the system were verified by experiments.

In addition, combined tillage equipment with the function of monitoring and controlling depth was researched and developed by some agricultural machinery enterprises such as OPICO, John Deere, and CASE. The HE-VA combined land preparation machine consists of groups of subsoiling components and disc harrows, of which the tillage depth is controlled by adjusting the height of the suppress roller through the hydraulic system [101]. John Deere developed a TruSet monitoring system for the land preparation equipment. Depth detection sensors, which are mounted on the supporting wheels and frame, are capable of accurately measuring the depth of the whole machine. Additionally, the depth of each group of tillage components is adjusted via the matched AccuDpthTM hydraulic system [102]. The AFS (Advanced Farming System) tillage depth intelligent monitoring system was developed by CASE, and is capable of precisely adjusting the tillage depth of the whole machine according to different soil conditions [103]. Before operation, the prescription map containing the predetermined tillage depth information of the field is input into the system. Owing to the tractor being equipped with a GPS positioning system, when the equipment moves to the corresponding position, the tillage depth is automatically adjusted by the hydraulic system to the required tillage depth. A stroke-detection sensor is installed in the hydraulic cylinder to provide real-time access to length changes of the hydraulic cylinder, and this is the key to the precise control of tillage depth.

5.2. Method of Independent Adjustment of Single Row

The abovementioned technologies in Section 5.1 are capable of adjusting the tillage depth of the whole machine or a group of tillage components of the combined land preparation equipment. However, due to surface relief or soil resistance variation, there exists the problem of lateral instability of tillage depth of the machine with several subsoiling shovels, which leads to the inconsistency of soil conditions in the lateral distribution. Therefore, it is necessary to develop subsoilers with special subsoiling assemblies that are capable of independently adjusting tillage depth of each subsoiling shovel. The method of independent adjustment of a single row is capable of avoiding undesired tillage depth of each subsoiling shovel to improve the tillage depth stability between rows.

A universal-type device which automatically monitors and controls the tillage depth of each subsoiling shovel was developed [104]. Laser-ranging sensors were used to measure the distance between the stand and surface. Then, the corresponding signal was sent to the Programmable Logic Controller (PLC) to be analyzed and calculated, and the real-time tillage depth was recorded. Comparing the real depth with the set value, the PLC sent an execution signal to the driver circuit of the electromagnetic valve to change the solenoid valve switch, and the hydraulic cylinder was adjusted to extend or shorten. Therefore, the tillage depth, in real time, was controlled to be consistent. Wang et al. [105] designed an electric-hydraulic control system which was capable of adjusting the tillage depth of subsoiling assemblies in a timely manner (Figure 4a). Two inclination sensors were used to detect the dip angle in the horizontal direction, and were mounted horizontally on the subsoiler frame and inclined frame of the suppress roller, respectively. Through internal algorithm, the tillage depth was calculated. The hydraulic cylinder performed the action, elongating or shortening to obtain desired tillage depth. Moreover, the performance of the system was evaluated in the aspect of the tillage depth stability between rows and within rows. Wu [106] developed a subsoiling machine to automatically monitor and control tillage depth based on ultrasonic sensors and hydraulic drivers. It was capable of independently monitoring and controlling the depth of each subsoiling shovel.

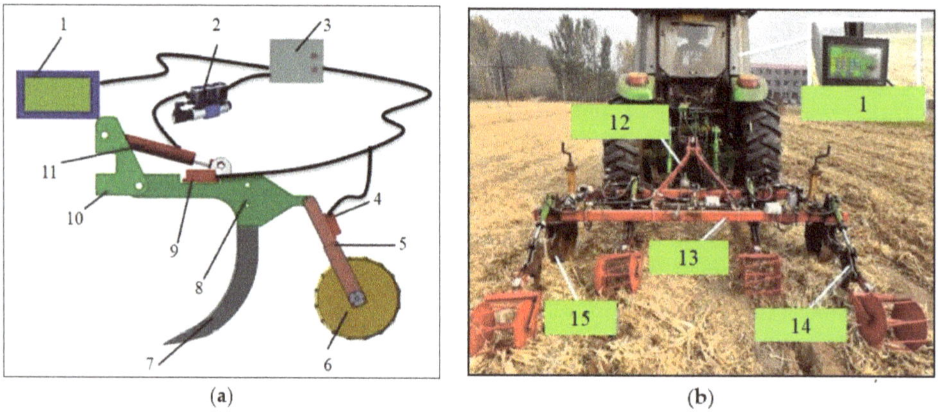

1. Touch screen 2. Proportional value and value amplifier 3. Controller box 4. Inclination sensor 5. Press roller frame 6. Press roller 7. Curved shovel 8. Shovel frame 9. Inclination sensor 10. Fixed frame 11. Hydraulic cylinder 12.Three-point suspension 13.Frame 14.Subsoiling assembly with the electric-hydraulic control system 15. Supporting wheel.

Figure 4. (**a**) Subsoiling assembly with the electric-hydraulic control system (Wang et al., 2018); (**b**) Subsoiler equipped with four subsoiling assemblies with the electric-hydraulic control system (Wang, 2019).

In conclusion, the control method of unified adjustment of each row adopts integrated adjustment and position adjustment combined with force adjustment. This mode not only ensures the tillage depth stability to a certain extent, but also takes the influence of tillage resistance into account. The control precision of tillage depth by this mode meets the operation requirements. However, relevant parameters in the proportional fuzzy control are difficult to determine, and the control process is more complicated. This control mode is suitable for the operations in flat fields. Furthermore, the single-row tillage depth cannot be independently adjusted according to undulation of local surfaces. By comparison, for the method of independent adjustment of a single row, the designed mechanism used to adjust depth has strong applicability. By adjusting the connecting parts of the control device, the adjustment mechanism can be installed on the different tillage machines. In addition, the main program of the automatic control system belonging to the separate adjustment mechanism is relatively simple. Moreover, tillage depth of subsoiling assemblies can be independently adjusted. Therefore, this control mode is excels at obtaining consistent depths of each subsoiling shovel.

6. Soil Mechanical Interaction

The interaction mechanism analysis of subsoiling shovels and soil is the key to the study of the shovel force and the effect of shovels on the dynamic changes of soil disturbance [107]. Wheeler [108] developed a mechanical model and studied the influence of speed on tillage resistance. Manuwa [109] set up models for the resistance prediction of different shovels and analyzed the effect of loosening depth on draft force. Based on the mechanical principle of two-side wedge and three-side wedge in soil, Ma and Zhou [110,111] studied the straight-legged shank and the circular shank and built mathematical models. In recent years, the Discrete Element Method (DEM) has been widely used to study the tillage mechanism [112,113]. Shmulevich [114] used PFC2D discrete-element software to research the interaction between the wide shovel and soil, and four types of shanks were chosen for simulation. Results showed that soil accumulation in front of shovels increased horizontal resistance during movement, and the soil flow below the tip influenced the vertical force applied to the shovel. A soil-tool-residue interaction model was developed by Zeng et al. [115] using the DEM and was used to investigate and compare the effects of forward speed on soil displacement, residue displacement, and residue cover reduction resulting from four types of shovels (Figure 5). Considering the normal tension generated among elements, Momozu et al. [116] used coefficients to modify the traditional discrete element model, and then to represent the adhesion generated by moisture among soil particles. Tamas et al. [117] developed a discrete element model to analyze the effects of the sweep rack angle and speed on soil loosening and porosity. Results showed that the comparison between the measured and simulated draught of subsoiling shovels with a 30° sweep rack angle behaved as a good match. In the selected speed interval of 0.5–2.4 m s^{-1}, the error range was 4–15%. A working model of subsoiling was established by Hang et al. [118], applying the DEM. Through simulation based on this model and an indoor soil-bin test, they compared and analyzed the micro-movement and macro-disturbance behavior of subsoil at different locations under the combined effect of two shovels (Figure 6).

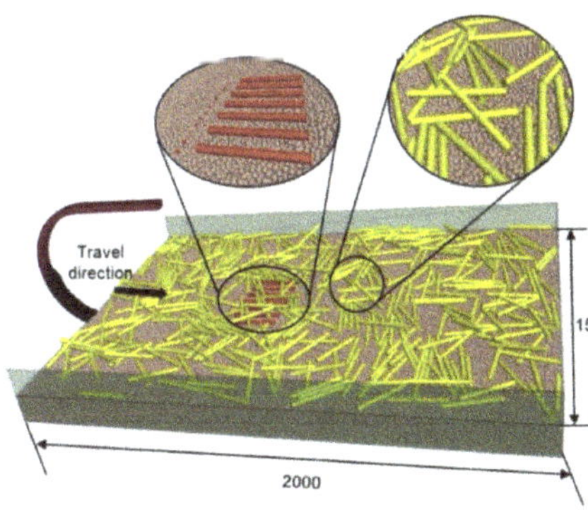

Figure 5. The soil-tool-residue interaction model (Zeng et al., 2020).

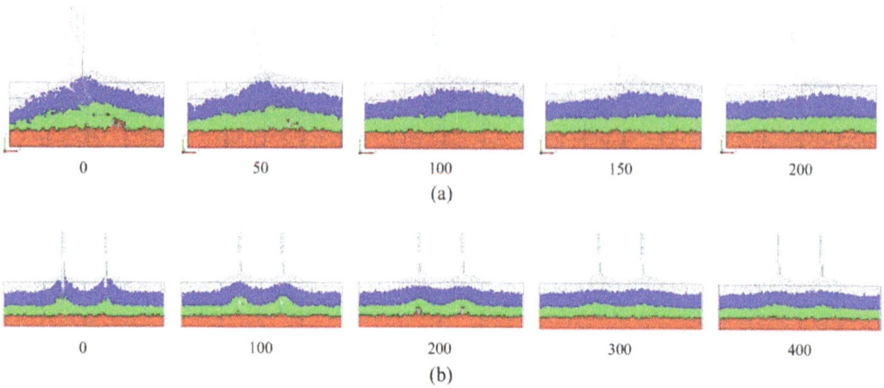

Figure 6. Analysis of the soil disturbance process (Hang et al., 2017): (**a**) Schematic diagram of the disturbance process in the longitudinal-sectional; (**b**) Schematic diagram of the disturbance process in the transverse-sectional.

In the subsoiling process, the soil is sheared, extruded, and uplifted by the subsoiling shovel, which provides the reacting force from the soil. Although the soil-mechanical interaction is a complex process, the disruption and deformation of soil under the force of subsoiling components still have some regularity. According to the soil structure, soil mechanics, as well as the failure and deformation regularity, the virtual soil model can be built in the simulation software, which is capable of replacing actual soil within a certain range of error. Furthermore, the simulation test of subsoiling machines and soil models can be carried out to study machines' performance and optimize structural parameters, which is conducive to the shortening of the development cycle of subsoiling machines, improving research efficiency, and saving time and resources. Additionally, the soil smashing mechanism during subsoiling and interaction between the shovel and soil in the process of adjusting tillage depth can be revealed.

7. Summaries and Recommendations

A large amount of research on subsoiling shovel design, anti-drag, tillage depth detection and control, and soil mechanical interaction have been done to improve subsoiling quality and reduce energy consumption. However, most soil models established in simulation software lacking plant root or straw are inconsistent with the field soil condition. There is insufficient information regarding the broken soil characteristics, stubble-breaking process analysis, and comprehensive effects of subsoiling. Further research should focus on the interaction mechanisms between soil, subsoiling shovels, and plant root or straw. The combination of tillage depth detection and control can effectively improve tillage depth stability, whereas the influence of different soil conditions on the control system is unclear. It is necessary to develop monitoring and control systems of tillage depth with good applicability. Apart from using the abovementioned methods to save energy, the performance of subsoiling components should be considered with more attention. As shovels directly make contact with soil, they wear severely. For this reason, studies on the wear resistance of subsoiling components should be conducted. Therefore, to develop high-performance subsoiling machines with good subsoiling quality and low energy consumption, this article proposes the following four recommendations.

(1) It is very important to improve simulation accuracy by optimizing the soil model in computer simulation. The discrete element model of a root-soil composite model with straw should be established, and the key parameters of soil models need to be calibrated by simulation and laboratory tests. Furthermore, the accuracy and reliability of soil models should be verified by subsoiling simulation and field experiments.

(2) Comprehensive methods combining computer simulation, field experiments, and theoretical analysis need be adopted to strengthen research on subsoiling mechanism and comprehensive effect. Thus, the interaction mechanism between root-soil composite models, subsoiling shovels, and straw, as well as the interactive relationship between energy consumption and yield, increase the benefits from subsoiling and can be clarified from macroscopic and microscopic perspectives.
(3) New tillage depth monitoring and control systems should be developed via mechanism innovation, algorithm optimization, and software development, which has the advantages of accurate detection, fast adjustment, convenient use, and suitability for different soil conditions. Moreover, by the combination with GPS positioning technology and data sharing technology, the tillage depth will be correlated to some information, such as nutrient, moisture, and yield, to provide guidance for the future management of farmland.
(4) Improving the wear-resisting properties of subsoiling shovels is a necessary method to save energy. According to Zhao [119], more than 80% of subsoiling shovels' failure is caused by wear, which is the main cause of low efficiency, poor quality, and high operating costs. Therefore, some wear-resistance improvements of subsoiling shovels should also be taken into account, such as applying advanced processing technologies, adopting wear resistant materials, and designing shovels with wear-resistant geometry of the biological surface.

Author Contributions: Conceptualization, S.L. and J.H.; writing—original draft preparation, S.L.; writing—review and editing, H.L. (Hongwen Li), Q.W., C.L., W.L., P.L., Z.Z., H.L. (Hui Li); funding acquisition, H.L. (Hongwen Li), J.H. All authors have read and agreed to the published version of the manuscript.

Funding: This research was funded by the China Agriculture Research System of MOF and MARA (Grant No.CARS-03); Innovative Research Team in University of China (Grant No. IRT13039).

Institutional Review Board Statement: Not applicable.

Informed Consent Statement: Not applicable.

Data Availability Statement: Not applicable.

Acknowledgments: This research was supported by The 2115 Talent Development Program of China Agricultural University.

Conflicts of Interest: The authors declare no conflict of interest.

References

1. Zulfiqar, U.; Hussain, S.; Ishfaq, M.; Ali, N.; Ahmad, M.; Ihsan, F.; Sheteiwy, M.S.; Rauf, A.; Hano, C.; El-Esawi, M.A. Manganese Supply Improves Bread Wheat Productivity, Economic Returns and Grain Biofortification under Conventional and No Tillage Systems. *Agriculture* **2021**, *11*, 142. [CrossRef]
2. Wolkowski, R.; Lowery, B. Soil compaction: Causes, concerns and cures. *Coop. Ext. Publ.* **2008**, *5*, 1–8.
3. Poehlitz, J.; Ruecknagel, J.; Schlueter, S.; Vogel, H.J.; Christen, O. Computed tomography as an extension of classical methods in the analysis of soil compaction, exemplified on samples from two tillage treatments and at two moisture tensions. *Geoderma* **2019**, *346*, 52–62. [CrossRef]
4. Ordonez-Morales, K.D.; Cadena-Zapata, M.; Zermeno-Gonzalez, A.; Campos-Magana, S. Effect of Tillage Systems on Physical Properties of a Clay Loam Soil under Oats. *Agriculture* **2019**, *9*, 62. [CrossRef]
5. European Commission. *The State of Soil in Europe-A Contribution of the JRC to the European Environment Agency's Environment State and Outlook Report-SOER 2010*; Publications Office of the European Union: Luxembourg, 2012.
6. Artemyeva, Z.S.; Kogut, B.M. The Effect of Tillage on Organic Carbon Stabilization in Microaggregates in Different Climatic Zones of European Russia. *Agriculture* **2016**, *6*, 63. [CrossRef]
7. Hargreaves, P.R.; Baker, K.L.; Graceson, A.; Bonnett, S.; Ball, B.C.; Cloy, J.M. Soil compaction effects on grassland silage yields and soil structure under different levels of compaction over three years. *Eur. J. Agron.* **2019**, *109*, 1–9. [CrossRef]
8. Nebo, G.L.; Manyevere, A.; Araya, T.; Tol, J.V. Short-Term Impact of Conservation Agriculture on Soil Strength and Saturated Hydraulic Conductivity in the South African Semiarid Areas. *Agriculture* **2020**, *10*, 414. [CrossRef]

9. Moradi, E.; Rodrigo-Comino, J.; Terol, E.; Mora-Navarro, G.; Silva, A.M.D.; Daliakopoulos, I.N.; Khosravi, H.; Fernandez, M.P.; Cerda, A. Quantifying Soil Compaction in Persimmon Orchards Using ISUM (Improved Stock Unearthing Method) and Core Sampling Methods. *Agriculture* **2020**, *10*, 266. [CrossRef]
10. Somerville, P.D.; May, P.B.; Livesley, S.J. Effects of deep tillage and municipal green waste compost amendments on soil properties and tree growth in compacted urban soils. *J. Environ. Manag.* **2018**, *227*, 365–374. [CrossRef]
11. Kristoffersen, A.O.; Riley, H. Effects of soil compaction and moisture regime on the root and shoot growth and phosphorus uptake of barley plants growing on soils with varying phosphorus status. *Nutr. Cycl. Agorecosyst.* **2005**, *72*, 135–146. [CrossRef]
12. Wang, M.J.; He, D.; Shen, F.; Huang, J.L.; Zhang, R.T.; Liu, W.B.; Zhu, M.J.; Zhou, L.; Wang, L.H.; Zhou, Q. Effects of soil compaction on plant growth, nutrient absorption, and root respiration in soybean seedlings. *Environ. Sci. Pollut. Res.* **2019**, *26*, 22835–22845. [CrossRef] [PubMed]
13. Haruna, S.I.; Nkongolo, N.V. Influence of Cover Crop, Tillage, and Crop Rotation Management on Soil Nutrients. *Agriculture* **2020**, *10*, 225. [CrossRef]
14. Kwiatkowski, C.A.; Harasim, E.; Feledyn-Szewczyk, B.; Antonkiewicz, J. Enzymatic Activity of Loess Soil in Organic and Conventional Farming Systems. *Agriculture* **2020**, *10*, 135. [CrossRef]
15. Wozniak, A. Chemical Properties and Enzyme Activity of Soil as Affected by Tillage System and Previous Crop. *Agriculture* **2019**, *9*, 262. [CrossRef]
16. Singh, K.; Choudhary, O.P.; Singh, H.P.; Singh, A.; Mishra, S.K. Sub-soiling improves productivity and economic returns of cotton-wheat cropping system. *Soil Tillage Res.* **2019**, *189*, 131–139. [CrossRef]
17. Wang, S.B.; Guo, L.L.; Zhou, P.C.; Wang, X.J.; Shen, Y.; Han, H.F.; Ning, T.Y.; Han, K. Effect of subsoiling depth on soil physical properties and summer maize (*Zea mays* L.) yield. *Plant Soil Environ.* **2019**, *65*, 131–137. [CrossRef]
18. Schneider, F.; Don, A.; Hennings, I.; Schmittmann, O.; Seidel, S.J. The effect of deep tillage on crop yield-What do we really know? *Soil Tillage Res.* **2017**, *174*, 193–204. [CrossRef]
19. Feng, X.M.; Hao, Y.B.; Latifmanesh, H.; Lal, R.; Cao, T.H.; Guo, J.R.; Deng, A.X.; Song, Z.W.; Zhang, W.J. Effects of subsoiling tillage on soil properties, maize root distribution, and grain yield on mollisols of northeastern China. *Soil Tillage Conserv. Manag.* **2018**, *110*, 1607–1615. [CrossRef]
20. Zhang, R.F.; Yang, H.S.; Gao, J.L.; Zhang, Y.Q.; Wang, Z.G.; Fan, X.Y.; Bi, W.B. Effect of subsoiling on root morphological and physiological characteristics of spring maize. *Trans. CSAE* **2015**, *31*, 78–84. [CrossRef]
21. Carr, P.M. Guest Editorial: Conservation Tillage for Organic Farming. *Agriculture* **2017**, *7*, 19. [CrossRef]
22. Wichelns, D. Achieving Water and Food Security in 2050: Outlook, Policies, and Investments. *Agriculture* **2015**, *5*, 188–220. [CrossRef]
23. Mullan, D. Managing Soil Erosion in Northern Ireland: A Review of Past and Present Approaches. *Agriculture* **2013**, *3*, 684–699. [CrossRef]
24. Pimentel, D.; Burgess, M. Soil Erosion Threatens Food Production. *Agriculture* **2013**, *3*, 443–463. [CrossRef]
25. 2016–2022 China's Subsoiling and Land Preparation Industry Special Investigation and 13th Five-Year Investment Prospect Forecast Report. Available online: http://baogao.chinabaogao.com/xumuye/245071245071.html (accessed on 6 August 2020).
26. Department of Farm Mechanization of Ministry of agriculture and rural affairs of the People's Republic of China. *National Agricultural Mechanization Statistics Annual Report*; Department of Farm Mechanization of MOA: Beijing, China, 2018; p. 14.
27. Ministry of Agriculture and Rural Affairs of the People's Republic of China. National Agricultural Machinery Subsoiling and Land Preparation Operation Implementation Plan (2016–2020). Available online: http://www.moa.gov.cn/nybgb/2016/disanqi/201711/t20171127_5920218.htm (accessed on 8 August 2020).
28. State Council of the People's Republic of China. Government Work Report. 2015. Available online: http://www.gov.cn/guowuyuan/2015-03/16/content_2835101.htm (accessed on 8 August 2020).
29. State Council of the People's Republic of China. Government Work Report. 2014. Available online: http://www.gov.cn/guowuyuan/2014-03/14/content_2638989.htm (accessed on 8 August 2020).
30. State Council of the People's Republic of China. Central Document No.1. 2014. Available online: https://wenku.baidu.com/view/7384313ceff9aef8941e06e8.html (accessed on 8 August 2020).
31. Ministry of Agriculture of the People's Republic of China. National Agricultural Machinery Subsoiling and Land Preparation Operation Implementation Plan (2011–2015). Available online: http://www.moa.gov.cn/nybgb/2011/deq/201805/t20180518_6142547.htm (accessed on 8 August 2020).
32. State Council of the People's Republic of China. Opinions of the State Council on Promoting Sound and Rapid Development of Agricultural Mechanization and Agricultural Machinery Industry. 2010. Available online: http://www.gov.cn/zhengce/content/2010-07/09/content_3009.htm (accessed on 8 August 2020).
33. State Council of the People's Republic of China. Central Document No.1. 2010. Available online: http://finance.sina.com.cn/g/20100131/17047342371.shtml (accessed on 8 August 2020).
34. Ministry of Finance of the People's Republic of China. Interim Measures for the Administration of the Centralized use of the Newly Increased Central Funds for Comprehensive Agricultural Subsidies for the Building of Basic Grain Capacity. 2009. Available online: http://www.chinalawedu.com/falvfagui/fg22016/514036.shtml (accessed on 8 August 2020).
35. State Council of the People's Republic of China. Central Document No.1. 2009. Available online: http://www.jingbian.gov.cn/xiwzx/yw/319.htm (accessed on 8 August 2020).

36. He, J.; Li, H.W.; Chen, H.T.; Lu, C.Y.; Wang, Q.J. Research progress of conservation tillage technology and machine. *Trans. Chin. Soc. Agric. Mach.* **2018**, *49*, 1–18. [CrossRef]
37. Li, B.F. *Agricultural Mechanics;* China Agriculture Press: Beijing, China, 2003; pp. 35–37.
38. Arable Land (Hectares per Person). Available online: https://data.worldbank.org/indicator/AG.LND.ARBL.HA.PC (accessed on 11 August 2020).
39. Maschio DIABLO Subsoiling Combined Machine. Available online: http://www.maschio.com/catalog/product/diablo/it_IT (accessed on 3 September 2020).
40. John Deere 2730 Subsoiling Combined Machine. Available online: https://www.deere.com/en/tillage/2730-combination-ripper/ (accessed on 3 September 2020).
41. Agrowplow AP31 Subsoiling Plough. Available online: http://www.agrowplow.com.au/agrowplow-ap31-plough (accessed on 3 September 2020).
42. Salfordgroup 9200 Subsoiling Plough. Available online: https://salfordgroup.com/9200-in-line-ripper (accessed on 5 September 2020).
43. Shandongdahua 1S-300C Omni-Directional Subsoiler. Available online: http://www.dhbl.net/product/73.html (accessed on 5 September 2020).
44. Shandongaolong 1SQ 330 Omni-Directional Subsoiler. Available online: http://www.sdaljx.com/zhsh.asp?id=520 (accessed on 5 September 2020).
45. Zhang, J.B. Bionic Drag Reduction and Wear-Resistant Theory and Techniques of Subsoiler. Ph.D. Thesis, Jilin University, Jilin, China, 2014.
46. Khalilian, A.; Han, Y.J.; Marshall, M.W.; Gorucu, S.; Abbaspour, G.Y.; Kirk, K.R. Evaluation of the clemson instrumented subsoiler shank in coastal plain soils. *Comput. Electron. Agric.* **2014**, *109*, 46–51. [CrossRef]
47. Liu, J.A.; Wang, X.Y.; Li, H.W.; He, J.; Wang, Q.J.; Li, W. Optimization of structural parameters of subsoiler based on soil disturbance and traction resistance. *Trans. Chin. Soc. Agric. Mach.* **2017**, *48*, 60–67. [CrossRef]
48. Lee, J.; Yamazaki, M.; Oidab, A.; Nakashimac, H.; Shimizub, H. Electro-hydraulic tillage depth control system for rotary implements mounted on agricultural tractor Design and response experiments of control system. *J. Terramech.* **1998**, *35*, 229–238. [CrossRef]
49. Garrido, M.; Concciçäo, L.A.; Baguena, E.M.; Valero, C.; Barreiro, P. Evaluating the need for an active depth-control system for direct seeding in Portugal. In Proceedings of the 8th European Conference Precision Agriculture, Prauge, Czech Republic, 11–14 July 2011; Wageningen Academic Publishers: Wageningen, The Netherlands.
50. Suomi, P.; Oksanen, T. Automatic working depth control for seed drill using ISO 11783 remote control messages. *Comput. Electron. Agric.* **2015**, *116*, 30–35. [CrossRef]
51. Wang, S.L.; Liang, H. Development and application of intelligent supervision system for agricultural machinery subsoiling. *Hebei Agric. Mach.* **2016**, *3*, 10–11.
52. Pereira, D.P.; Fiedler, N.C.; Juliao, S.L. Efficiency of subsoiling depth according to the slope of the land. *Cerne* **2012**, *18*, 607–612. [CrossRef]
53. Tian, S.Z.; Wang, Y.; Zhang, F.Y.; Bian, W.F.; Dong, L.; Luo, J.F.; Guo, H.H. Residue returning with subsoiling replacing rotary tillage improving aggregate and associated carbon. *Trans. CSAE* **2017**, *33*, 133–140. [CrossRef]
54. Kayad, A.; Paraforos, D.S.; Marinello, F.; Fountas, S. Latest Advances in Sensor Applications in Agriculture. *Agriculture* **2020**, *10*, 362. [CrossRef]
55. Yu, Y.C.; Liu, W.Y.; Zhao, Y.C.; Sun, J.Q. Force mathematical model and examination analysis of the column subsoiler. *Trans. CSAE* **2007**, *23*, 109–113. [CrossRef]
56. Zhang, Q.; Zhang, L.; Yu, H.Y.; Xiao, Y.K. Finite element analysis and experiment of soil resistance of multiplex-modality subsoiler. *Trans. Chin. Soc. Agric.Mach.* **2012**, *43*, 61–65. [CrossRef]
57. Lemken 1S-225C Subsoiler. Available online: https://lemken.com/en/soil-cultivation/stubble-cultivation/subsoiler/labrador/ (accessed on 15 September 2020).
58. Hedongxiongfeng 1S-150A Subsoiler. Available online: http://www.hedongxiongfeng.cn/pd.jsp?id=52#_pp=103_366 (accessed on 15 September 2020).
59. Maschio PINOCCHIO200 Subsoiler. Available online: http://www.maschio.com/catalog/category/dissodatori-di-profondita/zh_cmn (accessed on 15 September 2020).
60. Zhongnongboyuan 1S-200 Subsoiler. Available online: http://www.boyomac.com/product/17.html (accessed on 14 May 2021).
61. Woye 1SQ-340 Subsoiler. Available online: https://www.nongjitong.com/product/3072.html (accessed on 15 September 2020).
62. DALBO Ratoon Subsoiler. Available online: https://www.cnmasike.com/agriculture/used-other-agricultural-machines/dal-bo-ratoon/be6ohtgr.html (accessed on 15 September 2020).
63. Shandongdahua 1S-250 Subsoiler. Available online: https://www.dhbl.net/product/53.html (accessed on 15 September 2020).
64. Kostencki, P.; Stawicki, T.; Bialobrzeska, B. Durability and wear geometry of subsoiler shanks provided with sintered carbide plates. *Tribol. Int.* **2016**, *104*, 19–35. [CrossRef]
65. Chen, K. Study on Structural Optimization of Extroverted arc Subsoiler. Master's Thesis, Jilin Agriculture University, Jilin, China, 2017.
66. Zhao, S.H.; Wang, J.Y.; Chen, J.Z.; Yang, Y.Q.; Tan, H.W. Design and experiment of fitting curve subsoiler of conservation tillage. *Trans. Chin. Soc. Agric. Mach.* **2018**, *49*, 82–92. [CrossRef]

67. Wang, X.Z.; Yue, B.; Gao, X.J.; Zheng, Z.Q.; Zhu, R.X.; Huang, Y.X. Discrete element simulations and experiments of disturbance behavior as affected by mounting height of subsoiler's wing. *Trans. Chin. Soc. Agric. Mach.* **2018**, *49*, 124–136. [CrossRef]
68. Peng, Z.M.; Ding, Y.; Zhu, J.P.; Yuan, D. Technical analysis and research of detection in energy-saving emission reduction to farming machinery. *Chin. Agric. Mech.* **2009**, *225*, 69–75. [CrossRef]
69. Sun, Y.P.; Dong, X.Q.; Song, J.N.; Liu, C.L.; Wang, J.C.; Zhang, C. Parameter optimization of vibration subsoiler test bed for reducing resistance and vibration. *Trans. CSAE* **2016**, *32*, 43–49. [CrossRef]
70. Li, X.; Zhang, D.X.; Wang, W.X.; Cui, T.; Tang, M.J. Performance parameter optimization and experiment of forced-vibration subsoiler. *Trans. CSAE* **2015**, *31*, 17–24. [CrossRef]
71. Li, X.; Fu, J.F.; Zhang, D.X.; Cui, T.; Zhang, R. Experiment analysis on traction resistance of vibration subsoiler. *Trans. CSAE* **2012**, *28*, 32–36. [CrossRef]
72. Shahgoli, G.; Fielke, J.; Desbiolles, J.; Saunders, C. Optimising oscillation frequence in oscillation tillage. *Soil Tillage Res.* **2009**, *106*, 202–210. [CrossRef]
73. Linde, J.V.D. Discrete Element Modeling of a Vibratory Subsoiler. Ph.D. Thesis, Stellenbosch University, Stellenbosch, South Africa, 2007.
74. Zhang, J.C.; Yan, X.L.; Lin, Z.K.; Zhu, R.X. Design and experiment of self-exciting vibration deep-loosening and sub-soiling machine. *Trans. Chin. Soc. Agric. Mach.* **2016**, *47*, 44–49. [CrossRef]
75. Ma, X.; Qiu, L.C.; Milecki, Z. Experimental study on the dynamic characteristics of spring-tooth vibration sub-soiling shovel. *J. Mech. Eng. Res. Dev.* **2016**, *39*, 127–133. [CrossRef]
76. Soeharsono, S.; Radite, P.A.M.; Tineke, S.; Asep, H.; Wawan. Penurunan Draft Dan Energy Pembajakan Pada Model Subsoiler Getar Dengan Menggunakan Metode Self-Excited Vibration. Ph.D. Thesis, Bogor Agriculture University, Bogor, Indonesia, 2012.
77. Qiu, L.C.; Li, B.F. Experimental study on the self-excited vibration subsoiler for reducing draft force. *Trans. CSAE* **2000**, *16*, 72–76. [CrossRef]
78. Tong, J.; Sun, J.Y.; Chen, D.H.; Zhang, S.J. Geometrical features and wettability of dung beetles and potential biomimetic engineering applications in tillage implements. *Soil Tillage Res.* **2005**, *80*, 1–12. [CrossRef]
79. Zhang, J.B.; Tong, J.; Ma, Y.H. Design and experiment of bionic anti-drag subsoiler. *Trans. Chin. Soc. Agric. Mach.* **2014**, *45*, 141–145. [CrossRef]
80. Li, B.; Chen, Y.; Chen, J. Modeling of soil-claw interaction using the discrete element method (DEM). *Soil Tillage Res.* **2016**, *158*, 177–185. [CrossRef]
81. Zhang, X.R.; Wang, C.; Chen, Z.X.; Zeng, Z.W. Design and experiment of a bionic vibratory subsoiler for banana fields in southern China. *Int. J. Agric. Biol. Eng.* **2016**, *9*, 75–83. [CrossRef]
82. Wang, T.H.; Zhang, W.A.; Wang, L.B.; Yu, M.; Song, D.Q.; Yu, L. Stress analysis and subsoiling quality research of layered deep loosening shovel. *Farm Mach.* **2012**, *32*, 94–96.
83. Kasisira, L.L.; Plessis, H.L.M.D. Energy optimization for subsoilers in tandem in a sandy clay loam soil. *Soil Tillage Res.* **2006**, *86*, 185–198. [CrossRef]
84. Wang, H.L.; Zhang, W. Optimization design of the subsoiling shovel structural parameters based on Pro/E. *J. Agric. Mech. Res.* **2011**, *1*, 141. [CrossRef]
85. Kichler, C.M.; Fulton, J.P.; Raper, R.L.; Mcdonald, T.P.; Zech, W.C. Effects of transmission gear selection on tractor performance and fuel costs during deep tillage operations. *Soil Tillage Res.* **2011**, *113*, 105–111. [CrossRef]
86. Zhu, F.W. Analysis of the Morphology of Scarabaeoidea and Bionic Design of Subsoiling Components. Ph.D. Thesis, Jilin University, Jilin, China, 2005.
87. Yang, Z.J.; Zheng, K.; Geng, C.J.; Xia, J.Q. Design of multi-channel ultrasonic wave sensor ranging system based on RS–485 bus. *Transd. Microsyst. Technol.* **2018**, *37*, 79–82. [CrossRef]
88. Mouazen, A.M.; Anthonis, J.; Saeys, W.; Ramon, H. An automatic depth control system for online measurement of spatial variation in soil compaction 1. Sensor design for measurement of frame height variation from soil surface. *Biosyst. Eng.* **2004**, *89*, 139–150. [CrossRef]
89. Li, Z.X. Research and design of operating depth measurement apparatus for farm implement. *Trans. Chin. Soc. Agric. Mach.* **2000**, *31*, 8–91. [CrossRef]
90. Adamchuk, V.I.; Hummel, J.W.; Morgan, M.T.; Upadhyaya, S.K. On-the-go soil sensors for precision agriculture. *Comput. Electron. Agric.* **2004**, *44*, 71–91. [CrossRef]
91. Yuan, Y.W.; Fang, X.F.; Yang, B.N.; Liang, X.X.; Dong, X.; Zhou, L.M.; Zhang, J.N. A Suspended Subsoiler and Its Online Detection Device and Method. CHN CN104977586-A, 14 October 2015.
92. Zheng, Y.H. Study on Sensing Mechanism of Gas Pendulum Tilt Sensor and the Key Factors of Affecting its Property. Ph.D. Thesis, Beijing University of Post and Telecommunications, Beijing, China, 2008.
93. Zhao, J.; He, P.X.; Li, Q.D.; Jiang, M.; Kang, J.; Liu, H.B.; Zhu, K.X. Design of depth display system for micro-tiller. *J. Agric. Mech. Res.* **2016**, *10*, 83–86. [CrossRef]
94. Li, Q.L.; Sun, Y.J.; Sun, Y.T.; Shen, J.X.; Chen, G.; Dou, Q.Q. Development of DSP-based joint operations subsoiling machine monitoring system. *J. Agric. Mech. Res.* **2016**, *11*, 118–122. [CrossRef]
95. Xie, B.; Li, H.; Zhu, Z.X.; Mao, E.R. Measuring tillage depth for tractor implement automatic using inclinometer. *Trans. CSAE* **2013**, *39*, 15–21. [CrossRef]

96. Ayiding, K.; Wu, M.T.; He, P.X.; Liu, X.R.; Sun, B. The control system of automatic adjust for plowing depth. *J. Agric. Mech. Res.* **2013**, *3*, 160–163. [CrossRef]
97. Jia, H.L.; Guo, M.Z.; Yu, H.B.; Li, Y.; Feng, X.Z.; Zhao, J.L.; Qi, J.T. An adaptable tillage depth monitoring system for tillage machine. *Biosyst. Eng.* **2016**, *151*, 187–199. [CrossRef]
98. Shang, G.G.; Peng, H.J.; Xia, Y. Study on auto-control method and simulation for tractor depth based on fuzzy control. *Res. Agric. Mod.* **2014**, *35*, 825–829. [CrossRef]
99. Xie, B.; Mao, E.R. Development of CAN intelligent nodes of tractor electronic hydraulic hitch system. *Trans. Chin. Soc. Agric. Mach.* **2006**, *37*, 1–3. [CrossRef]
100. Nie, Y.H.; Kang, J.; He, J.H.; He, P.X.; Li, Q.R. Deep tillage automatically adjust the design and test of control system. *J. Agric. Mech. Res.* **2015**, *2*, 143–145. [CrossRef]
101. HE-VA Combined Land Preparation Equipment Equipped with Hydraulic System. Available online: https://www.he-va.co.uk/he-va-products/cultivation/combi-disc/ (accessed on 16 December 2020).
102. John Deere 2633VT with TruSet™ Tillage Technology. Available online: https://www.deere.com/en/tillage/truset/ (accessed on 16 December 2020).
103. CASE IH Tillage Tools Equipped with Advanced Farming System. Available online: https://www.caseih.com/northamerica/en-us/products/advanced-farming-systems/field-solutions/afs-soil-command (accessed on 16 December 2020).
104. He, J.; Wu, Y.H.; Li, H.W.; Wang, Q.J.; Lu, C.Y.; Zhang, Z.Q.; Zheng, K.; Liu, W.Z.; Zhao, H.B.; Ju, J.W.; et al. A Device, a Universal Type, Automatically Monitoring and Controlling Tillage Depth of Each Subsoiler Shank. CHN CN107980261-A, 4 May 2018.
105. Wang, Y.X.; Jing, H.R.; Zhang, D.X.; Cui, T.; Zhong, X.J.; Yang, L. Development and performance evaluation of an electric-hydraulic control system for subsoiler with flexible tines. *Comput. Electron. Agric.* **2018**, *151*, 249–257. [CrossRef]
106. Wu, Y.H. The Design of Depth Automatic Monitoring Control Subsoiler Based on Ultrasonic Sensor and Hydraulic Drive. Master's Thesis, China Agriculture University, Beijing, China, 2018.
107. Chen, K.; Hu, X.L.; Zhao, X.Z.; Yuan, H.Y. Research status and prospects of subsoilier at home and abroad. *Agric. Technol.* **2010**, *30*, 30–34.
108. Wheeler, P.N.; Godwin, R.J. Soil dynamics of single and multiple tines at speeds up to 20 km/h. *J. Agric. Eng. Res.* **1996**, *63*, 243–250. [CrossRef]
109. Manuwa, S.I. Performance evaluation of tillage tines operating under different depths in a sandy clay loam soil. *Soil Tillage Res.* **2009**, *103*, 399–405. [CrossRef]
110. Ma, S.F. Force Analysis and Computer Simulation of the Subsoiler in Working State. Master's Thesis, Henan Agriculture University, Henan, China, 2004.
111. Zhou, Y.Q. Force Mathematic Model and Computer Simulation of the Subsoiler. Master's Thesis, Henan Agriculture University, Henan, China, 2006.
112. Mak, J.; Chen, Y.; Sadek, M.A. Determining parameters of a discrete element model for soil-tool interaction. *Soil Tillage Res.* **2012**, *118*, 117–122. [CrossRef]
113. Chen, Y.; Munkholm, L.J.; Nyord, T. A discrete element model for soil-sweep interaction in three different soils. *Soil Tillage Res.* **2013**, *126*, 34–41. [CrossRef]
114. Shmulevich, I.; Asaf, Z.; Rubinstein, D. Interaction between soil and a wide cutting blade using the discrete element method. *Soil Tillage Res.* **2007**, *97*, 37–50. [CrossRef]
115. Zeng, Z.W.; Ma, X.; Chen, Y.; Qi, L. Modelling residue incorporation of selected chisel ploughing tools using the discrete element method (DEM). *Soil Tillage Res.* **2020**, *197*, 1–14. [CrossRef]
116. Momozu, M.; Oida, A.; Yamazaki, M.; Koolen, A.J. Simulation of a soil loosening process by means of the modified distinct element method. *J. Terramechanics.* **2003**, *39*, 207–220. [CrossRef]
117. Tamas, K.; Jori, I.J.; Mouazen, A.M. Modelling soil–sweep interaction with discrete element method. *Soil Tillage Res.* **2013**, *134*, 223–231. [CrossRef]
118. Hang, C.G.; Huang, Y.X.; Zhu, R.X. Analysis of the movement behaviour of soil between subsoilers based on the discrete element method. *J. Terramech.* **2017**, *74*, 35–43. [CrossRef]
119. Zhao, J.G.; Li, J.C.; Wang, A.; Ma, L.P.; Hao, J.J.; Zhang, J.G.; Ma, Y.J. Improvement of wear resistance of deep-shovel tip with Fe-based alloycoating by flame spray welding residual temperature quenching. *Trans. CSAE* **2018**, *34*, 65–71. [CrossRef]

agriculture

MDPI

Article

Development of Thermal Performance Metrics for Direct Gas-Fired Circulating Heaters

Benjamin C. Smith, Brett C. Ramirez * and Steven J. Hoff

Department of Agricultural and Biosystems Engineering, Iowa State University, Ames, IA 50011, USA; bcsmith1@iastate.edu (B.C.S.); hoffer@iastate.edu (S.J.H.)
* Correspondence: bramirez@iastate.edu; Tel.: +1-515-294-0468

Abstract: Many climate-controlled agricultural buildings use direct gas-fired circulating heaters (DGFCH) for supplement heat. There is no standardized test to calculate thermal efficiency for these heaters. This study aimed to develop a measurement system and analytical analysis for thermal efficiency, quantify the measurement uncertainty, and assess economics of DGFCH efficiency. The measurement system developed was similar to the ASHRAE 103 standard test stand with adaptations to connect the apparatus to the DGFCH. Two different propane measurement systems were used: input ratings < 30 kW used a mass flow system and input ratings > 30 kW used a volumetric gas meter. Three DGFCHs (21.9, 29.3, 73.3 kW) were tested to evaluate the system. Thermal efficiencies ranged from 92.4% to 100.9%. The resulting uncertainty (coverage factor of 2; ~95% Confidence Interval) ranged from 13.1% to 30.7% for input ratings of 56.3 to 11.4 kW. Key sources of uncertainty were propane and mass flow of air measurement. The economic impact of 1% difference in thermal efficiency ranged from USD $61.3 to $72.0 per heating season. Refinement of the testing system and procedures are needed to reduce the uncertainty. The application of this system will aid building designers in selection of DGFCHs for various applications.

Keywords: heating cost; thermal efficiency; uncertainty

Citation: Smith, B.C.; Ramirez, B.C.; Hoff, S.J. Development of Thermal Performance Metrics for Direct Gas-Fired Circulating Heaters. *Agriculture* **2021**, *11*, 588. https://doi.org/10.3390/agriculture11070588

Academic Editor: José Pérez-Alonso

Received: 26 April 2021
Accepted: 22 June 2021
Published: 24 June 2021

1. Introduction

The use of Direct Gas-Fired Circulating Heaters (DGFCH) is common in climate-controlled agricultural buildings (e.g., livestock, poultry, and greenhouses) as the heating component of the thermal environment modification system. Supplemental heat is needed to achieve the environmental control goal in an agricultural setting, that is, maintain an ideal environment for the growth of the agriculture commodity with minimal cost [1]. For both animal confinement buildings and greenhouses, heating during the winter months is necessary to maintain the optimal environment and presents a significant utility cost [1,2]. In most building types, a forced air, circulating style heater is used to provide circulation of the heated air to the occupied zone within the building. Gas-fired heaters are ubiquitous due to low operating costs and fuel availability to remote locations [2]. While DGFCH are popular for agricultural applications, there are downsides in implementing them.

DGFCH are simple heaters in comparison to residential units based on the construction and operation principles. DGFCHs use a propane burner manifold placed on the intake of a blower. This does not the vent the combustion gases out of the occupied space. The output of the blower can be equipped with various directional vanes to aid in the distribution of heated air. The placement of such heaters in agricultural buildings is highly variable with no standardized placement design currently available in literature [1,3]. Thermal efficiencies of DGFCHs are commonly reported with assumed values of approximately 99%; however, the water vapor and ventilation rates of a building can decrease the thermal efficiency to 80% [4]. In comparison to residential combustion heaters, which indirectly transfer heat to the air stream via a heat exchanger and combustion gases are vented outdoors, [5]. The use of the heater exchanger reduces the thermal efficiency to a range of 65% to 80%.

DGFCHs are direct-fired, resulting in combustion gases (i.e., CO and CO_2) formed directly in the heated air. In agricultural applications, the occupied (animal or plant) spaces where DGFCHs are present are continuously ventilated, unlike residential applications. The use of DGFCHs in greenhouses could result in safety concerns for workers, if prolonged operation throughout a day is necessary due to lower ventilation rates per floor area compared to animal buildings [4]. ANSI/ASHRAE [5] define the residential furnace thermal efficiency standard; however, there is limited standard thermal efficiency testing standards and guidelines for DGFCHs. Standardization of thermal efficiency testing and calculation allow for accurate comparison of different DGFCHs designs and capacities as well as generate informative equipment design specifications to support improved supplement heat capacity sizing and layout for agricultural buildings.

There are two types (direct and indirect) of combustion tests used to determine thermal efficiency. The direct test is a measure of the enthalpy gain of the air heated by the heater. Thermal efficiency can then be determined by dividing enthalpy gain by the sum of the heating fuel gross energy and electrical energy of the blower. This ratio of energy output over energy input for a given heater is the most accurate definition of thermal efficiency as the unit must obey the conservation of energy. The indirect test is a measure of the heating value of the fuel to determine the heat output less the flue gas or combustion gas enthalpy gain [5]. There are several challenges associated with applying the indirect test methodology to evaluate DGFCHs as the flue gas is mixed with the output air. The thermal efficiency calculated with the indirect methods typically is less accurate due to measurement error. Thus, DFGCHs are most appropriately evaluated by direct test methodology.

The objectives of this study were to (1) develop and validate a testing system and analytical analysis for determining the steady-state thermal efficiency of a DGFCH through a direct testing method; (2) evaluate the testing system across a range of DGFCH units; and (3) evaluate the economics of various DGFCH for agricultural uses.

2. Materials and Methods

2.1. Thermal Efficiency

The following sections will describe the methodology used to develop a DGFCH thermal efficiency measurement system. The system was designed to accommodate a range of rated heater capacities. For development purposes, three different DGFCHs (21.9 kW (75,000 BTU h^{-1}), 29.3 kW (100,000 BTU h^{-1}), 73.3 kW (250,000 BTU h^{-1})) were tested, procedural differences for each capacity are discussed in the testing procedure section. All tests were performed in the Air Dispersion Laboratory (ADL). See [6] for specifics of the laboratory.

2.2. Instrumentation

A Data Acquisition and Control (DAQC) system was developed to measure all necessary parameters to calculate the thermal efficiency of a DGFCH (analytical analysis to be discussed in a later section). The DAQC system consisted of two sub-systems, one to measure the input energy into the DGFCHs and one to measure the output enthalpy of the output air.

The instrumentation system for measuring input energy consisted of sensors to measure the propane input and electrical energy used by the DGFCH. Two different set-ups were created for propane input measurement. For DGFCHs rated <30 kW, a mass flow system was used, and for DGFCHs rated >30 kW, a volumetric device was used. This was performed as the minimum flow rates for traditional style diaphragm gas flowmeters exhibit high uncertainty for 30 kW rated heaters and below based on manufacturer specifications due to the extremely low flowrates. The mass flow system was not used for all heaters as the uncertainty in propane input was high, discussed in later section.

The mass flow system consisted of a 45.4 kg (100 lb) propane tank placed on a steel platform with a single-point 150 kg load cell with a 2 mV V^{-1} linear response signal (see Table 1 for specifications). The load cell signal was conditioned by a programmable-gain instrumentation amplifier (see Table 1 for specifications). A dual-stage regulator rated for 58.6 kW at 2.74 kPa was used (see Table 1 for specifications) with a 7.9 mm diameter flexible hose to connect the gas supply to the heater connection manifold.

Table 1. Energy input instruments used in the study in order as described in the text. FS = Full Scale.

Input Energy Instrumentation	Manufacturer's Rating	Manufacturer's Accuracy	Model	Manufacturer
Propane mass	150.0 kg	2.0×10^{-2}% FS	RLPWM12-150 kg	Rice Lake Weighing Systems, Rice Lake, WI, USA
Programmable amplifier	NA	NA	PGA 204	Texas Instruments, Dallas, TX, USA
Propane regulator	58.6 kW	NA	Model 9059398	Char-Broil, Columbus, GA, USA
Propane flowmeter	Minimum flow = 2.8E−4 m^3 s^{-1}	4.0×10^{-5} m^3 s^{-1}	Model PGM-150	EKM Metering, Santa Barbra, CA, USA
Propane temperature	−40.0 to 150.0 °C	±0.22 °C	Model TE-IBN-D044-14	Dwyer Instruments, Inc., Michigan City, IN, USA
Propane pressure	0.0 to 103.4 kPa	1.0% FS	Model 628	Dwyer Instruments, Inc., Michigan City, IN, USA
Precision resistor	1 kΩ	$\pm 5.0 \times 10^{-2}$% FS	Model 1k Tick	Labjack Corporation, Lakewood, CO, USA
Electrical power meter	Max 3000 A at 830 V AC	±0.7% of W	Model 1735	Fluke, Everett, WA, USA

The gas flowmeter (see Table 1 for specifications) was connected to the propane utility supply (two 3.79 m^3 propane tanks with a line pressure of 2.74 kPa) using a 12.7 mm diameter flexible hose. The flowmeter's output was connected to the heater connection manifold with a 12.7 mm Internal Diameter (ID) flexible hose. Flowrate was determined by dividing the difference between the recorded initial and final positions of the dial on the flowmeter by the duration of the test.

The heater connection manifold consisted of a sediment trap (following manufacturer's recommendation) and two taps that contained a silvered PT100 Resistance Temperature Detector (RTD; see Table 1 for specifications) and a pressure transducer (see Table 1 for specifications) to monitor supply propane temperature and pressure, respectively. A precision resistor (see Table 1 for specifications) was used to create a voltage divider circuit to condition the RTD output signal with a supply voltage of 2.5 Volt DC (VDC) and the 4 to 20 mA signal from the pressure transducer. The RTD temperature was determined using the alpha equation (Equations (1) and (2)) for RTDs, with alpha determined using the resistance at 0 °C and 100 °C. Heater electrical power consumption was measured using a portable power meter (see Table 1 for specifications).

$$T_i = \frac{\left(\frac{R_T}{R_0} - 1\right)}{\alpha} \tag{1}$$

where

T_i = temperature (°C); where i is RTD, propane, or exiting air
R_T = RTD resistance at T_i (Ω) see Equation (2)
R_0 = RTD resistance at 0 °C (100 Ω)
α = temperature coefficient of resistance (3.851E−4 Ω °C^{-1})

$$R_T = R_{d1} * \left(\frac{V_s}{V_{out}} - 1\right) \tag{2}$$

where

R_{d1} = resistance of voltage divider resistor (1 k Ω)
V_s = supply voltage (2.5 VDC)
V_{out} = analog voltage output measured by DAQC board (VDC)

Heater combustion exhaust measurement system consisted of a 0.2 m diameter, 2.44 m long round, galvanized steel duct that contained a PT100 RTD and a velocity pressure pitot tube (Figures 1 and 2, Tables 2 and 3). The same conditioning circuit and temperature equations were used as previously discussed. The supply end of the duct featured a flange fitting that was sealed with foil tape and secured to the face of the heater around the exhaust outlet. For heaters with larger output openings, a reduction flare fitting, 0.3 m to 0.2 m diameter and 0.3 m in length, was included. All seams on the inside of the duct were sealed with mastic caulking, and the outside of the duct was sealed with foil tape. The duct was wrapped with 51 mm (2 in.) thick, foil-backed fiberglass insulation. Additional insulation was placed on the seams on the fiberglass insulation, and all seams were sealed with foil tape. Velocity pressure was measured using a multi-range, differential pressure transducer (see Table 3 for specifications).

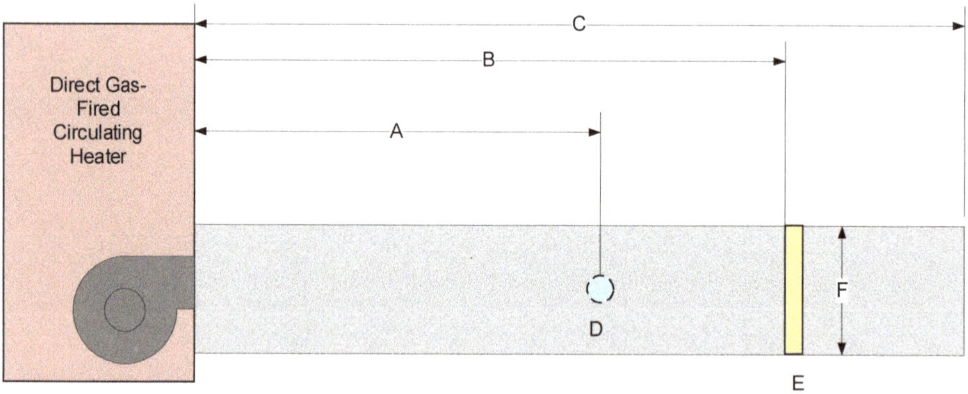

Figure 1. Test duct with locations of temperature measurement taps (D) and velocity pressure pitot tube (E). Measurements for the specific labels are in Table 1. Note diagram is not to scale.

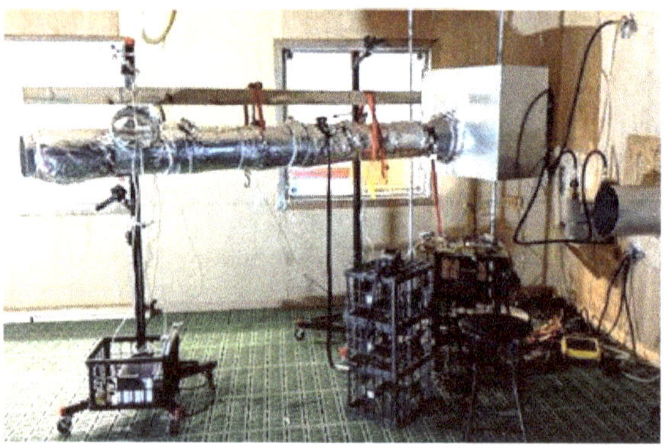

Figure 2. Test stand configuration and DAQC system in ADL for DGFCH test.

Table 2. Dimensions of the test duct in Figure 1.

Label	Description	Measurement (m)
A	Inlet to temperature tap	1.3
B	Inlet to velocity pressure pitot tube	1.6
C	Overall length of duct	2.4
F	Duct diameter	0.2

Table 3. Energy output instruments used in the study in order as described in the text.

Output Energy Instrumentation	Manufacturer's Rating	Manufacturer's Accuracy	Model	Manufacturer
Output air temperature	−40.0 to 150.0 °C	±0.2 °C	Model TE-IBN-D044-14	Dwyer Instruments, Inc., Michigan City, IN, USA
Output air velocity pitot tube	0.5 to 51 m s^{-1}	±2.0% FS	Model FLST-C8	Dwyer Instruments, Inc., Michigan City, IN, USA
Pitot tube pressure transducer	0.0 to 248.8 Pa	±1.0% FS	Model 267	Setra Systems, Inc., Boxborough, MA, USA
Incoming air temperature	−40.0 to 105.0 °C	±1.0% at 25 °C	Model TT05-10KC8-15-T105-1500	TEWA Sensors, LLC, Albuquerque, NM, USA
Precision resistor	10 k Ω	±5.0 × 10^{-2}% FS	Model 10K Tick	Labjack Corporation, Lakewood, CO, USA
Incoming air temperature and relative humidity	40.0 to 80.0 °C, 0.0 to 100.0% RH	±0.1 °C, ±1.5% RH	Model HMT120	Vaisala, Helsinki, Finland

Heater supply air (room) Dry-Bulb Temperature (Tdb) was measured using a grouping of four plastic-coated thermistors (see Table 3 for specifications) installed 0.25 m below the intake of the heater on the side opposite of the Liquid Propane (LP) intake and burner manifold. The four thermistors were multiplexed (Model: CD4052BE, Texas Instruments, Dallas, TX, USA) and conditioned with a voltage divider circuit consisting of a precision resistor (see Table 3 for specifications). Temperature from the thermistors was calculated using the β equation (Equation (3)) and then converted to Celsius for analysis. The thermistors were calibrated in a variable environment chamber (Series 7064-3140, Parameter Generation & Control, Black Mountain, NC, USA) at three unique temperatures (25 °C, 35 °C, and 45 °C) using the exact same DAQC setup prior to testing. The calibration regression was developed by regressing the chamber temperature recoded from the control panel on the x-axis and the average thermistor temperature on the y-axis. The final calibration equation was determined by taking the inverse of the previous calibration equation [7].

$$T_{TRi} = 1/\left[(1/298.15) + (1/\beta) * ln((R_{d10} * (V_{s1}/V_{out1} - 1))/R_{ref})\right] \quad (3)$$

where

T_{TR} = thermistor temperature i (K)
B = thermistor coefficient (3435)
R_{d10} = resistance of voltage divider resistor (Ω)
V_{s1} = voltage supply (5.0 VDC)
V_{out1} = analog voltage output measured by DAQC (VDC)
R_{ref} = thermistor reference resistance at 25 °C (Ω)

Supply air Tdb and Relative Humidity (RH) were measured in the same location using a portable hygrometer with a Liquid Crystal Display (LCD) screen (see Table 3 for specifications). The remote hygrometer was recorded manually at the start and end of each test.

All sensors, excluding the remote sensing unit, were recorded with a custom software interface (Visual Basic v19, Microsoft, Redmond, WA, USA). The software allowed for user-selectable sampling frequency and a calibration procedure to calibrate the load cell.

2.3. Testing Procedure

Data were collected on the three nominally rated DGFCHs (21, 29, and 73 kW) at the full rated output and half output setting. For the 21 and 29 kW rated heaters, three replicates of each output were measured. For the 73 kW heater, two replicates of each output were completed. This was due to issues with the test duct retaining residual heat between tests. Prior to test initiation, the load cell was verified with a 45.4 kg (100 lb) standard set of weights. During a test, the DAQC was configured to sample continuously at 1 Hz. For DGFCH ratings > 18 kW, the total sampling duration was 500 s with the DAQC system sampling at 1 Hz. For DGFCH ratings < 18 kW, the total sampling duration was 900 s, with the DAQC sampling at 1 Hz. The time duration was different for the heater and settings rated for <18 kW based on the anticipated propane consumption to have a minimum change in the propane tank mass double the load cells accuracy. Electrical energy consumption was recorded from the power meter at 3 min into the test and assumed constant for the duration of the test. The three-minute mark was selected based on ANSI/ASHRAE Standard 103, considering steady-state combustion at three minutes [5]. Between performance tests, the ADL was ventilated for 60 min to maintain similar room temperatures at the start of each test.

2.4. Thermal Efficiency Development

The energy balance of a DGFCH is the balance of the energy input and output of the heater (Equation (4)). Output DGFCH energy can be defined as the sum of the combustion heat in the output airflow and the radiative and convective heat loss of the heater jacket (Equation (5)). For agriculture buildings, the main objective is to warm the occupied zones of the house. Thus, the goal of a DGFCH thermal efficiency is to quantify the efficiency of heating the output air (Equation (6); [5]).

$$Q_{out} = Q_{in} \quad (4)$$

where

Q_{out} = energy output of a direct gas-fired circulating heater (kW)
Q_{in} = energy input (propane and electrical) of a direct gas-fired circulating heater (kW)

$$Q_{out} = Q_a + Q_j \quad (5)$$

where

Q_a = energy output via air (kW)
Q_j = energy output via heater jacket (kW)

$$\hat{Q}_{out} = Q_{in} \times n_{ss} \quad (6)$$

where

$\hat{Q}_{out}$ = measured energy output of heater output air (full or half load; kW)
n_{ss} = steady-state sensible energy output efficiency (%)

Energy input into a DGFCH can be quantified as the energy from the liquid propane plus the electrical energy (Equation (7)). Electrical energy is included as most of the DGFCH supply air is drawn in over the motor and electronic components for cooling purposes. For the 21 and 29 kW rated heaters, input propane was measured gravimetrically, while input propane for the 73 kW rated heater was measured volumetrically, with a temperature and pressure compensated gas meter.

$$Q_{in} = \left(\dot{x}_{LPG} \times HHV_{prop}\right) + W_{elec} \quad (7)$$

where

$\dot{x}_{LPG}$ = mass flow rate of propane to heater (kg s^{-1}) or volume flow rate of propane (m^3 s^{-1})
HHV_{prop} = higher heating value of propane (kJ kg or kJ m^{-3})
W_{elec} = electrical energy consumption of heater (kW)

$$\hat{Q}_{out} = \dot{m}_a \times (h_{out} - h_{in}) \quad (8)$$

where

$\dot{m}_a$ = mass flow rate of exhaust air from heater (kg s^{-1})
h_{out} = enthalpy of output air from heater (kJ kg^{-1})
h_{in} = enthalpy of oncoming air to heater (kJ kg^{-1})

$$\dot{m}_a = \vec{V} \times \left(\pi r^2\right) \times \rho \tag{9}$$

where
$\vec{V}$ = fluid velocity (m s^{-1})
r = radius of the duct (0.10 m)
ρ = density of the air exiting heater (kg m^{-3})

$$\vec{V} = \sqrt{\frac{2\delta p}{\rho}} \tag{10}$$

where
δp = velocity pressure measurement (Pa)

The output energy was quantified by the enthalpy change of the air moving through the heater (Equation (8)). The mass flow of the DGFCH exhaust air was calculated using the density measurements and the pitot tube measurement (Equations (9) and (10)). The specific volume and enthalpy were calculated using the humidity ratio of the starting room conditions [1,8]

By combining the input energy (Equation (4)) and the quantified energy output (Equation (5)), the steady-state thermal efficiency can be solved for the DGFCH (Equations (11) and (12)). The thermal efficiency equation is similar in nature to efficiency equations used to evaluate solar air heaters [9,10]

$$\dot{m}_a \times (h_{out} - h_{in}) = \left(\left(\dot{x}_{prop} \times HHV_{prop}\right) + W_{elec}\right) \times n_{ss} \tag{11}$$

$$n_{ss} = \left(\dot{m}_a \times (h_{out} - h_{in})\right) \div \left(\left(\dot{x}_{prop} \times HHV_{prop}\right) + W_{elec}\right) \tag{12}$$

2.5. Thermal Efficiency Calculations and Analysis

From each replicate, the mean of six random sub-samples of 30 s (n = 30 per sub-sample) of consecutive data was used to calculate the steady-state thermal efficiency (Equation (12)). A moving average of five data points was used to reduce fluctuations in the output air temperature. This processing step was justified due to the random error caused by turbulence and non-constant combustion of the heater. The mass flow rate of the propane was calculated by determining the weight change of the propane cylinder divided by the experiment duration. For DGFCH's rated >18 kW, a 4 s average before the heater solenoid was opened and after the test was used to determine the pre- and post-weight. For DGFCH's rated <18 kW, a 3 s average was used. The averaging of the cylinder weight was necessary to account for variation in the weight caused by vibration of the cylinder due to propane flow.

A sensitivity analysis of the steady-state thermal efficiency was performed with respect to the Higher Heating Value (HHV) of propane. Four sources of the HHV were used in the analysis due to the slight differences among sources (Table 4).

Table 4. Higher Heating Value (HHV) values from different sources used in the sensitivity analysis of the thermal efficiency equation.

Source	HHV (kJ kg^{-1})	References
A	50,009	[5]
B	50,411	[8]
C	50,656	[11]
D	50,129	[12]

2.6. Uncertainty Analysis

The final standard uncertainty of the thermal efficiency (Δn_{ss}) was calculated from the standard uncertainty obtained from all key measurement inputs propagated through (Equation (13) through Equation (20), Figure 3).

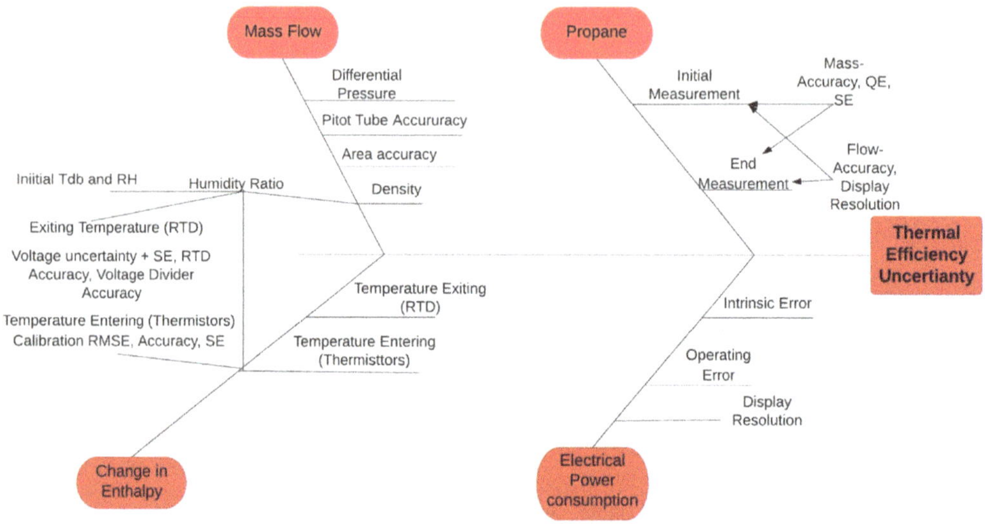

Figure 3. Fishbone diagram of the final standard uncertainty in the thermal efficiency calculation. Each line represents the integration of a measurement and its respective uncertainty into a standard uncertainty calculation.

A zeroth-ordered uncertainty budget was created for each measured input: voltage measurement for exiting air temperature (T_{ex}; Table 5), a Differential Pressure Transducer (DPT; Table 6), dry-bulb temperature, and relative humidity for humidity ratio calculation (Tdb and RH; Table 7), propane flow for mass and volumetric ($\dot{x}$; Table 8), and electrical energy consumption (W_{elec}; Table 9) and included the manufacturer's accuracy and long term stability, quantization error from the 14-bit Analog-to-Digital Converter (ADC), and the Standard Error (SE) from the experimental data (SE was removed from the budget as it changes with each experiment; [13]).

Table 5. Uncertainty budget analog voltage measurement by DAQC board for RTD measurement.

.	Value (VDC)	Probability Distribution	Divisor	Standard Uncertainty (Pa)
Accuracy [a]	3.7×10^{-3}	Rectangular	$\sqrt{3}$	2.1×10^{-3}
Quantization error [b]	3.1×10^{-4}	Rectangular	$\sqrt{3}$	1.8×10^{-4}
	Combined standard sensor uncertainty, ΔVDC			2.1×10^{-3}

[a] ±0.5 Manufacturer's stated accuracy with ±5. VDC input range. [b] ±0.5 ADC resolution = 14-bit ADC resolution, ±5 VDC reference range = 3.1E−4 V BL^{-1}. BL = Binary Level.

Table 6. Uncertainty budget for differential pressure transducer.

Source	Value (Pa)	Probability Distribution	Divisor	Standard Uncertainty (Pa)
Accuracy RSS [a]	2.5	Rectangular	$\sqrt{3}$	1.4
Long-term stability	5.0	Rectangular	$\sqrt{3}$	2.9
Quantization error [b]	3.2×10^{-2}	Rectangular	$\sqrt{3}$	1.8×10^{-2}
	Combined standard sensor uncertainty, $\Delta\delta p$			3.2

[a] Root sum square (at constant t_{db}), ±1.0% full scale ($\delta p1$) 0 to 248.8 Pa; [b] ±0.5 sensor resolution = (14-bit ADC resolution, ±10 VDC reference range = 6.1E−4 V BL^{-1}) (sensor sensitivity)$^{-1}$.

Table 7. Uncertainty budget for temperature and relative humidity sensor used to calculate humidity ratio.

Source	Value (°C, %RH)	Probability Distribution	Divisor	Standard Uncertainty (°C, %RH)
Accuracy RSS [a]	0.1, 1.5	Rectangular	$\sqrt{3}$	5.7×10^{-2}, 8.7×10^{-1}
Long-term stability	N/A [c], 1.0	Rectangular	$\sqrt{3}$	N/A [c], 5.7×10^{-1}
Display Resolution [b]	5.0×10^{-3}, 5.0×10^{-3}	Rectangular	$\sqrt{3}$	2.9×10^{-3}, 2.9×10^{-3}
Combined standard sensor uncertainty, Δt_{db}, Δrh				5.7×10^{-2}, 1.0

[a] Manufacturer stated accuracy; [b] ±0.5 smallest display value = 0.01; [c] long-term stability is included in accuracy term from manufacturer specifications.

Table 8. Uncertainty budget for propane for both mass and flow basis.

Source	Value (kg, m^3 s^{-1})	Probability Distribution	Divisor	Standard Uncertainty (kg, m^3 s^{-1})
Accuracy RSS [a]	1.0×10^{-2}, 4.4×10^{-4}	Rectangular	$\sqrt{3}$	5.7×10^{-2}, 2.6×10^{-4}
Long-term stability	N/A [c], N/A [c]	Rectangular	$\sqrt{3}$	N/A [c], N/A [c]
Display Resolution [b]	N/A, 1.4×10^{-3}	Rectangular	$\sqrt{3}$	NA, 2.5×10^{-3}
Quantization Error [d]	6.1×10^{-3}, N/A	Rectangular	$\sqrt{3}$	3.5×10^{-3}, N/A
Combined standard sensor uncertainty, $\Delta \dot{x}_m$, $\Delta \dot{x}_f$				5.8×10^{-2}, 2.4×10^{-3}

[a] Manufacturer stated accuracy; [b] ±0.5 smallest display value = 0.01; [c] long-term stability is included in accuracy term from manufacturer specifications; [d] ±0.5 sensor resolution = (14-bit ADC resolution, 1 VDC reference range = 1.2E−4 V BL^{-1}) (sensor sensitivity)$^{-1}$.

Table 9. Uncertainty budget for electrical energy consumption of the heater.

Source	Value (W)	Probability Distribution	Divisor	Standard Uncertainty (W)
Intrinsic error	241.5	Rectangular	$\sqrt{3}$	139.4
Operating error	517.5	Rectangular	$\sqrt{3}$	298.8
Display Resolution [a]	50	Rectangular	$\sqrt{3}$	28.9
Combined standard sensor uncertainty, ΔW_{elec}				331.0

[a] ±0.5 smallest display value = 50.

The uncertainty budget for the Tdb and RH includes a display resolution component as it was manually recorded (Table 7). The propane volumetric measurement also had a display resolution component from the dial on the meter display from manually recording measurements.

The standard uncertainty associated with the area of the duct was determined by propagating the standard uncertainty of the duct radius ($\Delta r = 2.70 \times 10^{-3}$ m, one-half the reading resolution set by the manufacturer) through the area equation of a circle resulting in a standard uncertainty of the area $\Delta a = 3.39 \times 10^{-3}$ m^2.

$$\Delta VDC_i^2 = \left(\frac{\Delta VDC}{n}\right)^2 + SE^2 \tag{13}$$

where

ΔVDC_i = combined standard uncertainty in analog voltage measurement (T_{ex} VDC)
SE = standard error of the voltage measurement (VDC)

$$\Delta T_{ex}^2 = \left(\frac{\partial T_{ex}}{\partial VDC}\Delta VDC_i\right)^2 + \left(\frac{\partial T_{ex}}{\partial R_{d1}}\Delta R_{d1}\right)^2 + \left(\frac{\partial T_{ex}}{\partial R_{ref}}\Delta R_{ref}\right)^2 + RMSE^2 \tag{14}$$

where

ΔT_{ent} = combined standard uncertainty in entering air temperature (°C)
ΔR_{d1} = 1 kΩ resistor in divider circuit (±0.05%; Ω; rectangular distribution)
ΔR_{ref} = 100 Ω at 0 °C (±0.12%; Ω; rectangular distribution)

$RMSE$ = root mean square error from nonlinear regression (°C)

The standard uncertainty of the entering temperature measured by the thermistor did not include propagation of uncertainty through (Equation (3)) as the DAQC system was identical in both calibration and testing. Thus, it was assumed that the RMSE of the calibration encompasses the uncertainty associated with the measurements propagated through the equation.

$$\Delta T_{ent}^2 = ACC^2 + RMSE^2 + SE^2 \tag{15}$$

where

ΔT_{ent} = combined standard uncertainty in entering air temperature (°C)
ACC = manufacturer's accuracy (±0.2 °C; rectangular distribution)
$RMSE$ = root mean square error from linear calibration regression (°C)
SE = standard error of the mean (°C)

$$\Delta W^2 = \left(\frac{\partial W}{\partial t_{db}}\Delta t_{db}\right)^2 + \left(\frac{\partial \rho}{\partial rh}\Delta rh\right)^2 \tag{16}$$

where

ΔW = combined standard uncertainty in humidity ratio ($kg_{H2O}\ kg_{da}{}^{-1}$)

$$\Delta \rho^2 = \left(\frac{\partial \rho}{\partial t_{ex}}\Delta t_{ex}\right)^2 + \left(\frac{\partial \rho}{\partial W}\Delta W\right)^2 \tag{17}$$

where

$\Delta\rho$ = combined standard uncertainty in moist air density (kg m^{-3})

The standard uncertainty in the mass flow (Equation (18)) was determined from the propagation of the velocity pressure uncertainty, density uncertainty, area uncertainty, and accuracy of the pitot tube system.

$$\Delta \dot{ma}^2 = \left(\frac{\partial \dot{ma}}{\partial dp}\Delta dp\right)^2 + \left(\frac{\partial \dot{ma}}{\partial \rho}\Delta \rho\right)^2 + \left(\frac{\partial \dot{ma}}{\partial a}\Delta a\right)^2 + ACC_v{}^2 \tag{18}$$

where

$\Delta\dot{ma}$ = combined standard uncertainty in mass flow moist air (kg s^{-1})
ACC_v = accuracy of the velocity measurement (1.0%, [5])

$$\Delta h_{ex-ent}{}^2 = \left(\frac{\partial h_{out-in}}{\partial W}\Delta W\right)^2 + \left(\frac{\partial h_{out-in}}{\partial T_{ent}}\Delta T_{ent}\right)^2 + \left(\frac{\partial h_{out-in}}{\partial T_{ex}}\Delta T_{ex}\right)^2 \tag{19}$$

where

Δh_{out-in} = combined standard uncertainty in change in enthalpy (kJ kg^{-1})

The standard uncertainty of $\dot{x}_{LPG}$ included in (Equation (20)) included the SE term for the initial and final measurement from the gravimetric system as a multi-second average was used. It was assumed that there was no uncertainty in the run time of each test factored into $\dot{x}_{LPG}$.

$$\Delta n_{ss}{}^2 = \left(\frac{\partial n_{ss}}{\partial \dot{ma}}\Delta \dot{ma}\right)^2 + \left(\frac{\partial n_{ss}}{\partial h_{ex-ent}}\Delta h_{ex-ent}\right)^2 + \left(\frac{\partial n_{ss}}{\partial \dot{x}}\Delta \dot{x}_{LPG}\right)^2 + \left(\frac{\partial n_{ss}}{\partial W_{elec}}\Delta W_{elec}\right)^2 \tag{20}$$

where

Δn_{ss} = combined standard uncertainty in thermal efficiency (%)

The combined standard uncertainty was calculated using the data from each replicate of the heater model and output setting. Thus, for each replicate and output, a standard deviation was calculated for the uncertainty due to variations in measurements between each replicate.

2.7. Economics of Heater Efficiency

Economic implications of heating agricultural buildings were assessed by modifying the degree-day heating method (Equation (21); [3]). The method is simpler than other heating cost methods that use binned ambient weather data, with the consequence of overestimating the heating season costs. The efficiency represents the Annual Fuel Energy Use Efficiency (AFEUE). The AFEUE is based on the ASHRAE standard 103 [5]. For DGFCH without a pilot light, the AFEUE is the same as the steady-state thermal efficiency. For a DGFCH with a pilot light, the calculation of AFEUE considers the proportion of time with the pilot light on and its efficiency over the duration of the heating season.

$$E = C_D \times \left[\left(H_L \times D \times S \right) \div \left(\Delta t \times n_{ss/AEU} \times HHV_{prop} \right) \right] \tag{21}$$

where

E = propane usage per season (kg)
C_D = correction factor for degree-days (0.6)
H_L = design heating load at minimum ventilation (kW)
D = degree-days (°C day)
S = seconds per day (86,400 s d^{-1})
Δt = design temperature difference, inside/outside (°C)
HHV_{prop} = higher heating value for propane (kJ kg^{-1})

The calculation of the heating load for a building (heat gain/heat loss) was developed to account for the wide variety of building specifications (Equation (22)). This was performed by accounting for the heat gain on the basis of floor area. The heating load is to be calculated at the wintertime 99% design temperature for the given location.

$$HL = \left[(UA \times \Delta T) + \left(\dot{V_b} \div v \times C_p \times \Delta T \right) \right] + \left[q_g \times A_f \right] \tag{22}$$

where

UA = overall building heat loss coefficient (W °C^{-1})
$\dot{V_b}$ = minimum ventilation rate (m^3 s^{-1})
v = specific volume of air/water mixture (kg m^{-3})
C_p = specific heat of air (J kg^{-1})
Q_g = heat production per unit of floor area (W m^{-2})
A_f = floor area of building (m^2)

A sensitivity analysis of the energy usage (Equation (21)) was performed based on recommended thermal resistance (R values) for various building components, internal design temperature, two different minimum ventilation rates, and heater efficiency. This analysis used a high heat gain per unit floor area of 1.6 W m^{-2}. It was assumed that the outside design temperature was −29 °C (99% heating value for Minnesota; [8]). For this economic analysis, the building size was 45.7 m (L), 15.2 m (W) with a 2.5 m ceiling (150 ft L, 50 ft W with 8 ft ceiling). The building's thermal analysis assumed no windows and doors and did not include any perimeter or floor heat loss. The R values, internal building temperature, and minimum ventilation rates (Table 10). For the analysis of the first three components (level of insulation, design internal Tdb, and minimum ventilation rate), the heater efficiency was 99%, and the cost of propane was assumed at USD $0.53 kg^{-1} ($1 gallon^{-1}).

Table 10. Sensitivity analysis inputs for R values, internal building temperature, and minimum ventilation rate. Recommended levels of insulation for homes based on climate [3].

Level of Insulation	Wall (R m^2 °C W^{-1})	Ceiling (R m^2 °C W^{-1})
UA_{low}	2.1	3.5
UA_{medium}	2.5	4.4
UA_{high}	3.5	5.8
$UA_{ultra\text{-}high}$	3.5	6.7
Design internal temperature		
Ti_{high} = 28 °C	Ti_{medium} = 25 °C	Ti_{low} = 20 °C
Minimum ventilation rate		
	V_{high} = 6 ACH [a]	V_{low} = 3 ACH [a]

[a] ACH = air changes per hour.

3. Results and Discussion

3.1. Thermal Efficiency

The average values of the operational parameters and results for the three DGFCHs tested at each output performed as expected (Table 11). The procedures developed were capable of detecting the relatively small differences in input and output energy, although the 73 kW did not reach rated operation. The time to steady-state assumed from ASHRAE 103 was represented in the results of this study (Figure 4), with all heaters and replicates showing similar trends in temperature change over time. Future tests should involve a high accuracy CO and O_2 sensor to verify the thermal efficiency through stoichiometry (indirect combustion test). A unique trend was noted in the comparison of the half and full output thermal efficiencies. Heater models A and B exhibited higher thermal efficiencies at full output, while heater model C exhibited a higher thermal efficiency at half output. The thermal efficiency of over 100% is related to the system's uncertainty, discussed in the next section. The general trend of efficiency increasing with heater output could be influenced by the test stand gaining heat during the test and affecting measurements. This was the reason behind two replicates with heater model C. The heater models used in the study had limited thermal efficiency values reported, model A had no reported values, and models B and C reported 99.99% thermal efficiency. The reported thermal efficiencies in this study are higher than what is assumed for DGFCH in buildings 80%, with high humidity and low ventilation [4]. However, it was not within the scope of this study to evaluate the heaters under such conditions that are believed to drastically reduce the performance of these heaters in buildings. The thermal efficiency of the DGFCH used in this study is significantly higher (>90%) compared to the thermal efficiency of common gas-fired residential units (63% to 44%; [13]). This drastic difference is mainly contributed to the difference in the heaters as the residential heaters.

Table 11. Average (standard deviation) of test conditions and measured performance values for all three DGFCH models and outputs.

Measured Input [a] (kW)	Heater Model	Manifold Pressure (kPa)	Output Air Temperature (°C)	Enthalpy Gain (kJ kg da^{-1})	Volumetric Output (L min^{-1})	Thermal Efficiency [b] (%)
11.43	A	2.5 (3.0×10^{-2})	67.2 (2.0)	51.2 (1.1)	13,198 (51)	94.3% (1.5)
17.00	B	2.3 (3.0×10^{-2})	80.0 (1.2)	61.9 (0.4)	16,352 (33)	92.4% (1.5)
21.39	A	2.1 (6.0×10^{-2})	118.3 (1.4)	102.1 (1.1)	14,631 (401)	96.5% (1.5)
29.89	B	1.9 (5.0×10^{-2})	140.8 (1.0)	125.5 (0.6)	17,169 (242)	94.98% (1.1)
42.20	C	1.7 (3.0×10^{-2})	131.0 (1.6)	109.2 (0.5)	29,019 (93)	100.9% (0.3)
56.27	C	1.4 (3.0×10^{-2})	180.4 (2.3)	159.1 (3.6)	29,439 (200)	98.8% (2.0)

[a] Measured input calculated using HHV from [5]. [b] Thermal efficiency calculated using HHV from [5].

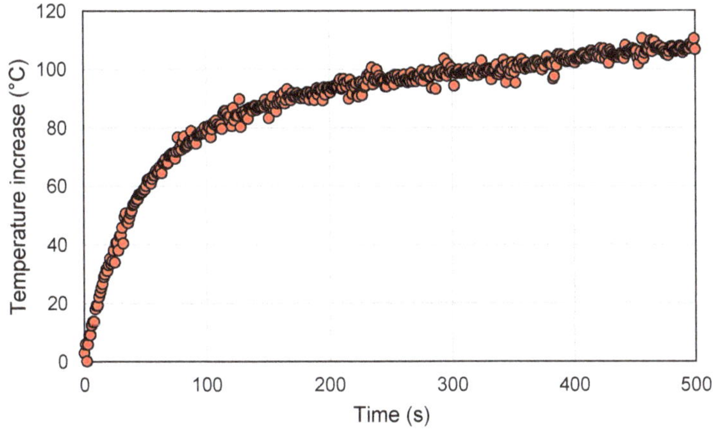

Figure 4. Temperature gain of the exiting air over the course of a test for heater model A at full output.

The measured input energy to heater model C at full output was significantly lower than its nominal rating of 73.3 kW. Upon investigation of the operational parameters of the tests, it was noted that manifold pressure exhibits a negative trend in relation to the measured input (Figure 5). The temperature of the propane did not vary between any heaters and replicates. This is most likely caused by an improper design of the propane supply of ADL that limits the maximum output due to propane flow and not the heater reaching its rated output.

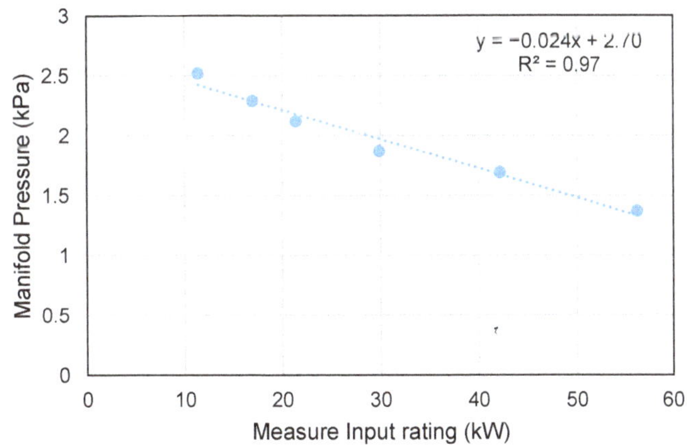

Figure 5. Measured input of propane (x-axis) versus the manifold propane pressure (y-axis).

The sensitivity analysis of the thermal efficiency equation in relation to the HHV was performed for heater model A at full output. The maximum difference noted in the analysis is between HHV source A (97.3%) and D/C (96.0%), with source B (96.5%) falling in-between. While all four sources are relatively similar in value, this highlights the differences in reported values among groups (sources A, B, C) and the federal government (source D). A unified HHV between all organizations is imperative to avoid discrepancies in design and in the rating of all types of propane combustion heaters.

3.2. Uncertainty Analysis

The expanded (coverage factor = 2; ~95% confidence interval) uncertainty ranged from 13.1% (Standard Deviation; SD = 0.1%) to 30.7% (SD = 8.0%) for the input energy rating 56.3 to 11.4 kW; the SD is from the multiple replicates of each heater output tested. There is a distinct difference in the variability of the uncertainty between the replications for a heater based on the propane measurement system. The mass flow system had significantly greater variability (Figure 6). The breakdown of the uncertainty contributions (Table 12) demonstrates that for the mass flow system, the propane mass flow and air mass flow are the largest contributors while using volumetric flow measurement, the air mass flow, and change in enthalpy are the largest contributors. Overcoming the variability in the propane mass flow for future tests could require a hardware-based signal conditioning filter to account for the signal noise due to vibrations. Alternatively, extended test duration could be used to reduce variability.

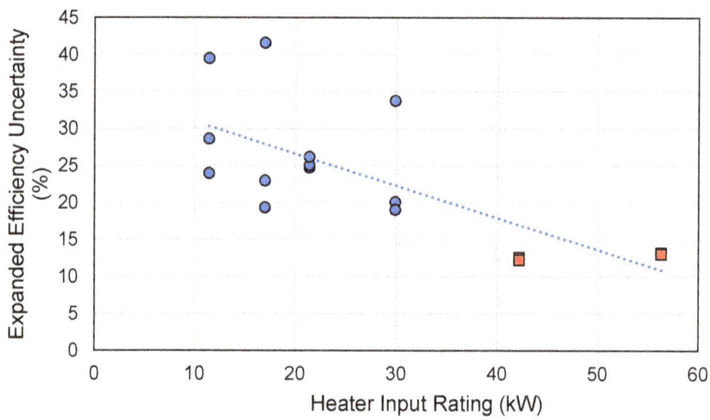

Figure 6. Expanded thermal efficiency (y-axis) versus heater input rating (x-axis). The blue circles represent the measurements using the mass flow system, and the red squares represent the volumetric flow measurements.

Table 12. Expanded uncertainty component contributions for Equation (20) based on the propane measurement system. The values in parentheses represent 1 standard deviation of the mean.

	Airflow	Change in Enthalpy	Propane Flow	Electrical Consumption
Mass flow	38.0% (14.3)	4.0% (3.0)	55.9% (16.2)	2.1% (1.4)
Volumetric Flow	80.8% (4.9)	17.5% (4.5)	0.6% (0.2)	1.2% (0.3)

The expanded uncertainty component contributions for the volumetric propane measurement system suggest that if a high accuracy, low flow rate option was feasible, it would reduce the uncertainty considerably. To reduce uncertainty in future experiments, a higher-accuracy pressure transducer and RTD would be warranted.

3.3. Economics of Heater Efficiency

The economic estimates for heating the specified building using design conditions show a wide heating cost for a heating season (Table 13). There are a few limitations to the degree-day method for estimating this cost. First, it assumes a constant balance temperature for a building. Balance temperature is the ambient temperature that separates the need for heating or cooling, which is different for the various types of agricultural buildings (livestock versus greenhouse). For livestock buildings, this approach cannot account for the increase in the animal's heat production through a growth cycle and the change in sensible heat production with indoor temperature. This economic assessment confirms that the greatest heating cost factor is the minimum ventilation rate, with building internal temperature and insulation level being the other factors impacting cost. It should be cautioned against decreasing the ventilation rate of a building to reduce the seasonal heating cost as air quality and moisture levels will be directly impacted and could negatively impact production within the building. The costs and corresponding energy inputs are higher than reported values from field studies for animal agricultural buildings with similar heat production values per floor area [14–16]. These differences are most likely attributed to the method used and the local weather data differences from the reported study's locations.

Table 13. Economic analysis (values reported in USD) for a general agricultural building for a heating season with a heat production of 1.6 W m^{-2}.

Level on Insulation	V_{high}			V_{low}		
	Ti_{high}	Ti_{medium}	Ti_{low}	Ti_{high}	Ti_{medium}	Ti_{low}
UA_{low}	$7059	$6756	$6260	$3844	$3676	$3398
UA_{medium}	$6941	$6644	$6159	$3727	$3564	$3297
UA_{high}	$6794	$6505	$6032	$3579	$3424	$3170
$UA_{very\ high}$	$6762	$6475	$6005	$3548	$3395	$3144

The sensitivity analysis for the scenario regarding heater efficiency (Table 14) shows the economic impact of reducing the DGFCH thermal efficiency by 1% for the combinations of internal temperature and building envelope insulation level. The range of costs is caused by the different range of supplemental heat needs based on internal temperature and insulation value. The magnitude of the costs would be significantly higher if an AEU efficiency for a heater with a pilot light.

Table 14. Economic cost (values reported in USD) of reducing the DGFCH thermal efficiency by 1% for a heating season.

Level on Insulation	Ti_{high}	Ti_{medium}	Ti_{low}
UA_{low}	$72.0	$68.9	$63.9
UA_{medium}	$70.8	$67.8	$62.8
UA_{high}	$69.3	$66.4	$61.6
$UA_{very\ high}$	$69.0	$66.1	$61.3

The methods used to perform this economic analysis could be adapted to a range of scenarios to evaluate utility costs for building types, climates, and construction for a cost estimate. Other methods for estimating heating costs are binning methods, where data is binned by day for heating demand (time and temperature difference). This method has advantages in being able to incorporate livestock heat production curves and variable ventilation rates. The downside is that this method requires extensive calculations for each time period within each day and is not a simple calculation for agriculture professions to estimate heating costs for a building.

4. Conclusions

A custom testing apparatus and instrumentation system were developed and evaluated for Direct Gas-Fired Circulating Heaters (DGFCH) commonly used in agricultural buildings. Three DGFCH were used for the evaluation testing of the system. The steady-state thermal efficiency of the tests was determined using data from the testing system and based on calculations from a standard for residential systems [5]. An uncertainty analysis was performed on the thermal efficiency calculation for the given measured inputs. An economic analysis for heating common agricultural buildings was performed. An inadequate propane service limited the systems testing ability for 73 kW rated heaters. A sensitivity analysis of the thermal efficiency in respect to the Higher Heating Value (HHV) results in a maximum difference between sources of HHV of 1.3% efficiency. This highlights the need for standardization between all organizations involved. The uncertainty of the system ranged from 13.1% to 30.7% across the input range of 56.2 to 11.4 kW, respectively. The propane mass was the key source of uncertainty at the lower ratings. A more robust hardware system and longer run times would reduce this key source of uncertainty. At higher input ratings, the mass flow of air was the key source of uncertainty. Reducing this would require the use of higher-accuracy sensors. The economic analysis showed that for a 1% difference in heater efficiency across the combination of inputs (insulation and inside temperature), the costs per heating season ranges from USD $61.3 to $72.0. Future evolutions of this system are needed to reduce the overall uncertainty in thermal efficiency. It is suggested that a flowmeter with lower minimum flow ratings be used or longer test durations to reduce the main source of uncertainty. Other instrumentation improvements are needed for the output air and velocity measurement in the duct. Overall this system shows a great prospect for testing DGFCH commonly used in agricultural buildings. This study has highlighted the need for standardized procedures for quantifying DGFCH thermal efficiency specific to agricultural applications. The unique nature of these heaters presents challenges for testing and has a large economic impact on the agricultural operation during heating seasons.

Author Contributions: Conceptualization, B.C.S., B.C.R., and S.J.H.; methodology, B.C.S., B.C.R., and S.J.H.; software, B.C.S. and S.J.H.; validation, B.C.S.; formal analysis, B.C.S.; resources, B.C.R.; data curation, B.C.S.; writing—original draft preparation, B.C.S.; writing—review and editing, B.C.S., B.C.R., and S.J.H.; supervision, B.C.R.; project administration, B.C.R.; funding acquisition, B.C.R. All authors have read and agreed to the published version of the manuscript.

Funding: This research is a product of the Iowa Agriculture and Home Economics Experiment Station, Ames, Iowa. Project Number IOW04100 is sponsored by the Hatch Act and State of Iowa funds. The content of this article is, however, solely the responsibility of the authors and does not represent the official views of the USDA.

Institutional Review Board Statement: Not applicable.

Informed Consent Statement: Not applicable.

Data Availability Statement: The data presented in this study are available on request from the corresponding author.

Acknowledgments: The authors would like to thank the contributions of undergraduate assistants, Devri Cliburn and Julia Bowman, for their assistance with the calibration and construction of the testing apparatus.

Conflicts of Interest: The authors declare no conflict of interest. The funders had no role in the design of the study; in the collection, analyses, or interpretation of data; in the writing of the manuscript; or in the decision to publish the results.

References

1. Albright, L.D. *Environment Control for Animals and Plants*; American Society of Agricultural and Biological Engineers: St. Joesph, MI, USA, 1990; ISBN 0-929355-08-3.
2. Cui, Y.; Theo, E.; Gurler, T.; Su, Y.; Saffa, R. A Comprehensive Review on Renewable and Sustainable Heating Systems for Poultry Farming. *Int. J. Low Carbon Technol.* **2020**, *15*, 121–142. [CrossRef]
3. Midwest Plan Service. *Structures and Environment Handbook*; MWPS-1-TSM; Midwest Plan Service: Ames, IA, USA, 1983.
4. Sanford, S. *Greenhouse Unit Heaters—Types, Placement & Efficiency*; University of Wisconsin Cooperative Extension: Madison, WI, USA, 2011; p. 6.
5. ANSI/ASHRAE. *Method of Testing for Annual Utilization Efficiency of Residential Central Furnances and Boilers*; ASHRAE: Atlanta, GA, USA, 2017.
6. Hoff, S.J.; Utrecht, D.V.; Harmon, J.D.; Mangold, D.W. A General Purpose Laboratory for Evaluating Livestock Ventilation Systems. *Appl. Eng. Agric.* **2000**, *16*, 701–710. [CrossRef]
7. Doebelin, E.O. *Measurement Systems: Application and Design*, 5th ed.; McGraw-Hill: New York, NY, USA, 2004; ISBN 0-07-243886-X.
8. ASHRAE. *ASHRAE Handbook*; ASHRAE: Atlanta, GA, USA, 2013.

9. Parker, B.F.; Lindley, M.R.; Colliver, D.G.; Murphy, W.E. Thermal Performance of Three Solar Air Heaters. *Sol. Energy* **1993**, *51*, 467–479. [CrossRef]
10. Saxena, A.; Verma, P.; Srivastava, G.; Kishore, N. Design and Thermal Performance Evaluation of an Air Heater with Low Cost Thermal Energy Storage. *Appl. Therm. Eng.* **2020**, *167*, 114768. [CrossRef]
11. Energy Information Administration (EIA). EIA Energy Units and Calculators Explained—U.S. Available online: https://www.eia.gov/energyexplained/units-and-calculators/ (accessed on 21 January 2021).
12. REGO. *LP—Gas Serviceman's Manual*; REGO: Elon, NC, USA, 2020.
13. Juanicó, L.E.; González, A.D. Thermal Efficiency of Natural Gas Balanced-Flue Space Heaters: Measurements for Commercial Devices. *Energy Build.* **2008**, *40*, 1067–1073. [CrossRef]
14. Sharpe, K.T.; Reese, M.H.; Buchanan, E.S.; Tallaksen, J.E.; Janni, K.A.; Johnston, L.J. Electrical and Thermal Energy Consumption in Midwest Commercial Swine Facilities. *Appl. Eng. Agric.* **2018**, *34*, 857–864. [CrossRef]
15. Barber, E.M.; Classen, H.L.; Thacker, P.A. Energy Use in the Production and Housing of Poultry and Swine—An Overview. *Can. J. Anim. Sci.* **1989**, *69*, 7–21. [CrossRef]
16. Fernandez, M.D.; Losada, E.; Ortega, J.A.; Arango, T.; Ginzo-Villamayor, M.J.; Besteiro, R.; Lamosa, S.; Barrasa, M.; Rodriguez, M.R. Energy, Production and Environmental Characteristics of a Conventional Weaned Piglet Farm in North West Spain. *Agronomy* **2020**, *10*, 902. [CrossRef]

agriculture

MDPI

Article

Design and Experimental Research on Soil Covering Device with Linkage and Differential Adjustment of Potato Planter

Zhiqi Zheng, Zuoli Fu *, Chenyang Wang, Yuxiang Huang and Jinpu He

College of Mechanical and Electronic Engineering, Northwest A&F University, Yangling 712100, China; zhiqizheng@nwsuaf.edu.cn (Z.Z.); wangchenyang333@nwsuaf.edu.cn (C.W.); hyx@nwsuaf.edu.cn (Y.H.); hjp@nwsuaf.edu.cn (J.H.)

* Correspondence: fuzuoli@nwsuaf.edu.cn

Abstract: When the potato planter works on sloping field, it will cause problems such as poor film mulching quality due to the difference in volume of soil covering both sides of the discs and the inconvenient adjustment of the soil covering disc. The soil covering device with linkage and differential adjustment was designed to improve the mulching quality. The main research content includes explaining the structure and principle of the soil covering device and analyzing the structural parameters of the adjustment mechanism. The field experiment was completed to verify the performance of soil covering device, which takes the stability coefficient and uniformity coefficient of the volume of covering soil as factors. The result shows the following: (1) The volume of covering soil changes exponentially with the angle of the disc through data fitting, which can standardize the angle of covering disc; and (2) when the angle of disc is 30° and 60°, respectively, the uniformity coefficient of volume of covering soil is lower than 1.4, which has premium soil covering quality. When the angles of the discs on both sides differ greatly, the stability coefficient of volume of covering soil is 0.41, which can meet the requirements of the mulching quality of potato planter. This research provides the technical support for high-quality potato planting.

Keywords: potato planter; soil covering device; design; experiment

Citation: Zheng, Z.; Fu, Z.; Wang, C.; Huang, Y.; He, J. Design and Experimental Research on Soil Covering Device with Linkage and Differential Adjustment of Potato Planter. *Agriculture* **2021**, *11*, 665. https://doi.org/10.3390/agriculture11070665

Academic Editor: José Pérez-Alonso

Received: 30 May 2021
Accepted: 12 July 2021
Published: 14 July 2021

1. Introduction

In Ningxia, Gansu, Shaanxi and other regions of China, the plastic film mulching potato planter are widely used. The cultivated land in these areas is mainly hills and is mountainous and there are a lot of planters working on sloping fields. The main function of mulching is to maintain warmth and moisture levels, to promote the rapid germination of seeds and to achieve the purpose of increasing yield and reducing the growth period. The effect of covering soil is an important evaluation of the quality of mulching and there is a large impact on the volume of covering soil on both sides of mulch in sloping fields. Furthermore, it is necessary to make tentative adjustments based on experience and it is difficult to achieve best the covering quality. Therefore, it is of great significance to develop a soil covering device for potato planters that achieve the same volume of covering soil on both sides in sloping fields [1,2].

The research on the soil covering device of potato planter has been conducted earlier in foreign countries with advanced technology and high reliability. The typical equipment includes GL series potato planter produced by GRIMME in Germany and the LK series potato planter produced by STRUIK in Netherlands. These machines have high efficiency and quality, which are suitable for areas with large-scale fields rather than hills and mountainous areas. Recently, the development of potato planting machinery technology and equipment in China has become rapid. The potato planter was designed and tested and includes the sowing depth control device and wide ridge pattern suitable planter [3–5]. Liu and other scholars designed the film mulching potato planter, which applied to rice, quinoa, potato and so on [6–8]. Chen and other scholars put forward film mulching technology, soil conveying technology, etc., which provide technical support for the research and development of potato planters [9–13]. The study on the optimization of working parameters and improvement of the design structure of the potato planter and mulching device was conducted by Xin and others [14–18]. These technologies and equipment have been experimented and verified to basically suit the requirements of field working. When working on flat fields, the quality of covering soil is higher; however, in the sloping field, there will be inconsistencies in the volume of covering soil on both sides and the stability of cover will decrease. The covering quality is reduced, which is not conducive to heat and moisture preservation and will affect the high yield and premature delivery of potatoes [19–21].

Therefore, a soil covering device was designed to improve seeding quality by linkage and differential adjustments of the disc separately and the field experiment was completed to verify the performance of soil covering device, which takes the stability coefficient and uniformity coefficient of the volume of covering soil as factors.

2. Material and Methods

2.1. The Device of Experiment

As shown in Figure 1, the potato planter mainly includes fertilizer box, seeder, seed box, ridging machine, laminating machine and soil covering device. The prototype takes the Qingdao HONGZHU potato planter and the planting method utilizes double rows on a ridge. The fertilizer from the fertilizer box is applied to the middle position of the ridge and planted on both sides of the ridge.

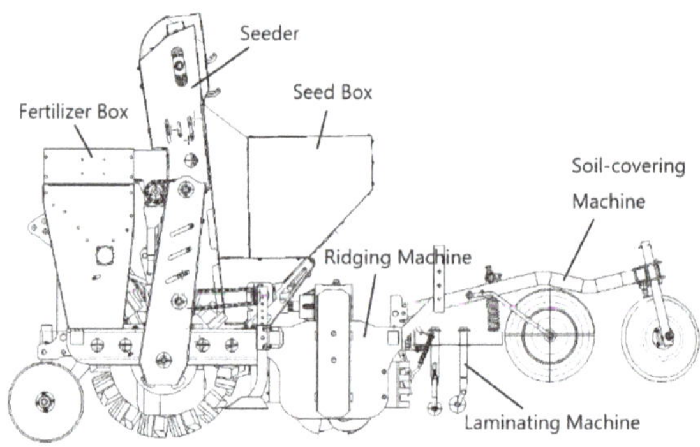

Figure 1. The structure scheme of potato planter.

A spoon-chain seeder metering device is used to take out the lumpy potato seeds from the seed box and to shake the excess potato seeds back into the seed box under the action of the vibrating seed cleaning device. The ridging machine trims the ridge shape according to the set value. The laminating machine covers the film on the ridge surface and the packer wheel and soil covering machine takes the soil covering both sides of film.

Figure 2 shows the structure diagram of soil covering device and the device mainly includes covered soil discs, adjusting mechanism, frame and so on. The distance between the selection centers of the covered soil discs on both sides is adjustable from 1100 to 1500 mm, which can suit the work of 0.8–1.2 m width film and the maximum adjustment of the rotation angle of the covering disc is 75°.

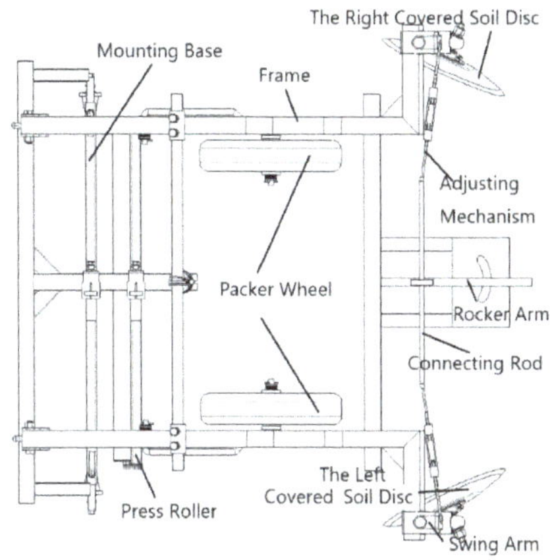

Figure 2. The structure scheme of soil covering device.

The tractor pulls the device forward and the press roller presses the film on the ridge surface. The packer wheel then presses down the film on the ridge side and the covered soil disc pushes the loose soil of the furrow to the film on the ridge side to complete the entire film.

The uniformity and stability of the volume of covering soil is an important factor for the quality of the soil covering device. When working on flat ground, the angle of the covered soil discs on both sides is generally required to be consistent and the covering quality can basically meet the operation requirements. However, as shown in Figure 3, when working on sloping field, the volume of soil on both sides will be affected by the slope angle and the amount of soil on the upper side is much larger than that on the lower side. Therefore, it is necessary to adjust the angle of the covered soil disc to keep the volume of covering soil on both sides consistent.

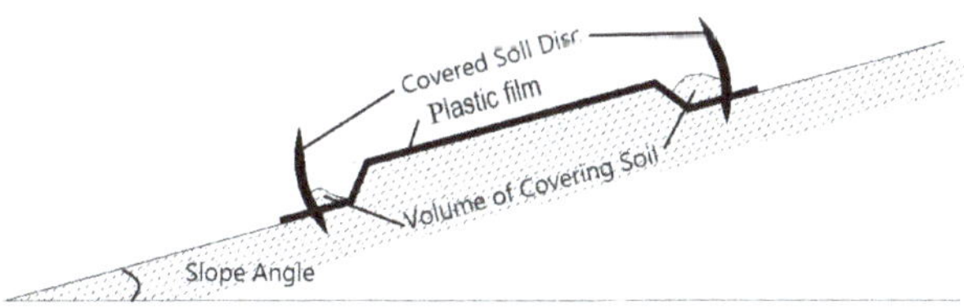

Figure 3. The soil covering at the sloping field.

2.2. The Data Analysis

The main function of the adjustment mechanism is to adjust the angle of the soil covering discs on both sides to improve the mulching quality.

As shown in Figure 4, the adjustment mechanism mainly includes the covered soil disks on both sides, the rocker arm, the four-bar mechanism composing the frame, the connecting rod and the swing arm. The connecting rod adopts a two-way screw structure and, by rotating the middle nut, the length of the connecting rod can be changed and the angle of the discs on both sides can be adjusted in linkage under the action of the four-bar mechanism. The two sides of the connecting rod and the swing arm are correspondingly connected by hinged shafts. On the other hand, the rocker arm can rotate around the end point. During the rotation, the connecting rod will rotate synchronously to realize the differential adjustment of the covered soil disc. The angles α and β of both side discs are correspondingly related to the mechanism parameters of the four-bar mechanism.

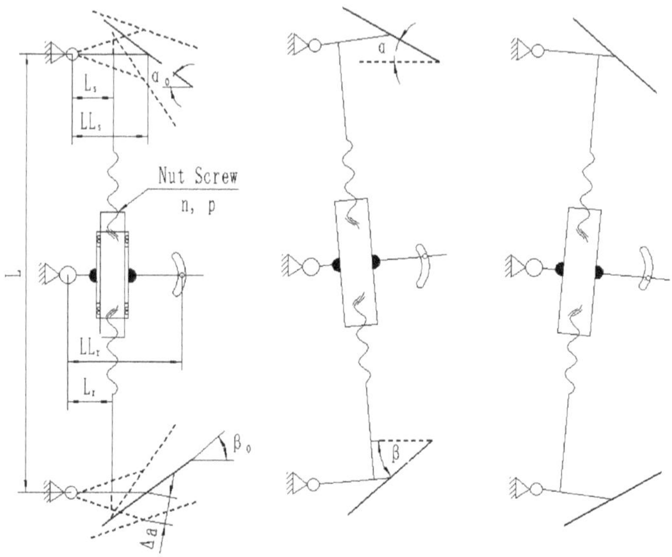

(**a**) The linkage adjustment (**b**) The differential adjustment

Figure 4. The schematic diagram of the adjusting mechanism.

Firstly, the model equations of the angle and related parameters are established through a theoretical calculation and then the corresponding relationships between different slope angles and adjustments are obtained based on the test data.

(1) The linkage adjustment

The intercept of the two-way nut screw is p, the length of the swing arm is Ls and LLs, respectively. When the function of the hinge axis is not considered, the Equation (1) exists and is described as follows:

$$\begin{cases} \frac{np}{\Delta a} = \frac{Ls}{LLs} \\ \alpha - \alpha_0 = 2arcsin\frac{np}{2Ls} \end{cases} \quad (1)$$

where n is number of rotations and Δa is the turning distance of the end of the swing arm in mm.

(2) The differential adjustment

When the four-bar mechanism is composed of the frame, the connecting rod and the swing arm is arranged in a parallel. If the rocker arm is deflected counterclockwise and the angle is γ, then there exists the following equation:

$$\alpha - \alpha_0 = \beta - \beta_0 = \gamma$$

where a_0 and β_0 are the initial angles of the discs on both sides.

When the four-bar mechanism is not arranged in a parallel, the calculation method of steering trapezoid is referred to and described in the following.

$$\begin{cases} \Delta\alpha = \gamma \\ \Delta\beta = \gamma - arcsin\frac{\sin((\gamma+\gamma_0)}{\sqrt{\left(\frac{L}{L_s}\right)^2+1-2\frac{L}{L_s}\cos(\gamma+\gamma_0)}} - arccos\frac{\frac{L}{L_s}[2cos\gamma_0-\cos(\gamma+\gamma_0)-cos2\gamma_0]}{\sqrt{\left(\frac{L}{L_s}\right)^2+1-2\frac{L}{L_s}\cos(\gamma+\gamma_0)}} \\ \gamma_0 = arcsin\frac{np}{L_s} \end{cases} \quad (2)$$

where γ_0 is the initial bottom angle of trapezoidal four-bar mechanism in degrees and L is the length of the connecting rod (short side) in mm.

The size of adjusting mechanism in potato planter is shown in Figure 5. The initial bottom angle of the trapezoidal four-bar mechanism is determined by the length of the swing arm, the intercept of the nut screw and the number of rotations. By using the Equation (3), the rotation angle of the swing arm can be quantified. It is also necessary to calibrate the inclination requirements of the upper and lower soil covering discs according to different slope angles. So far, the design parameters of the four-bar mechanism can be obtained.

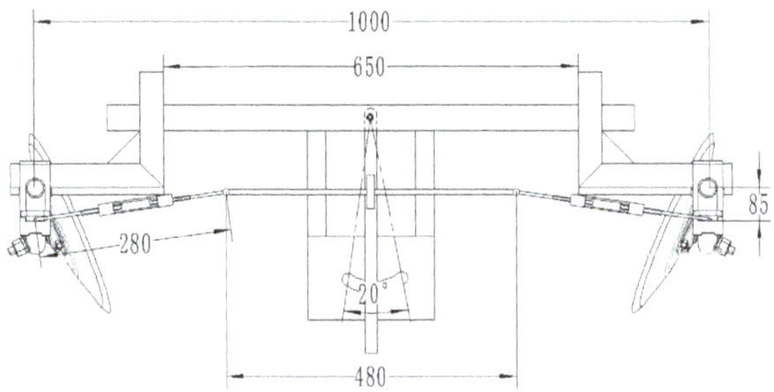

Figure 5. The size of adjusting mechanism.

The purpose of the experiment is to calibrate the volume of covered soil under different angles of the disc. On the other hand, when the angles of the discs on both sides are inconsistent, the experiment detects the difference and stability of the volume of covered soil on both sides. The flat fields have been used to complete the field experiment by using the stability coefficient of the difference in the volume of covered soil on both sides to evaluate the quality of the covering soil.

The potato planter used in the experiment adopts a self-made soil covering device with adjusting mechanism, which based on the HONGZHU-1.2 large ridge double-row planter. The adjusting mechanism was shown in Figure 6 and the mulch uses ordinary plastic film with a thickness of 0.0012 mm and a width of 1.2 m.

Figure 6. The adjusting mechanism.

The tractor uses DONGFANGHONG-504 from LUOTUO (about 36.8 kW) and the operating speed is about 3.6 km/h.

2.3. The Method of Experiment

On 9 April 2020 at the Northwest A&F University test farm, the operation performance test of soil covering device was carried out. The field of experiment has a length of 110 m and a width of 10 m. It is a typical columbine soil condition. Moreover, corn was planted in the first stubble and two rotary tillage treatments were carried out after harvest. The soil moisture content of the cultivated layer (5–10 cm) is 18.5%(w.b.) and the average firmness of soil is 2.05 MPa. The soil covering disc adopts a concave disc with a diameter of 45 cm and the depth of the disc is adaptively adjusted by spring preloading and force adjustment, the measured data shows that the average depth of the disc into the soil is 14.27 mm.

In order to calibrate the volume of soil covering under the adjustment of different disc angles, five points were randomly selected to measure in each level and 30°, 45° and 60°, respectively, were used as linkage adjustment angles of the covered soil disc. Moreover, the angles of the covering discs on both sides are selected to be 30° and 60°, respectively, as the difference adjustment. The Table 1 shows the method for adjusting the angle of the covered soil discs on both sides. The sampling points were selected according to the five-point method and randomly selected on a ridge in stable working area as the sampling points. Repeat three ridge operations under the same angle of discs and take the average for calculation.

Table 1. Method for adjusting the angle of covered soil disc.

Project	The Left Disc (°)	The Right Disc (°)
Linkage adjustment	30	30
	45	45
	60	60
Difference adjustment	30	60
	60	30

As shown in Figure 7, the volume of covered soil is calculated by the area of the section and the length of the sampling (take 1 m), as shown in Equation (3) as follows:

$$V = \frac{a+b-h}{2}hl \tag{3}$$

where V is the volume of covered soil, m^3; a is the length of the section, m; b is the width of the section, m; h is the height of the section, m; l is the length of the sampling, m.

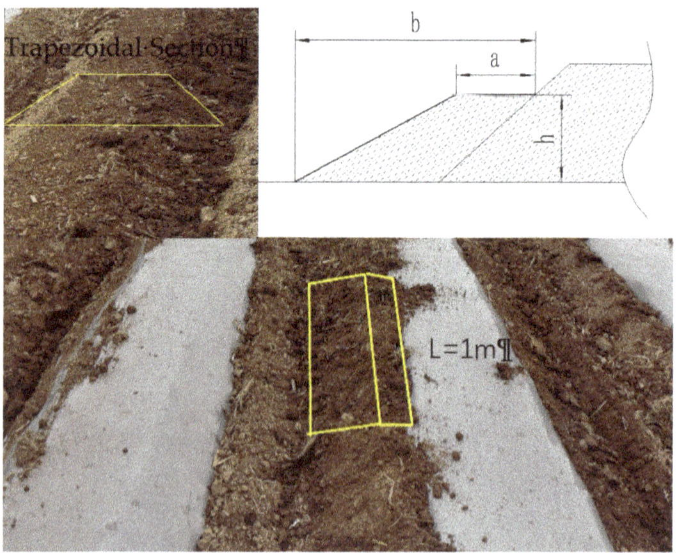

Figure 7. The diagram of the volume of covering soil and section.

Therefore, the main measurement factors include the width, height and length of the covered soil section. Considering that the ridge structure is relatively stable (the slope angle is about 45°), the area of the covering soil section can be formed by subtracting the area of an isosceles right triangle from a right-angled trapezoid for which the side lengths are a, b and h.

The area of the covering soil section can be formed by a right-angled trapezoid with width, length and height as three sides and minus the area of the isosceles right triangle with the height as the right side.

The stability coefficient and uniformity of the volume of covered soil are used as indicators to investigate the feasibility of the differential adjustment and linkage adjustment. Refer to the "uniformity determination" and "stability determination" methods in the test method of the quality evaluation technical specification for fertilization machinery (NY/T 1003-2006) for calculation. The uniformity mainly measures the uniformity coefficient of the volume of covered soil. When the angles of the soil covering discs on both sides are different, the uniformity coefficient of each side is measured separately, which can characterize the feasibility of the differential adjustment. The stability mainly measures the stability coefficient of the volume of covered soil. When the angle of the soil covering discs on both sides are same, the stability coefficient of each side is measured separately, which can characterize the stability of the linkage adjustment.

2.3.1. The Calculation of Uniformity Coefficient

The uniformity coefficient of the volume of covered soil includes standard deviation and the variation coefficient. Before the experiment, the parameters of the soil covering plate were adjusted and the tractor was moving forward steadily at an average speed of 3.6 km/h. Five points were randomly selected in the test area and the section parameters of the covered soil on both sides were correspondingly measured. The volume of covered soil was calculated according to the above method and, according to Equation (4), the standard deviation and the variation coefficient of the uniformity on both sides were calculated as follows:

$$\begin{cases} x = \frac{\sum_{i=1}^{n} x_i}{n} \\ S = \frac{\sum_{i=1}^{n} (x_i - x)^2}{n-1} \\ V = \frac{S}{x} \times 100\% \end{cases} \quad (4)$$

where x is average volume of covered soil, m^3; x_i is volume of covered soil of the point, m^3; S is the standard deviation of the uniformity; V is the variation coefficient of the uniformity, %; n is the number of measuring points, ($n = 5$).

2.3.2. The Calculation of Stability Coefficient

The stability coefficient of the volume of covered soil includes the standard deviation and the variation coefficient. The test method is the same as the calculation of the uniformity coefficient. Refer to the calculation in Equation (3), where, x is average value of the difference in the volume of covered soil on both sides, m^3; x_i is difference of volume of covered soil on both sides of the point, m^3; S is the standard deviation of the stability; V is the variation coefficient of the stability, %; n is the number of measuring points, ($n = 5$).

3. Results and Discussion

As shown in Figure 8, when the angles of the soil covering discs on both sides are 30° and 45°, the volume of covered soil at the measuring points is 0.28–0.73 $\times 10^{-4}$ m^3 and 1.07–1.48 $\times 10^{-4}$ m^3, respectively. The volume of covered soil is quite different between both the angles of discs, but the difference is relatively close at different points. The maximum difference of 0.79 $\times 10^{-4}$ m^3 appears at the first point and the minimum difference of 0.61 $\times 10^{-4}$ m^3 appears at the fourth point. When the angle of the covering disc is 30°, the average volume of covered soil is 0.55 $\times 10^{-4}$ m^3 and when the angle of the covering disc is 45°, the average volume is 1.12 $\times 10^{-4}$ m^3.

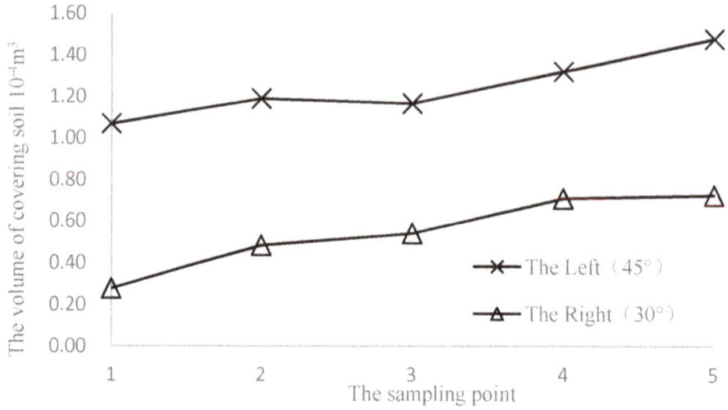

Figure 8. The volume of covered soil at different sampling points (Both the angles of left and right disc are 45°and 30°, respectively).

It shows that the volume of covered soil is different at each point but, at the same point, there is a relatively fixed difference of the soil covering discs on both sides.

As shown in Figure 9, when the angles of the soil covering discs on both sides are 45° and 60°, the volume of covered soil at the measuring points is 0.67–1.11 × 10^{-4} m^3 and 1.24–1.52 × 10^{-4} m^3, respectively. The maximum difference of 0.60 × 10^{-4} m^3 appears at the first point and the minimum difference of 0.40 × 10^{-4} m^3 appears at the second point. When the angle of the covering disc is 45°, the volume of covered soil is 0.85 × 10^{-4} m^3 and when the angle of the covering disc is 60°, the average volume is 1.37 × 10^{-4} m^3.

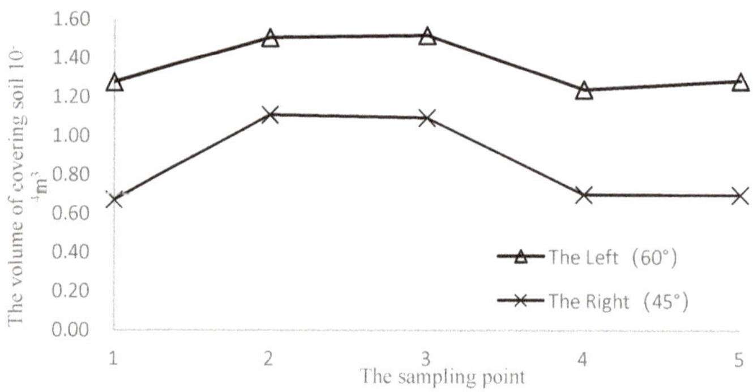

Figure 9. The volume of covered soil at different sampling points (Both the angles of left and right disc are 60°and 45°, respectively).

When the angle of covering disc is 45°, the volume of covered soil is about 31% smaller than that of the previous group. The reason may be that there will be some interference between adjacent rows during the operation. Specifically, the engaged previous row of soil covering will reduce the amount of soil in the ditch and, as a result, the amount of floating soil on the adjacent side becomes less while the volume of covered soil is reduced.

As shown in Figure 10, when the angles of the soil covering discs on both sides are 30° and 60° the volume of covered soil at the measuring points is 0.61–0.88 × 10^{-4} m^3 and 1.19–1.49 × 10^{-4} m^3, respectively. The maximum difference of 0.70 × 10^{-4} m^3 appears at the first point and the minimum difference of 0.58 × 10^{-4} m^3 appears at the fourth point. When the angle of the covering disc is 30°, the volume of covered soil is 0.77 × 10^{-4} m^3 and when the angle of the covering disc is 60°, the average volume is 1.40 × 10^{-4} m^3.

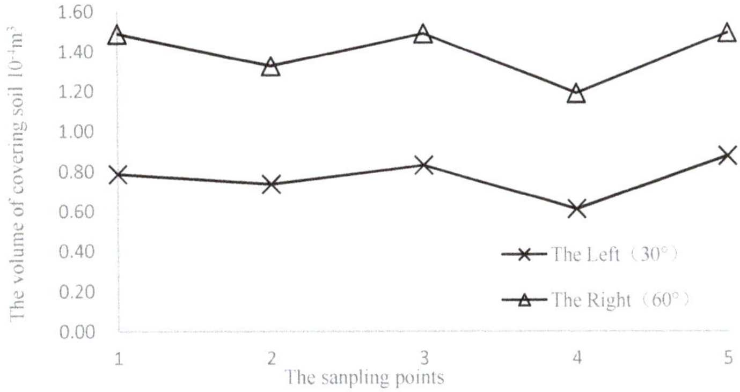

Figure 10. The volume of covered soil at different sampling point (Both the angles of left and right disc are 30°and 60°, respectively).

When the angle of the covering disc is 30°, the amount of soil covering is about 29% compared to that of the first group. The reason may be that the first row is close to the edge of the plot during operation and the soil is not as loose as the middle position and so the covering amount is lesser. When the inclination angle is 60°, the amount of soil covered is close to that of the second group with a difference of less than 2.2%.

Therefore, the reason for the larger change in the volume of covered soil may be caused by soil conditions and the curve can be explained as shown in Figures 8–10.

When the angles of covering discs 30°, 45° and 60° are, respectively, fitted and the relationship between the volume of covered soil and the angle of disc is obtained as a quadratic function, as shown in Equation (5). The angle of the disc can be calibrated through the following equation:

$$Y = 0.001X^2 - 0.0701x + 1.9597 \quad (5)$$

where X is the angle of disc, degrees; is the volume of covering soil, 10^{-4} m^3.

According to the test data, the uniformity coefficient and stability coefficient of the volume of covered soil under different angles of the soil covering discs are calculated, as shown in Table 2.

Table 2. Coefficient of uniformity and stability of covering soil at different angles.

Factor [1]	Uniformity Coefficient		Stability Coefficient	
	Standard Deviation [2]	Coefficient of Variation %	Standard Deviation [2]	Coefficient of Variation %
30°	105.49	13.7	/	/
45°	508.23	59.5	/	/
60°	184.28	13.2	/	/
45° (30°)	/	/	62.58	9.0
60° (45°)	/	/	87.79	17.2
30° (60°)	/	/	25.70	4.1

Note: [1] The factor column indicates the angle of the soil covering disc and the brackets indicate that the angles of the soil covering disc are different on the left and right sides; [2] the relevant dimensions of the covering soil section during the calculation of the standard deviation in mm.

According to the standard, the variation coefficient of uniformity should be less than 40% and the variation coefficient of stability should be less than 7.8%. When the coefficient of variation is less than the specified value, it indicates that the machine possesses good quality covering soil.

As shown in Table 2, when the angle of the soil covering disc is 30° and 60°, respectively, the uniformity coefficient is relatively small (less than 40%), indicating that the volume of covered soil is more consistent and that the quality is good. When the angle of the soil covering disc is 45°, the uniformity coefficient is relatively large (more than 40%), indicating that the uniformity of the volume of covered soil is not good. When the difference between the angles of the discs on both sides is large (30° and 60°), the standard deviation and variation coefficient of stability coefficient are the lowest (less than 7.8%), indicating that the greater the difference between the discs on the both sides, the easier it is to achieve differential overburden, which is beneficial under sloping fields.

4. Conclusions

This paper designs a soil covering device that can perform linkage and differential adjustments separately and the field experiment was completed to verify the performance of soil covering device. The results show that this device can obtain higher quality potato planting. However, due to the condition of test sloping fields being hard to obtain, in this paper, the flat fields were used to complete the field experiment, which needs verification in further research. Moreover, the Soil-bin Testing Platform for Sloping Field will be installed in the Northwest A&F University, which can be used to test the performance of device in sloping field.

In order to improve the mulching quality, a soil covering device that can realize linkage and differential adjustment was designed and field experiments were carried out to obtain the fitting equation of the volume of covered soil and the angle of the covering disc; the uniformity coefficient and stability coefficient are measured and calculated.

(1) The structure and principle of the soil covering device are put forward and the design calculation and analysis are focused on the four-bar adjustment mechanism and reasonable structural parameters are obtained. The linkage adjustment is carried out through the two-way screw and nut mechanism and the differential adjustment is carried out through the steering mechanism.

(2) By using data fitting, it was found that the volume of covered soil changes in a binomial manner with the angle of the discs. Field test results show the following: when the angles of discs on both sides are the same (30° or 60°), the uniformity coefficient of volume of covering soil is lower than 14% and when the angles of the discs are different (respectively, 30° and 60°), the stability coefficient of volume of covering soil is 4.1% (less than 7.8%), which indicates that the linkage and difference adjustment device can meet the quality requirements of the potato film side covering soil.

Author Contributions: Conceptualization, Z.F.; methodology, Z.Z. and Z.F.; validation, J.H.; investigation, Y.H. and Z.Z.; resources, Z.F.; writing—original draft preparation, C.W.; writing—review and editing, C.W. All authors have read and agreed to the published version of the manuscript.

Funding: This research was funded by the Key research and development projects of Shaanxi Province (2019NY-169 and 2018ZDCXL-NY-03-06).

Institutional Review Board Statement: Not applicable.

Informed Consent Statement: Not applicable.

Data Availability Statement: The data presented in this study are available upon requrest from the correspondence author at (fuzuoli@nwsuaf.edu.cn).

Acknowledgments: This work was financially supported by the Northwest A&F University. All supports and assistance are sincerely appreciated.

Conflicts of Interest: The authors declare no conflict of interest.

References

1. Li, Z.; Wen, X.; Lv, J.; Li, J.; Yi, S.; Qiao, D. Analysis and Prospect of Research Progress on Key Technologies and Equipments of Mechanization of Potato Planting. *Trans. Chin. Soc. Agric. Mach.* **2019**, *50*, 1–16.
2. Xue, J.F.; Yuan, Y.Q.; Zhang, H.L.; Ren, A.X.; Lin, W.; Sun, M.; Gao, Z.Q.; Sun, D.S. Carbon footprint of dryland winter wheat under film mulching during summer-fallow season and sowing method on the Loess Plateau. *Ecol. Indic.* **2018**, *95*, 12–20. [CrossRef]

3. Guo, D.; Zheng, Y.; Wang, Z.; Yang, R. Design and Test of 2CM-SF Potato Planter. *J. Agric. Mech. Res.* **2021**, *43*, 103–106.
4. Lyu, J.; Yang, X.; Feng, X.; Li, Z.; Li, J.; Liu, Z. Design and experiment of sowing depth control device of potato planter. *Trans. Chin. Soc. Agric. Eng.* **2020**, *36*, 13–21.
5. Qi, J.; Meng, H.; Kan, Z.; Chen, S.; Yan, H.; Lu, S. The Design of Wide Ridge Pattern Suitable for Potato Planter. *J. Agric. Mech. Res.* **2017**, *39*, 126–130.
6. Liu, G. Design and Test of Rice Mulching Seeder. *Agric. Sci. Technol. Equip.* **2020**, *1*, 21–22.
7. Wang, X.; Kang, J.; Ding, H.; Peng, Q.; Zhang, C.; Li, N. Design and Test of 2BM-4 Film Mulch Precision Planter for Quinoa. *Trans. Chin. Soc. Agric. Mach.* **2020**, *51*, 86–94.
8. Searle, C.L.; Kocher, M.F.; Smith, J.A.; Blankenship, E.E. Field slope effects on uniformity of corn seed spacing for three precision planter metering systems. *Appl. Eng. Agric.* **2018**, *24*, 581–586. [CrossRef]
9. Chen, J.; Liu, X. Development and Application of High-speed Hole Planter with Full Film Covering on Furrow and Ridge. *Farm Mach.* **2020**, *12*, 82–83.
10. Hu, Y.; Wang, X.; Chen, C.; Cai, J.; Pu, T.; Zhang, L.; Yang, W. Design and parameter optimization of maize hole irrigation and plastic film fertilization planter in hilly dry land of Sichuan Province. *J. N. Agric. Univ.* **2020**, *51*, 61–68.
11. Shao, Y.; Liu, Y.; Xuan, G.; Hu, Z.; Han, X.; Wang, Y.; Chen, B.; Wang, W. Design and Test of Multifunctional Vegetable Transplanting Machine. *IFAC Pap.* **2019**, *52*, 92–97. [CrossRef]
12. Shi, L.; Yang, X.; Zhao, W.; Sun, W.; Li, R.; Sun, B. Design and Test of Potato Combine Seeder with Throwing and Covering Soil on Film Edge. *Trans. Chin. Soc. Agric. Mach.* **2018**, *49*, 129–137.
13. Sun, W.; Liu, X.; Zhang, H.; Wang, H.; Tian, B. Design of potato casing soil planter in all-in-one machine combined with fertilizing, sowing, ridging, complete film mulching and planting line covering. *Trans. Chin. Soc. Agric. Eng.* **2017**, *33*, 14–22.
14. Xin, S.; Dai, F.; Shi, L.; Zhao, W.; Liu, F.; Xin, B.; Song, G. Optimization of working parameters of potato planter with fertilizer and mulching. *Agric. Res. Arid Areas* **2017**, *35*, 282–287.
15. Dai, F.; Zhao, W.; Song, X.; Xin, S.; Liu, F.; Xin, B. Operating Parameter Optimization and Experiment of Device with Elevating and Covering Soil on Plastic-film. *Trans. Chin. Soc. Agric. Mach.* **2017**, *48*, 88–96.
16. Cui, Y.; Jia, L.; Chen, W.; Zheng, D.; Du, W. Design and Test of the Rain-fed Potato Planter with Micro-ridge and Plastic Film Mulching. *J. Agric. Mech. Res.* **2016**, *38*, 62–66.
17. Chen, L.D.; Du, F.Y.; Du, F.B.; Wang, W.Z. Design on Multifunctional Water-Saving Mulch Planter. *Adv. Mater. Res.* **2014**, *912–914*, 628–632. [CrossRef]
18. Huang, J. Study on the mechanized technology of potato sowing, fertilization and film mulching. *Sci. Technol. Inf.* **2014**, *12*, 239.
19. Nyawade, S.O.; Karanja, N.N.; Gachene, C.K.K. Schulte-Geldermann, E.; Parker, M. Effect of potato hilling on soil temperature, soil moisture distribution and sediment yield on a sloping terrain. *Soil Tillage Res.* **2018**, *24*, 184.
20. Kozelko, Y.; Usenko, M. Definition of performances of conveyor potato-planter apparatus with pinning devices. *INMATEH—Agric. Eng.* **2016**, *49*, 47–52.
21. Karmakar, S.; Mandal, P.; Roy, S.; Majumder, R. Evaluation of tractor drawn potato planter in West Bengal state of India. *Agric. Mech. Asia Afr. Lat. Am.* **2016**, *47*, 36–40.

agriculture

MDPI

Article

An Investigation of Pull-Out Force of Semi-Buried Lotus Roots after Hydraulic Scouring

Rong Zeng [1,2], Yitao Lin [1,2], Zhihua Wan [1,2], Ming Tu [1,2], Jun Jiao [1,2] and Guozhong Zhang [1,2,*]

1 College of Engineering, Huazhong Agricultural University, Wuhan 430070, China; zengrong@mail.hzau.edu.cn (R.Z.); ouzuxiix1@163.com (Y.L.); wanzhihua_83@webmail.hzau.edu.cn (Z.W.); mingtu@mail.hzau.edu.cn (M.T.); jiaojun@webmail.hzau.edu.cn (J.J.)
2 Key Laboratory of Agricultural Equipment in Mid-Lower Yangtze River, Ministry of Agriculture and Rural Affairs, Wuhan 430070, China
* Correspondence: zhanggz@mail.hzau.edu.cn

Abstract: Hydraulic scouring is the most effective approach to harvest lotus roots, but its application is largely restricted by the low harvesting rate. After hydraulic scouring, some mature lotus roots are still partially buried in the soil. Therefore, it is highly necessary to develop an appropriate approach to harvest semi-buried lotus roots. In this work, we for the first time studied the interaction between semi-buried lotus roots and soil, as well as analyzing the pull-out process and the factors affecting the pull-out force of semi-buried lotus roots. Firstly, a simple testing platform was designed based on the virtual prototype technology and the tests on pull-out force were conducted on five lotus roots with similar shape but different weights, with the burial depth and pull-out speed as the experimental factors. The results revealed that the maximum pull-out force is not significantly affected by the pull-out speed, whereas it is significantly influenced by the burial depth and the surface area of lotus roots. The maximum pull-out force increased with increasing lotus root surface area and burial depth. In addition, the discrete element method was employed to simulate the pull-out process of lotus root at different pull-out speeds. The simulation results indicated that a higher pull-out speed would result in a greater pull-out force at the same displacement of the lotus root from the soil. Both experimental and simulation results revealed that soil adhesion contributes the most to the pull-out resistance. It was also observed that a slight loosening of semi-buried lotus roots could drastically reduce the pull-out force. These results suggest that some kind of mechanical structure or improvement of water flow can be applied to the existing lotus root harvester to reduce the adhesion between lotus roots and soil. Overall, our findings provide a novel direction for optimizing hydraulic harvesting machines of lotus roots.

Keywords: semi-buried lotus root; pull-out force; hydraulic scouring; soil adhesion

Citation: Zeng, R.; Lin, Y.; Wan, Z.; Tu, M.; Jiao, J.; Zhang, G. An Investigation of Pull-Out Force of Semi-Buried Lotus Roots after Hydraulic Scouring. *Agriculture* **2021**, *11*, 706. https://doi.org/10.3390/agriculture11080706

Academic Editor: José Pérez-Alonso

Received: 17 May 2021
Accepted: 26 July 2021
Published: 27 July 2021

1. Introduction

Lotus root is one of the largest aquatic vegetables in China. According to statistics, the planting area of lotus root in China in 2014 was about 500,000 to 700,000 hectares [1]. With crispy and refreshing flavor and rich nutrition, lotus roots are popular among Chinese people. Due to the huge economic benefits, the number of lotus root growers and the planting area have been increasing year by year. Lotus roots are mostly planted in low-lying and rotten fields with deep silt [2]. The special planting environment and growth conditions result in great difficulties in mechanized harvesting. As a result, manual harvest is still the main harvesting method. With the rapid decrease in labor resources and increase in labor cost in rural China, the production cost of lotus roots has increased dramatically, which seriously restricts the scale development of the lotus root industry. Hence, some scholars have developed several simple harvesting machines for lotus roots based on the planting characteristics, and the mechanized harvesting of lotus roots has been partly realized.

Currently, shovel digging and hydraulic scouring are the two most commonly used methods for mechanized harvesting of lotus roots. Due to the fragile texture and slender shape of lotus roots, harvesting with shovel digging causes high damage and loss. The harvesting machine with hydraulic scouring uses a high-pressure water jet to wash the mud around lotus roots. When the lotus roots are free from the mud, they will float to the water's surface so that the operators can easily pick them up. The pressure of the water jet can be adjusted to reduce damage to the lotus roots. Therefore, hydraulic harvesting machines are more widely applied to the harvesting of lotus roots [3–5]. The shape of lotus root is similar to a drum. Sludge attached to the upper surface of lotus root can be washed away by the water jet, whereas the sludge attached to the lower surface is obstructed and cannot be washed away by the water jet. Due to high viscosity of soil, some lotus roots are still partially bound by sludge, and will be in a semi-buried state and thus cannot float up. These lotus roots need to be harvested manually. To develop a reasonable harvesting method, it is essential to analyze the interaction between the semi-buried lotus roots and soil, which may be best reflected by the pull-out force of semi-buried lotus roots. Aiming to improve the working performance of harvesting machines for crops from soil, research has been focused on the pull-out force of crops from soil. For example, Li et al. designed a device to measure the pull-out force of cotton stalk and analyzed the relationship between the pull-out force and cotton root diameter, pull-out direction and pull-out speed [6]. Similarly, Song et al. studied the factors affecting the pull-out resistance of cotton stalk and proposed that the diameter of the root of cotton stalk and the soil properties had significant influence on the pull-out resistance [7]. Moreover, Xie et al. simulated the pull-out force of leaf vegetables from soil by EDEM software and determined the optimal pull-out force [8]. Based on the simulation results, they also proposed the optimal working parameters of the harvesting machine. In addition, according to the grip strength, Seongin et al. developed a harvesting robot for lettuce [9].

Understanding the pull-out force of semi-buried lotus roots from soil may help to develop effective methods for the subsequent harvesting of semi-buried lotus roots. At present, there has been no research on the pull-out force of lotus roots. Hence, in order to explore the factors affecting the pull-out force and analyze the pull-out process of lotus roots, this paper constructed a simple platform to test the pull-out force of semi-buried lotus roots. The pull-out force of different lotus roots was investigated at different pull-out speeds and burial depths, and the effect of these factors on the pull-out force was evaluated. In addition, the pull-out process of a testing lotus root was simulated by the discrete element method via the EDEM software. Experimental and simulation analysis results suggest the importance of finding an effective approach to reduce the pull-out resistance, so as to improve the harvesting machine with hydraulic scouring for lotus roots.

2. Materials and Methods

2.1. Analysis of Pull-Out Force of Semi-Buried Lotus Roots after Hydraulic Scouring

After washing by water jet, the lotus roots may naturally float out of the water surface. Taking one lotus root as an example, the floating process is demonstrated in Figure 1. Firstly, the lotus root is buried in a certain depth under the mud (a). Then, the hydraulic harvesting machine starts to wash the mud layer (b), and the mud surrounding the lotus root is washed away (c). Finally, the lotus root is separated from the mud layer and floats out of the water surface (d). Since the burial depth of each lotus root in mud layer is not consistent, the distance between the nozzle of the water jet and each lotus root is uncertain, resulting in the failure of deeper lotus roots to float out. In addition, the water jet of hydraulic scouring will generate a downward impact force. If the mud layer is soft, the impact force may press the lotus roots into a deeper layer, preventing the lotus roots from floating. These lotus roots will be semi-buried in the mud and require further measures to be harvested.

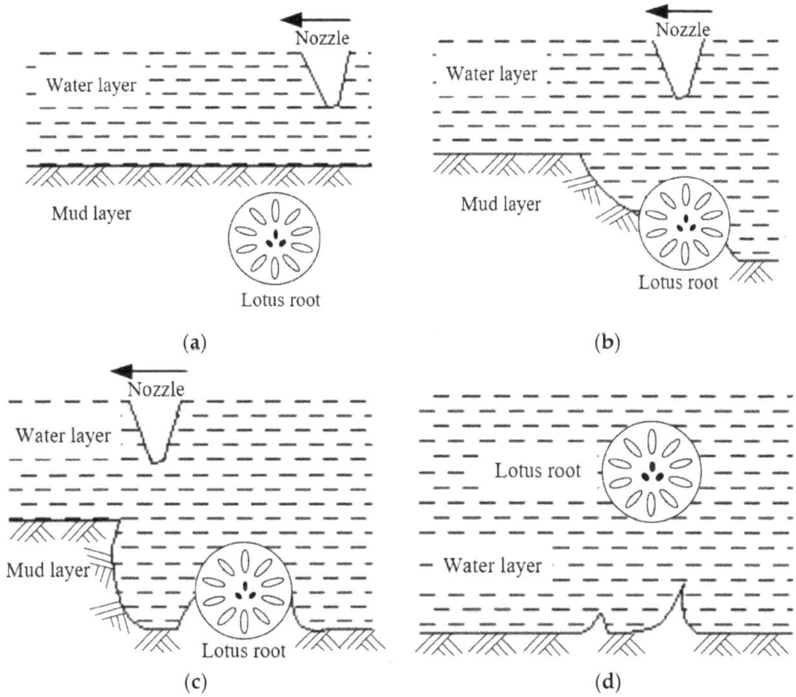

Figure 1. Floating process of a lotus root. (**a**) Initial state of the lotus root; (**b**) Start of floating; (**c**) Continuous floating; (**d**) Total floating.

In this section, the forces acting on the semi-buried lotus roots are analyzed. When the semi-buried lotus root tends to move upward, it is under the buoyancy of water, gravity, the soil adhesion force, the normal force of the mud layer and the pull-out force. Taking the cross-section center of the lotus root as the origin to establish the coordinates, the free-body diagram of the lotus root is depicted in Figure 2.

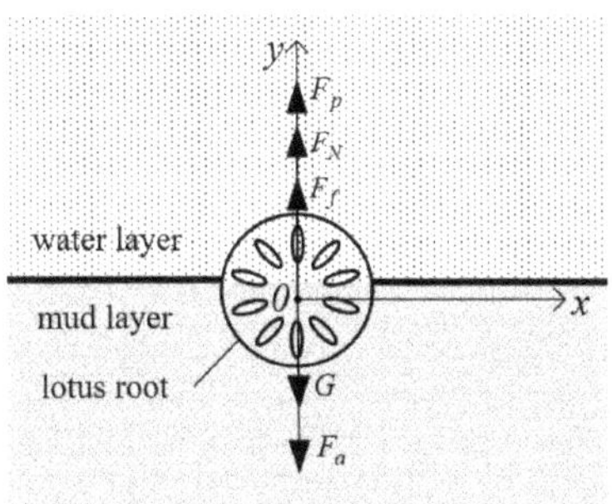

Figure 2. Free-body diagram of the semi-buried lotus root.

Figure 2 shows the vertical force acting on the lotus root when it is pulled out, and the resultant force can be expressed as:

$$\sum \vec{F} = \vec{F}_f + \vec{F}_p + \vec{F}_N + \vec{G} + \vec{F}_a, \quad (1)$$

where $\sum F$ represents the resultant force; F_f and F_p denote the buoyancy of water and pull-out force, respectively; F_N is the normal force of the mud layer; G is the gravitational force; and F_a describes the equivalent soil adhesion force in the vertical direction. When the pull-out force is lower than the vertical resistance, the lotus root will stay static in the mud layer, and thus, $\sum F$ is 0. Therefore, the equilibrium equation is given by:

$$0 = \vec{F}_p + \vec{F}_{pz}, \tag{2}$$

and the vertical resistance is:

$$\vec{F}_{pz} = \vec{F}_f + \vec{F}_N + \vec{F}_a + \vec{G}, \tag{3}$$

where F_{pz} is the pull-out resistance. If the pull-out resistance is overcome, the lotus root will start to float.

2.2. Pull-Out Force Test

2.2.1. Testing Platform

A testing platform for the measurement of pull-out force of semi-buried lotus roots is designed in this section. Figure 3 depicts the working principle.

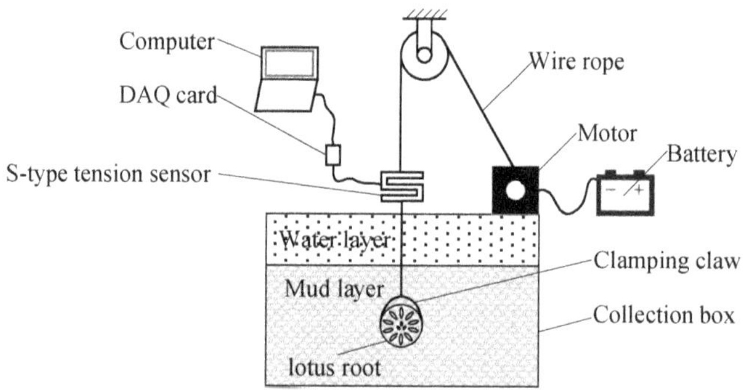

Figure 3. Sketch map of the testing platform.

A single lotus root is buried in the mud layer in the collection box. A clamping claw is employed to grasp the lotus root. The wire rope on the clamping claw is driven upward by the motor to provide a pull-out force. A S-type tension sensor is connected to the wire rope to measure the pull-out force. The output voltage of the tension sensor is acquired by a USB type data acquisition (DAQ) card and converted to pull-out force by a computer with the LabVIEW software. To establish the relationship between the output voltage and the pull-out force, the sensor was calibrated first, and the linear relationship was obtained as:

$$F_p = KU \tag{4}$$

where K denotes the linear coefficient, which was calibrated to be 872.15, and U is the output voltage of the S-type tension sensor.

2.2.2. Experimental Materials

Lotus roots generally present a drum shape, and there is certain regularity in the diameter distribution. In this experiment, five single lotus roots of the No. 5 Hubei variety were selected as the test objects according to the diameter distribution. The testing lotus roots showed similar shapes but different sizes without obvious damage in appearance. The average length of the five single lotus roots was 22.58 cm, and the minimum and maximum diameters were 55.1 mm and 94.5 mm, respectively. First, their weights were measured and the values are presented in Table 1.

Table 1. Weights of the testing lotus roots.

Lotus Root	No.1	No.2	No.3	No.4	No.5
Weight/(g)	789	855	734	696	657

The mud layer in the collection box was set as 30 cm in depth, and the water layer was set as 10 cm in depth. Soil samples were collected from a lotus root field located at Binhu fishing ground in Dongsheng Town, Shishou City, Hubei Province, China. The soil hardness and moisture content were measured and presented in Tables 2 and 3, respectively. Since the distribution of soil layer was not uniform, the soil hardness was measured five times at different positions to obtain the mean value (Table 2). Table 3 shows the moisture content in different layers of soil. As the surface layer is mainly solid with silt, it has higher fluidity and moisture content. With increasing soil depth, the soil hardness increases, accompanied by a decrease in moisture content.

Table 2. Soil hardness.

Number of Test	1	2	3	4	5	Mean Value
Soil hardness/(kPa)	330	350	370	300	330	336

Table 3. Soil moisture content (%).

Number of Test	Soil Depth	
	0~10 cm	Beneath 20 cm
1	40.25	20.82
2	48.25	22.05
3	43.11	23.07
Mean value	43.87	21.98

2.2.3. Experimental Methods

Since lotus roots are usually buried in a disordered state, different burial depths of lotus roots will require different pull-out force. Therefore, the burial depth was taken as one experimental factor. During the test, the burial depth was defined as shown in Figure 4. Before the test, the testing lotus root was slowly pressed into the mud layer, and the distance between the upmost point of the lotus root and the water layer surface was measured by a tape measure, which was recorded as the burial depth. Then, the mud attached to the upper surface was removed to form a semi-buried state. As the soil hardness increases with increasing soil depth, it is difficult to bury lotus roots to depths greater than 20 cm. Therefore, the burial depths of 15 cm and 20 cm were chosen to measure the pull-out force of the semi-buried lotus roots.

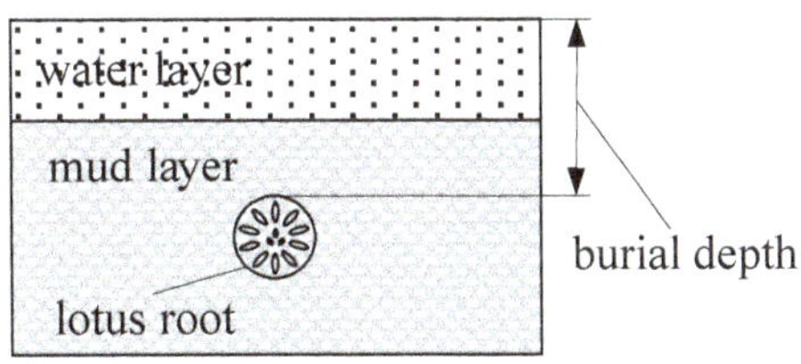

Figure 4. Definition of the burial depth.

Pull-out speed was another experimental factor to be considered in the test. The adjustment of the pull-out speed was fulfilled by the governor of the motor. The rotational speed of the motor was controlled at 10 r/min, 6.6 r/min and 3.3 r/min, which correspond to the pull-out speeds of 20.9 mm/s, 13.8 mm/s and 6.9 mm/s, respectively. Finally, by taking the buried depth, pull-out speed and different lotus roots as the factors, the pull-out force for the testing lotus roots was measured. The experimental factors are listed in Table 4.

Table 4. Factors for the experiment.

Level	Pull-Out Speed/(mm/s)	Burial Depth/(cm)	Lotus Root
1	6.9	15	No.1
2	13.8	20	No.2
3	20.9		No.3
4			No.4
5			No.5

2.2.4. Testing Procedure

The designed platform was employed to test the pull-out force, and the test site is depicted in Figure 5. Before operation of the motor, the sampling rate of the measurement system was set to 10 Hz. During the test, the motor was adjusted to the required speed and the lotus roots were pulled out by the clamping claw, during which the pull-out force was observed. When the pull-out force exceeded the pull-out resistance, that was, the lotus root broke away from the soil, the test was considered to be finished. The next test was carried out after recovery of the mud layer and five minutes of standing. Since the testing lotus root needed to be buried in the soil again in each test, in order to eliminate the influence of operation, each test was repeated three times, and the average pull-out force was recorded.

Figure 5. Test site.

2.3. Simulation of Pull-Out Process of Semi-Buried Lotus Root

The ultimate aim of the pull-out test is to obtain the value of the pull-out force which can pull lotus root out of the mud layer under a static state. To reveal the changes in pull-out force in the process of lotus root breaking away from the mud layer, the discrete element method (DEM) was used to simulate the pull-out process of semi-buried lotus root, and the simulation was carried out by EDEM 2.6 (DEM Solutions, Edinburgh, UK).

2.3.1. Lotus Root Model

A 3D scanner was used to model the lotus root. The surface information of the lotus root was read by the cameras of the 3D scanner and then converted into a '.stl' format file. Then, the SolidWorks 2016 software (Dassault Systemes, Paris, France) was employed to convert the file into a '.stp' file, which can be used in the EDEM 2.6 software. In practice, the cameras were set at the top and bottom of the testing lotus root hung by a string. The cameras ran along the length direction of the lotus root to scan the lotus root and generate the 3D model (Figure 6).

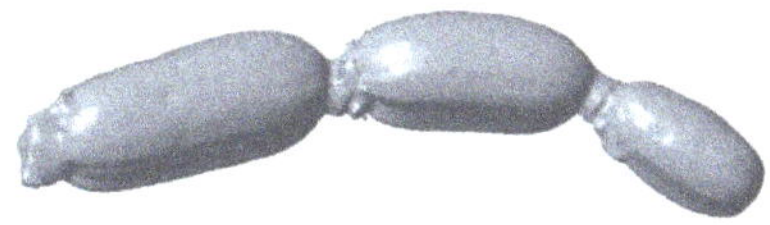

Figure 6. 3D model of the lotus root.

Density is a key parameter that determines the material property of the lotus root in EDEM simulation. Lotus root is not solid, and the inner part is porous and full of air, indicating that pulp density is more applicable to the simulation than bulk density. The pulp volume was estimated based on the pulp area and stomatal area of the lotus root, which can be used to determine the pulp density.

First, the total volume of the lotus root was measured by the drainage method. In order to measure the pulp area and stomatal area, the testing lotus root was cut into several slices along its axial direction, which were then placed on several black papers with a size of 10 × 10 cm. A camera was then used to capture the images of the slices. Each image was treated and shown in Figure 7. After that, an image with white pulp and another image with white stomata were obtained. From Figure 7a, the area of the pulp part could be determined. In addition, the area of the stomatal part could be measured from Figure 7b. Accordingly, the ratio of pulp part area to the total area of the section (the area of pulp + stomata) was calculated to be 0.817. Finally, the pulp volume and the pulp density of the five testing lotus roots could be obtained (Table 5).

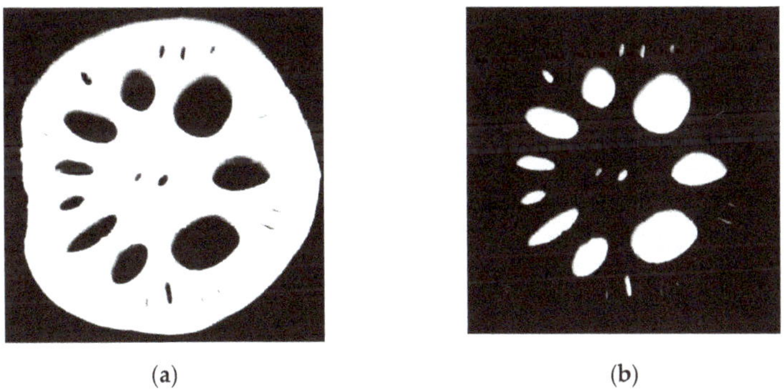

Figure 7. Images of a slice of the testing lotus root. (**a**) White pulp; (**b**) White stomata.

Table 5. Volume and density of pulp.

Lotus Root	No.1	No.2	No.3	No.4	No.5
Total Volume/(mm^3)	919	1089	877	853	849
Volume of pulp/(mm^3)	751	890	717	697	694
Density of pulp/(kg/m^3)	1050.60	960.67	1023.71	998.57	946.69

Finally, the pulp density and bulk density were calculated to be about 996.05 kg/m^3 and 818.78 kg/m^3, respectively. Since the bulk density is lower than the water density, the lotus roots can float out from water after being separated from the mud layer.

Xiao et al. and Guo et al. have measured several mechanical properties of lotus roots and demonstrated their anisotropy [10,11]. They also demonstrated that there are only small differences in mechanical properties between the axial and radial directions of lotus roots. Therefore, lotus roots can be regarded as isotropic to facilitate the calculation. The elastic modulus of lotus root was measured to be 0.451–2.529 Mpa [10,11]. Moreover, due to the low elastic modulus and poor compression performance, the Poisson's ratio of lotus roots was set to be 0.47 by referring to the mechanical properties of common materials. The shear modulus of lotus roots can be calculated by Equation (5):

$$G = \frac{E}{2(1+v)}, \tag{5}$$

where G is the shear modulus, E is the elastic modulus and v is the Poisson's ratio. Thus, the shear modulus of lotus roots was calculated to be 0.39 Mpa.

Currently, there has been no detailed research on the mechanical behavior of the interface between soil and lotus roots; moreover, it is difficult to calibrate the contact parameters. In this study, an approximate method was selected to determine the contact parameters in this simulation. The surface of lotus root is smooth, and its hardness is far greater than that of the sludge attached to the lotus root surface. For this reason, the contact between lotus roots and soil can be similarly regarded as the contact between a farm tool and soil. With reference to the parameters of soil–farm tool contact, we set the static friction coefficient, the dynamic friction coefficient and the recovery coefficient of the soil–lotus root contact as 0.5, 0.05 and 0.6, respectively [12].

2.3.2. Soil Model

The soil of lotus root field is characterized by sticky properties. Thus, the JKR model was selected to model the soil. Due to its ability to characterize the contact surface energy of soil particles, the JKR model has been proved to be suitable for modeling cohesive soil [13,14]. The soil parameters of the JKR model are presented in Tables 6 and 7 [15–17].

Table 6. Intrinsic parameters of soil.

Poisson's Ratio	Shear Modulus/(MPa)	Density/(kg/m^3)
0.4	408.3	1714

Table 7. Contact parameters of soil particles.

Dynamic Friction Coefficient	Static Friction Coefficient	Recovery Coefficient
0.2	1.16	0.15

To improve the calculation efficiency, the soil particles were generated in a 300 × 150 × 70 mm box as depicted in Figure 8. The soil particles were of a sphere shape and were generated based on a radius of 3 mm with 0.5–1.5 fold changes. The surface energy of JKR was 3.5–10.5 J/m^2 [15–17].

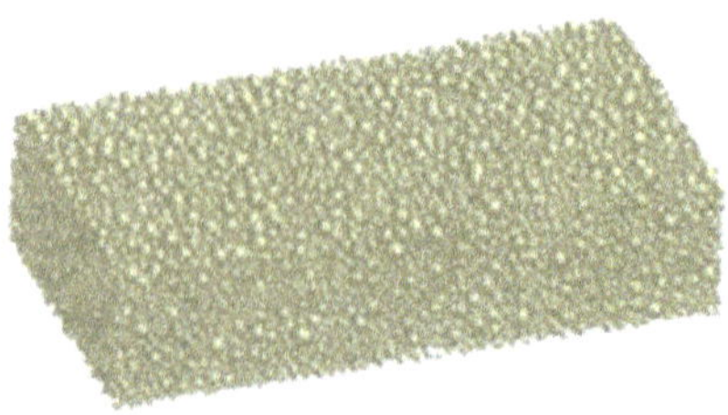

Figure 8. Soil box.

After setting the parameters, a single testing lotus root was semi-buried in the soil model with a burial depth of 15 cm. The EDEM simulation was carried out at the pull-out speeds of 10 mm/s and 5 mm/s. In the simulation, the effect of the water layer was neglected. When the lotus root was observed to be completely separated from the soil, the simulation was stopped, and the pull-out force varying with time and displacement was obtained through post processing.

3. Results and Discussion

3.1. Pull-Out Force Testing Results and Analysis

3.1.1. Force-Time Curves with Different Testing Factors

Figure 9 depicts the pull-out force of No.4 lotus root varying with pull-out speed and burial depth. Figure 10 displays the pull-out force of the five testing lotus roots at the burial depth of 20 cm and the pull-out speed of 6.9 mm/s.

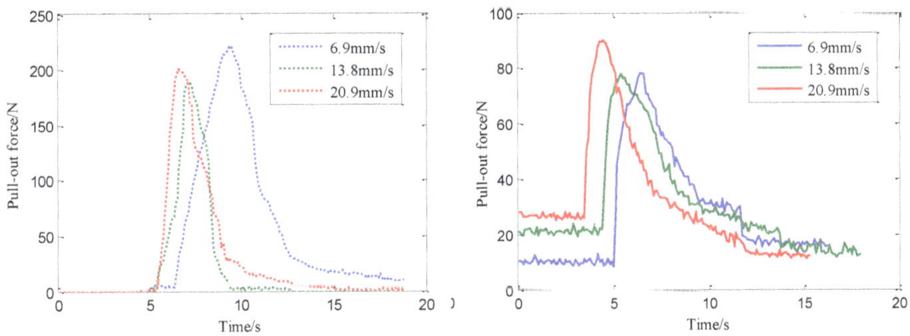

(a) (b)

Figure 9. Force-time curves of No.4 lotus root. (**a**) Burial depth of 20 cm; (**b**) Burial depth of 15 cm.

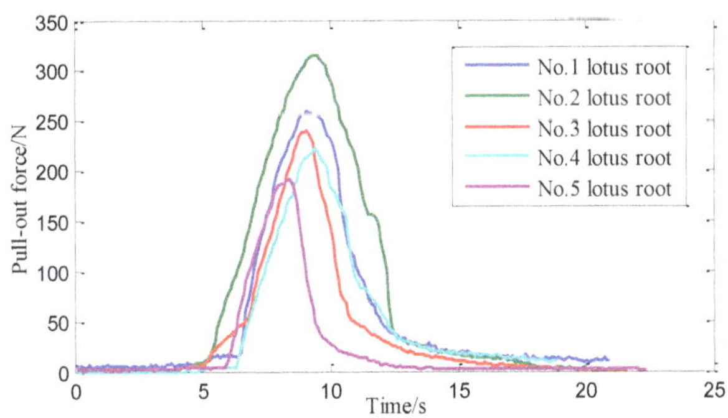

Figure 10. Force-time curves of five testing lotus roots with burial depth of 20 cm and pull-out speed of 6.9 mm/s.

Figures 9 and 10 show the pull-out process in the time domain. The pull-out force first increases slowly to reach the peak value with time, followed by a sharp decrease, probably because the lotus root needs to first overcome the resistance, and once the resistance is overcome, the lotus root starts to move and break away from the mud. It can also be discovered that the pull-out force increases with increasing burial depth. Since the moisture content of the mud layer decreases with increasing burial depth, the soil of the deeper layer is harder with a higher viscosity, resulting in a higher resistance due to soil adhesion. In addition, Figure 9 shows that the pulling speed does not affect the maximum pull-out force. Figure 10 indicates that a greater lotus root weight will require a greater pull-out force.

3.1.2. Average Maximum Pull-Out Force of Each Lotus Root under Different Testing Factors

Figures 11 and 12 present the average maximum pull-out force of each lotus root with different pull-out speeds at the burial depths of 15 cm and 20 cm, respectively.

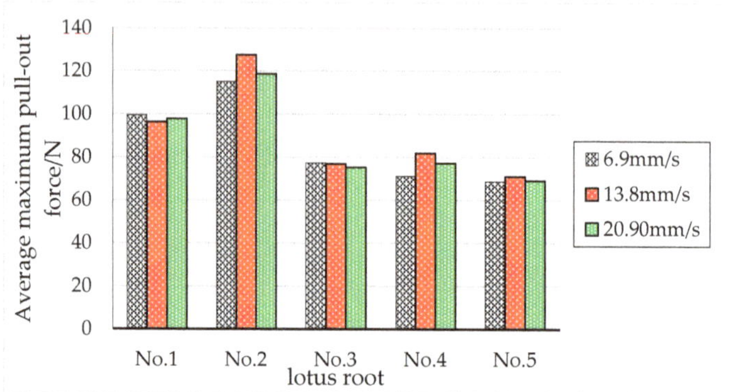

Figure 11. Average maximum pull-out force of different lotus roots at the burial depth of 15 cm.

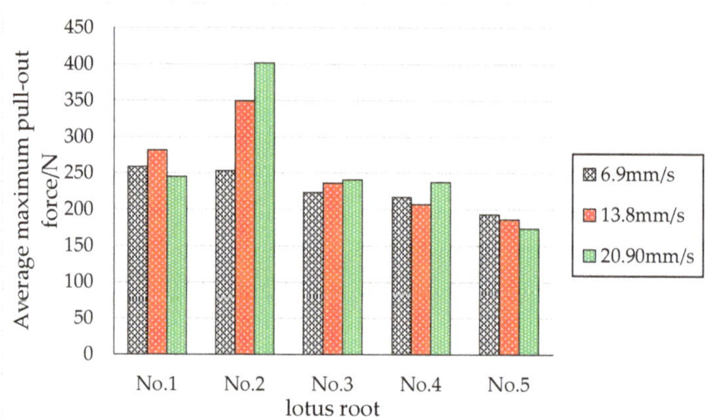

Figure 12. Average maximum pull-out force of different lotus root at the burial depth of 20 cm.

The results show that the average maximum pull-out force varies significantly among different lotus roots. Besides, the burial depth shows significant influence on the average maximum pull-out force, and pull-out speed exhibits no effect.

3.1.3. Analysis of Variance

Repeated analysis of variance (ANOVA) was carried out on the average maximum pull-out force of the testing lotus roots under different burial depths and pull-out speeds. The analysis results of No. 1 lotus root are shown in Table 8.

The ANOVA results of other lotus roots are consistent with those in Table 8. According to Table 8, burial depth has a significant impact on the pull-out force. Since the above tests were carried out under a static state, the pull-out speed shows no significant influence on the pull-out force.

Considering the individual differences of the testing lotus roots, the average maximum pull-out force with the burial depth of 15 cm was analyzed by ANOVA, and the results are presented in Table 9.

Table 8. ANOVA of the average maximum pull-out force under different burial depths and pull-out speeds.

Source of Differences	SS	df	MS	F	Significance
Buried depth	121,054.96	1	121,051.96	138.22	significance
Pull-out speed	993.06	2	496.53	0.57	
Interaction	1179.08	2	589.54	0.67	
Error	10,510.03	12	875.84		
Sum	133,737.14	17			

Table 9. AVOVA of the maximum pull-out force of different lotus roots at the burial depth of 15 cm.

Source of Differences	SS	df	MS	F	Significance
Different lotus root	15,697.32	4	3924.33	22.75	significance
Error	6898.61	40	172.47		
Sum	22,595.93	44			

The results obtained at the burial depth of 20 cm are consistent with those in Table 9. The results revealed that different lotus roots have significantly different maximum pull-out forces.

3.1.4. Effect of Individual Differences of the Testing Lotus Roots on the Maximum Pull-Out Force

Since individual lotus roots have different surface areas, the contact area between lotus roots and the soil is different, which will result in different soil adhesion force. It is difficult to measure the surface area of each lotus root. Considering the similar shape of lotus roots, the volume of the testing lotus roots can be used to indicate the surface area and the mass. Thus, the relationship between the mass of each lotus root and the maximum pull-out force was analyzed (Figure 13).

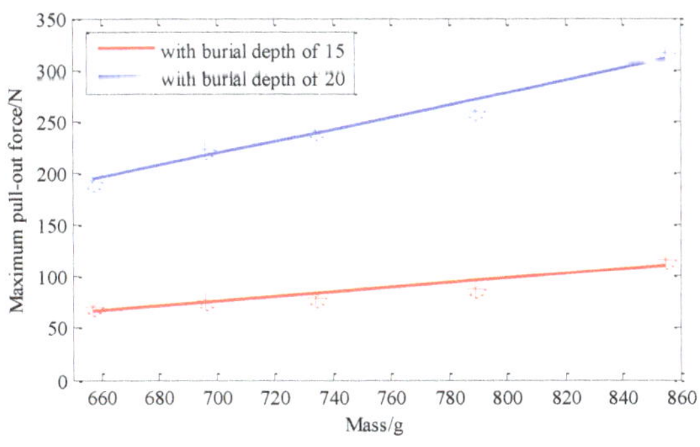

Figure 13. Curve fitting of the maximum pull-out force at different buried depths.

As shown in Figure 13, the maximum pull-out force of each lotus root linearly increases with the mass, indicating that the maximum pull-out force is positively correlated with the surface area of the lotus root. The linear fitting equations illustrating the relationship between the mass and the maximum pull-out force at the burial depth of 15 cm and 20 cm can be given by:

$$y_{15\text{cm}} = 0.2236\, m - 80.72, \quad (6)$$

$$y_{20\text{cm}} = 0.5903\, m - 194.30 \quad (7)$$

where y denotes the maximum pull-out force and m represents the mass of the lotus root. The determination coefficients of the equations are 0.9005 and 0.9770, respectively.

3.1.5. Summary and Discussion

In summary, when the pull-out force is greater than the resistance, the semi-buried lotus root can break away from the mud layer, and the instant pull-out force reaches the maximum value at the moment of separation. In addition, the maximum pull-out force is not affected by the pull-out speed, but is significantly influenced by the burial depth and surface area of the lotus root: it increases with increasing burial depth and surface area of the lotus root.

During the test, it was also observed that the maximum pull-out force greatly decreased when the testing lotus root was slightly loosened in the mud layer. The maximum pull-out force was detected to be only 20 N when the lotus root was loosened beforehand and then pulled out from the soil at a burial depth of 15 cm. The water penetrates between the loosened lotus root and the soil attached on the lotus root, disrupting the originally closed environment and weakening the adhesion. Furthermore, further pressing the lotus root causes a large increase in the pull-out force, which is similar to the impact of the water jet. The direct impact of the water jet on the lotus root results in an additional downward pressure, making the lotus root further adhere to the soil, which is not conducive to the floating of lotus roots.

Among the components of pull-out resistance, the buoyancy and normal force play a role in helping lotus root to float up, both of which are related to the gravity of buried lotus roots. Since the gravity only takes a very small proportion in the pull-out resistance, the main factor affecting the pull-out force of semi-buried lotus roots is the adhesion between lotus roots and the soil. In this experiment, significant factors such as burial depth and surface area of lotus roots affect the adhesion of lotus roots to the soil. The results show that once the adhesion between the lotus root and soil is destroyed, the lotus root can be pulled out easily.

Due to the uneven burial depth of lotus roots in the soil, some lotus roots are still difficult to float out of the sludge after being washed by the water jet under a certain pressure during harvesting. Too high jet pressure will cause damage to lotus root skin. It is difficult to balance the floating rate and damage rate. According to the above analysis, some mechanical methods can be used to vibrate and loosen the soil after the first hydraulic scouring, or the water jet can be adjusted to weaken the adhesion between lotus roots and the soil. After that, a second small pressure scouring can be used to improve the floating rate of lotus roots on the premise of reducing the damage.

3.2. Simulation Results Analysis and Discussion

The pull-out process is shown in Figure 14. During the pull-out process, the pull-out force varying with time and displacement was recorded (Figure 15).

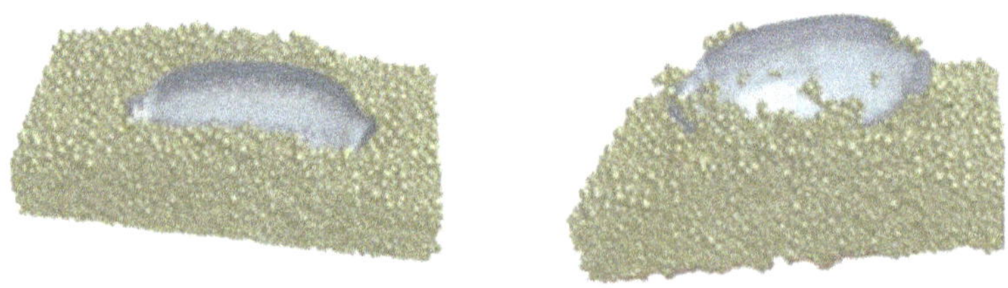

Figure 14. Pull-out process of semi-buried lotus roots in EDEM simulation.

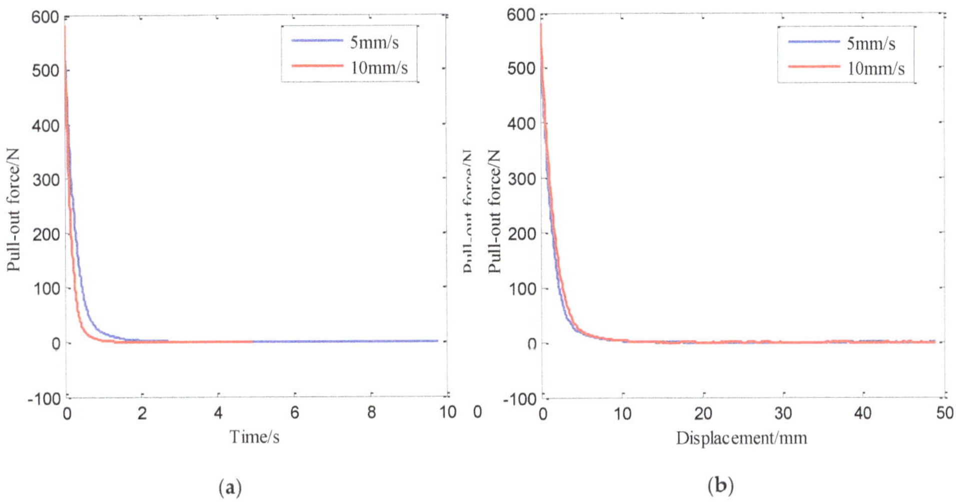

Figure 15. Pull-out force in simulation. (**a**) Force–time curve; (**b**) Force–displacement curve.

Figure 15 shows that the maximum pull-out force of the lotus root in the simulation is greater than that in the test. In the simulation, the water layer is not involved in the soil model, which would result in a larger soil adhesion force. The parameters of soil–soil contact and soil–lotus root contact are the main factors influencing the simulation results. In this study, the soil–lotus root contact parameters were set based on the soil–tool contact in cohesive soil [12]. Furthermore, the soil–soil contact parameters were set based on the properties of the cohesive soil [15–17]. Due to the influence of the water layer, the soil of a lotus root field has certain rheological properties. Hence, the parameters may not accurately simulate the real situation.

The force-displacement curve (Figure 15b) indicates that the pull-out force at the speed of 10 mm/s is greater than that at the speed of 5 mm/s for the same displacement. For cohesive soil, the adhesion is proportional to the velocity. Similarly, when lotus roots are being moved from the mud layer, a higher velocity usually requires a greater pull-out force. When the lotus root begins to break away from the soil, the gap between the lotus root and the soil increases, which will reduce the adhesive force and the pull-out resistance. In this case, the pull-out force decreases rapidly at first, followed by mild decreases until disappearance.

The simulation results also show that the adhesion force of the mud layer mainly contributes to the pull-out resistance. The increase in burial depth and surface area of the lotus root substantially increases the pull-out resistance, resulting in an increase in the maximum pull-out force. Since the adhesion force decays rapidly with the separation of the lotus root from the soil, the required pull-out force will sharply decrease with a slight loosening of the lotus root. The results of simulation are consistent with those of the test.

3.3. Limitations of the Study

This research explores the pull-out process of semi-buried lotus roots and analyzes the factors influencing the pull-out force based on experiment and simulation, which can provide a theoretical basis for improving the lotus root harvester with hydraulic scouring. However, there are still several limitations in this study, which need to be further addressed in the future.

In this experiment, the testing lotus root was pulled out in the vertical direction. At this time, the soil adhesion hinders the separation of lotus root and soil, which is also in the vertical direction. If the pull-out direction is adjusted, the direction of soil adhesion will also be changed, and the pull-out effect should be different from that of vertical pull-out.

However, the test platform proposed in this study can only pull lotus roots out from the vertical direction. Hence, the pull-out angle should be considered in the experimental study to determine the optimal pull-out direction.

Soil samples were collected from a real lotus field in the test and reconstructed according to the actual soil state of the lotus field. However, in the test process, the method of burying the lotus root might cause soil compaction, which may lead to certain deviation of the pull-out force value from that in the actual field. In addition, with the progress of the test, the operation might repeatedly compact and disrupt the soil layer, which may influence the accuracy of the test results.

In the simulation analysis, the parameters for lotus root–soil and soil–soil contact were set by referring to the existing literature, and these contact parameters were not calibrated, which may affect the accuracy of the simulation results.

In the follow-up research, the pull-out force test platform can be improved to make the pull-out angle adjustable and consider hydraulic scouring. In addition, the physical and chemical properties of lotus roots and the mechanical properties of soil in lotus field will be studied to explore the calibration method for the lotus root–soil and soil–soil contact parameters. In the future, we will further explore the interaction among water jet, soil, lotus root, water layer and other relevant factors, so as to provide better solutions for lotus root harvesting with high harvest rate and low damage rate.

4. Conclusions

1. A testing platform was designed to observe the pull-out process and acquire the pull-out force for semi-buried lotus roots. The testing platform was proved to be highly efficient and convenient for the real-time measurement of the pull-out force.
2. The testing results show that the pull-out force increases with increasing burial depth and surface area of the lotus root, and the adhesion between lotus roots and soil contributes the most to the pull-out resistance.
3. The discrete element method was employed to simulate the pull-out process of lotus roots from soil. The results also show that the adhesion of soil layer to lotus root is the main factor affecting the pull-out force, and the adhesion decreases rapidly with the separation of lotus roots from the soil.
4. Both experimental and simulation results provide important implications for the improvement of the existing lotus root harvester with hydraulic scouring. To improve the floating rate of lotus roots after hydraulic scouring, proper mechanical structure can be applied after hydraulic scouring, or the water jet can be adjusted.

Author Contributions: Conceptualization, R.Z. and G.Z.; methodology, R.Z. and Y.L.; software, Y.L.; validation, R.Z., Y.L. and J.J.; investigation, R.Z. and Z.W.; resources, R.Z. and M.T.; data curation, G.Z.; writing—original draft preparation, R.Z.; writing—review and editing, R.Z.; visualization, R.Z.; supervision, G.Z.; project administration, R.Z.; funding acquisition, G.Z. All authors have read and agreed to the published version of the manuscript.

Funding: This research was funded by China Agriculture Research System of MOF and MARA, grant number CARS-24-D-02.

Institutional Review Board Statement: Not applicable.

Informed Consent Statement: Not applicable.

Data Availability Statement: The data presented in this study are available on demand from the first author at (zengrong@mail.hzau.edu.cn).

Acknowledgments: G.Z. and the Technical System of National Characteristic Vegetable Industry.

Conflicts of Interest: The authors declare no conflict of interest.

References

1. Xie, J.; Han, D.; Wang, J. Development Status Quo and Prospect of China's Lotus Root Industry. *Agric. Outlook* **2017**, *13*, 42–45, 51.
2. Liu, M.; Mao, D.; Wang, Z. Research Status of Lotus Root Digging Technology and Prospect of Lotus Root Harvesting Technology in Yellow River Delta Region. *Agric. Eng.* **2018**, *8*, 1–5.
3. Wang, C.; Liu, H.; Wang, L. Overall Design of Water Chestnut Harvester. *J. Chin. Agric. Mech.* **2017**, *38*, 22–26.
4. Du, J.; Xia, J.; Wu, H. An Investigation of the Performance of Water Jet for Lotus Root Digging Device: Simulation and Experiment. *Int. J. Fluid Mach. Syst.* **2020**, *13*, 160–166. [CrossRef]
5. He, J.; Chen, J.; Wang, A. Study on Flow Field of a New Digging Lotus Based on the Vortex Theory. *Comput. Simul.* **2015**, *32*, 184–187, 329.
6. Li, Y.; Zhang, G.; Zhou, Y. Design and field experiment of drawing resistance measurement system for cotton stalk. *Trans. Chin. Soc. Agric. Eng. (Trans. CSAE)* **2013**, *29*, 43–60.
7. Chen, M.; Song, D.; Wang, Z. Research on the Cotton-Stalk Uprooting Resistance. *J. Agric. Mech. Res.* **2016**, *38*, 64–68.
8. Gao, G.; Xie, H.; Wang, T. EDEM simulation and experiment of pullout force of protected vegetable harvester. *Trans. Chin. Soc. Agric. Eng. (Trans. CSAE)* **2017**, *33*, 24–31.
9. Cho, S.; Chang, S.; Kim, Y. AE—Automation and emerging technologies: Development of a three degrees of freedom robot for harvesting lettuce using machine vision and fuzzy logic control. *Biosyst. Eng.* **2002**, *82*, 143–149. [CrossRef]
10. Xiao, K.; Xia, J.; Chen, Z. Biomechanics of lotus root. *J. Huazhong Agric. Univ.* **2016**, *35*, 125–130.
11. Guo, Y.; Xia, J.; Xiao, K. Determination and analysis on the yield strength and compressive strength of lotus roots. *J. Huazhong Agric. Univ.* **2015**, *34*, 148–151.
12. Yang, N.; An, Z. Simulation and Analysis on Ridging Process of Ridging Machine. *Mech. Res. Appl.* **2018**, *31*, 41–43.
13. Gilabert, F.; Roux, J.; Castellanos, A. Computer simulation of model cohesive powders. *Influ. Assem. Proced.* **2007**, *1*, 184–194.
14. Johnson, K.; Kendall, K.; Roberts, A. Surface energy and the contact of elastic solids. *Proc. R Soc. Lond.* **1971**, *3*, 301–313.
15. Wu, T.; Huang, W.; Chen, X. Calibration of discrete element model parameters for cohesive soil considering the cohesion between particles. *J. South China Agric. Univ.* **2017**, *38*, 93–98.
16. Ucgul, M.; John, M.; Chris, S. 3D DEM tillage simulation: Validation of a hysteretic spring (plastic) contact model for a sweep tool operation in a cohesionless soil. *Soil Tillage Res.* **2014**, *144*, 220–227. [CrossRef]
17. Wu, H.; Xia, J.; Zhang, G. Design and experiment of spin-jet flow type lotus root digging machine based on EDEM-Fluent. *Trans. Chin. Soc. Agric. Eng. (Trans. CSAE)* **2018**, *34*, 9–14.

agriculture

MDPI

Article

Experimental Evaluation of Rootstock Clamping Device for Inclined Inserted Grafting of Melons

Kang Wu [1], Jianzhong Lou [2], Chen Li [1] and Jianping Li [1,*]

[1] College of Biosystems Engineering and Food Science, Zhejiang University, Hangzhou 310058, China; kang_wu@zju.edu.cn (K.W.); lichen2020@zju.edu.cn (C.L.)

[2] School of Additive Manufacturing, Zhejiang Institute of Mechanical & Electrical Engineering, Hangzhou 310053, China; loujianzhong@zime.edu.cn

* Correspondence: jpli@zju.edu.cn

Abstract: A grafting machine is a kind of machine that can quickly graft scion to rootstock instead of manual grafting. Currently, an inclined inserted grafting machine uses the mechanical clamping method, which can easily damage the rootstock seedlings due to its high stiffness, thus, reducing the success rate of grafting. This study proposed an effective, flexible clamping device for grafting. The suction hole diameter (HD), the negative pressure (NP), and the surface inclination angle (IA) of the clamping device were studied via a single factor experiment to obtain optimal parameter ranges. Optimal HD, NP, and IA were 2–3 mm, 4–8 kPa, and 10–20°, respectively. The orthogonal experiment results showed that the optimal parameter combination for maximum holding success rate was HD, 2.5 mm; NP, 6 kPa; and IA, 10°. The experimental verification was carried out using the optimal parameter combination, with a holding success rate of 98.3% and no damage. This study provides a reference for the optimal design of an inclined inserted grafting machine.

Keywords: inclined inserted grafting; mechanical damage; flexible holding; orthogonal experiment; optimal design

Citation: Wu, K.; Lou, J.; Li, C.; Li, J. Experimental Evaluation of Rootstock Clamping Device for Inclined Inserted Grafting of Melons. *Agriculture* **2021**, *11*, 736. https://doi.org/10.3390/agriculture11080736

Academic Editors: José Pérez-Alonso and Massimo Cecchini

Received: 25 June 2021
Accepted: 30 July 2021
Published: 2 August 2021

1. Introduction

Watermelon, belonging to *Citrillus lanatus*, is a cash crop grown worldwide, especially in China. China planted watermelon on about 1.85 million hectares (ha) in 2017, yielding 74.84 million tons, accounting for 53.3% and 67.4% of the global planting and production, respectively [1]. Watermelon contains rich nutrients, such as minerals, vitamins, and amino acids [2,3]. Because of the limited cultivated land available and the low payoffs of crop rotation, watermelons are commonly continuously grown in the same field to meet the consumer demand. However, the continuous planting leads to the establishment of *Fusarium* wilt, significantly affecting watermelon production [4,5]. *Fusarium* wilt can cause rapid withering of plant cotyledons and fruits [6–8]. Although *Fusarium* wilt is controllable by crop rotation and not using land for a period, the farmers need the income. Developing resistant varieties against *Fusarium* wilt could be an effective and friendly measure to control *Fusarium* wilt disease in watermelon [9]. However, no commercial resistant varieties have been developed currently [5]. Grafting, an innovative and friendly technology [10], can produce high yield and strong resistance to *Fusarium* wilt disease [11], and it is commonly used in continuous planting worldwide. Planting grafted watermelons can increase the yield and production due to the strong rooting system [12], about 30–50% higher than that of ungrafted watermelons [13].

Traditional watermelon grafting was by hand, which is labor-intensive and inefficient. The emergence of grafting robots has attracted attention of many scholars due to continuous development of science and technology [14–16]. In China, the watermelon grafting machine mainly uses two methods [17]: splice and inserted methods, as shown in Figure 1. The splice method is to cut both the rootstock and the scion into bevels, then fix the two bevels together with a grafting clamp. The inserted method is first removing the growth point of rootstock, then punching a hole in the rootstock using a needle. In order to insert the scion into the hole smoothly and ensure that the scion will not loosen after being inserted into the hole without a grafting clamp, the size of the needle should be as thick as the scion. Next, cut the stem at 10–15 mm below the scion cotyledon into a splice of 70°, and then insert the scion into the rootstock [18]. When the punching needle is extracted, the diameter of the hole in the rootstock becomes smaller due to the rebound of the tissue in the hole. After inserting the scion into the rootstock, since the diameter of the scion is close to the hole diameter, the scion fits closely with the hole in the rootstock. Therefore, compared with the splice method, the inserted method does not require a grafting clamp during grafting. The inserted method consists of the straight inserted method and the inclined inserted method. Compared with the straight inserted method, the scion is not easy to insert into the cavity (a stem structure of melons) of the rootstock using the inclined inserted method. The inclined method can prevent self-rooting and also can prevent the nutrients loss caused by the splice method. This study is based on the inclined inserted method. The design of the rootstock clamping device directly affects the success of punching a hole in the rootstock and the union of the rootstock and the scion, thus, the rootstock clamping device is a key component in the inclined inserted melon vegetables grafting machine. Lou et al. [19] designed an adaptive pressing mechanism for rootstock cotyledons using the inclined inserted method. The springs are arranged at both ends of the clamping mechanism, automatically adjusting the position during the clamping process. However, clamping parts in the middle of the clamping mechanism are rigid and can damage the rootstock stems. Yang et al. [20] proposed a kind of air suction clamp for rootstock. The mechanism uses negative pressure to hold the rootstock cotyledon for making flat; however, too many suction holes and larger negative pressure cause greater energy consumption.

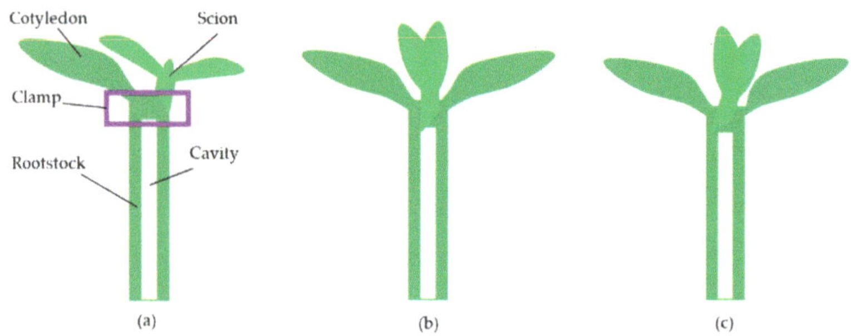

Figure 1. Grafting methods of melon vegetables: (**a**) Splice grafting method; (**b**) Straight inserted grafting method; (**c**) Inclined inserted grafting method.

This study proposed an effective, flexible clamping device for successful inclined inserted rootstock clamping. The device was equipped with silicone rubber to prevent the rootstock stem damage during clamping. The rootstocks were then held on the clamping block via a seedling pressing mechanism and negative pressure. The rootstock clamping device test-bed was designed, and the structural parameters and working parameters of the key components were optimized through experimental analysis. The device was experimentally verified for the optimal parameter combination. This study provides a reference for the rootstock clamping analysis in an inclined inserted grafting machine.

2. Materials and Methods

2.1. Inclined Inserted Grafting Method

The process of inclined inserted grafting method shown in Figure 2 are as follows. Firstly, the scion seedling is manually placed in the scion feeding mechanism and held by the negative pressure inside the air suction fixing device (Figure 2a). Then, the stem at 10 mm below the scion cotyledon is quickly cut off the splice with a cutter (Figure 2b,c).

Meanwhile, in another position of the grafting machine, the rootstock seedling is manually placed in the clamping blocks and then held by the clamping device (Figure 2d). The clamping device mainly consists of a seedling pressing mechanism and a clamping mechanism. The growth point of rootstock is completely exposed when removing the seedling pressing mechanism (Figure 2e). The growth point removal mechanism moves according to the designed trajectory to remove the rootstock growth point (also known as apical meristem, can produce new leaves and buds.), and the root of the rootstock is cut off quickly using a cutter (Figure 2e). As new roots after root cutting grow more vigorously, the root of rootstock is generally cut off during grafting. After rotating the rootstock clamping device by 90°, a pneumatic cylinder drives a needle (2.5 mm diameter in this study) that is as thick as the scion to punch a hole in the rootstock (Figure 2f). Next, the air suction fixing device drives the scion to the union position and inserts the scion into the rootstock diagonally downward as the needle is extracted (Figure 2g). Due to the close diameters between the scion and the hole in the rootstock, the scion fits closely in the hole in the rootstock and will not loosen after being inserted into the rootstock. Therefore, the grafted seedling does not require a grafting clamp. The grafted seedling is removed from the grafting machine and the grafting action is completed (Figure 2h).

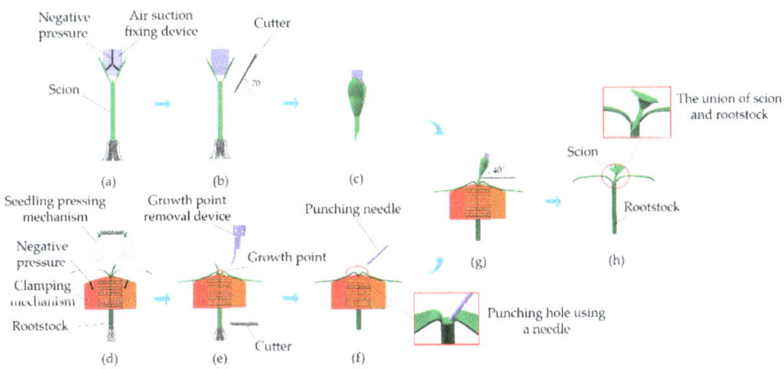

Figure 2. Grafting process of inclined inserted grafting method: (**a**) holding the scion; (**b**) cutting the stem of scion; (**c**) the scion after cutting; (**d**) holding the rootstock; (**e**) cutting the root and removing the growth point of rootstock; (**f**) punching a hole in the rootstock; (**g**) inserting the scion into the rootstock; (**h**) seedling after grafting.

The rootstock clamping device is one of the essential components of the grafting machine. Inaccurate positioning of the rootstock and the inability to fully expose the growth point during the clamping process will increase the difficulty of removing the growth point. Therefore, it is necessary to study the rootstock clamping device.

2.2. Structure and Working Principle of Rootstock Clamping Device

2.2.1. Structure of Rootstock Clamping Device

The 3D model of the rootstock clamping device is shown in Figure 3. The device consists of cylinders, seedling pressing mechanism, clamping blocks, frame, and other components. In pneumatic transmission, the external air pressure pushed the piston in the cylinder to reciprocate and then drives the actuator fixed at one end of the piston rod to move linearly or swing back and forth [21]. Pneumatic cylinder "a", fixed on the frame, can drive the pneumatic cylinder "b" and the seedling pressing mechanism to move directly above the clamping blocks. The function of the seedling pressing mechanism is to press the cotyledons of the rootstock seedling to the upper surface of the clamping blocks to facilitate holding the cotyledons. The rotary table cylinder is fixed on the cylinder fixing frame, and the clamping block "a" and the clamping block "b" are connected with the rotary table cylinder through the connecting frame. The rotary table cylinder can drive the clamping blocks to rotate to another direction so as to facilitate the union of scion and rootstock. Silicone rubber, shown in Figure 4, is fixed by glue at the middle of the clamping block, and it is food grade and belongs to low hardness silicone rubber with a hardness of 10°. Squeezing silicone rubber causes its deformation to cushion the clamping impact.

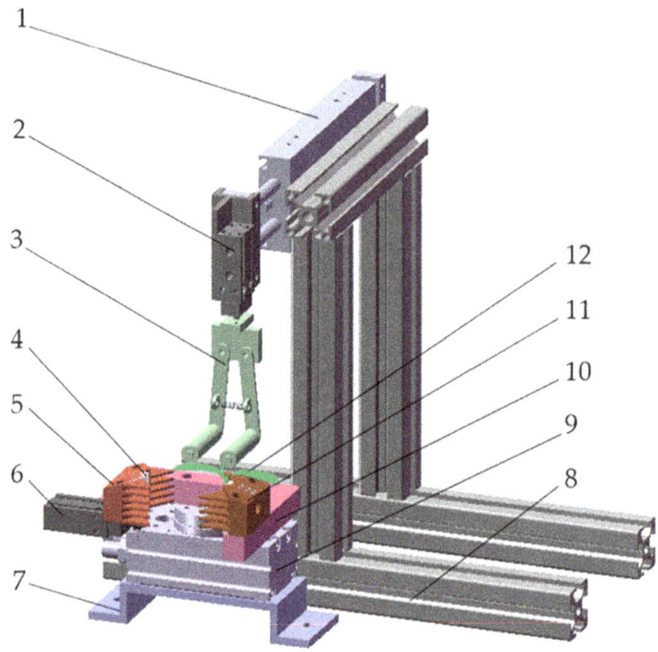

Figure 3. 3D model of rootstock clamping device. 1. Pneumatic cylinder "a". 2. Pneumatic cylinder "b". 3. Seedling pressing mechanism. 4. Silicone rubber. 5. Clamping block "a". 6. Pneumatic cylinder "c", 7. Cylinder fixing frame, 8. Frame. 9. Rotary table cylinder. 10. Connecting frame. 11. Clamping block "b". 12. Rootstock seedling.

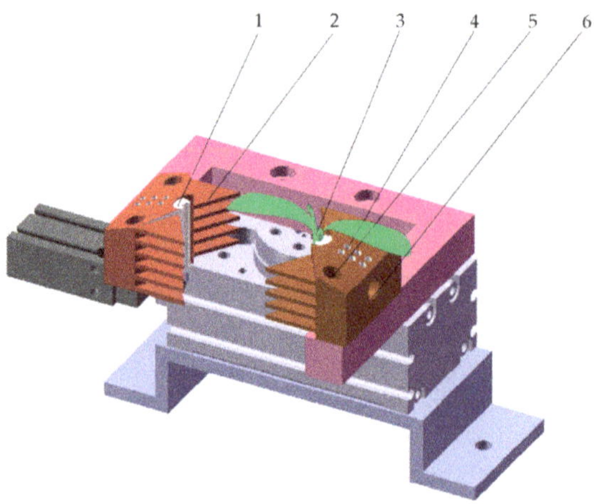

(**a**)

Figure 4. *Cont.*

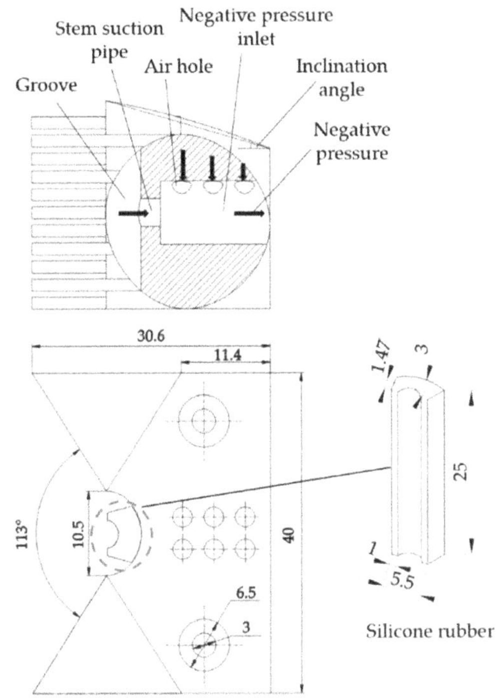

(**b**)

Figure 4. Clamping mechanism: (**a**) 3D model of Clamping mechanism; (**b**) engineering drawing of clamping block "b" (unit: mm). 1. Silicone rubber. 2. Positioning piece. 3. Seedling. 4. Air hole. 5. Connecting hole. 6. Negative pressure inlet.

2.2.2. Working Principle

The rootstock seedling is manually placed on the silicone rubber fixed inside the clamping block "b", and then the negative pressure (in Figure 4b) generated by the vacuum pump holds the stem of the rootstock. The pneumatic cylinder "c" pushes the clamping block "a" towards the rootstock seedling. The v-shaped positioning pieces (in Figure 4a) on two clamping blocks can cross together and gather the stem in the middle, realizing the clamping positioning. After fixing the stem of the rootstock, the pneumatic cylinder "a" pushes the pneumatic cylinder "b" directly above the rootstock seedling, making the seedling pressing mechanism press down the rootstock cotyledons. The negative pressure holds the cotyledons when the cotyledons are pressed on the upper surface of the clamping blocks, and the seedling pressing mechanism completes seedling pressing and returns to its original position. The growth point is fully exposed and the clamping action completes, waiting for the follow-up growth point removal operation.

2.2.3. Design of Clamping Mechanism

The clamping mechanism is shown in Figure 4. The fit degree between the cotyledon and the upper surface of the clamping block is one of the key factors for the holding success. The upper surface of the clamping block is curved at an inclination angle of 5–25°. The two rows of air holes (diameter, 1.5–3.5 mm) are arranged on the upper surface. The distances between the two rows of air holes are larger than the leaf vein width in the middle of the cotyledons to ensure that the cotyledons are effectively held. At the same time, to avoid the waste of negative pressure, the total width of two rows of air holes is less than the cotyledon width. The negative pressure inlet (diameter, 8 mm) at one end is connected with the negative pressure pipe. Compared to clamping block "a", a stem suction pipe is set in the clamping block "b", and other structural parameters of clamping block "b" are the same as those in clamping block "a". Silicone rubber is then put in the middle groove of the clamping block. A hole connected with the stem suction pipe of the clamping block "b" is designed inside the silicone rubber. The silicone rubber is smaller than the groove size to achieve silicone rubber deformation space after squeezing.

The straight stems can be fixed at the middle of the clamping block only by suction. It is necessary to locate the curved stems using other methods since negative pressure can not completely hold and fix the stems in the middle position [22]. In this study, the multi-layer positioning pieces are arranged on the clamping block to realize the curved stems positioning.

The force of the rootstock stem in the clamping process is shown in Figure 5.

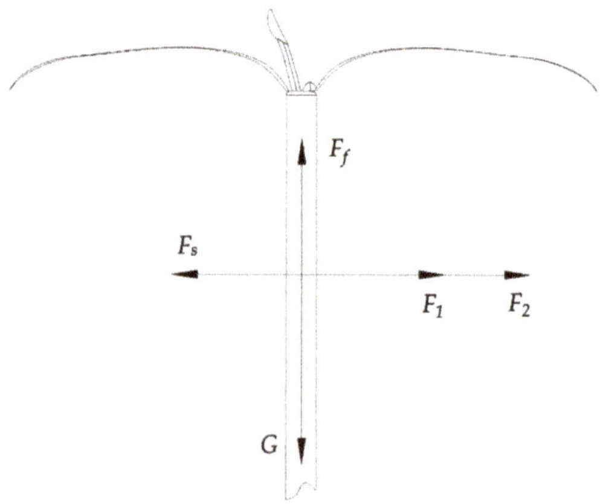

Figure 5. The force diagram of the rootstock stem.

Assuming that the force of the stem reaches an instant balance, the force equation can be expressed as:

$$\begin{cases} F_s = F_1 + F_2 \\ G = F_f \end{cases} \tag{1}$$

$$\begin{cases} F_f = \mu F_s \\ F_1 = PS = \pi P r^2 \end{cases} \tag{2}$$

Formulas (1) and (2) can be obtained using the following formula:

$$\frac{F_1}{S} + \frac{F_2}{S_0} = P + \frac{G - \pi\mu P r^2}{\mu S_0} \tag{3}$$

F_s—The supporting force of silicone rubber to the stem, N;
F_1—The air suction pressure, N;
F_2—The impact force of clamping block "a" to the stem, N;
G—The rootstock seedlings gravity, N;
F_f—The friction force between silicone rubber and the stem, N;
μ—The friction coefficient between silicone rubber and the stem;
P—Negative pressure, Pa;
S—The contact area between the suction hole and the stem, m^2;
r—The radius of suction hole, r = 1.75 mm;
S_0—Pressure area of the stem, m^2;
π—Pi, π = 3.14.

Stress generated by F_1 and F_2 should not be greater than yield stress of the stem to ensure that the rootstock stem is not pinched during the clamping process. Therefore, at constant μ, G, and r, it is necessary to study the air suction negative pressure change in the Equation (3).

2.2.4. Design of Seedling Pressing Mechanism

The structure of the rootstock seedling pressing mechanism is shown in Figure 6. The seedling pressing mechanism contains connecting block, connecting rod, spring, bearing, shaft, bearing retaining ring, and seedling pressing rod. Pneumatic cylinder "b" is fixed on one end of the rootstock seedling pressing mechanism through the connecting block, controlling the up and down movement of the seedling pressing rod. The seedling pressing rod is installed on the shaft through the bearing, and the outer end of the bearing is equipped with a bearing retaining ring. Pneumatic cylinder "b" drives the seedling pressing mechanism downwards during operation, pushing the seedling pressing rod down after touching the cotyledons of the rootstock. The friction between the seedling pressing rod and the cotyledons rotates the seedling pressing rod, moving along the extension direction of the cotyledons. The spring in the middle of the two connecting rods adjusts the horizontal movement distance of the seedling pressing rod. It quickly restores the seedling pressing rod to its original position after seedling pressing.

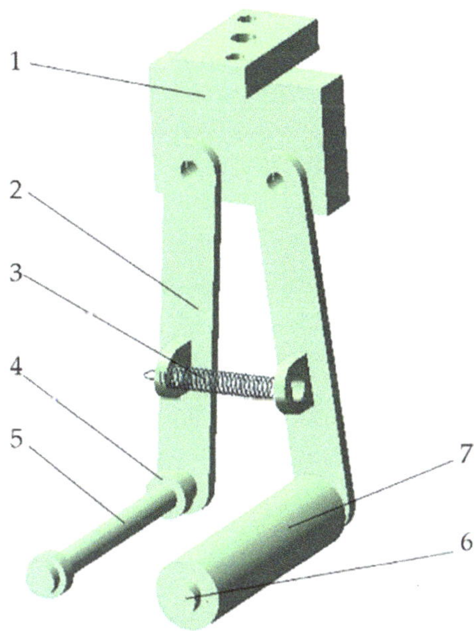

Figure 6. 3D model of seedling pressing mechanism. 1. Connecting block. 2. Connecting rod. 3. Spring. 4. Bearing. 5. Shaft. 6. Bearing retaining ring. 7. Seedling pressing rod.

The force on a single cotyledon during seedling pressing is shown in Figure 7. The cotyledon force equation at equilibrium moment can be expressed as:

$$\begin{cases} G\cos\alpha + F_p + F_a = F_n \\ G\sin\alpha + F_{f2} = F_{f1} \end{cases} \tag{4}$$

$$\begin{cases} F_{f1} = \mu_1 F_p \\ F_{f2} = \mu_2 F_n \\ F_a = P_0 S_1 = n P_0 \pi r_0{}^2 \end{cases} \tag{5}$$

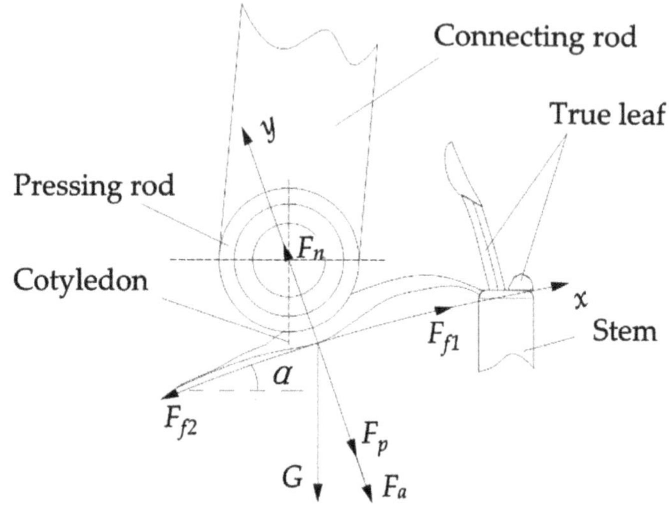

Figure 7. Stress analysis of a single cotyledon.

Then,

$$\frac{F_a}{S_1}+\frac{G\cos\alpha+F_p}{S_2}=P_0+\frac{G\sin\alpha+\mu_1 G\cos\alpha+n\mu_2 P_0\pi {r_0}^2}{S_2(\mu_1-\mu_2)} \tag{6}$$

G—The cotyledon gravity, N;
α—The surface inclination angle of the clamping block, °;
F_p—The force of the pressing rod on the cotyledon, N;
F_a—The air suction pressure, N;
F_n—The supporting force of the clamping block to the cotyledon, N;
F_{f1}—The friction force between the pressing rod and cotyledon, N;
F_{f2}—The friction force between the clamping block surface and the cotyledon, N;
μ_1—The friction coefficient between cotyledon and pressing rod;
μ_2—The friction coefficients between cotyledon and clamping block surface;
P_0—Negative pressure, Pa;
S_1—The contact area between suction hole and cotyledon, m^2;
r_0—The radius of suction holes on the surface of the clamping block, m;
n—The number of suction holes, n = 6;
S_2—Pressure area of the cotyledon, m^2;

Stress of cotyledon should not be greater than the yield stress of cotyledon to ensure that cotyledons are not crushed. Equation (6) indicates that the pressure of the pressing rod on the cotyledon is related to the surface inclination angle of the clamping block, the negative pressure, and the suction hole diameter. Therefore, it is necessary to investigate the above parameters to reduce the seedling damage and further optimize the design of the clamping device.

2.3. Experiment Materials

Yongzhen No. 5 cucurbit, a special rootstock for watermelon (Top-Yield Seed Technology Co., Ltd., Ningbo, China), was selected as the experimental material. The rootstock seedling shown in Figure 8 was planted in a greenhouse at 22–26 °C for 9–12 days, and the characteristic parameters of rootstock seedlings are shown in Table 1. The watermelon variety "Zaojia 8424" was used as the scion. The experiment was conducted when the cucurbit seedlings had two leaves (one true leaf unfolded and the second true leaf exposed).

(a)

(b)

Figure 8. Rootstock seedlings: (**a**) rootstock seedlings in a greenhouse; (**b**) structure of single rootstock seedling.

Table 1. Characteristic parameters of rootstock.

Long Axis of Stem (mm)	Short Axis of Stem (mm)	Cotyledon Width (mm)	Single Cotyledon Length (mm)
3.52 ± 0.29	2.94 ± 0.29	20.23 ± 1.84	32.37 ± 3.29

The 3D model of clamping blocks were designed in SolidWorks (Dassault Systemes, Concord, MA, USA) and then imported into a 3D printer (CR5080, Creality 3D Technology Co., Ltd., Shenzhen, China). The clamping blocks were made using 3D printing with polylactic acid (PLA). The fabrication process of silicone rubber was that the liquid silicone rubber was poured in a mold and cured at a room temperature of 28 °C for 4–5 h. The experiment gas sources included a Boge air compressor (pressure range 0–10 bar) and vacuum pump (pressure range −8–0 bar).

2.4. Experiment Conditions and Methods

The experiment was conducted on the self-made rootstock clamping test-bed to analyze the influence of the actual working environment and optimize the operating parameters. The vacuum pump produced negative pressure to hold the stem of rootstock on the clamping block. Another clamping block was pushed by the pneumatic cylinder to clamp the rootstock. The seedling pressing mechanism pressed down the cotyledons of rootstock, and then the cotyledons were held by the negative pressure of air holes. The rootstock holding process is shown in Figure 9. The negative pressure pipe was connected with a digital display meter to display the negative pressure value. The positive air pressure produced by the air compressor moved the pneumatic cylinder to control the clamping block and the seedling pressing mechanism. The air pressure pipe had a pressure regulating valve, which controlled the pressure.

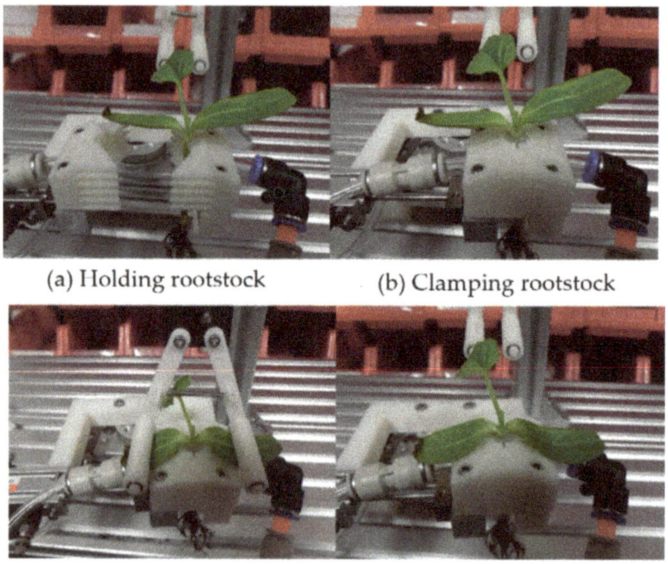

(a) Holding rootstock (b) Clamping rootstock

(c) Pressing cotyledon (d) Experiment completed

Figure 9. Diagram of rootstock holding process.

2.5. Experimental Factors and Indicators

The clamping test-bed was used to investigate the adsorption effect of the rootstock clamping device. Previous experiments and studies [23] have shown that air suction hole diameter (HD), negative pressure value (NP), and surface inclination angle (IA) are the main experimental factors. The holding success rate and the damage rate of rootstock were used as the experimental indicators. Rootstock seedlings held on the clamping blocks with no damage were treated as successful experiments; otherwise, they were recorded as failed. Obvious extrusion marks or breakage on the stems or cotyledons of rootstock seedlings were regarded as experimental damage. The damage rate included the damage to the stems and cotyledons of rootstocks. The indicators were calculated as follows.

$$\begin{cases} P = \frac{n_0}{N} \\ Q = \frac{n_1}{N} \end{cases} \tag{7}$$

P—The holding success rate, %;
n_0—The number of failed cotyledons of rootstocks;
N—The number of rootstocks;

Q—The damage rate of rootstocks, %;
n_1—The number of damaged rootstocks.

2.6. Experimental Design

This experiment selected HD, NP, and IA as influencing factors and the success rate and the damage rate as experimental indicators. Each factor had five levels: HD from 1.5 to 3.5 mm in increments of 0.5 mm, NP from 2 to 10 kPa in increments of 2 kPa, and IA from 5 to 25° in increments of 5°. The influence of the experimental factors on clamping was studied via a single factor experiment and orthogonal experiment. The orthogonal experiment could analyze the significant influence of each experiment factor on the experiment indicators through a small number of representative experiments and found out the optimal combination of factor levels [24], which has been applied in many fields [25,26]. The factor level table of the orthogonal experiment is shown in Table 2. A total of 15 groups of single factor experiments and nine groups of orthogonal experiments were carried out (60 rootstock seedlings in each group).

Table 2. Factors level.

Level	HD (mm)	NP (kPa)	IA (°)
1	2	4	10
2	2.5	6	15
3	3	8	20

2.7. Data Analysis

The orthogonal experiment results were processed and analyzed using the software SPSS 19.0 (IBM, Armonk, NY, USA). The analysis of variance (ANOVA) was conducted to assess the significance ($p < 0.05$) of experimental factors. The software Origin 8.5 (OriginLab, Northampton, MA, USA) was used for graphing and data analysis.

3. Results and Discussion

All the experiments were conducted in the laboratory of the College of Biosystems Engineering and Food Science, Zhejiang University, China. However, there were potential limitations in this study. First, compared with the actual facility agriculture grafted operation, the sample of this study is relatively small. Second, this study just investigates the rootstock clamping device, and no attempt has been made to study the device in the grafting machine; therefore, there could be other factors that affect the desired properties of the rootstock clamping device. Thus, a profound study is warranted in the future. Nonetheless, since fewer studies are available on the rootstock clamping device currently, these results could be helpful for the research of flexible clamping devices.

3.1. Single-Factor Experiment Analysis and Discussion

The single-factor experimental results of HD are shown in Figure 10. The results showed that with the increase of HD, the holding success significantly increased at first, then slowed down, finally turning into a downward trend. Decreased HD increased the velocity in the air hole, reducing the adsorption effect on rootstock cotyledons due to the small suction force between the air hole and cotyledon. Similarly, increased HD increased the suction force between the air hole and cotyledons. This improved the holding success rate while gradually decreasing the flow velocity and negative pressure. The negative pressure and flow rate rapidly decreased at certain HD increments, decreasing the adsorption success rate. Similar conclusions were reported by other researchers, who found that the airflow velocity and pressure gradually decreased with increasing outlet diameter of the suction chamber [20,27,28]. The clamping device had a better holding effect at an HD range of 2–3 mm.

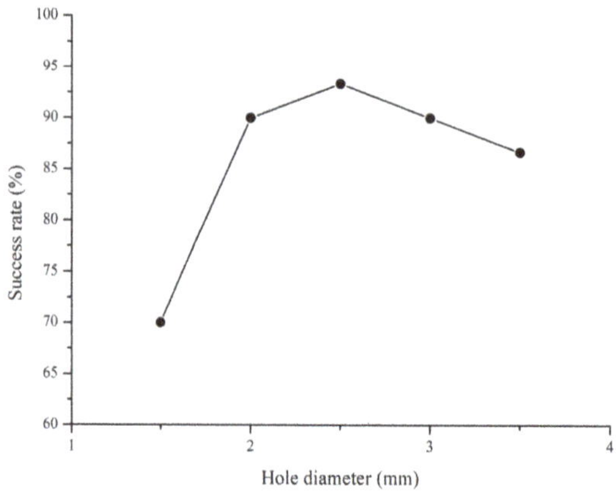

Figure 10. Relationship between hole diameter (HD) and success rate. Note: negative pressure (NP) was 4 kPa and inclination angle (IA) was 20°.

The single factor experimental results of NP are shown in Figure 11. Negative pressure flowed into the stem suction pipe and air hole from the negative pressure inlet, and the seedling pressing mechanism assisted in holding the rootstock on the clamping block. It can be seen in Figure 11 that the holding success rate of rootstock gradually increased with increasing NP. The cotyledons cannot be held on the clamping block at an NP of 2 kPa because low pressure leads to a small suction force. However, with the gradual increase of NP, this situation changed significantly. NP was caused by the pressure difference between the inside and outside of air hole. Increased NP led to increased pressure difference and decreased pressure inside the air hole. According to Bernoulli's principle, the flow velocity increased when the pressure decreased, so the flow velocity inside the air hole increased. At constant HD, increasing NP promoted the flow rate and suction force of the air hole, thus, improving the holding success rate. Jiang et al. (2019, 2020) obtained a similar conclusion that the increase of negative pressure increased the vacuum in the gas chamber, thus, increasing the flow rate and improving the adsorption capacity [29,30]. In this experiment, the injury of rootstock seedlings occurred mainly at the junction between cotyledons and stems when NP was 10 kPa, possibly due to the high NP increasing suction force. When HP ranged 4–8 kPa, the holding effect had satisfying values.

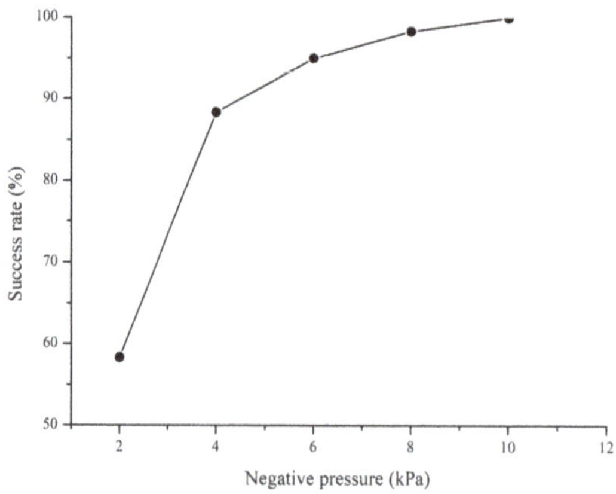

Figure 11. Relationship between negative pressure (NP) and success rate. Note: hole diameter (HD) was 2.5 mm and inclination angle (IA) was 20°.

At a constant of NP and HD, inclination angle (IA) reflected the fit of cotyledon and the upper surface of clamping block. Excessive IA led to damage in the junction between the cotyledon and the stem during the pressing cotyledons process; additionally, the negative pressure of air holes close to the stem could not effectively hold the cotyledon. The single factor experimental results of IA are shown in Figure 12. As can be seen from Figure 12, the holding success rate showed a downward trend with increasing IA. Increased IA increased the distance of the cotyledons pressed to the surface of the clamping block, promoting the resistance of the cotyledons to the clamping block. Furthermore, increased IA reduced the fit degree between the cotyledons and the clamping block surface, resulting in some negative pressure loss, thus, reducing the adsorption effect. The holding success rate was higher when the IA was 5°. However, the growth point of the rootstock could not be fully exposed at 5°, thus, it was not conducive for the subsequent growth point removal operation. Therefore, the holding effect can be improved to an acceptable level at an IA range of 10–20°.

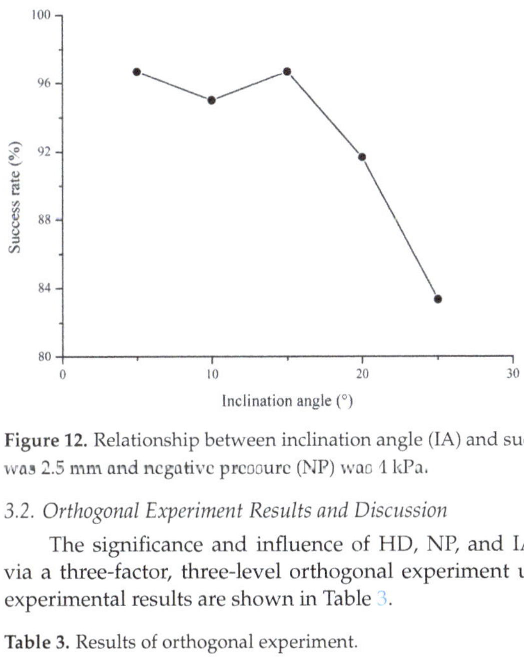

Figure 12. Relationship between inclination angle (IA) and success rate. Note: hole diameter (HD) was 2.5 mm and negative pressure (NP) was 4 kPa.

3.2. Orthogonal Experiment Results and Discussion

The significance and influence of HD, NP, and IA on holding were investigated via a three-factor, three-level orthogonal experiment using the clamping device. The experimental results are shown in Table 3.

Table 3. Results of orthogonal experiment.

No.	HD (mm)	NP (kPa)	IA (°)	Success Rate (%)	Damage Rate (%)
1	1	1	1	93.33	0
2	1	2	2	95	0
3	1	3	3	91.67	0
4	2	1	2	96.67	0
5	2	2	3	95	0
6	2	3	1	98.33	0
7	3	1	3	90	1.67
8	3	2	1	96.67	0
9	3	3	2	95	5

The success rate of the clamping device was not less than 90% with a low damage rate, indicating an effective clamping device (Table 3). Through the experiment, it was found that the rootstock stem was deflected during clamping, inhibiting cotyledon absorption, losing some negative pressure, which led to the adsorption failure. This observation was in agreement with the results obtained by others [31]. The holding damage was due to the uneven cotyledons of the rootstock, and the distal end of some cotyledons were lower than the position of clamping blocks.

This study mainly focused on the holding success rate analysis since the damage rate was low. The analysis results are shown in Table 4. Table 4 showed that the p-values for HD, NP, and IA were 0.032, 0.072, and 0.023, respectively, indicating that HD and IA had a significant effect on the success rate ($p < 0.05$). By contrast, NP had no significant influence on the success rate since its p-value for NP was greater than 0.05.

Table 4. ANOVA results for adsorption effect.

Source	Sum of Squares	df	Mean Square	F-Value	p-Value
HD	19.131	2	9.565	30.502	0.032 *
NP	8.031	2	4.015	12.804	0.072
IA	26.523	2	13.262	42.289	0.023 *
Error	0.627	2	0.314		
Total	80,647.845	9			

* means this term is significant ($p < 0.05$).

The F-values were compared to determine the influence order of the three factors on the experiment results. Higher F-values indicated a greater influence of the corresponding factor on the experiment results [32]. The order of influence of the three factors in this study was IA > HD > NP. Therefore, more attention should be on adjusting IA and HD than NP during the experiment since the influence order of NP was the last and had no significant effect on the experimental results.

As can be seen from Figure 13, the optimal combination based on achieving the maximum holding success rate was HD2, NP2, and IA1. The experimental verification was conducted using the parameter combination with 60 rootstock seedlings. The optimal parameter combination achieved a holding success rate of 98.3% with no damage, indicating an effective clamping device designed in this study, achieving grafting requirements.

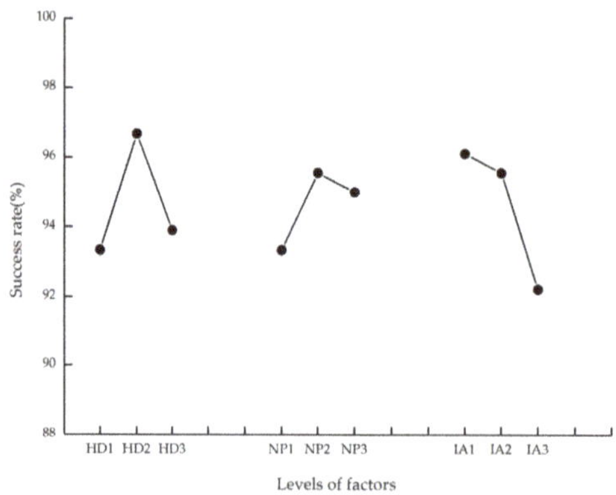

Figure 13. Comparison of effects of influencing factors on holding success rate.

4. Conclusions

This study proposed a flexible clamping device to reduce the holding damage of rootstock seedlings during grafting. The clamping device was evaluated to determine its performance. The following conclusions were made:

(1) The single factor experiment was used to determine the optimal factor ranges of the clamping device: HD, 2–3 mm; NP, 4–8 kPa; and IA, 10–20°.

(2) The orthogonal experiment was used to assess the influence of structural parameters and working parameters on holding. The ANOVA analysis indicated the influence order as follows: IA > HD > NP.

(3) The optimal parameter combination was determined by orthogonal experiment: HD, 2.5 mm; NP, 6 kPa; and IA, 10°. The verification experiment was conducted using the parameter combination. A holding success rate of 98.3% with no damage was obtained. This study can provide a reference for the development of an inclined inserted grafting machine.

Author Contributions: Methodology, K.W. and J.L. (Jianping Li); software, K.W. and C.L.; validation, K.W.; formal analysis, K.W. and C.L.; investigation, K.W.; data curation, K.W.; writing—original draft preparation, K.W.; writing—review and editing, J.L. (Jianzhong Lou) and J.L. (Jianping Li); visualization, K.W. and C.L.; supervision, J.L. (Jianping Li); project administration, J.L. (Jianping Li); funding acquisition, J.L. (Jianping Li). All authors have read and agreed to the published version of the manuscript.

Funding: This research was funded by National Natural Science Foundation of China (Grant No. 51775490).

Institutional Review Board Statement: Not applicable.

Informed Consent Statement: Not applicable.

Data Availability Statement: The data presented in this study are available on demand from the first author at (kang_wu@zju.edu.cn).

Acknowledgments: The authors are very grateful to the members of the research group for their help in the experiment. We also appreciate the work of the editors and the reviewers of the paper.

Conflicts of Interest: The authors declare no conflict of interest.

References

1. Yang, X.; Yang, F.; Liu, Y.; Li, J.; Song, H.L. Identification of key off-flavor compounds in thermally treated watermelon juice via gas chromatography-olfactometry-mass spectrometry, aroma recombination, and omission experiments. *Foods* **2020**, *9*, 227. [CrossRef]
2. Collins, J.K.; Wu, G.; Perkins, V.P.; Spears, K.; Claypool, P.L.; Baker, R.A.; Clevidence, B.A. Watermelon consumption increases plasma arginine concentrations in adults. *Nutrition* **2007**, *23*, 261–266. [CrossRef] [PubMed]
3. Tarazona-díaz, M.; Viegas, J.; Moldao-martins, M.; Aguayo, E. Bioactive compounds from flesh and by-product of fresh-cut watermelon varieties. *J. Sci. Food Agric.* **2011**, *91*, 805–812. [CrossRef]
4. Martyn, R.D. Fusarium wilt of watermelon: 120 years of research. *Hortic. Rev.* **2014**, *42*, 349–442. [CrossRef]
5. Wu, Y.C.; Zhou, J.Y.; Li, C.G.; Ma, Y. Antifungal and plant growth promotion activity of volatile organic compounds produced by Bacillus amyloliquefaciens. *MicrobiologyOpen* **2019**, *8*, e813. [CrossRef]
6. Cohen, R.; Pivonia, S.; Burger, Y.; Edelstein, M.; Gamliel, A.; Katan, J. Toward integrated management of monosporascus wilt of melons in Israel. *Plant Dis.* **2000**, *84*, 496–505. [CrossRef] [PubMed]
7. Castro, G.; Perpiñá, G.; Esteras, C.; Armengol, J.; Picó, B.; Pérez-de-Castro, A. Resistance in melon to Monosporascus cannonballus and M. eutypoides: Fungal pathogens associated with Monosporascus root rot and vine decline. *Ann. Appl. Biol.* **2020**, *177*, 101–111. [CrossRef]
8. Xu, L.H.; Nicolaisen, M.; Larsen, J.; Zeng, R.; Gao, S.G.; Dai, F.M. Pathogen infection and host-resistance interactively affect root-associated fungal communities in watermelon. *Front. Microbiol.* **2020**, *11*, 3256. [CrossRef]
9. Keinath, A.P.; Coolong, T.W.; Lanier, J.D.; Ji, P.S. Managing Fusarium wilt of watermelon with delayed transplanting and cultivar resistance. *Plant Dis.* **2019**, *103*, 44–50. [CrossRef]
10. Aslam, A.; Zhao, S.; Azam, M.; Lu, X.; He, N.; Li, B.; Dou, J.; Zhu, H.; Liu, W. Comparative analysis of primary metabolites and transcriptome changes between ungrafted and pumpkin-grafted watermelon during fruit development. *Peer J.* **2020**, *8*, e8259. [CrossRef]
11. Rouphael, Y.; Venema, J.; Edelstein, M.; Savvas, D.; Colla, G.; Ntatsi, G.; Ben-Hur, M.; Kumar, P.; Schwarz, D. Grafting as a tool for tolerance of abiotic stress. In Proceedings of the Vegetable Grafting Principles and Practices, CAB International, Wallingford, UK, 5 June 2017; pp. 171–215. [CrossRef]
12. Aloni, B.; Cohen, R.; Karni, L.; Aktas, H.; Edelstein, M. Hormonal signaling in rootstock-scion interactions. *Sci. Hortic.* **2010**, *127*, 119–126. [CrossRef]
13. Johnson, G. Grafted Watermelons Revisited. Weekly Crop Update. University of Delaware Cooperative Extension. 2017. Available online: https://sites.udel.edu/weeklycropupdate/?p=10563 (accessed on 6 May 2021).
14. Chiu, Y.C.; Chen, S.; Chang, Y.C. Development of a circular grafting robotic system for watermelon seedlings. *Appl. Eng. Agric.* **2011**, *10*, 95–102. [CrossRef]
15. Lee, J.M.; Kubota, C.; Tsao, S.J.; Bie, Z.; Echevarria, P.H.; Morra, L.; Oda, M. Current status of vegetable grafting: Diffusion, grafting techniques, automation. *Sci. Hortic.* **2010**, *127*, 93–105. [CrossRef]

16. Wang, X.Y.; Gu, S. The production experiment of 2jc-500 grafting machine carrying on the watermelon to graft grows seedlings. *J. Agric. Mech. Res.* **2008**, *1*, 148–149. [CrossRef]
17. Jiang, K.; Zhang, Q.; Chen, L.P.; Guo, W.Z.; Zheng, W.G. Design and optimization on rootstock cutting mechanism of grafting robot for cucurbit. *Int. J. Agric. Biol. Eng.* **2020**, *13*, 117–124. [CrossRef]
18. Ma, Z.Y.; Mu, Y.H.; Gu, S.J. Techniques of automatic grafting of cucurbitaceous vegetables. *J. ZhongKai Univ. Agric. Technol.* **2012**, *1*, 48–51. [CrossRef]
19. Lou, J.Z.; Wu, K.; Chen, J.Y.; Ma, G.Y.; Li, J.P. Design and test of self-adaptive stock cotyledons pressing and clamping mechanism for oblique inserted grafting of Cucurbitaceous vegetables. *Trans. CSAE* **2018**, *34*, 76–82. [CrossRef]
20. Yang, Y.L.; Li, K.; Chu, Q.; Zhong, L.X.; Jia, D.D.; Gu, S. Air suction clamp structure of rootstock cotyledons for inclined inserted grafting machine and its optimized experiment of operation parameters. *Trans. CSAE* **2014**, *30*, 25–31. [CrossRef]
21. Zhu, X.C.; Jin, X.; Yao, B.; Cao, J. Modeling and Design of a frictionless pneumatic cylinder system with air bearings. In Proceedings of the 5th Annual IEEE International Conference on Cyber Technology in Automation, Control and Intelligent Systems, Shenyang, China, 8–12 June 2015.
22. Lou, J.Z.; Li, J.P.; Zhu, P.A. Design and test on growing point removal mechanism of melon vegetable grafting stock. *Trans. CSAE* **2016**, *32*, 64–69. [CrossRef]
23. Lou, J.Z.; Li, J.P.; Zhu, P.A.; Lv, G.L.; Wang, M. Optimization of suction head of scion clamping mechanism for vegetable grafting machine. *Trans. CSAM* **2013**, *44*, 63–67. [CrossRef]
24. Zheng, W.K.; Dong, J.K.; Zhang, L.; Chen, Z.H. Heating performance for a hybrid radiant-convective heating terminal by orthogonal test method. *J. Build. Eng.* **2020**, *33*, 101627. [CrossRef]
25. Bai, J.; Ma, S.C.; Wang, F.L.; Xing, H.N.; Ma, J.Z.; Hu, J.W. Field test and evaluation on crop dividers of sugarcane chopper harvester. *Int. J. Agric. Biol. Eng.* **2021**, *14*, 118–122. [CrossRef]
26. Li, H.; Zeng, S.; Luo, X.; Fang, L.; Liang, Z.; Yang, W. Design, DEM simulation, and field experiments of a novel precision seeder for dry direct-seeded rice with film mulching. *Agriculture* **2021**, *11*, 378. [CrossRef]
27. Zhang, X.; Bai, S.; Jin, W.; Yan, J.; Shi, Z.; Yu, N.; Yuan, P.; Zhu, X. Design and parameter optimization of an air suction jujube picking and conveying device. *Trans. ASABE* **2020**, *63*, 943–954. [CrossRef]
28. Chen, H.T.; Wang, H.F.; Wang, Y.C.; Shi, N.Y.; Wei, Z.P.; Dou, Y.K. Design and experiment of three-leaf type air-suction seed meter with automatic clear and replace seeds features for soybean plot test. *Trans. CSAM* **2020**, *51*, 75–85. [CrossRef]
29. Jiang, K. Study on Mechanism and Experimental Device of Splice Mechanical Grafting of Cucurbit. Ph.D. Thesis, Northeast Agricultural University, Harbin, China, 2019.
30. Jiang, K.; Zhang, Q.; Chen, L.P.; Guo, W.Z.; Mou, Y.Q. Simulation design and performance experiment of adsorption block in feeding and positioning mechanism for rootstock. *Trans. CSAE* **2020**, *36*, 73–80. [CrossRef]
31. Lou, J.Z. Mechanism Study and Optimization Design of Inclined-Insert Grafting Device of Cucurbita Vegetable. Ph.D. Thesis, Zhejiang University, Hangzhou, China, 2014.
32. Wang, F.; Ma, S.; Xing, H.; Bai, J.; Hu, J. Base cutting energy consumption for sugarcane stools using contra-rotating basecutters. *Trans. ASABE* **2021**, *64*, 221–230. [CrossRef]

agriculture

MDPI

Article

Study on the Distribution Pattern of Threshed Mixture by Drum-Shape Bar-Tooth Longitudinal Axial Flow Threshing and Separating Device

Jianwei Fu [1,2], Gan Xie [1,2], Chao Ji [1,2], Weikang Wang [1,2], Yong Zhou [1,2,*], Guozhong Zhang [1,2], Xiantao Zha [1,2] and Mohamed Anwer Abdeen [1,3]

1 College of Engineering, Huazhong Agricultural University, Wuhan 430070, China; fjwtap@mail.hzau.edu.cn (J.F.); xieg6@webmail.hzau.edu.cn (G.X.); chaoji@webmail.hzau.edu.cn (C.J.); wangwk@webmail.hzau.edu.cn (W.W.); zhanggz@mail.hzau.edu.cn (G.Z.); zhaxiantao@webmail.hzau.edu.cn (X.Z.); mohamed.anwer2010@yahoo.com (M.A.A.)

2 Key Laboratory of Agricultural Equipment in Mid-Lower Yangtze River, Ministry of Agriculture and Rural Affairs, Wuhan 430070, China

3 Agricultural Engineering Department, Zagazig University, Zagazig 44519, Egypt

* Correspondence: zhyong@mail.hzau.edu.cn

Abstract: To determine the distribution pattern of the threshing and separating device, the simulation experiment on the distribution pattern of our self-designed drum-shape bar-tooth longitudinal axial flow threshing and separating device was carried out with the help of the EDEM software, by which the axial and radial distribution curve of the threshed mixture along the cylinder was acquired. The three-dimensional distribution of the mass of the threshed mixture was drawn by using the Matlab software, and the bench test was carried out on the self-built small-scale longitudinal axial flow threshing cylinder performance test platform, which was consistent with the simulation conditions. The results showed that the axial and radial distribution of the threshed mixture was uneven, and the axial distribution of the threshed mixture decreased gradually, which was mainly distributed in the first third section of the cylinder. The distribution of the threshed mixture along the radial area of the cylinder was gradually decreasing at first and then increasing, i.e., the total mass of the threshed mixture on the left and right sides was higher than that of the middle area, which was basically consistent with the simulation results. The research can provide reference for the optimization of structural parameters of threshing and separating device and cleaning system.

Keywords: threshing and separating device; cleaning system; threshed mixture; discrete element method (DEM); Matlab

Citation: Fu, J.; Xie, G.; Ji, C.; Wang, W.; Zhou, Y.; Zhang, G.; Zha, X.; Abdeen, M.A. Study on the Distribution Pattern of Threshed Mixture by Drum-Shape Bar-Tooth Longitudinal Axial Flow Threshing and Separating Device. *Agriculture* **2021**, *11*, 756. https://doi.org/10.3390/agriculture11080756

Academic Editor: José Pérez-Alonso

Received: 11 June 2021
Accepted: 3 August 2021
Published: 9 August 2021

1. Introduction

In the actual operation of the combine harvesters, the crops enter the threshing cleaning system through the header and conveying channel, and are threshed under the action of threshing elements, and then the threshed mixture falls onto the cleaning screen through the grain layer and concave screen under the centrifugal force [1]. The research on the distribution features of the threshed mixture can provide a reference for the parameter design of the cleaning system, and conditions for reducing the cleaning load and improving the grain cleanliness.

A large number of theoretical and experimental studies have been carried out on the distribution pattern of threshed mixture. Miu et al. [2] established a mathematical model for the distribution of the threshed mixture by an axial flow threshing device. The model described the distribution characteristics of the unthreshed grain, threshed grain, and free grain with the length of the cylinder in the threshing process, and analyzed the effects that the cylinder speed, feed rate, and crop moisture content had on grain separation. Li et al. [3] theoretically analyzed the working mechanism of the threshing and separating device and established a comprehensive mathematical model of the threshing and separating process from the angle of probability theory. Zhang et al. [4] studied the distribution pattern of the threshed mixture along the axial direction of the cylinder by using the self-developed axial flow threshing and separating device with helical blade and plate teeth. The results showed that the axial distribution of the threshed mixture was greatly affected by the grain distribution, and the grain distribution curve was prominent. The distribution pattern was similar under different feeding rates, and the peak value of the curve was higher when the feeding rate was large. Yi et al. [5–7] conducted an experimental study on the distribution pattern of the threshed mixture along the axial and tangential directions of the spiral blade strip-tooth and spike-tooth axial flow threshing and separating devices, and carried out a regression analysis on the distribution curve of each threshed mixture. The results showed that the degree of impurity and stem breakage of the spike-tooth axial flow threshing and separating device was higher than those of the combined device. Li et al. [8,9] conducted a comparative experimental study on the axial distribution of the threshed mixture from the short-striped bar-plate teeth and the spike-tooth cylinder. The results showed that the distribution of the threshed mixture from the short-striped bar-plate teeth was more even than that from the spike-tooth cylinder, and the impurity content of the threshed mixture was lower. Guo et al. [10] also conducted a comparative experimental study on the radial distribution of the threshed mixture from the trapezoidal plate tooth cylinder and spike-tooth cylinder. The results suggested that the threshed mixture from spike-tooth cylinder was more evenly distributed along the radial direction of the cylinder. Tang et al. [11] used self-made rectangular tooth plate, short grain rod-plate tooth, spike tooth, blade tooth, and trapezoidal plate tooth to test the threshing and separating performance of grain under the feeding rate of 7 kg/s and 8 kg/s, and compared the initial threshing and separating rate of tangential flow device under different threshing elements and the axial and radial distribution of threshing mixture of longitudinal axial flow re-threshing device under different threshing elements. Chen et al. [12] carried out a comparative experiment on the distribution pattern of the threshed mixture between the semi-feeding coaxial differential threshing cylinder and the single-speed threshing cylinder, and established a 3D image and its mathematical model with Matlab for the measured data of all kinds of threshed mixture. The results indicated that the distribution of all kinds of threshed mixture by the differential threshing cylinder was more evenly distributed on the screen. Chen et al. [13] conducted a single factor experiment on the axial and radial distribution changes of the longitudinal axial flow flexible threshing device. The results showed that the distribution of threshed mixture and seeds of longitudinal axial flow flexible threshing device conformed to the exponential function distribution, the distribution of pods and light impurities conforms to cubic function distribution, and the distribution of threshed mixture, seeds, pods, and light impurities conforms to the multinomial form with different coefficients.

The above researches show that the distribution of the threshed mixture was affected by the feeding mode of the crops, the structure of the threshing and separating device, the movement parameters, the feeding amount, the material characteristics, etc. Most of the existing researches on the distribution pattern of threshed mixture were conducted by combining theoretical models with bench test or field test. However, in the process of experiment, it is complex to collect and process the threshed mixture, which not only increases the difficulty of experiment, but also prolongs the experiment period.

Aiming at the above problems, this paper presents a mathematical model of threshing and separating based on the probability theory, which was based on the self-designed drum-shape bar-tooth longitudinal axial flow threshing and separating device. The distribution pattern of threshing was simulated and analyzed by using the EDEM (2018, DEM Solutions Ltd., 6th Floor, 1 Rutland Court, Edinburgh, EH3 8FL UK) software, and a bench test was also set up to verify the results, which could create conditions for optimizing the structural parameters of the threshing and separating device and cleaning system.

2. Materials and Methods

2.1. Structure of Longitudinal Axial-Flow Threshing and Separating Device

A drum-shape bar tooth longitudinal axial-flow threshing and separating device was designed with the reference to existing researches [14–16], which mainly consisted of a drum-shape cylinder, a concave screen, a top cover with guide plate, a rack, etc., as shown in Figure 1.

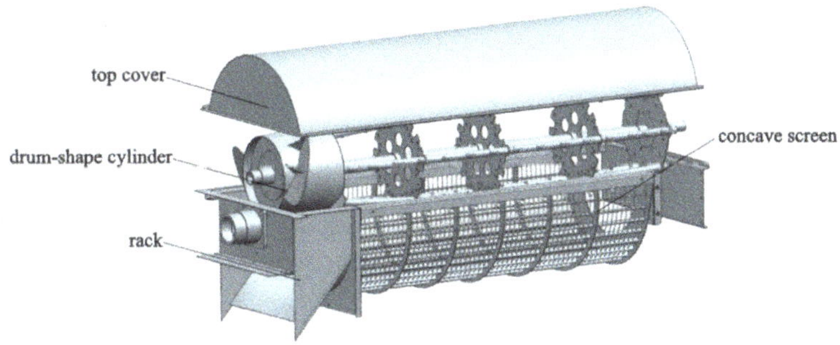

Figure 1. Structure diagram of drum-shape bar-tooth longitudinal axial-flow threshing and separating device.

Its working principle is that the crops enter the threshing and separating device from the conveying channel, get into the threshing section under the grasping of the conical spiral feeding head, and take off and separate the rice grains from the earhead under continuous hitting, brushing, and rubbing of the bar tooth. At the same time, under the action of the cylinder and the guide plate, the rice plants make a spiral movement into the separating section. Under the centrifugal force generated by the high-speed rotation of the cylinder, the detached grains are continuously separated from the concave sieve, and the straw is discharged from the discharge section. Most of the grains are removed and separated in the first half of the threshing cylinder, and the second half mainly separates the free grains from the concave screen.

The traditional threshing cylinder with single longitudinal flow mainly consists of two types: cylindrical and conical. On the basis of the existing threshing cylinder structure, we designed a kind of drum-shape bar-tooth longitudinal axial flow threshing cylinder combined with the theoretical research and experimental analysis of scholars at home and abroad, which aimed at reducing drag and consumption. The cylinder was mainly composed of spiral feeding head, bar-tooth, wheel, and threshing teeth. The structure is shown in Figure 2.

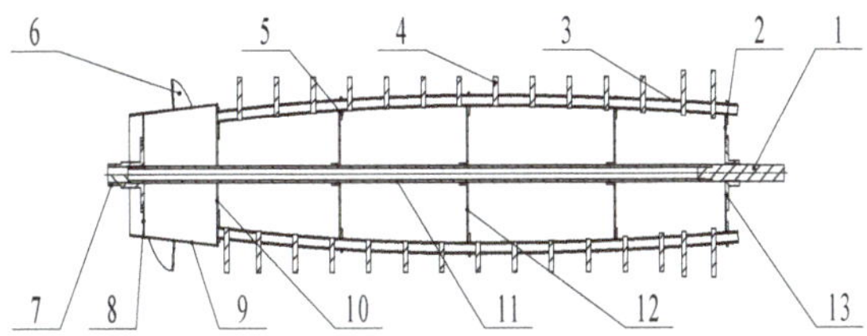

Figure 2. Structure of the drum-shape bar-tooth longitudinal axial flow threshing cylinder. 1. Shaft head; 2. Trailer wheel; 3. Drum-shape bar-tooth; 4. Threshing teeth; 5. Bar-tooth mounting plate; 6. Spiral blade; 7. Front flange; 8. Front disc; 9. Conical cylinder; 10. Rear disc; 11. Shaft tube; 12. Middle wheel; 13. Rear flange.

2.2. Mathematical Model of the Threshing and Separating Process

The function of the threshing and separating device is to thresh the grain from the ear under the action of the threshing elements, and then allow the free grain to be separated from the concave screen through the grain layer under the centrifugal force generated by the high-speed rotation of the cylinder. It was hypothesized that it is equally likely for the grain to be taken off at any part of the cylinder. From the perspective of probability, the probability of grain being threshed depends on the amount of unthreshed grain, and the probability of grain being separated depends on the number of free grains in the process of threshing and separating. It was based on this idea that the mathematical model of the threshing and separating process was established and the working process of drum-shape bar-tooth longitudinal axial flow threshing and separating devices was studied.

It was hypothesized that:

(1) There is no threshing and separating phenomenon before the crops entering the threshing and separating device for single longitudinal flow threshing cylinder.
(2) In the process of threshing, the chance of grain being threshed at any position is equal, and there is a proportional relationship between the number of grain being threshed and the number of grain not being threshed, which is recorded as threshing coefficient k.
(3) In the process of separating, the chance of grain being separated at any position is also equal, and there is a proportional relationship between the number of grain being separated and the number of grain not being separated, which is recorded as separating coefficient p.

2.2.1. Mathematical Model of Threshing Process

Take a point x in the axial direction of concave screen, take a small distance Δx near x, and express the probability of threshing here with frequency. Using the idea of limit, let $\Delta x \to 0$, the equation of probability density function $f(x)$ of threshing in axial flow threshing cylinder can be obtained as follows [17,18]:

$$\frac{f(x)}{[1 - F(x)]} = k \tag{1}$$

where x is distance of axial flow cylinder from threshing and separating starting point of concave screen to tail along axis, mm; $F(x)$ is cumulative threshing probability from threshing separation starting point of concave screen to x.

It can be obtained from the probability theory that:

$$F(x) = \int_0^x f(\xi)d\xi \tag{2}$$

Take the derivation from both sides of Equation (2) and substitute it into Equation (1):

$$\frac{dF(x)}{1 - F(x)} = kdx \tag{3}$$

By integrating both sides of Equation (3), we can get the following results:

$$F(x) = 1 - e^{-kx} \tag{4}$$

By deriving Equation (4), the probability density function of grain threshing at the axis x of concave sieve is obtained as follows:

$$f(x) = ke^{-kx} \tag{5}$$

Assuming that the feed of grain is continuous and even, the unthreshed rate at concave x is:

$$S_n(x) = 1 - \int_0^x ke^{-kx}dx \quad (6)$$

when $x = L$, $Sn(L)$ which indicates the unthreshed rate is:

$$S_n(L) = e^{-kL} \quad (7)$$

where L is length of the concave screen, mm.

2.2.2. Mathematical Model of Separating Process

In the same way, the probability density function of the separated free grain at the axis x of the concave screen in the single longitudinal axial flow threshing and separating device can be obtained as follows:

$$g(x) = pe^{-px} \quad (8)$$

Separation is a process in which the threshed free grain passes through the crop layer under centrifugal force and is separated from the concave screen. The threshing of the grain is the precondition of the separating of the grain. According to the probability theory, the probability of the grain to be separated is the convolution of the probability of grain being threshed and the probability of the free grain being threshed but not being separated. Therefore, the probability density function of grain being separated at x can be expressed as:

$$h(x) = \int_0^x f(x)g(x - \xi)d\xi \quad (9)$$

By integrating the two sides of Equation (9) at the same time, it can be concluded that the cumulative amount of separated grains $H(x)$ at concave x is:

$$H(x) = \frac{k(1 - e^{-px}) - p(1 - e^{-kx})}{k - p} \quad (10)$$

The above mathematical model was simulated by the MATLAB (R2016b, The MathWorks. Inc, 1 Apple Hill Drive Natick, MA 01760 USA) software, in which, $k = 3.4$, $p = 3.2$. The axial distribution curve of the unthreshed rate of grain and the cumulative separating rate of threshed grain along the concave screen were obtained as shown in Figure 3.

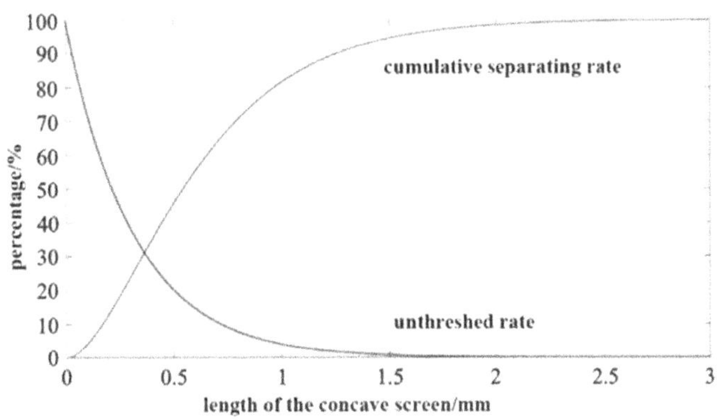

Figure 3. The axial distribution curve of the unthreshed rate of grain and the cumulative separating rate of threshed grain along the concave screen.

According to Figure 3, when the crops enter the threshing cylinder, the amount of the total grain increases sharply at first and then slows down to a steady level. Threshing and separating mainly happen in the first half of the threshing cylinder. Thus, the length of the axial threshing cylinder depends on the effect of separation. In the separating section of the threshing cylinder, it is the threshing cylinder that separates the free grain which has been threshed but not separated from the crops. Some grain that is hard to be threshed can be taken off at this section. The longer the cylinder is, the longer the separating time is, the better the separation effect will be. However, it may increase the break rate and the power dissipation at the same time. Under the premise of ensuring the effect of threshing and separating the length of threshing cylinder should be as small as possible. Its length on the existing axial flow drum combine harvester is generally 1~3 m.

2.3. Discrete Element Simulation Test

It is an important method to analyze the threshing performance of the cylinder to study the movement pattern and force condition of grain mixture in threshing and separating device. However, its movement and force in threshing and separating device are complex, which makes it difficult to analyze by traditional theory or experimental research. Fortunately, the continuous development of discrete element method and the application of EDEM software provide a new choice for simulating the movement of grain mixture in the threshing and separating device.

The mixture from threshing and separating device includes grain, short straw, glume, light debris, etc. However, its main components are grain and short straw, which takes up 95% of the mixture mass [19–21]. In order to simplify the types of particle models and improve the simulation efficiency, only the effects of grain and short straw were considered in the simulation process. The rice grain was simplified as an ellipsoid of homogeneous linear elastic material, with the length of 6.5 mm and the diameter of the largest section circle in the length direction of 3.5 mm. According to the actual size of the short straw in the mixture, the rice grain was simplified as a cylinder composed of 34 spherical particles with a diameter of 4 mm and a length of 70 mm, as shown in Figure 4.

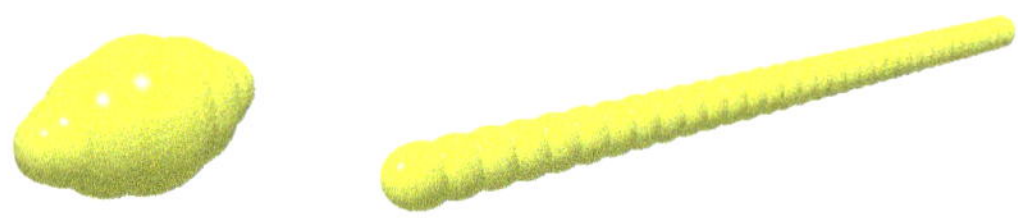

Figure 4. Model of rice grain and short straw.

The physical parameters of the rice grain, short straw, and threshing device are shown in Table 1 [22].

Table 1. Mechanical characteristic parameters of materials and threshing device.

Model	Poisson's Ratio	Shear Modulus (MPa)	Density (kg/m^3)
Grains	0.3	26	1350
Short straw	0.4	10	100
Threshing device	0.3	70,000	7800

The contact model is an important basis of the discrete element method. The analysis and calculation of the contact model directly determine the magnitude of the force and moment on the particle. In this paper, the Hertz–Mindlin non-sliding contact model was used to analyze the contact between the material and threshing device [23–25]. The contact stiffness of this model varied with the contact overlap of particles, and it was a nonlinear model. The contact parameters between materials and materials and between materials and threshing device are shown in Table 2 [22].

Table 2. Interaction parameters of materials and threshing device.

Contact Object	Coefficient of Restitution	Coefficient of Static Friction	Coefficient of Rolling Friction
Between grains and grains	0.2	1.0	0.01
Between grains and short straw	0.2	0.8	0.01
Between grains and Threshing device	0.5	0.58	0.01
Between short straw and short straw	0.2	0.9	0.01
Between short straw and Threshing device	0.2	0.8	0.01

Two pellet factories were set up at the feeding entrance cylinder, namely rice grain pellet factory and short straw pellet factory. Since the mass of the short straw among the threshed mixture processed by full feeding longitudinal axial flow threshing and separating device took up 30% of the total mass [19], the feeding rate was set to be 1.2 kg/s (the rate of rice grain production was 0.84 kg/s, the rate of rice straw production was 0.36 kg/s), the particle generation time was 2 s, and the total simulation time was 5 s. According to the actual situation where the particles were produced at the feeding inlet of the threshing and separating device, the initial speed was set as 2.8 m/s according to the conveying speed of the intermediate conveying device of the combine harvester. The threshing cylinder was set to be working in the simulation with a speed of 1100 r/min. The calculation time step was set as 10% of Rayleigh time step and the output time step as 0.01 s.

The total length of the designed threshing cylinder was 1360 mm, in which the feeding section is 190 mm, the outlet section was 170 mm, the threshing and separating section was 1000 mm, and the width of the plates on both sides of the threshing cylinder was 500 mm. The threshing and separating device were divided into 10 areas along the axial direction and 5 areas along the radial direction, with a total of 10 areas × 5 = 50 areas, each of whose size was 100 mm × 100 mm, as shown in Figure 5.

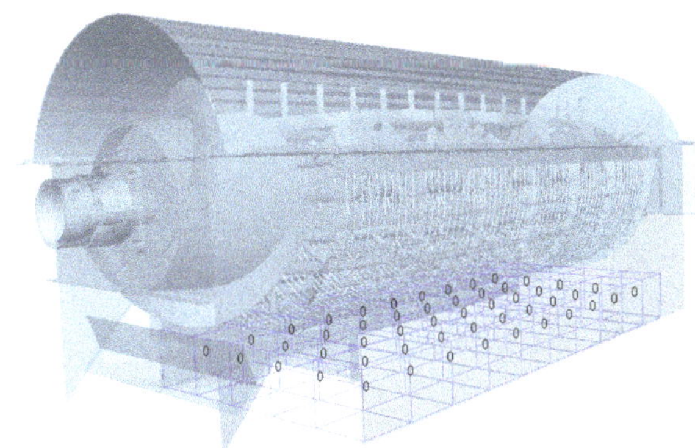

Figure 5. Regional distribution of statistical domain.

In order to explain the axial and radial distribution of the threshed mixture, the statistical fields in Figure 5 are numbered as shown in Figure 6.

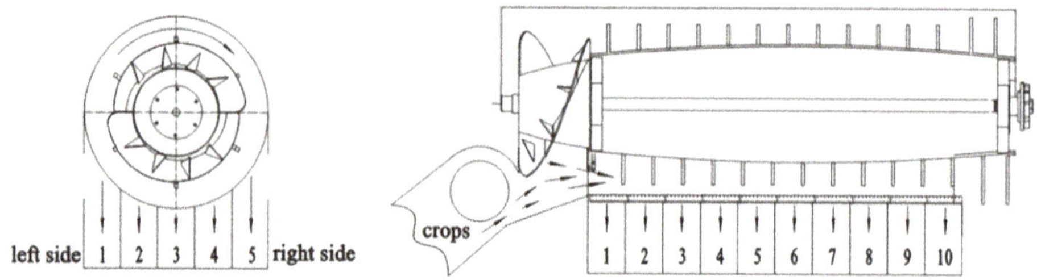

Figure 6. Regional distribution of the threshing and separating device along the axial and radial directions.

The numbers of each area are shown in Table 3.

Table 3. Area distribution of threshing and separating performance test.

	Axial									
radial	1-1	1-2	1-3	1-4	1-5	1-6	1-7	1-8	1-9	1-10
	2-1	2-2	2-3	2-4	2-5	2-6	2-7	2-8	2-9	2-10
	3-1	3-2	3-3	3-4	3-5	3-6	3-7	3-8	3-9	3-10
	4-1	4-2	4-3	4-4	4-5	4-6	4-7	4-8	4-9	4-10
	5-1	5-2	5-3	5-4	5-5	5-6	5-7	5-8	5-9	5-10

2.4. Bench Test

The performance test platform of small longitudinal axial flow threshing and separating device designed in this paper was modified from the small longitudinal axial flow combine harvester "4LZ-1.6" produced by Hunan Nongyou Machinery Group Co., Ltd., and its structure is shown in Figure 7.

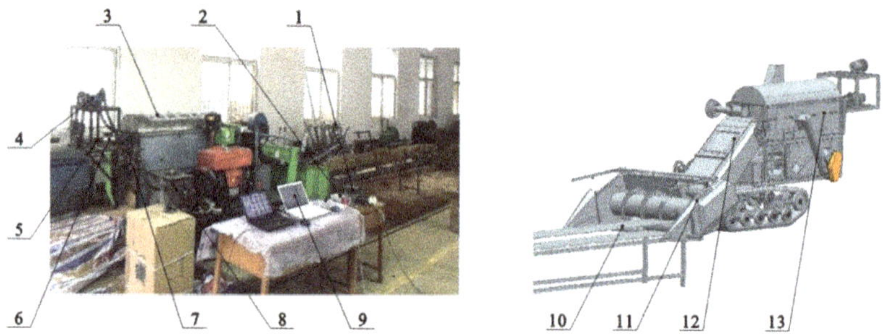

Figure 7. Performance test platform of longitudinal axial flow threshing and separating device. 1. Frequency converter; 2. Torque speed power tester; 3. Acrylic board; 4. Hydraulic motor; 5. Elastic coupling; 6. Torque sensor; 7. Hydraulic control valve; 8. Hydraulic tank; 9. Computer; 10. Conveyor belt; 11. Header; 12. Conveying channel; 13. Threshing and separating device.

The total length of the conveyor belt is 6.0 m, and the head end is 1.0 m free. Firstly, 6 kg rice straw was evenly laid at 5.0 m behind the conveyor belt, so that the conveyor belt could feed rice after obtaining a stable transmission speed to ensure accurate and uniform feeding [26,27]. When the frequency of the frequency converter was adjusted to make the belt speed 1.0 m/s, the designed feeding amount of the threshing cylinder is 1.2 kg/s, which was consistent with the feeding amount of the simulation test. The experiment was repeated three times.

In each test, the threshing cylinder was started to rotate through the hydraulic valve. When the rotation was stable, the transmission of header feeding auger and conveying channel was started through the clutch. Finally, the transmission of conveyor belt was started through the frequency converter. The material receiving device under the concave screen collected the threshed mixture, and the oil cloth was used under the straw outlet to collect the discharged mixture. After feeding, the transmission of conveyor belt was stopped first, and then the transmission of header auger and conveying channel were stopped. When all the rice straw in the cylinder was discharged from the straw outlet, the transmission of threshing cylinder was stopped. Refer to the above test evaluation index for the treatment method of prolapse, and finally take the average value of three tests for the mass of threshed mixture in each area. After stopping the machine, we took out the material receiving device, collected the threshed mixture in each small box respectively, and weighed it with a BT457 electronic balance produced by Suzhou Shunqiang Electromechanical Equipment Co., Ltd., with an accuracy of 0.1 g. A total of three tests were conducted to obtain the average value of the mass of the mixture in each area.

The experiment was conducted in the Electromechanical Training Center of Huazhong Agricultural University. The rice variety was Huanghuazhan. The average plant height was 1076.50 mm. The rice was manually harvested, the stubble was 200–300 mm, the rice length was 876.50–776.50 mm, the 1000 grain weight was 28.25 g, the moisture content of grain was 22.35%, and the moisture content of straw was 68.31%. The rice was planted in the rice experimental base of Huazhong Agricultural University.

3. Results and Discussion

3.1. Simulation Process Analysis

The whole process of threshing and separating of threshed mixture in the cylinder could be simulated by software simulation. The threshing simulation model is shown in Figure 8. The mixed material enters the threshing cylinder under the action of the spiral feeding head. Under the action of threshing teeth and guide plate, the mixed material flows forward along the spiral of the inner arc surface composed of concave plate and top cover, and then it is discharged from the straw outlet. During this period, a large number of rice grains are rapidly threshed and separated in the front part of the cylinder under the action of centrifugal force generated by the high-speed rotation of the cylinder.

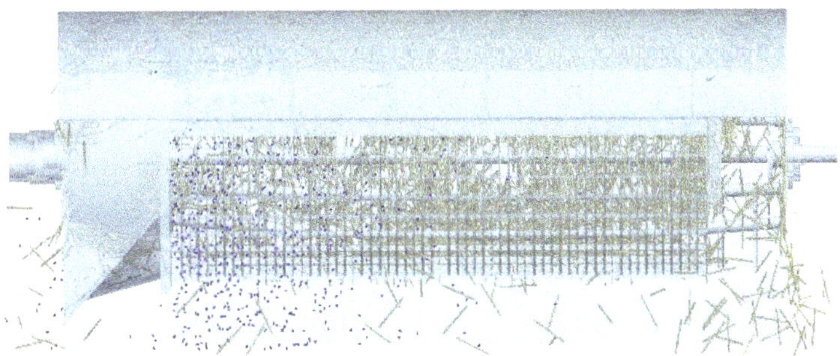

Figure 8. Simulation model of threshing.

3.2. Simulation Results and Analysis

The distribution of the threshed mixture in the statistical domain is shown in Figures 9 and 10. Axial zone 1 refers to 1-1 to 1-10 in Table 3, and there are 5 zones in the axial direction; radial region 1 refers to 1-1 to 5-1 in Table 3, and there are 10 radial regions. Considering that the subsequent distribution of the threshed mixture is relatively small, the radial distribution of the threshed mixture is analyzed in the radial 1–5 region.

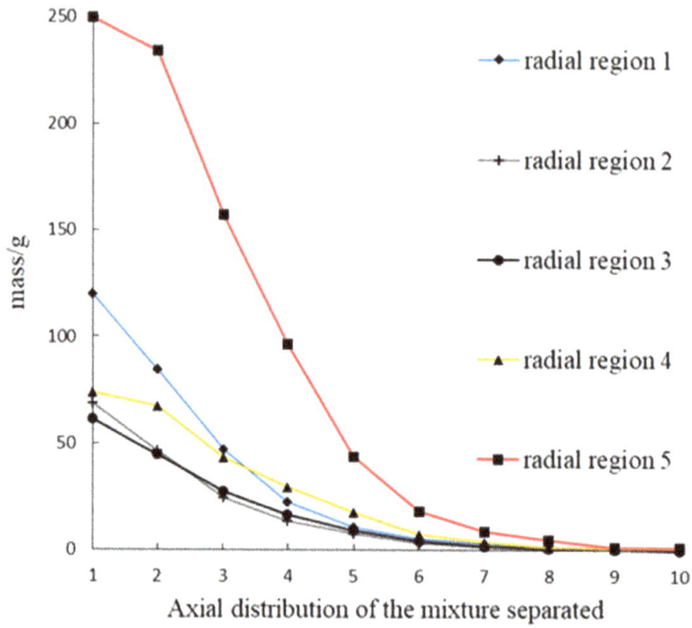

Figure 9. Axial distribution of the mixture separated by the drum-shape cylinder.

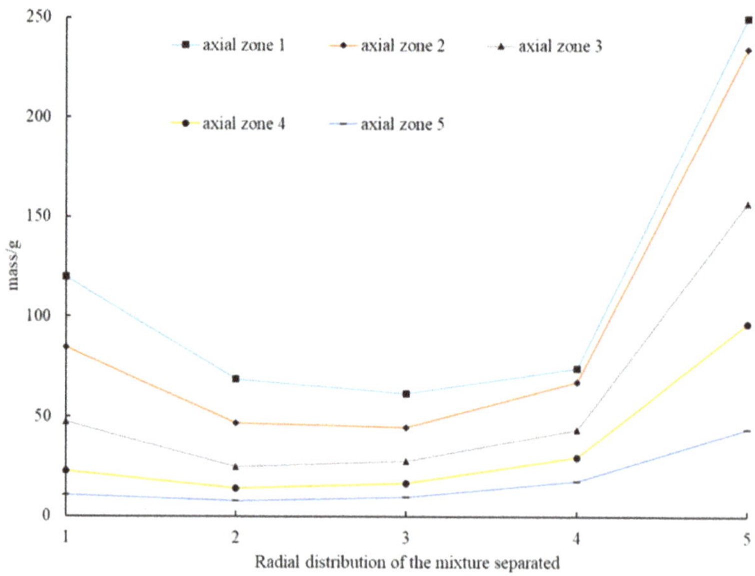

Figure 10. Radial distribution of the mixture separated by the drum-shape cylinder.

It can be seen from Figure 9 that for any radial region, the mass of the threshed mixture decreases gradually along the axial direction of the cylinder, and mainly distributes in the first 1/3 section of the cylinder. Axial zone 1 and axial zone 5 correspond to the left and right sides of the threshing cylinder respectively, and the total mass of the threshed mixture is higher than that of the middle zone.

Figure 10 indicates that for any radial region, the mass of the threshed mixture along the axial region of the cylinder decreases gradually at first and then increases. The mass of the threshed mixture in the radial region 5 is always higher than that in the region 1. The results also show that the spike tooth reaction force and bar tooth impact force of threshing device are complex periodic functions of angular displacement of grain motion [28–30]. Because of the rotation direction of the threshing cylinder, the "compaction area" is formed between the threshing cylinder and the right side of the concave screen. In this area, the difference between the linear speed of the material and the threshing cylinder is large, and the impact force of the rod teeth is large.

In order to observe the axial and radial variation trend of the threshed mixture along the cylinder more comprehensively, the three-dimensional distribution of the mass of threshed mixture is drawn by using MATLAB software, as shown in Figure 11.

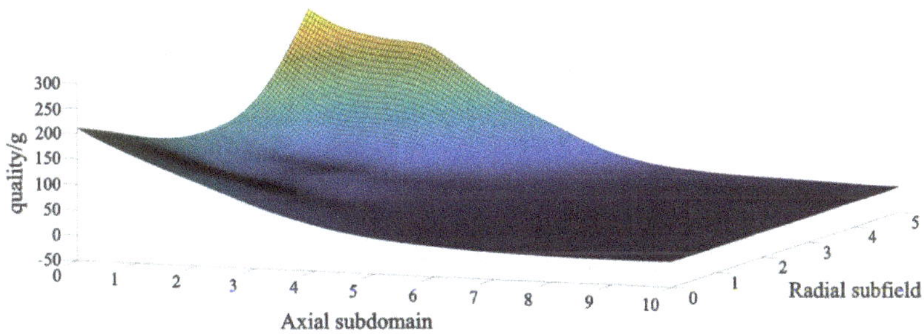

Figure 11. Grain distribution diagram in each area of the vertical axial-flow threshing cylinder.

It can be seen from Figure 11 that the distribution of the threshed mixture along the axial and radial direction of the cylinder is uneven. Its mass distribution is like a "skateboard", which is high on both sides, low in the middle, steep in the front, and flat in the back.

3.3. Bench Test Results and Analysis

The threshed mixture from the drum-shape threshing cylinder is shown in Figure 12.

Figure 12. Physical picture of mixture distribution.

We weighed the mixture of each small box in Figure 12, and obtained the change pattern of the mass of the mixture along the axial and radial direction of the cylinder, as shown in Figures 13 and 14. According to the simulation test method, the mixture was divided into 1~5 zones along the radial direction, and the 1~5 zones near the feeding end in axial direction were taken for the experimental analysis.

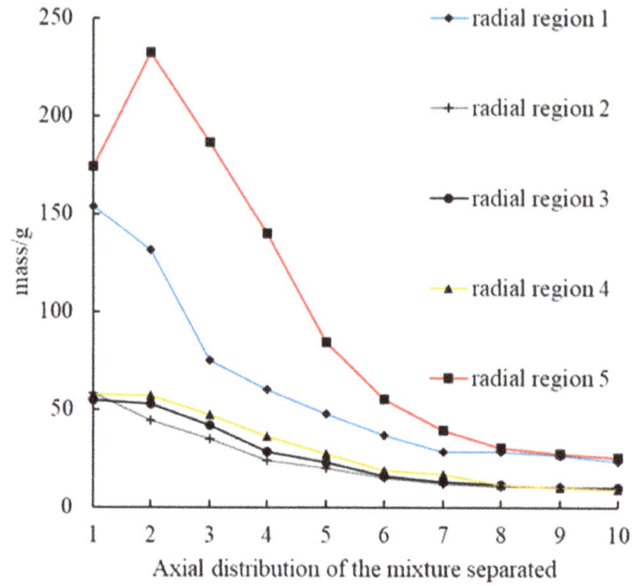

Figure 13. Axial distribution of the mixture separated by the drum-shape cylinder.

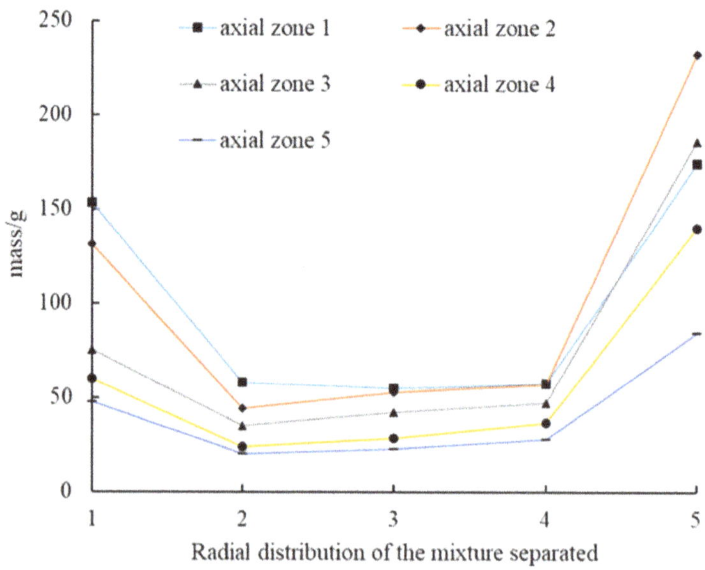

Figure 14. Radial distribution of the mixture separated by the drum-shape cylinder.

It can be seen from Figure 13 that along the axial direction of the cylinder, the mass of the mixture in zone 1~4 decreases gradually from the feeding inlet to the straw discharge outlet, and the maximum peak value appears in zone 5–2. Among the five radial regions, the axial region 5 had the highest mass of the mixture, followed by region 1. In addition, the mass of the mixture is mainly concentrated in the first half of the threshing cylinder. This result was basically consistent with the simulation test, but the radial zone 5 appeared the maximum value in the axial zone 2 in the bench test, which may be caused by the uneven initial moment when the material enters the cylinder. In the simulation test, there was basically no threshed mixture after axial zone 6, while in the bench test, there was still some mixture left after axial zone 6. This is because the materials set in the simulation test were only grains and short straw, while in the actual work, there were also leaves, ear heads and branches, etc. Although these parts account for less mass, they will affect the collision between particles, thus affecting the separating effect.

According to Figure 14, the mass of the mixture in axial 1–5 region decreases first and then increases along the radial direction of the cylinder. For the same radial region, the mass of the mixture in region 5 is always higher than that in region 1, which is consistent with the simulation results. It indicates that it is feasible to use the EDEM software to simulate the distribution pattern of the threshed mixture from the longitudinal axial flow threshing and separating device. However, as the simulation parameters in this paper all refer to the existing research settings, there may be some deviation from the physical parameters of the actual test materials, thus affecting the test effect. In a future study, the key parameters in the simulation process can be further calibrated to make it more accurate.

4. Conclusions

(1) In this paper, a drum-shape bar-tooth longitudinal axial flow threshing and separating device was designed, and its working principle was studied.
(2) Based on the probability theory, the threshing and separating model was established, and the threshing and separating characteristic curve of the drum-shape bar-tooth longitudinal axial flow threshing cylinder was obtained. According to the curve, threshing and separating basically occur in the first half of the threshing cylinder.
(3) The length of the cylinder could be selected according to the distribution pattern of the axial threshed mixture. On the premise of meeting the threshing performance, the length of the threshing cylinder should be as short as possible.
(4) The distribution of the threshed mixture along the axial region of the cylinder was gradually decreasing, and mainly distributed in the first one-third section of the cylinder. The mass of the threshed mixture along the radial region of the cylinder decreases gradually at first and then increases, and the total mass of threshed mixture on the left and right sides was higher than that in the middle area. The bench test results were basically consistent with the simulation results. This study can provide a reference for optimizing the structure parameters of the threshing and separating device and cleaning system.

5. Patents

Two patents have been applied in China for a drum-shape bar-tooth threshing cylinder for combine harvester in this manuscript (Patent No. CN201921966457.2 and Application No. CN201911113778.2).

Author Contributions: Conceptualization, J.F., Y.Z. and G.X.; methodology, J.F. and G.X.; software, J.F., G.X. and C.J.; validation, J.F., Y.Z., C.J., G.X., W.W., G.Z., X.Z. and M.A.A.; formal analysis, J.F.; investigation, J.F., C.J., G.X., W.W., X.Z. and M.A.A.; resources, J.F.; data curation, J.F.; writing—original draft preparation, J.F.; writing—review and editing, J.F.; visualization, J.F.; supervision, Y.Z.; project administration, G.Z.; funding acquisition, G.Z. All authors have read and agreed to the published version of the manuscript.

Funding: This work was financially supported by the Project of Scientific and Technological Innovation Team for the Excellent Young and Middle-aged in Hubei Province (T201934), the Fundamental Research Funds for the Central Universities (2662018PY038) and the National Key Research and Development Program of China (Technological Innovation of Regenerated Rice Mechanization in the North of the Middle and Lower Reaches of the Yangtze River, 2017YFD0301404-05).

Institutional Review Board Statement: Not applicable.

Data Availability Statement: The data presented in this study are available on demand from the first author at (fjwtap@mail.hzau.edu.cn).

Conflicts of Interest: The authors declare no conflict of interest.

References

1. Li, Y.M. Study on Separation of Rethreshed Stripped Mixtures and Their Cleaning Characteristics for Rice. Ph.D. Thesis, Nanjing Agricultural University, Nanjing, China, 2004.
2. Miu, P.; Kutzbach, H. Mathematical model of material kinematics in an axial threshing unit. *Comput. Electron. Agric.* **2007**, *58*, 93–99. [CrossRef]
3. Li, J.; Yan, C.L.; Yang, F.F. Theoretical model and simulation of threshing of axial unit with axial feeding. *J. Jiangsu Univ.* **2006**, *4*, 299–302.
4. Zhang, Y.F.; Lai, Y.J.; Zhang, K.; Sun, J.J. Research on The Axial Direction Distribution Regularities of Emerging Object with Axial Flow Threshing Installation. *J. Heilongjiang August First Land Reclam. Univ.* **2006**, *5*, 34–36.
5. Zhang, W.; Yi, S.J. The Establishment and Simulation of Mathematical Model on the Axial Threshing and Separating Unit. *J. Agric. Mech. Res.* **2007**, *7*, 40–42+47.
6. Yi, S.J.; Tao, G.X.; Mao, X. Comparative experiment on the distribution regularities of threshed mixtures for two types of axial flow threshing and separating installation. *Trans. CSAE* **2008**, *6*, 154–156.
7. Tao, G.X.; Yi, S.J. Distribution of mixture assembled axial flow device. *J. Agric. Mech. Res.* **2009**, *31*, 134–136+140.
8. Li, Y.M.; Li, H.C.; Xu, L.Z. Comparative experiments on threshing performance between short-rasp-bar tooth cylinder and spike tooth cylinder. *Trans. CSAE* **2008**, *3*, 139–142.
9. Li, Y.M.; Li, H.C.; Xu, L.Z.; Zhao, Z. Performance Test of Short-rasp-bar of Axial Flow Threshing and Separating Unit. *Trans. Chin. Soc. Agric. Mach.* **2009**, *40*, 88–92.
10. Guo, Y.; Li, Y.M.; Li, H.C.; Xu, L.Z. The Radial Distribution Regularities of Emerging Object with Longitudinal Axial Flow Threshing and Separating Device. *J. Agric. Mech. Res.* **2011**, *33*, 110–112.
11. Tang, Z.; Li, Y.M.; Xu, L.Z.; Zhao, Z.; Li, H.C. Effects of different threshing components on grain threshing and separating by tangential-axial test device. *Trans. CSAE* **2011**, *27*, 93–97.
12. Chen, N.; Yu, H.J.; Chen, D.J.; Gong, Y.J.; Zhang, Z.Z. Design and Test on Coaxial Differential Threshing Rotor of Head-feed Combine Harvester. *Trans. Chin. Soc. Agric. Mach.* **2011**, *42*, 39–42.
13. Chen, Y.P.; Kang, Y.; Wang, T.E.; Ning, X.J.; Jin, C.Q.; Yin, X. Distribution regularities of the threshed mixtures in longitudinal axial flow flexible thresher of soybean harvester. *J. China Agric. Univ.* **2020**, *25*, 104–111.
14. Xie, G.; Zhang, G.Z.; Fu, J.W.; Zhou, Y.; Wang, Y.; Gao, Y.; Wang, W.K.; Mohamed, A. Comparing power consumption of drum and cylindrical bar-tooth longitudinal axial flow threshing roller. *J. Huazhong Agric. Univ.* **2021**, *40*, 202–209.
15. Xie, G. Design and Experimental Research of Drum-Shape Bar-Tooth Longitudinal Axial Flow Threshing and Separating Device. Master's Thesis, Huazhong Agricultural University, Wuhan, China, 2020.
16. Abdeen, M.A.; Salem, A.E.; Zhang, G.Z. Longitudinal Axial Flow Rice Thresher Performance Optimization Using the Taguchi Technique. *Agriculture* **2021**, *11*, 88. [CrossRef]
17. Xu, L.Z. Simulation and Test Research on Threshing Unit of Stripping Combine. Master's Thesis, Jiangsu University, Zhenjiang, China, 2003.
18. Wang, X.W.; Xie, F.P.; Ren, S.G.; Wang, X.S.; Zhang, Z.Z. A mathematical model and test of the horizontal axial flow threshing separation device. *J. Hunan Agric. Univ.* **2020**, *46*, 480–487.
19. Lu, K. Design and Performance Experiment of a Small Horizontal-Axial Threshing and Separating Device for Ratoon Rice. Master's Thesis, Huazhong Agricultural University, Wuhan, China, 2017.
20. Zhao, S.H.; Zhang, G.Z.; Zhang, S.J.; Fu, J.W.; Xie, G.; Mohamed, A. Designing a real-time feed measurement system for horizontal axial flow threshing drum based on thin film sensor. *J. Huazhong Agric. Univ.* **2020**, *39*, 160–169.
21. Zhao, S.H. Design of Measurement System for Horizontal Axial Flow Threshing Drum Based on Film Sensors. Master's Thesis, Huazhong Agricultural University, Wuhan, China, 2019.
22. Wang, D.Z. Simulation Analysis and Threshing Performance of Horizontal-Axial Threshing Device. Master's Thesis, University of Jinan, Jinan, China, 2018.
23. Zha, X.T.; Zhang, G.Z.; Han, Y.H.; Salem, A.E.; Fu, J.W.; Zhou, Y. Structural Optimization and Performance Evaluation of Blocking Wheel-Type Screw Fertilizer Distributor. *Agriculture* **2021**, *11*, 248. [CrossRef]
24. Wang, W.Z.; Liu, W.R.; Yuan, L.H.; Qu, Z.; He, X.; Lv, Y.L. Simulation and Experiment of Single Longitudinal Axial Material Movement and Establishment of Wheat Plants Model. *Trans. Chin. Soc. Agric. Mach.* **2020**, *51*, 170–180.
25. Su, Z.; Li, Y.M.; Dong, Y.H.; Tang, Z.; Liang, Z.W. Simulation of rice threshing performance with concentric and non-concentric threshing gaps. *Biosyst. Eng.* **2020**, *197*, 270–284. [CrossRef]
26. Fu, J.W.; Zhang, G.Z.; Xie, G.; Wang, Y.; Gao, Y.; Zhou, Y. Development of double-channel feeding harvester for ratoon rice. *Trans. Chin. Soc. Agric. Eng.* **2020**, *36*, 11–20.

27. Fu, J.W. Development of Double-Channel Full Feeding Harvester for Ratoon Rice. Ph.D. Thesis, Huazhong Agricultural University, Wuhan, China, 2020.
28. Zhang, R.C.; Sang, Z.Z. Dynamic Simulation of the Axial Threshing Process of Combine Harvesters. *Trans. Chin. Soc. Agric. Mach.* **2001**, *47*, 58–60.
29. Tian, L.Q.; Li, H.Y.; Hu, H.D.; Chen, L.B.; Xiong, Y.S.; Jin, R.D. Design and Experiment on Coaxial Double Speed Threshing for Combine Harvester. *Trans. Chin. Soc. Agric. Mach.* **2020**, *51*, 139–146.
30. Huang, X.N. Design and Experimental Research on Tangential and Transverse Axial Flow Threshing System of Windrow Harvesting for Buckwheat. Master's Thesis, Northwest A&F University, Xianyang, China, 2020.

agriculture

Article

Coupled Bionic Design Based on Primnoa Mouthpart to Improve the Performance of a Straw Returning Machine

Jiale Zhao [1], Xiaogeng Wang [2], Jian Zhuang [1], Huili Liu [1], Yijia Wang [3,*] and Yajun Yu [2,*]

1 Key Laboratory of Bionics Engineering, Ministry of Education, Jilin University, Changchun 130025, China; zhaojl@jlu.edu.cn (J.Z.); Zhuangjian@jlu.edu.cn (J.Z.); congyj1970@outlook.com (H.L.)
2 College of Biological and Agricultural Engineering, Jilin University, Changchun 130025, China; wxg20@mails.jlu.edu.cn
3 Department of Industrial and Manufacturing Systems Engineering, The University of Hong Kong, Hong Kong, China
* Correspondence: yijiawang@connect.hku.hk (Y.W.); yuyajun@jlu.edu.cn (Y.Y.); Tel.: +86-166-297-4902 (Y.W.); +86-151-4317-3701 (Y.Y.)

Abstract: The high energy consumption and low crushing length qualification rate of traditional straw returning machines in the main maize-growing regions of northeast China make it difficult to promote straw returning operations in the region. The primnoa locust mouthpart is extremely efficient in cutting maize rootstocks. In this paper, it was found that there are significant differences between the primnoa locust mouthpart and the conventional machine, these exist mainly in the cutting edge structure and cutting motion. Thus, this paper develops a coupled bionic design for structural and kinematic coupling elements to develop a bionic straw returning machine. This paper found that the operating performance of the bionic straw returning machine was mainly affected by the blade rotation radius and the output rotation speed of the drive mechanism through DEM (discrete element method) simulation, and the optimal combination of the two parameters was 248 mm rotation radius and 930 r/min output rotation speed. Finally, this paper finds that the most obvious operational performance difference of the bionic straw returning machine compared with the traditional straw returning machine is that it can reduce the cutting power consumption by 9.4–11.7% and improve the crushing length qualification rate by 10.4–14.7% through the operational performance comparison test. Based on the above findings, this paper suggests that in future research and development of straw returning machines, more attention can be focused on finding suitable bionic prototypes and improving bionic design methods.

Keywords: straw returning; coupling bionic; cutting energy consumption; crushing length qualification rate; maize straw

Citation: Zhao, J.; Wang, X.; Zhuang, J.; Liu, H.; Wang, Y.; Yu, Y. Coupled Bionic Design Based on Primnoa Mouthpart to Improve the Performance of a Straw Returning Machine. *Agriculture* **2021**, *11*, 775. https://doi.org/10.3390/agriculture11080775

Academic Editor: José Pérez-Alonso

Received: 7 July 2021
Accepted: 11 August 2021
Published: 14 August 2021

1. Introduction

The soil type in the maize-growing region of northeastern China is a very rare cold black soil [1,2], which forms at a rate of 1 cm/400 years [3,4] and is a precious resource common to all mankind [5]. However, the rate of black soil erosion has reached 1 cm/year due to the difficulty of promoting straw returning operations in the region [6,7]. Straw returning means that crop straw is crushed into pieces no larger than 10 cm and then retained in the farmland [8], which is currently recognized as one of the most effective means to inhibit soil erosion [9]. Therefore, if the straw returning operations can be promoted on a large scale in the region, it can effectively protect the black soil resources in the region.

The main reason why straw returning cannot be rapidly promoted in the maize-growing region of northeast China is that traditional straw returning machines are difficult to apply to this region [10–12]. The planting cycle is long (6 months) in the maize growing region of northeast China [13]. Maize rootstocks are extremely thick and high in water content after maturity [14,15], resulting in extremely high power consumption [10,11] and low qualification rate of crushing length during the operation of traditional straw returning machines [12]. Related studies have shown that, compared to other maize-growing regions in China, traditional machines will improve the operational power consumption by 6–10% when operating in the maize-growing regions of northeast China [10,11,16], and the crushing length qualification rate will be reduced by 3–5% [12,17]. Higher operational power consumption can significantly increase the cost of maize production [17]. Lower crushing length qualification rate will enhance the difficulty of sowing operations in the second year, which in turn will result in lower maize yield [18]. The above-mentioned problems have seriously affected the motivation of agricultural producers to promote straw returning operations [19]. Obviously, if a brand-new machine with efficient cutting capacity can be developed, it can significantly enhance the progress of straw returning in the region.

The unique functions of organisms have been a source of creativity for humans to achieve technological breakthroughs [20]. Researchers studying an insect unique to northeastern China (primnoa, which feeds mainly on maize stalks) found that its mouthparts were far more efficient at cutting maize stalks than traditional straw returning blades [21]. The power consumption of the primnoa mouthparts gnawing on the corn rootstocks was only 23% of that of the traditional return blade [22]. The crushing length qualification rate of the traditional straw returning blades can only be maintained at 80% [23], which is due to its cut-off success rate of only about 70% for a single cut [24], while the cut-off success rate of the primnoa mouthpart is nearly 100% for each gnawing action [25]. Coupled bionics is an emerging design approach [20], which can replicate the unique functions of organisms onto mechanical equipment [20]. Therefore, it is highly likely that the operational performance of straw returning machines can be substantially improved by coupled bionic design if the bionic coupling element with efficient cutting ability of the primnoa mouthparts can be clarified.

In summary, this paper develops a completely new bionic straw returning machine by specifying the bionic coupling elements that can improve the efficiency of cutting operation, and integrating the coupled bionic design and DEM simulation parameter optimization test. The effect of different bionic coupling elements on the operational performance and the actual operational performance of the bionic straw returning machine were verified by soil bin tests. The research in this paper can provide a new research idea and method for the design of straw returning machines. At the same time, it can promote the promotion of conservation tillage in northeastern China and slow down the rate of soil erosion of black soil resources in the region, which is of great ecological significance.

2. Materials and Methods

2.1. Bionic Coupling Element Selection and Coupled Bionic Design

The search for the bionic coupling element of the primnoa mouthpart with efficient cutting ability should start from the main differences between it and the traditional straw returning machine. As shown in Figure 1a, the main difference between the primnoa mouthpart and the traditional straw returning machine is the cutting edge geometry and the cutting motion. The cutting edge structure of the primnoa mouthpart is a segmented-serrated structure, while the cutting edge of the traditional straw returning blade is smooth curve. The primnoa uses a motion with two mouthparts rotating in equal and opposite directions to gnaw on the corn rootstalk. Traditional straw returning machines commonly use a motion in which all blades rotate in the same direction to cut maize rootstocks. In this paper, based on the uniqueness of the structure and motion coupling elements of the primnoa mouthpart, the bionic straw returning blade based on the bionic structure coupling elements and the isokinetic reverse bionic drive mechanism based on the bionic motion coupling elements are designed respectively.

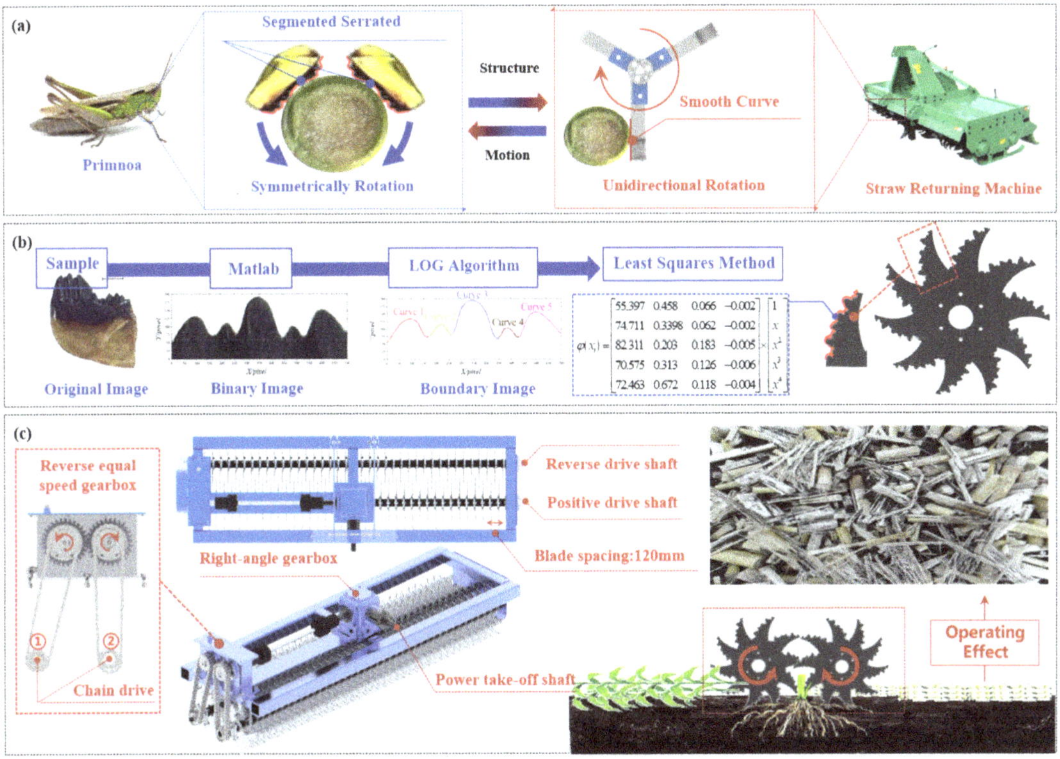

Figure 1. The search for the bionic coupling element of the primnoa mouthpart with efficient cutting ability should start from the main differences between it and the traditional straw returning machine. (**a**) Analysis of the difference between the primnoa mouthpart and the traditional straw returning machine. (**b**) Design process of bionic straw returning blade based on structural coupling elements. (**c**) Operating principle of bionic isokinetic reverse drive system based on motion coupling elements.

As shown in Figure 1b, the bionic straw returning blade achieves a high degree of reduction of the primnoa mouthpart. In the first step, 30 primnoa mouthparts samples were collected, and thus the original images were obtained. In the second step, the 30 original images were imported into MATLAB software (The Mathworks, Natick, MA, USA), and the multi-image fitting process was performed on the contour map of locust mouthparts using the function commands of rgb2gray, imerode, imdilate, im2bw, Imfill and edge, respectively, and then the fitted binary images were obtained. In the third step, the binary image is plotted as a boundary image by LOG algorithm. In the fourth step, the binary image is divided into five bionic curves according to the image curve continuity characteristics, and the bionic curve parameter equation $\varphi(x_i)$ is obtained on Origin software using LS algorithm. Finally, the laser cutting operation is carried out by CNC (Shandong Ruiji CNC Machine Tool Co., Ltd., Tengzhou, China) machine according to the affine curve $\varphi(x_i)$ to obtain the affine straw returning blade.

As shown in Figure 1c, the isokinetic reverse bionic drive mechanism consists of a power input shaft, a right-angle gearbox, an isokinetic reverse gearbox, a forward drive shaft, and a reverse drive shaft. During operation, the power input shaft delivers the torque to the equal-speed reverse gearbox through the right-angle reduction gearbox. The equal speed reverse gearbox converts the torque into forward and reverse torque at the same time, and transmits the forward and reverse torque to the forward and reverse blade shafts respectively through chain drive ① and chain drive ②. The bionic straw returning blades on the forward and reverse drive shafts are arranged in a staggered manner with a spacing of 120 mm between adjacent blades on the same blade shaft. The above design allows the bionic straw returning machine to highly recreate the way the locust mouthparts cutting the maize rootstalk.

2.2. Parameter Optimization Experiment of Bionic Straw Returning Machine Based on EDEM2018

The core operational performance of the straw returning machine is the cutting power consumption and crushing length qualification rate [21], so both were selected as test indicators. Numerous studies have shown that the two test indicators mainly depend on the rotational radius of the blade and the output rotation speed of the drive mechanism [26], so they are mainly optimized for both.

A total of three 3D simulation models of the bionic straw returning machine, soil and maize straw were required for the DEM (discrete element method) simulation experiments. As shown in Figure 2a, a single-sphere model was chosen as the soil particle model, and the Hertz-Mindlin with bonding contact model was used between each soil particle. A vertical load was applied on the surface of all soil particles to provide a compaction effect. Each physical parameter of the soil model after the above treatment is highly similar to the actual situation in the region [27]. As shown in Figure 2b, the single-sphere particle model was selected as the straw model, and the Hertz-Mindlin with bonding contact model was used between the single-sphere particle models, and each physical parameter setting of the straw model was highly similar to the actual situation in the region [14]. As shown in Figure 2c, CATIA V5R21 software (Dassault Systèmes, Paris, France) was used to build the 3D solid model of the bionic straw returning machine, and its material properties were all the same as the solid prototype. As shown in Figure 2d, the operation process was simulated by the Creator and Simulator modules in officially licensed genuine EDEM2018 software (Altair Engineering, Inc., Troy, MI, USA), and the power consumption and crushing length qualification rate of the operation were calculated by the Total Energy and Torque modules. The amount of straw returning is generally 30%, 50%, 80% and 100% in the maize growing regions of Northeast China [8,28]. In order that the results obtained from the parameter optimization experiments could maximize all the returning amounts, the straw returning amount was taken as the middle value of 65% in the DEM simulation experiments.

Since the DEM simulation test significantly reduces the test cost, a 2-factor, 5-level full-scale test is used in this paper. Five levels of rotation radius R were selected: 200, 225, 250, 275 and 300 mm. The output rotation speed was selected at 600, 900, 1200, 1500 and 1800 r/min. All the experiments were conducted in a total of 25 groups, and each group was repeated 5 times. Finally, according to the test results, the influence law of the 2 test factors on the test index was analyzed, and a regression analysis model was established, and then the optimal design parameter value combination was optimized.

2.3. Bionic Straw Returning Machine Operational Performance Test

The purpose of this test was to verify the effects of different bionic coupling elements on the operating performance and the operating effect of the bionic straw returner under different return volume conditions. Therefore, the experiment selected the bionic coupling element and straw return volume as the test factors, and the operating power consumption and crushing length qualification rate as the test indexes.

The bionic coupling element was set up in four levels: bionic structure coupling element + bionic motion coupling element (BB + BD), bionic structure coupling element + traditional drive method (BB + TD), bionic motion coupling element + traditional blade structure (TB + BD), and traditional blade structure + traditional drive method (TB + TD). The machine used for this experiment was a bionic straw returning machine with 1JGH-2 straw returning machine (the most common model in the region [29]). Before the start of each experiment, the selection of the test machine was made according to the set level of the bionic coupling element, based on Table 1.

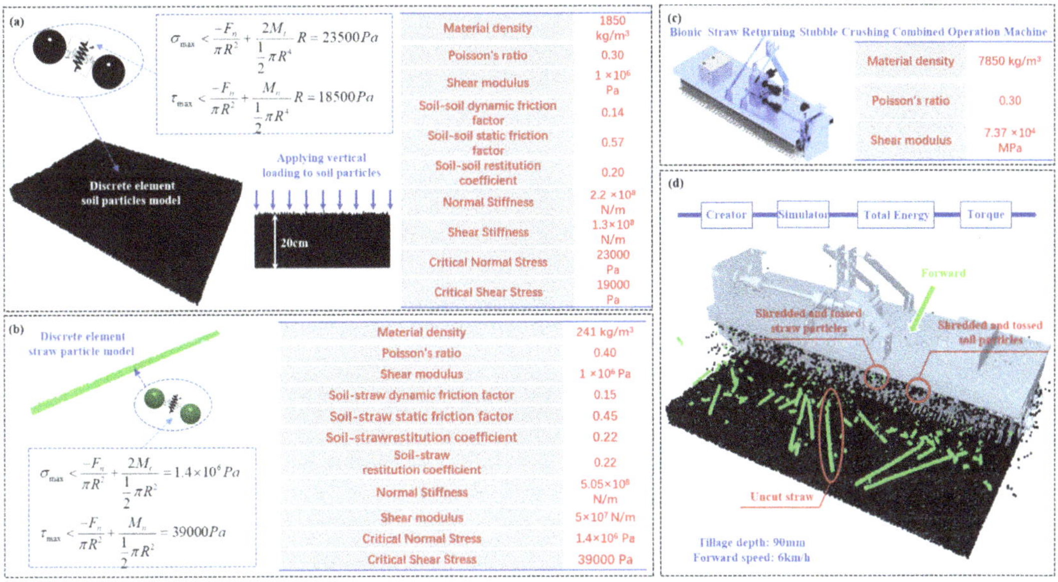

Figure 2. DEM simulation model construction and simulation operation process. (**a**) Discrete element soil model and parameter construction. (**b**) Discrete element maize straw model and parameter construction. (**c**) Construction of 3D model and parameters of the bionic straw returning machine. (**d**) The DEM-based simulation of the operation process of the bionic straw returning machine.

Table 1. Bionic coupling element level and the test machines comparison table.

Bionic Coupling Element Levels	Test Machine Selection
BB + BD	Soil bin test using a bionic straw returning machine
BB + TD	Installing the Bionic straw returning-stubble cleaning universal blade on the 1JGH-2 Straw returning machine for soil bin test
TB + BD	Installing the traditional straw returning-stubble cleaning universal blade on the bionic straw returning machine for soil bin test
TB + TD	Soil bin test with 1JGH-2 straw returning machine

The maize planting density in the northeast China maize growing region is 75,000 plants/hm^2 [30]. The effective test area of the soil bin (Agricultural Machinery Soil Bin Laboratory, Jilin University, Changchun, China) used in this paper was 80 m^2. Therefore, four levels of straw returning were set for the soil bin tests, 180, 300, 480 and 600 straws, according to the common return levels (30%, 50%, 80% and 100%) in the area. Before the start of each test, straw and root stubble were evenly spread or buried in the bins according to the set straw returning level.

The environmental conditions of the soil bin were made similar to the field environment by manual soil preparation, watering and compaction before the test, and the moisture content and compactness of the soil trough before the test are shown in Table 2. Post-harvest maize rootstocks were collected and sealed before the test. During the test, the maize rootstocks were removed from the plastic sealing bag to minimize water dissipation.

Table 2. Indicators of the soil in the soil bin before the test.

Soil Parameters	Soil Bin for Test	Average Maize-Growing Region in Northeast China
soil type	black clay	black clay
Soil compactness (MPa)	0.99 ± 0.04	1.03 ± 0.06
Soil moisture content (% d. b.)	19.6 ± 0.9	20.1 ± 1.2

Note: The data in the Table is from Inner Mongolia Statistical Yearbook 2020, Heilongjiang Province Statistical Yearbook 2020 and Jilin Province Statistical Yearbook 2020. The data in the table are the average of the corresponding test sites for nearly 20 years [31–33].

As shown in Figure 3a, the machine is connected to the bench test vehicle, and the test relies on the input power of the bench test vehicle (produced by Harbin Aoshen Technology Co., Harbin, China, Power output range: 0–25 kW, rotation speed output range: 2000 r/min), and the forward speed is selected as the most common 6 km/h [34]. The TQ-660 torque sensor (produced by Beijing Shitong Technology Co., Beijing, China, range: 0 ± 500 Nm, accuracy: 0.5 V) is connected to the drive shaft of the blade through a coupling, and then the torque of the machine operation is measured in real time. As shown in Figure 3b, the torque sensor can transmit the cutting resistance torque to the computer in real time, and the test data in the effective test area is fitted to the time-torque curve g(t) by MATLAB software, and the operating power consumption is derived according to Equation (1). As shown in Figure 3c, the straw crushing length was measured manually, and the crushing length qualification rate was derived according to Equation (2).

$$Q = \int_0^{L/V} g(t)dt \tag{1}$$

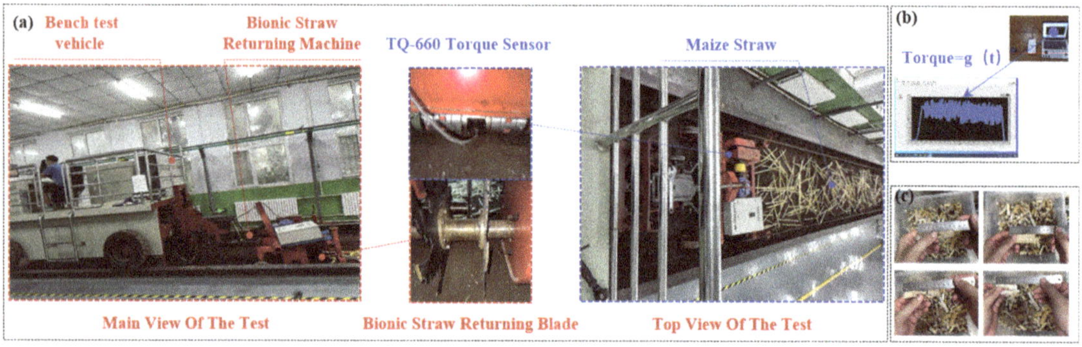

Figure 3. Soil bin test procedure. (**a**) Test equipment linking method. (**b**) Cutting energy consumption test data output. (**c**) Manual test results of crushing length qualification rate.

Note: Q means cutting energy consumption, L means Effective length of soil bin, V means forward speed of bench test vehicle, g(t) means instantaneous torque-time function equation, t means operation time.

$$Y = \frac{N_1}{N_1 + N_2} \times 100\% \tag{2}$$

Note: Y means crushing length qualification rate, N_1 means crushing length less than 10 cm pieces, N_2 means crushing length greater than 10 cm pieces.

2.4. Test Data Statistics

Based on the results of the parameter optimization test, a regression model was established using regression analysis, and the accuracy of the model was evaluated by ANOVA analysis and lack-of-fit term analysis. Both the parameter optimization test and the operational performance comparison test used LSD to determine whether there was a significant difference between the different levels, and ANOVA to determine whether the test factors had a significant effect on the test index. All data were processed and analyzed using MATLAB and Design-expert software, and plotted using Origin software.

3. Results

3.1. Parameter Optimization Test Results

As shown in Figure 4a,b, under the condition of fixed rotation radius, the cutting energy consumption first decreases and then increases as the output rotation speed accelerates, while the crushing length qualification rate becomes a gradually increasing trend, and the increasing trend gradually decreases. As shown in Figure 4c,d, under the condition of the fixed output rotation speed, the cutting energy consumption decreases and then rises with the increase of the rotation radius, while the crushing length qualification rate realizes a gradually rising trend, and the rising trend gradually decreases. The regression models between the two test factors and the two test indicators were derived separately using Design-Expert software (Equations (3) and (4)). As shown in Table 3, all parameters in the two regression models (R (Blade rotation radium), ω (Blade rotation speed), $R\omega$, R^2, ω^2) have significant effects on the cutting energy consumption and crushing length qualification rate. Meanwhile, the p-values of the lack-of-fit term test for both regression models were greater than 0.1.

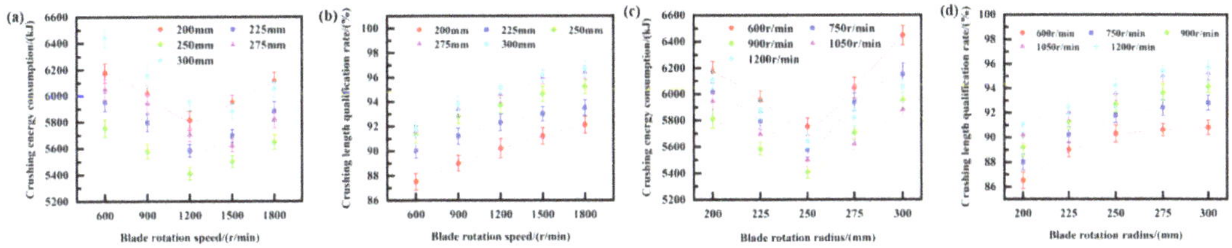

Figure 4. Parameter optimization test results. (**a**) Effect of blade rotation speed on cutting energy consumption. (**b**) Effect of blade rotation speed on crushing length qualification rate. (**c**) Effect of blade rotation radius on cutting energy consumption. (**d**) Effect of blade rotation radius on crushing length qualification rate.

Table 3. Test data.

Resources	Regression Model on Crushing Energy Consumption (Equation (3))				Regression Model on Crushing Length Qualification Rate (Equation (4))			
	Sum of Squares	df	F-Value	p-Value	Sum of Squares	df	F-Value	p-Value
Model	2.05×10^6	5	86.32	<0.0001 **	138.80	5	584.39	<0.0001 **
R	22,285.06	1	30.34	0.0389 *	58.75	1	1236.85	<0.0001 **
ω	1.41×10^5	1	4.80	<0.0001 **	70.09	1	1475.57	<0.0001 **
$R\omega$	53,592.95	1	11.54	0.0025 **	0.25	1	5.26	0.0312
R^2	9.35×10^5	1	106.35	<0.0001 **	1.84	1	38.71	<0.0001 **
ω^2	4.94×10^5	1	201.24	<0.0001 **	6.47	1	136.11	<0.0001 **
Residual	1.07×10^5	23	/	/	1.09	23	/	
Lack of fit	96,392.45	19	1.94	0.2746	1.00	19	2.29	0.2196
Pure error	10,455.22	4	/	/	0.092	4		
Cor total	2.11×10^6	28	/	/	139.89	28		

Note: ** indicates highly significant ($p < 0.01$), * indicates significant ($0.01 < p < 0.05$).

The above experimental results show that the two regression models can provide accurate estimates of the optimal values of the two design parameters. Using MATLAB software, the minimum crushing energy consumption/acre under the condition that the crushing length qualification rate is greater than 95% is used as the solution condition.

The optimal combination of operating parameters of the machine was obtained as follows: when the blade shaft rotation speed is 930 r/min and the blade rotation radius is 248 mm, the estimated crushing length qualification rate can reach 92.89% and the crushing energy consumption can reach 5431.23 kJ.

$$Q = 5436.96 - 68.55\omega + 29.85R - 42.27R\omega + 165.00R^2 + 215.96\omega^2 \tag{3}$$

$$Y = 92.78 + 2.168\omega - 2.368R + 0.21R\omega - 0.601R^2 - 1.126\omega^2 \tag{4}$$

3.2. *Performance Verification Test Results of Bionic Straw Returning Machine*

As shown in Figure 5a–d, when the straw is returned in the same amount, the crushing energy consumption from largest to smallest is TB + TD > BB + TD > TB + TD > BB + BD. As shown in Figure 5e–h, when the straw is returned in the same amount, the crushing length qualification rate from high to low is: BB + BD > TB + BD > BB + TD > TB + TD.

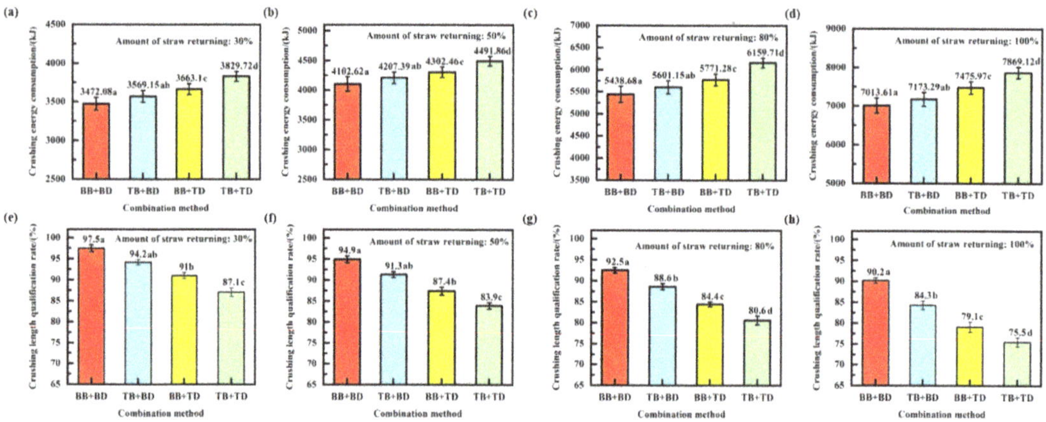

Figure 5. Parameter optimization test results. (**a**) Effect of different mechanism collocation methods on operating power consumption at 30% of straw returning amount. (**b**) Effect of different mechanism collocation methods on operating power consumption at 50% of straw returning amount. (**c**) Effect of different mechanism collocation methods on operating power consumption at 80% of straw returning amount. (**d**) Effect of different mechanism collocation methods on operating power consumption at 100% of straw returning amount. (**e**) Effect of different mechanism collocation methods on crushing length qualification rate at 30% of straw returning amount. (**f**) Effect of different mechanism collocation methods on crushing length qualification rate at 50% of straw returning amount. (**g**) Effect of different mechanism collocation methods on crushing length qualification rate at 80% of straw returning amount. (**h**) Effect of different mechanism collocation methods on crushing length qualification rate at 100% of straw returning amount. Averages followed by different lowercase letters are significantly different according to LSD's multiple range experiment at the significance level of 0.05. Error bars are standard deviation.

When the amount of straw returning is 30%, compared with TB + TD, BB + BD, TB + BD and BB + TD improve the crushing length qualification rate by 10.4%, 7.1% and 3.9%, respectively, and reduce the crushing energy consumption by 9.4%, 6.8% and 4.4%. When the amount of straw returning is 50%, compared with TB + TD, BB + BD, TB + BD and BB + TD improve the crushing length qualification rate by 11%, 7.4% and 3.5%, respectively, and reduce the crushing energy consumption by 9.5%, 6.3% and 4.2%. When the amount of straw returning is 80%, compared with TB + TD, BB + BD, TB + BD and BB + TD improve the crushing length qualification rate by 11.9%, 8% and 3.8%, respectively, and reduce the crushing energy consumption by 11.7%, 9.1% and 6.3%. When the amount of straw returning is 100%, compared with TB + TD, BB + BD, TB + BD and BB + TD improve the crushing length qualification rate by 14.7%, 8.8% and 3.6%, respectively, and reduce the crushing length consumption by 10.9%, 8.8% and 5.0%.

The above experimental results show that both the bionic blade structure and the bionic drive method can reduce the crushing energy consumption and improve the crushing length qualification rate, and the effect is more significant as the amount of straw returning increases. Compared with the bionic blade structure, the bionic drive method has a more significant effect on the improvement of operating performance.

4. Discussion

4.1. Discussion on the Law of Influence of Design Parameters on Operational Performance

The crushing length qualification rate of the straw returning machine depends mainly on the crushing success rate and the number of cutting times [35]. Maize straw can be cut off only when the driving force of the blade is greater than the cutting resistance [36]. Obviously, when the driving force of the blade is the same, the lower the cutting resistance, the higher the crushing success rate. Under the condition of the same crushing success rate, the more the number of cutting times, the higher the qualification rate of crushing length. Numerous studies have shown that the maximum cutting resistance is inversely proportional to the cutting linear speed of the blade [37]. From Equation (5), it can be seen that the larger the rotation radius R and output rotation speed ω, the larger the cutting linear speed of the blade. The higher the output rotation speed, the more times the straw is cut. Therefore, the crushing length qualification rate showed a gradual increase with increasing the rotation radius and output rotation speed. This is similar to the findings of He et al. in the maize-growing region of central China [17].

$$V_c = \omega R \frac{R-h}{R} \pm V = \omega R - \omega h \pm V \tag{5}$$

Note: V_c means cutting linear velocity, ω means output rotation speed, R means rotation radius, h means blade entry depth, V means Forward speed of bench test vehicle.

The crushing energy consumption Q of the bionic straw returning machine mainly includes the energy consumption Q_1 means the energy consumption generated by cutting straw Q_2 generated by disturbing soil. Lu et al. showed that Q_1 mainly depends on the cutting resistance, and the smaller the cutting resistance the lower the Q_1 [38]. Ma et al. showed that Q_2 mainly depends on the cutting pitch and the entry depth, and the deeper the entry depth or the shorter the cutting pitch the greater the Q_2 [39]. From the previous discussion, it is clear that the larger the rotation radius R and the output rotation speed, the smaller the Q_1. From Equation (6), the larger the rotation radius R and output rotation speed ω, the smaller the cutting pitch. Therefore, the larger the rotation radius R and output rotation speed ω, the greater the energy consumption Q_2 for cutting the soil. Since the entry depth (90 mm) of the straw returning operation is small, the soil is less disturbed by the bionic straw returning machine when the cutting pitch is larger, and the magnitude of Q_1 is larger than Q_2 at this time. Therefore, when the rotation radius R and the output rotation speed ω are small, Q decreases as both increase. However, when the cutting pitch is small, the amount of soil disturbance by the bionic straw returning machine is larger, and the magnitude of Q_2 exceeds Q_1 at this time. Therefore, at a larger rotation radius R and output rotation speed, Q realizes a gradual increase as both increase.

$$\begin{bmatrix} Q = Q_1 + Q_2 \\ S = \frac{3000\omega^2(R-h)}{\pi Z} \end{bmatrix} \tag{6}$$

Note: Q means crushing energy consumption, Q_1 means energy consumption generated by crushing straw, Q_2 means energy consumption from soil disturbance, S means cutting pitch, ω means output rotation speed, R means rotation radius, h means blade entry depth, Z means Number of bionic straw returning blades.

4.2. A Discussion on the Impact of Bionic Design on Operational Performance

Maize straw consists of the epidermis and pith [40]. The epidermis is composed of plant fibers, the pith is composed of lignin and spongy mesophyll [21]. Zhao et al. showed that the resistance generated by cutting the epidermis accounted for more than 90% of the total cutting resistance [41]. Zhao et al. showed that the straw epidermis was in a pulled-up state before the blade pierced the straw epidermis [21]. As the tensile deformation of the epidermis gradually increases, the tensile stress σ of the fiber gradually increases, resulting in a continuous increase in cutting resistance [42]. The maximum value of cutting resistance occurs at the moment when the straw epidermis is pierced [42]. Obviously, the longer it takes for the blade to pierce the straw epidermis, the greater the deformation generated by the straw epidermis and the greater the maximum cutting resistance. As shown in Figure 6, the contact area between the bionic blade structure and the straw epidermis is smaller compared to the traditional returning blade. As a result, the bionic blade structure produces less cutting resistance and allows for a higher straw crushing length qualification rate. Compared with the traditional blade structure, the bionic blade structure will significantly reduce the energy consumption of straw crushing energy operation Q_1, and thus reduce the total energy consumption of returning operations.

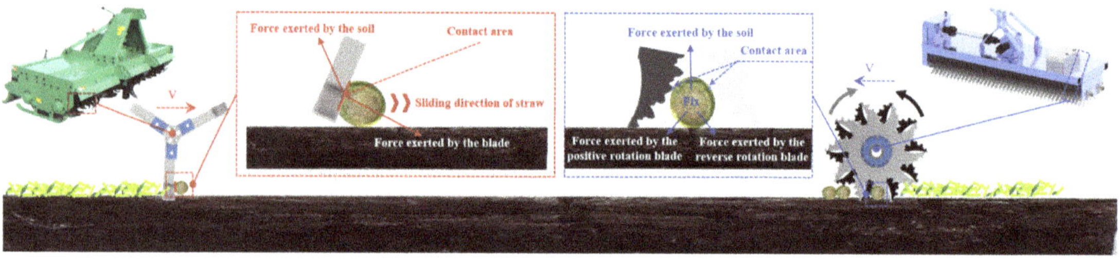

Figure 6. Comparison of the operating principle between traditional straw returning machine and bionic straw returning machine. The straw slips during the operation of traditional straw returning machine, and the contact area between the blade and the straw is large. During the operation of the bionic straw returning machine, the straw is fixed and the contact area between the blade and the straw is smaller.

As shown in Figure 6, when adopting the bionic drive method, the straw can realize the force balance under the joint action of forward and reverse rotation return blades and soil, and the cutting method at this time belongs to fixed cutting. When adopting the traditional driving method only one side of the straw is stressed in the horizontal direction, and the straw will move in the horizontal direction, and the cutting method at this time belongs to dynamic cutting. Numerous studies have shown that the operational resistance of fixed cutting is much lower than dynamic cutting [17]. Therefore, the bionic drive method produces less cutting resistance and can obtain higher straw crushing length qualification rate. The traditional driving method will throw the straw up and also cause the straw to translate on the soil surface, thus generating frictional energy and potential energy consumption. Therefore, the bionic drive method can significantly reduce the energy consumption of the straw returning operation compared to the traditional drive method.

In recent years, many scholars in the field of agricultural machinery engineering have conducted a large number of studies on methods to improve the operation performance of straw returning machines, but have focused more on the optimization of design parameters or operational parameters of existing machines [8,10–13,15]. The bionic straw returning machine developed by coupled bionic design with primnoa as the bionic prototype in this paper has made significant improvements in the cutting edge structure and cutting motion, thus improving the quality of straw returning operation more significantly than the previous studies.

5. Conclusions

In this paper, we found that the efficient gnawing ability of the primnoas on maize rootstocks was due to the segmented serrated structure of their mouthparts and the isokinetic reverse rotational movement. In this paper, we found that the blade rotation radius and the drive mechanism output rotation speed had a significant effect on the crushing energy consumption and crushing length qualification rate of the bionic straw returning machine. The optimized combination of design parameter values most suitable for the maize production region in northeast China should be 930 r/min and 248 mm. In this paper, it was found that the bionic blade structure is easier to pierce the epidermal skin of maize straw, thus reducing the cutting resistance. The bionic drive method can transform the cutting method of straw returning operation into fixed cutting, which effectively limits the displacement of straw during the cutting process while reducing the cutting resistance. The findings of this paper showed that the bionic straw returning machine can improve the straw crushing length qualification rate by 10.4–14.7% and reduce the straw cutting energy consumption by 9.4–11.7% compared with the traditional straw returning machine. The findings of this paper showed that the coupled bionic design can replicate the efficient cutting ability of the primnoa mouthpart to the straw returning machine. Therefore, this design method can be an effective means to provide future agricultural machinery design field.

It should also be noted that the current fuel cost of straw returning operations in the main maize-growing regions of northeast China is about 30 RMB/acre, and the bionic straw returning machine can reduce power consumption by up to 10.4–14.7%, thus potentially reducing that cost to 25.59–26.88 RMB/acre. The improvement of the straw crushing length qualification rate will also reduce the difficulty of sowing operations, which will lead to an increase in production and also bring greater economic benefits.

There are two limitations to the findings of this paper. First, the findings of this paper are mainly applicable to the main maize producing regions in northeast China, and the applicability to other regions or other crops needs further in-depth study. Secondly, due to the more severe epidemic situation at present, large-scale field trials were not conducted in this paper, but soil bin tests were used instead of field trials, so the findings of this paper may not fully reflect the influence of field environment on the operating effect of the machine.

Author Contributions: Conceptualization, J.Z. (Jiale Zhao); software, X.W.; validation, Y.W.; investigation, J.Z. (Jian Zhuang); resources, Y.Y.; visualization, H.L. All authors have read and agreed to the published version of the manuscript.

Funding: The research was financially supported by the National Key Research and Development Project of China, grant number 2017YFD0700701, the National Natural Science Foundation of China, grant number 52075215, the Science and Technology Development Program of Jilin Province, grant number 20200403153SF, and the Thirteenth Five-Year Plan Science and Technology Project of Jilin Provincial Education Department, grant number JJKH20201009KJ, and Innovation and Entrepreneurship Training Program for College Students of Jilin University 202010183616.

Institutional Review Board Statement: Not applicable.

Informed Consent Statement: Not applicable.

Data Availability Statement: Not applicable.

Conflicts of Interest: The authors declare that there is no conflict of interest.

References

1. Sun, B.J.; Jia, S.X.; Zhang, S.X.; McLaughlin, N.; Zhang, X.P.; Liang, A.Z.; Chen, X.W.; Wei, S.C.; Liu, S.Y. Tillage, seasonal and depths effects on soil microbial properties in black soil of Northeast China. *Soil Till. Res.* **2016**, *155*, 421–428. [CrossRef]
2. Zhang, S.X.; Chen, X.W.; Jia, S.X.; Liang, A.Z.; Zhang, X.P.; Yang, X.M.; Wei, S.C.; Sun, B.J.; Huang, D.D.; Zhou, G.Y. The potential mechanism of long-term conservation tillage effects on maize yield in the black soil of Northeast China. *Soil Till. Res.* **2015**, *154*, 84–90. [CrossRef]

3. Liang, A.Z.; Yang, X.M.; Zhang, X.P.; Chen, X.W.; Mclaughlin, N.B.; Wei, S.C.; Zhang, Y.; Jia, S.X.; Zhang, S.X. Changes in soil organic carbon stocks under 10-year conservation tillage on a Black soil in Northeast China. *J. Agr. Sci.* **2016**, *154*, 1425–1436. [CrossRef]
4. Feng, H.Y.; Li, Q.Y.; Sun, L.Y.; Cai, Q.G. Using Cs-137 to study spatial patterns of soil ero-sion and soil organic carbon (SOC) in an agricultural catchment of the typical black soil region, Northeast China. *J. Environ. Radioact.* **2012**, *112*, 125–132. [CrossRef]
5. Li, H.Q.; Yao, Y.F.; Zhang, X.J.; Zhu, H.S.; Wei, X.R. Changes in soil physical and hydraulic properties following the conversion of forest to cropland in the black soil region of North-east China. *CATENA* **2021**, *198*, 104986. [CrossRef]
6. Wang, J.W.; Tang, H.; Wang, J.F. Comprehensive Utilization Status and Development Analysis of Crop Straw Resource in Northeast China. *Trans. CSAM* **2017**, *48*, 1–21.
7. Cheng, S.L.; Fang, H.J.; Zhu, T.H.; Zheng, J.J.; Yang, X.M.; Zhang, X.P.; Yu, G.R. Effects of soil erosion and deposition on soil organic carbon dynamics at a sloping field in Black Soil region, North-east China. *Soil Sci. Plant Nutr.* **2012**, *56*, 521–529. [CrossRef]
8. Zhao, J.L.; Lu, Y.; Tian, H.L.; Jia, H.L.; Guo, M.Z. Effects of Straw Returning and Residue Cleaner on the Soil Moisture Content, Soil Temperature, and Maize Emergence Rate in China's Three Major Maize Producing Areas. *Sustainability* **2019**, *11*, 5796. [CrossRef]
9. Horning, L.B.; Strtler, L.D.; Saxton, K.E. Surface residue and soil toughness for wind ero-sion protection. *Trans. ASAE* **1998**, *41*, 1061–1065. [CrossRef]
10. Zhou, H.; Zhang, J.M.; Xia, J.F.; Tahir, H.M.; Zhu, Y.H.; Zhang, C.L. Effects of subsoiling on working quality and total power consumption for high stubble straw returning machine. *Int. J. Agric. Biol. Eng.* **2019**, *12*, 56–62. [CrossRef]
11. Bao, X.B.; Zhao, X.Y.; He, J.; Li, H.W.; Wang, Q.J.; Liu, W.Z. Design and performance test of plowing and rotaty tillage combined machine. *INMATEH Agric. Eng.* **2019**, *58*, 213–222.
12. Wang, W.W.; Li, J.C.; Chen, L.Q.; Qi, H.J.; Liang, X.T. Effects of key parameters of straw chopping device on qualified rate, non-uniformity and power consumption. *Int. J. Agric. Biol. Eng.* **2018**, *11*, 122–128. [CrossRef]
13. Yin, X.G.; Olesen, J.E.; Wang, M.; Ozturk, I.; Chen, F. Climate effects on crop yields in the Northeast Farming Region of China during 1961–2010. *J. Agr. Sci.* **2016**, *154*, 1190–1208. [CrossRef]
14. Zhang, T.; Zhao, M.Q.; Liu, F.; Tian, H.Q.; Wulan, T.Y.; Yue, Y.; Li, D.P. A Discrete Element Method Model of Corn Stalk and Its Mechanical Characteristic Parameters. *Bioresources* **2020**, *15*, 9337–9350. [CrossRef]
15. Liao, N.; Han, L.J.; Huang, G.Q.; Chen, L.J.; He, C. Effects of moisture content and compression frequency on straw open compression energy consumption. *Trans. CSAE* **2011**, *27* (Suppl. S1), 318–322, (In Chinese with English Abstract).
16. Yu, C.Y.; Liu, J.; Zhang, J.; Xue, K.; Zhang, S.; Liao, J.; Tai, Q.L.; Zhu, D.Q. Design and opti-mization and experimental verification of a segmented double-helix blade roller for straw returning cultivators. *J. Chin. Inst. Eng.* **2021**, *44*, 379–387. [CrossRef]
17. Zheng, Z.Q.; He, J.; Li, H.W.; Diao, P.S.; Wang, Q.J.; Zhang, X.C. Design and Experiment of Straw-chopping Device with Choppingand Fixed Knife Supported Slide Cutting. *Trans. CSAM* **2016**, *47*, 108–116.
18. Shen, Y.; Zhang, T.; Cui, J.; Chen, S.; Han, H.; Ning, T. Subsoiling increases aggre-gate-associated organic carbon, dry matter, and maize yield on the North China Plain. *PeerJ* **2021**, *9*, e11099. [CrossRef]
19. Zhang, C. Study on the Dynamic of Soil Organic Carbon Accumulation and Maize Pro-Duction in Long Term Straw Returning. Master's Thesis, Gansu Agricultural University, Lanzhou, China, 2017.
20. Ren, L.Q.; Liang, Y.H. Biological couplings: Classification and characteristic rules. *Sci. China Ser. E Technol. Sci.* **2009**, *52*, 2791–2800. [CrossRef]
21. Zhao, J.L.; Guo, M.Z.; Lu, Y.; Huang, D.Y. Design of bionic locust mouthparts stubble cutting device. *Int. J. Agric. Biol. Eng.* **2020**, *13*, 20–28. [CrossRef]
22. Jia, H.L.; Li, C.Y.; Zhang, Z.H.; Wang, G. Design of Bionic Saw Blade for Corn Stalk Cut-ting. *J. Bionic. Eng.* **2013**, *10*, 497–505. [CrossRef]
23. Liang, Y.C. Design and Test of Combined Straw and Stubble Crushing and Returning Ma-Chine. Master's Thesis, Shandong University of Technology, Zibo, China, 2020.
24. Wiedermann, A.; Harms, H.H. Straw-cutting machine Investigations on combine har-vester straw-cutting machine with exact reaping. In Proceedings of the Conference on Agricultural Engineering, Limenas Hersonissou, Greece, 23–25 June 2008; Volume 2045, pp. 257–262.
25. Li, C.Y. Bionic Blade of Corn Harvester for Leaving High Stubble and Its Cutting Mechanism. Ph.D. Thesis, Jilin University, Changchun, China, 2014.
26. Hu, J.P.; Zhao, J.; Pan, H.R.; Liu, W.; Zhao, X.S. Prediction Model of double axis rotary power consumption based on discrete element method. *Trans. CSAM* **2020**, *51*, 9–16.
27. Zhao, S.H.; Liu, H.P.; Yang, C.; Yang, L.L.; Gao, L.L.; Yang, Y.Q. Design and discrete ele-ment simulation of interactive layered subsoiler with maize straw returned to filed. *Trans. CSAM* **2021**, *52*, 75–87.
28. Lin, J.; Qian, W.; Niu, J. Design and Experiment of Stubble-cutting and Anti-blocking Mechanism for Ridge-till and No-till Planter. *J. Shenyang Agric. Univ.* **2015**, *46*, 691–698.
29. Pei, Y.; Bian, S.F.; He, Z.; Cao, Y.; Ma, X.; Yu, J.Q. Study on a New Tilling Method of Striped Deep Loosening of Wide-Narrow Row Alternate Planting and Its Attached Machines. *Trans. CSAE* **2020**, *5*, 67–70.
30. Wang, F.L.; Dong, Z.G.; Wu, Z.H.; Fang, K. Optimization of maize planting density and fer-tilizer application rate based on BP neural network. *Trans. CSAE* **2017**, *33*, 92–99. (In Chinese with English abstract)

31. Inner Mongolia Province National Bureau of Statistics. In *China Statistical Yearbook*; China Statistical Publishing House: Beijing, China, 2020.
32. Heilongjiang Province National Bureau of Statistics. In *China Statistical Yearbook*; China Statistical Publishing House: Beijing, China, 2020.
33. Jilin Province National Bureau of Statistics. In *China Statistical Yearbook*; China Statistical Publishing House: Beijing, China, 2020.
34. Chen, L.Q.; Liang, X.T.; Cao, C.M. Virtual Simulation and Power Test of Straw Counters-field Based on Multi-body Dynamics. *Trans. CSAM* **2016**, *47*, 106–111.
35. He, J.; Zhang, Z.Q.; Li, H.W.; Wang, Q.J. Development of small/medium size no-till and minimum-till seeders in Asia: A review. *Int. J. Agric. Biol. Eng.* **2014**, *7*, 1–12.
36. Liao, Q.X.; Gao, H.W.; Shu, C.X. Design of sawing anti-blocking mechanism for no-tillage planter and its cutting mechanism. *Trans. CSAE* **2003**, *5*, 64–70.
37. Igathinathane, C.; Womac, A.R.; Sokhansanj, S. Corn stalk orientation effect on mechanical cutting. *Biosyst. Eng.* **2010**, *107*, 97–106. [CrossRef]
38. Lu, C.Y.; He, J.; Li, H.W.; Wang, Q.J.; Zhang, X.C.; Liu, J.A. Finite Element Analysis and Experiment on Anti-blocking Device Based on Support Cutting. *Trans. CSAE* **2013**, *44*, 61–66.
39. Ma, H.L.; Gao, H.W.; Li, H.W.; Wei, S.Y. Experimental Study on Corn Stalk and Rootstalk Cutting by Driven Disc. *Trans. CSAE* **2007**, *5*, 47–50+54.
40. Maraldi, M.; Molari, L.; Regazzi, N.; Molari, G. Analysis of the parameters affecting the mechanical behaviour of straw bales under compression. *Biosyst. Eng.* **2017**, *160*, 179–193. [CrossRef]
41. Zhao, J.L.; Wang, X.G.; Zhuang, J.; Cong, Y.J.; Lu, Y.; Guo, M.Z. Fine-Crush Straw Returning Enhances Dry Matter Accumulation Rate of Maize Seedlings in Northeast China. *Agronomy* **2021**, *11*, 1144. [CrossRef]
42. Wu, H.X. Research and Simulated Analysis for Corn Stalk Harvesting Technology and Equipment. Ph.D. Thesis, Jilin University, Changchun, China, 2014.

agriculture

MDPI

Article

Optimization of the Process Parameters of an Air-Screen Cleaning System for Frozen Corn Based on the Response Surface Method

Ning Zhang [1,2], Jun Fu [1,2,3,*], Zhi Chen [2,3], Xuegeng Chen [4] and Luquan Ren [1,2]

1 Key Laboratory of Bionic Engineering, Ministry of Education, Jilin University, Changchun 130022, China; nzhang19@mails.jlu.edu.cn (N.Z.); lqren@jlu.edu.cn (L.R.)
2 College of Biological and Agricultural Engineering, Jilin University, Changchun 130022, China; chenzhi@caamm.org.cn
3 Chinese Academy of Agricultural Mechanization Sciences, Beijing 100083, China
4 College of Mechanical and Electrical Engineering, Shihezi University, Shihezi 832000, China; chenxg431@163.com
* Correspondence: fu_jun@jlu.edu.cn

Abstract: The threshing of frozen corn is accompanied by breakage and adherence, which influence the cleaning performance when the corn-cleaning mixture is separated and cleaned. In order to reduce the impurity ratio and loss ratio during frozen corn cleaning and provide theoretical support for frozen corn combine harvesting, this study employed a self-made air-screen cleaning system with adjustable parameters. The optimal process parameters of frozen corn cleaning were determined by using the response surface method (RSM). The influences of the fan speed (FS), vibrational frequency (VF), and screen opening (SO) on the cleaning performance were explored. The results showed that all three process parameters had significant effects on the impurity ratio (IR) and loss ratio (LR). The fan speed had the most significant impact. The cleaning performance was optimal when the fan speed was 102.7 rad/s, the vibration frequency was 6.42 Hz, and the screen opening was 21.9 mm, corresponding to a 0.80% impurity ratio and a 0.61% loss ratio. The predicted values of the regression models were consistent with the experimental results with a relative error of less than 5%. The reliability and accuracy of regression models were established and confirmed.

Keywords: frozen corn; freeze adhesion; corn combine harvester; cleaning performance; parameter optimization

Citation: Zhang, N.; Fu, J.; Chen, Z.; Chen, X.; Ren, L. Optimization of the Process Parameters of an Air-Screen Cleaning System for Frozen Corn Based on the Response Surface Method. *Agriculture* **2021**, *11*, 794. https://doi.org/10.3390/agriculture11080794

Academic Editor: José Pérez-Alonso

Received: 1 August 2021
Accepted: 18 August 2021
Published: 19 August 2021

Publisher's Note: MDPI stays neutral with regard to jurisdictional claims in published maps and institutional affiliations.

1. Introduction

Corn is the most widely produced crop in the world, and it is an important food and feed source. Corn production has great significance in ensuring food security [1–4]. In Canada, Ukraine, and northeast China, owing to high latitudes, the temperature is already below zero when corn is harvested. In particular, owing to the long harvesting period, a large amount of corn is harvested after frost and snowfall [5]. The physical properties of corn change after freezing, which results in a high loss ratio in the combine harvesting operation. The combine harvester can simultaneously complete the processes of ear picking, threshing, cleaning, and collection, resulting in high operational efficiency and low operating costs [6]. Therefore, it is widely used for frozen corn and harvesting of other grains.

In combine harvesting, cleaning is an important procedure. The performance of the cleaning system in a combine harvester directly affects the loss ratio and the impurity ratio [7]. In order to improve the performance of the cleaning system, several researchers have carried out numerous analyses on cleaning system in harvesters. Li et al. [8] simulated and analyzed the motion of rice particles in cleaning device utilizing the discrete element method (DEM) coupled with computational fluid dynamics (CFD). Their simulation results revealed that the longitudinal velocity of the short straw was significantly affected by inlet airflow velocity. Badretdinov et al. [9] established a kinematic model of the linkage structure of a grain combine harvester cleaning system by deriving coordinates, velocity, and acceleration of the nodal points. For wind speed distribution, Ueka et al. [10] analyzed the turbulent flow characteristics of the cleaning airflow. The results found that the main factors affecting the airflow distribution were the friction and pressure change of material particles. Gebrehiwot et al. [11] simulated and compared the airflow distributions of three forward curved centrifugal fans. The results indicated that adding a cross-flow opening in the width direction of the centrifugal fan outlet could enhance the utilization efficiency of airflow. To investigate cleaning screen types, Wang et al. [12,13] designed several corn cleaning screens, such as a curved screen, combined screen, and rubber screen. Sabashkin et al. [14] proposed a cylinder screen with a screw dispenser for grain cleaning. Ivanov et al. [15] established a mathematical model of grain movement in cylindrical screen, and discussed the impact of feed rate and rotation speed on grain screening. Krzysiak et al. [16] invented a new conical rotary screen and investigated the effect of the drum inclination angle on cleaning performance. To address the blocking problem of the cleaning screen, Cheng et al. [17] proposed an accumulation rule for the corn cob blockage mass in a chaff screen as the operating hours increased. The response surface method (RSM) was used to obtain the optimal vibration parameters with minimal corn cob blockage. For the real-time monitoring of loss during the cleaning process, Xu et al. [18] developed a sieve loss sensor based on the signal analysis of impacts. In addition, Craessaerts et al. [19,20] proposed a multivariate input selection methodology and a fuzzy control system. By selecting and controlling the variables of the sieve, the cleaning performance of the combine harvester was optimized under different operation conditions. In general, all of these studies were conducted under conditions in which the temperature was above zero at the time of grain harvesting.

However, the physical properties of corn change after freezing, and some corn kernels adhere to each other, which introduce challenges to the cleaning process. Unfortunately, no experimental studies of air-screen cleaning systems have been reported for such conditions, and to date, no optimal process parameters have been specified. Therefore, the objective of this study is to optimize the process parameters of the air-screen cleaning system for frozen corn to reduce the impurity ratio (IR) and loss ratio (LR). The characteristic dimensions and the physical properties of components in the cleaning mixture were measured. A single-factor design and the Box–Behnken design (BBD) were implemented. The effects of the fan speed (FS), vibrational frequency (VF), the screen opening (SO) on cleaning performance were analyzed, respectively. The experiments were conducted to determine the combination of process parameters to obtain the optimal IR and LR. Using the impurity ratio and loss ratio as response values, the optimal combination of process parameters was determined and verified by experiments. The results may have potential to use for setting frozen corn combine harvesting parameters.

2. Materials and Methods

In this section, we provide a brief introduction of the materials used for the cleaning study. Then, we describe the experimental apparatus. Next, we present the experimental designs, including the single-factor and the Box–Behnken experimental designs. Finally, we present the method used to analyze experimental data.

2.1. Materials

For the cleaning study on frozen corn, a corn-cleaning mixture was prepared. The corn cultivar Feitian 358 was selected as the sample. First, corn ears with the husks were picked by hand in Changchun city (N 43°56′, E 125°14′), Jilin province. The picking time was from 10 to 15 December 2020, while the corn ears were frozen. Then, the ears were threshed on a longitudinal axial threshing cylinder test device [5]. Referring to the operating parameters of corn grain harvesters in Northeast China, the feed rate of corn ears was 8.5 kg/s, the drum speed was 40.3 rad/s, and the concave clearance was 40 mm [5]. The corn-cleaning mixture was collected and stored outdoors. The mass of all preparative mixture was 3135 kg. Finally, a corn cleaning test was performed. The whole experimental process, from threshing to cleaning, was completed outdoors within 7 days after picking the corn ears. Over the experimental periods, the outdoor average temperature ranged from −10.2 to −12.6 °C. This means that the experimental materials were in a native frozen state throughout the experimental periods.

2.2. Experimental Apparatus

The air-screen cleaning system used in this study is shown in Figure 1. The air-screen cleaning system separates corn kernels from impurities under the action of airflow and vibration. It mainly consists of frame, feeding hopper, oscillating plate, fan, fan drive motor, crank, crank drive motor, swing, upper screen, upper screen box, lower screen, lower screen box, tailing screen, and collection box. The upper screen and tailing screen are bolted onto the upper screen box. Similarly, the lower screen is bolted onto the lower screen box. In Figure 1, the nearside plate of the lower screen box is hidden so as to observe the position of the lower screen. The oscillating plate, upper screen box, and lower screen box were powered for reciprocation with a 7.5 kW electric motor through the crank and swing. A FR500-4T-7.5G frequency converter (Frecon Electric Co., Ltd., Shenzhen, China) was used to adjust the vibrational frequency of the oscillating plate, upper screen box, and lower screen box in a range from 0 to 25 Hz. The fan was driven by a 2.2 kW electric motor and was adjusted by a FR150-2S-2.2B frequency converter (Frecon Electric Co., Ltd., Shenzhen, China), the speed of which ranged from 0 to 150.7 rad/s. All screens were chaffer sieves with adjustable opening. The screen slices of the chaffer sieve were arranged in parallel. The vertical distance between two adjacent parallel screen slices is the screen opening (SO).

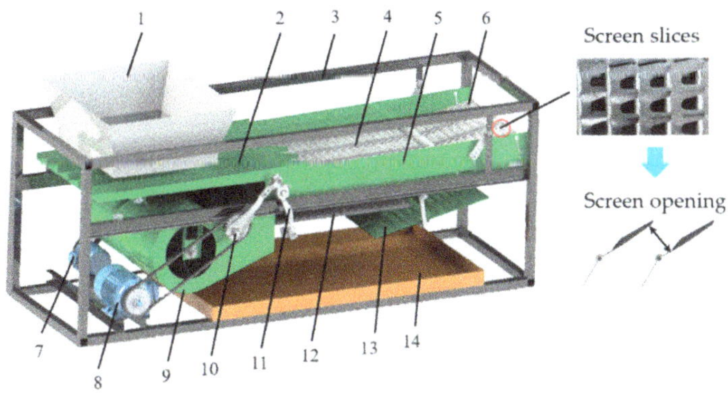

Figure 1. Air-screen cleaning system: 1. feeding hopper, 2. oscillating plate, 3. frame, 4. upper screen, 5. upper screen box, 6. tailing screen, 7. fan drive motor, 8. crank drive motor, 9. fan, 10. crank, 11. swing, 12. lower screen, 13. lower screen box, and 14. collection hopper.

2.3. Experimental Design

2.3.1. Single-Factor Experiment

According to the material properties and the operating principle of the air-screen cleaning system, the impurity ratio (IR) and loss ratio (LR) depend on several factors, such as fan speed (FS), vibration frequency (VF), and screen opening (SO) [21]. A single-factor experiment was designed to analyze the effects of these three factors [22,23]. The single-factor experiment was arranged with six values for each factor with the feed rate of the corn-cleaning mixture set to 5.5 kg/s. For each set of parameters, a single trial was repeated three times. The experimental scheme is shown in Table 1. Given that air flow plays an important role in the cleaning process, the fan outlet airflows at different fan speeds were measured with an AS866 hot-film anemometer (SMART SENSOR, Hong Kong, China) and the airflows corresponding to the six fan speed values were 7.9, 9.4, 10.8, 12.7, 14.3, and 16.8 m/s.

Table 1. Experimental scheme design of the single-factor experiment.

Numbers	Factors	Values	Condition
1–5	Fan speed (FS) (rad/s)	73.2, 83.7, 94.2, 104.7, 115.2, 125.6	VF = 5 Hz SO = 22 mm
6–10	Vibration frequency (VF) (Hz)	3, 4, 5, 6, 7, 8	FS = 104.7 rad/s SO = 22 mm
11–15	Screen opening (SO) (mm)	18, 20, 22, 24, 26, 28	FS = 104.7 rad/s VF = 5 Hz

2.3.2. Box–Behnken Experiment

A Box–Behnken design (BBD) with three factors and three levels was implemented to explore the interaction between factors [24,25]. The coded levels are shown in Table 2. The impurity ratio and loss ratio were used as evaluation indexes. As shown in Table 3, the experiment was performed with seventeen groups of trials including twelve combinations of factors and five replicates at the center point [26]. The feed rate of the corn-cleaning mixture was set to 5.5 kg/s.

Table 2. Levels of each experimental factor.

Levels	Fan Speed (rad/s)	Vibration Frequency (Hz)	Screen Opening (mm)
−1	94.2	5	20
0	104.7	6	22
1	115.2	7	24

Table 3. The experimental design and results.

	Factors			Evaluation Indexes	
Numbers	Fan Speed (rad/s)	Vibration Frequency (Hz)	Screen Opening (mm)	Impurity Ratio (%)	Loss Ratio (%)
1	−1	−1	0	1.39	0.69
2	1	−1	0	0.85	1.47
3	−1	1	0	0.94	0.71
4	1	1	0	0.79	1.24
5	−1	0	−1	1.05	0.64
6	1	0	−1	0.9	1.65
7	−1	0	1	1.32	0.82
8	1	0	1	1.03	1.19
9	0	−1	−1	0.99	1.03
10	0	1	−1	0.89	1.01
11	0	−1	1	1.25	1.08
12	0	1	1	0.9	0.82
13	0	0	0	0.73	0.73
14	0	0	0	0.85	0.66
15	0	0	0	0.83	0.69
16	0	0	0	0.75	0.73
17	0	0	0	0.87	0.74

After cleaning, the mixture was collected in the collection box, and the impurities therein were manually selected and weighed to calculate the impurity ratio. All discharged materials at the end of the tailing screen were gathered in a net bag, and the corn kernels therein were picked out and weighed to calculate the loss ratio. The impurity ratio and loss ratio were calculated by using the following equations [7]:

$$IR = \frac{m_i}{m_h} \times 100\%, \quad (1)$$

$$LR = \frac{m_l}{m_t} \times 100\%, \quad (2)$$

where m_i is the mass in kg of impurities in the collection box, m_h is the mass in kg of corn kernels in the collection box, m_l is the mass in kg of lost kernels, and m_t is the total mass in kg of the corn-cleaning mixture.

2.4. Data Analysis Method

In this study, the experimental results of the BBD were statistically analyzed using Design-Expert 2021 software (Stat-Ease Inc., Minneapolis, MN, USA). The response surface method (RSM) was applied to analyze the experimental data. Quadratic regression models were evaluated through the coefficient of determination (R^2) [27]. The significance of each factor for the experimental evaluation indexes was determined using the analysis of variance (ANOVA) [27], and the significance level was $p = 0.05$. Subsequently, response surface plots for interactions were generated using RSM. In addition, the optimal values of process parameters and the predicted IR and LR values were determined. The optimal parameters obtained from the regression analysis were further verified by experiments.

3. Results

In this section, the physical properties of materials are described, and the effects of process parameters on the cleaning performance are analyzed. Regression models and the optimal parameters combination are presented.

3.1. Physical Properties of Components

The physical properties of materials are an important basis for the design and determination of parameters for an air-screen cleaning system. After freezing, the physical properties of corn components change, which affects the cleaning performance [28]. Therefore, in order to provide a data reference, the physical properties of each component in the corn-cleaning mixture were measured. The corn-cleaning mixture included five crop components, namely, corn kernels, corn cobs, corn stalks, corn husks, and corn stigma, together with the noncrop component, ice. Five hundred grams corn kernels, 100 g corn cobs, 50 g corn stalks and 50 g corn husks were randomly selected from the mixture as a group of samples to determine the moisture contents. Samples were prepared in triplicate and were dried in a DZF–6050 thermostatic drying oven (Rongshida Electric Equipment Co., Ltd., Kunshan, China) at 105 ± 1 °C for 24 h [29]. The moisture contents of kernels, cobs, stalks, hulks, and stigma, measured on a wet basis method, were 24.8 ± 0.19%, 54.5 ± 0.22%, 68.3 ± 0.15%, 26.6 ± 0.17%, and 25.7 ± 0.11%, respectively.

The shapes of corn kernels were horse-toothed, conical, and spherical [30]. Taking the horse-toothed kernel as an example, the characteristic dimensions comprise the upper width (W_1), bottom width (W_2), height (H_h), and thickness (T). The dimensions were measured by a digital caliper with 0.01 mm accuracy (Prokits Industries Co., Ltd., Shanghai, China). The characteristic dimensions of corn kernel in three shapes are shown in Table 4.

Table 4. Shapes and characteristic dimensions of corn kernels in three shapes.

Shape	Characteristic Dimensions	Value	Proportion (%)
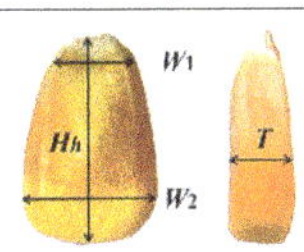 Horse-toothed kernel	W_1 (mm)	4.44–6.39	88.2
	W_2 (mm)	7.32–9.61	
	H_h (mm)	10.18–14.20	
	T (mm)	4.01–6.46	

Table 4. *Cont.*

Shape	Characteristic Dimensions	Value	Proportion (%)
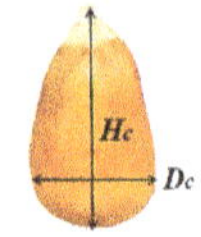	D_c (mm)	4.65–6.23	7.5
Conical kernel	H_c (mm)	9.26–13.65	
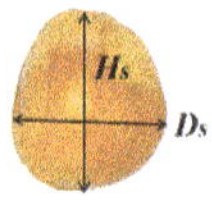	D_s (mm)	3.98–6.27	4.3
Spherical kernel	H_s (mm)	3.65–6.51	

The length (L_c), radius (R_c), and broken angle (α) were determined as the characteristic dimensions of corn cobs. The length (L_s) and radius (R_s) were determined as the characteristic dimensions of corn stalks. The dimensions were measured with the digital caliper and an angle gauge with 0.01° accuracy (Dongmei Instruments Ltd., Shenzhen, China). The characteristic sizes and the measurement results are listed in Tables 5 and 6.

Table 5. Characteristic dimensions of corn cobs.

	Characteristic Dimensions	Value	Proportion (%)
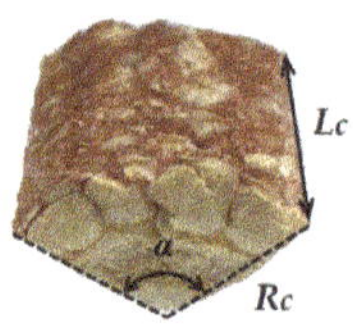	R_c (mm)	10–20	38.1
		20–30	36.4
		30–40	15.4
		40–50	5.5
		50–60	2.7
		60–70	1.9
	L_c (mm)	8–10	22.2
		10–12	37.8
		12–14	26.4
		14–16	4.5
Corn cob	α (°)	0–90	51.2
		90–180	29.0
		180–270	9.1
		270–360	10.7

Table 6. Characteristic dimensions of corn stalks.

	Characteristic Dimensions	Value	Proportion (%)
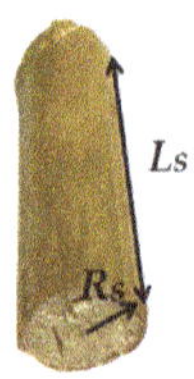	L_s (mm)	10–20	11.9
		20–30	38.5
		30–40	31.7
		40–50	12.8
		50–60	5.1
	R_s (mm)	5–6	6.9
Corn stalk		6–7	44.2
		7–8	35.7
		8–9	13.2

The coefficient of static friction was measured by applying a self-made test platform (Figure 2). A corn kernel was placed on the wall; the slope was then lifted by adjusting the screw and was stopped just as the kernel began to slide [31]. The angle displayed on the digital protractor was recorded. When the coefficient of friction between kernels was measured, a corn kernel was placed on the wall that was glued with kernels, and then the previous step was repeated. The coefficient of static friction can be calculated by Equation (3):

$$\mu = \tan\theta \quad (3)$$

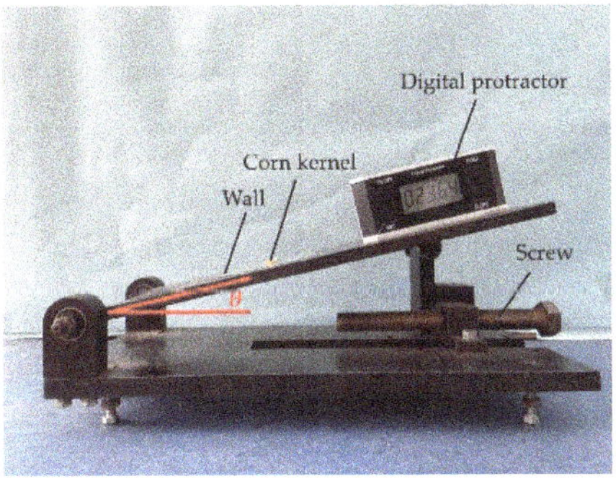

Figure 2. Platform to test the static friction coefficient.

The coefficient of rolling friction was measured by using the slope method [32]. As shown in Figure 3, a corn stalk was released at no initial velocity at the top of the slope and rolled on the horizontal surface until static. The vertical distance h and rolling distance d were measured, and coefficient of rolling friction was calculated using Equation (4):

$$f = \frac{h}{d} \quad (4)$$

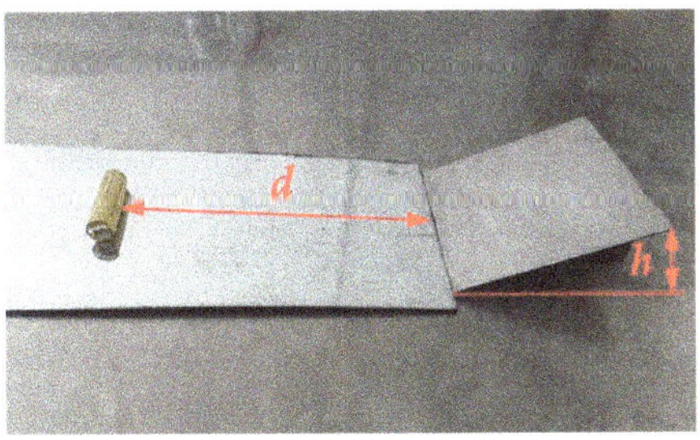

Figure 3. Platform to test the roll friction coefficient.

The coefficient of restitution was measured by the self-made restitution coefficient test platform (Figure 4). We used a dropping test to determine the coefficient of restitution between each component and screen [33]. A corn kernel was released from the position of height H_0 and collided with the horizontal collision plane. A high-speed camera was used to record the rebound height of the kernel H_1 after the collision. The coefficient of restitution can be calculated by using Equation (5). We used a pendulum test to determine the coefficient of restitution between components [34]. As shown in Figure 4, two corn kernels were glued and connected with fishing line. The kernel on the right was released at the height H_0 with no initial velocity and collided with another kernel. After collision, kernels reached the heights H_1 and H_2. The coefficient of restitution was calculated by using Equation (6). The results of the physical properties characterizing the contact are shown in Table 7.

$$e_w = \sqrt{\frac{H_1}{H_0}} \quad (5)$$

$$e_p = \frac{\sqrt{H_2} - \sqrt{H_1}}{\sqrt{H_0}} \quad (6)$$

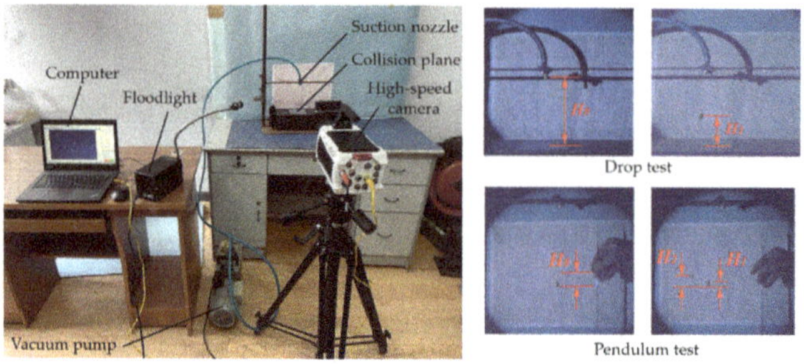

Figure 4. Platform to test the restitution coefficient.

Table 7. The physical properties of characterizing contact between components.

Property	Coefficient of Restitution	Coefficient of Static Friction	Coefficient of Rolling Friction
Corn kernel–corn kernel	0.37	0.36	0.04
Corn kernel–corn cob	0.28	0.62	0.02
Corn kernel–corn stalk	0.26	0.34	0.05
Corn kernel–screen	0.58	0.39	0.05
Corn cob–corn cob	0.25	0.78	0.02
Corn cob–corn stalk	0.24	0.39	0.04
Corn cob–screen	0.35	0.65	0.02
Corn stalk–corn stalk	0.23	0.38	0.06
Corn stalk–screen	0.30	0.34	0.05

When corn was harvested in the frozen state, some corn kernels after threshing presented the phenomenon of freeze-adhesion, as shown in Figure 5. Adhering kernels had difficulty penetrating the screen due to the overall increase in dimensions. Therefore, the characteristic dimensions of adhering kernels were measured, as shown in Table 8.

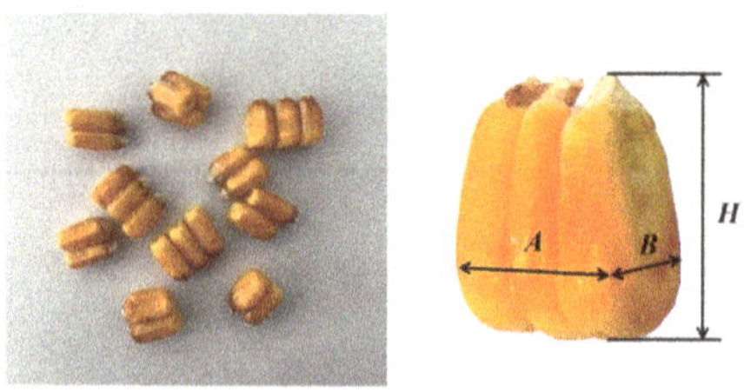

Figure 5. Adhering kernels and their characteristic dimensions.

Table 8. Types and characteristic dimensions of adhesive kernels.

Types	A (mm)	B (mm)	H (mm)
Two-kernel adhesion	6.9–8.8	7.3–9.6	12.2–14.5
Three-kernel adhesion	11.9–14.3	7.5–10.1	12.0–15.4
Four-kernel adhesion	15.5–17.2	7.2–10.5	11.8–15.9

3.2. Results of Single-Factor Experiment

3.2.1. Fan Speed

The nonlinear fitting curve in Figure 6 indicates the influence of fan speed on the impurity ratio and loss ratio. With the increase in fan speed, the IR decreased from 1.45% to 0.86%. Conversely, the LR increased from 0.75% to 1.42%. It can be seen in Figure 6 that the reduction in IR was more significant, ranging from 73.2 to 104.7 rad/s. The IR maintained a small decline from 104.7 to 125.6 rad/s. The LR increased slightly at first and then experienced a sharp increase when the fan speed exceeded 104.7 rad/s. As the fan speed increased, more impurities, such as corn cobs and husks. were blown out from the cleaning system, resulting in a decrease in the IR. However, the blowing effect of airflow on corn kernels was strengthened with the increase in the fan speed. Some kernels intermingled within impurities were blown out without being screened, which caused a large loss.

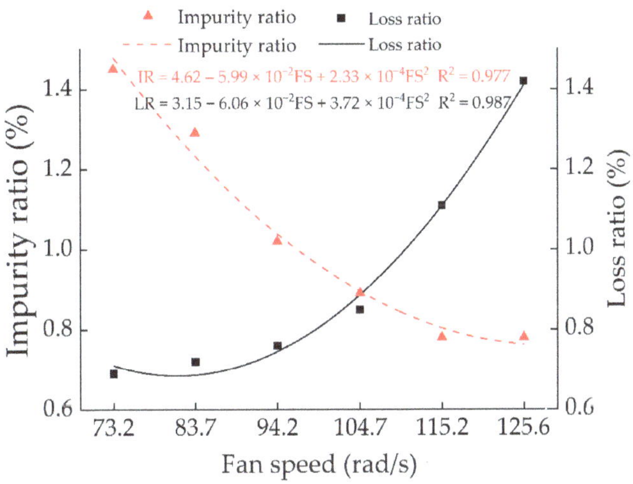

Figure 6. Influence of fan speed on the impurity ratio and loss ratio.

3.2.2. Vibration Frequency

The nonlinear fitting curve in Figure 7 indicates the influence of vibration frequency on the impurity ratio and loss ratio. With the increase in the vibration frequency, the IR decreased from 1.44% to 0.80%. Conversely, the LR increased from 0.79% to 1.18%. In particular, the IR decreased mainly in the range of 3–6 Hz, and it leveled off within the range of 6–8 Hz. The rising tendency of the LR caused by the vibration frequency was similar to that caused by the fan speed. The LR increased significantly when the vibration frequency exceeded 6 Hz. The increase in vibration frequency facilitated layering and dispersal of material from the cleaning mixture on the screen surface, which was conducive to kernel penetration. The airflow on the screen surface was evenly distributed, which facilitated the backward movement of impurities. Therefore, the IR continuously decreased. However, the kernel loss increased as the vibration frequency increased. The reason for this is that the screening time of the cleaning-mixture decreased as the vibration frequency increased. Some corn kernels were discharged from the chaff screen before passing through it.

3.2.3. Screen Opening

The nonlinear fitting curve in Figure 8 indicates the influence of the screen opening on the impurity ratio and the loss ratio. With the increase of the screen opening, the LR decreased from 1.39% to 0.72%. Conversely, the IR increased from 0.73% to 1.42%. It can be seen that the IR increased slightly at first. After reaching 22 mm, the rise in the IR became steep. The LR showed a sharp downward trend from 18 to 24 mm and a slow downward trend from 24 to 28 mm. As the screen opening increased, corn kernels were more likely to pass through the screen rather than exit the cleaning system, explaining the decline in the LR. With a small screen opening, a large number of corn stalks and corn cobs were unable to penetrate the screen due to the size limitation. When the size of the screen opening increased until it exceeded that of impurities, the impurities passed through the screen and entered the collection box, which caused an increase in the IR.

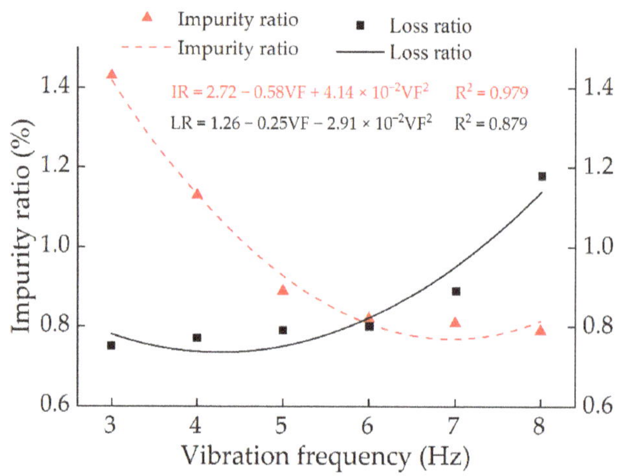

Figure 7. Influence of vibration frequency on the impurity ratio and loss ratio.

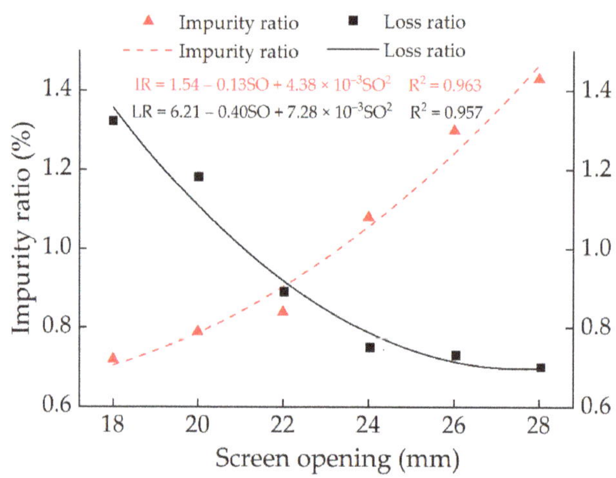

Figure 8. The influence of screen opening on the impurity ratio and the loss ratio.

3.3. Analysis of Variance

The results of ANOVA on the impurity ratio are shown in Table 9. In the ANOVA results, a *p*-value of less than 0.05 indicates that the model term had a significant influence. The *p*-value of this model was less than 0.001, which indicates that the regression model of the IR was significant. For the linear terms, the results clearly show that FS, VF, and SO all had a significant influence on the IR. Moreover, FS was the most significant factor, with an F-value of 43.15, and VF was the second most significant. For the interaction terms, only FS–VF had a significant influence on the IR. Moreover, the quadratic terms $(FS)^2$ and $(SO)^2$

also had a significant influence on the IR. Therefore, after excluding nonsignificant terms, the regression model of the IR is shown as follows:

$$IR = 31.834 - 0.274FS - 1.119VF - 1.158SO + 9.286 \times 10^{-3} FS \times VF + 1.152 \times 10^{-3} (FS)^2 + 3.5500 \times 10^{-2} (SO)^2 \quad (7)$$

Table 9. ANOVA of the impurity ratio.

Cause of Variance	Sum of Squares	Freedom	Mean Square	F-Value	*p*-Value	Significant
Model	0.57	9	0.064	17.25	0.0006	*
FS	0.16	1	0.16	43.15	0.0003	*
VF	0.12	1	0.12	31.14	0.0008	*
SO	0.056	1	0.056	15.17	0.0059	*
FS–VF	0.038	1	0.038	10.28	0.0149	*
FS–SO	0.0049	1	0.0059	1.32	0.2876	
VF–SO	0.016	1	0.016	4.22	0.0789	
$(FS)^2$	0.068	1	0.068	18.36	0.0036	*
$(VF)^2$	0.015	1	0.015	4.03	0.0847	
$(SO)^2$	0.085	1	0.085	22.95	0.0020	*
Residual	0.026	7	0.003699			
Lack of Fit	0.010	3	0.003458	0.89	0.5186	
Pure Error	0.016	4	0.003.88			
Total	0.60	16				

* Significant ($p < 0.05$)

The results of ANOVA on the loss ratio are shown in Table 10. The *p*-value of this model was less than 0.0001, implying that the regression model of the LR was extremely significant. For the linear terms, it was clear that FS, VF, and SO all had a significant influence on changing the LR. Moreover, FS was the most significant factor. Furthermore, the F-value of FS far outweighed that of VF and SO. For the interaction terms, only FS–VF had a significant influence on changing the LR. Moreover, the quadratic terms of $(FS)^2$, $(VF)^2$, and $(SO)^2$ had a significant influence on changing the LR. Therefore, the regression model of the LR after excluding the nonsignificant items is shown as follows:

$$LR = 34.508 - 0.317FS - 2.787VF - 0.935SO + 6.429 \times 10^{-3} FS \times VF + 1.73 \times 10^{-3} (FS)^2 + 0.176 (VF)^2 + 2.594 \times 10^{-2} (SO)^2 \quad (8)$$

Table 10. ANOVA for the loss ratio.

Cause of Variance	Sum of Squares	Freedom	Mean Square	F-Value	*p*-Value	Significant
Model	1.69	9	0.19	58.86	<0.0001	*
FS	1.24	1	1.24	388.90	<0.0001	*
VF	0.023	1	0.023	7.25	0.0310	*
SO	0.034	1	0.034	10.60	0.0140	*
FS–VF	0.018	1	0.018	5.71	0.0481	*
FS–SO	0.0081	1	0.0081	2.54	0.1550	
VF–SO	0.0004	1	0.0004	0.031	0.8645	
$(FS)^2$	0.15	1	0.15	48.29	0.0002	*
$(VF)^2$	0.13	1	0.13	41.01	0.0004	*
$(SO)^2$	0.045	1	0.045	14.21	0.0070	*
Residual	0.022	7	0.003189	58.86		
Lack of Fit	0.018	3	0.005908		0.0739	
Pure Error	0.0046	4	0.00115			
Total	1.71	16				

* Significant ($p < 0.05$).

3.4. Response Surface Analysis

The response surface plots shown in Figure 9 depict the effects of interactions between FS, VF, and SO on the IR. The IR presents a similar trend in Figure 9a–c; it first decreases and then increases with the increase in FS, VF, and SO. However, comparing the curve gradient of the response surface shows that the changing trend shown in Figure 9a is the strongest. The *p*-value of FS–VF listed in Table 9 is greater than those of FS–SO and VF–SO. Therefore, the interaction of FS–VF for the IR had the most significant effect on the IR.

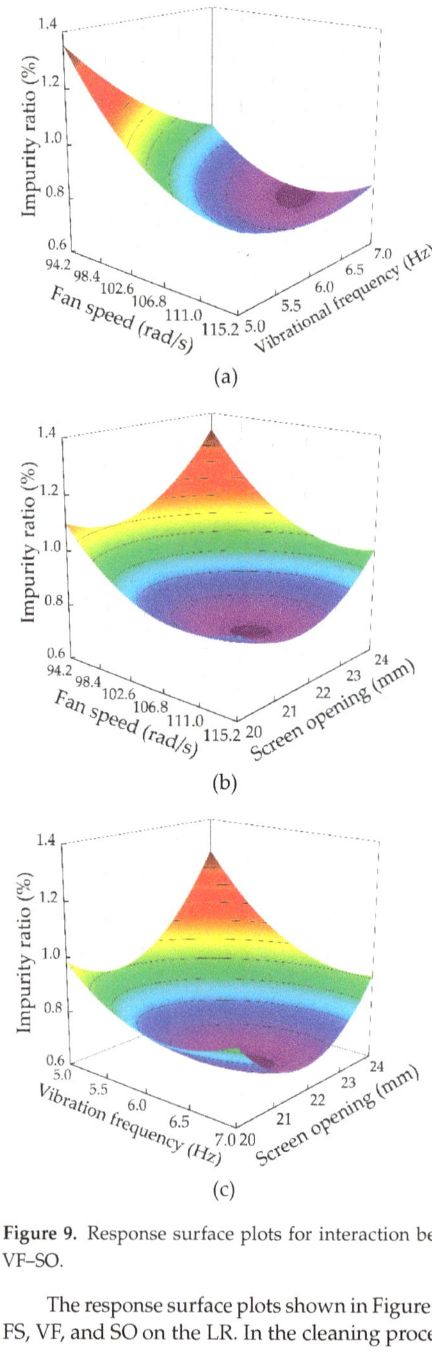

Figure 9. Response surface plots for interaction between the pairs (**a**) FS–VF, (**b**) FS–SO, and (**c**) VF–SO.

The response surface plots shown in Figure 10 depict the effects of interactions between FS, VF, and SO on the LR. In the cleaning process, the LR significantly increased with the increase in FS. The enhancement of the LR can be explained by the stronger carrying effect of the airflow on the corn kernels. The increase in VF led to a corresponding decline in the LR when the VF was low. However, a further increase in VF resulted in a rise of the LR. The trend of the LR caused by the increase of SO was similar to that of VF. Comparing the curve gradient of the response surfaces in Figure 10 and the *p*-values in Table 10 confirms that the interaction of FS–VF had the most significant effect on the LR.

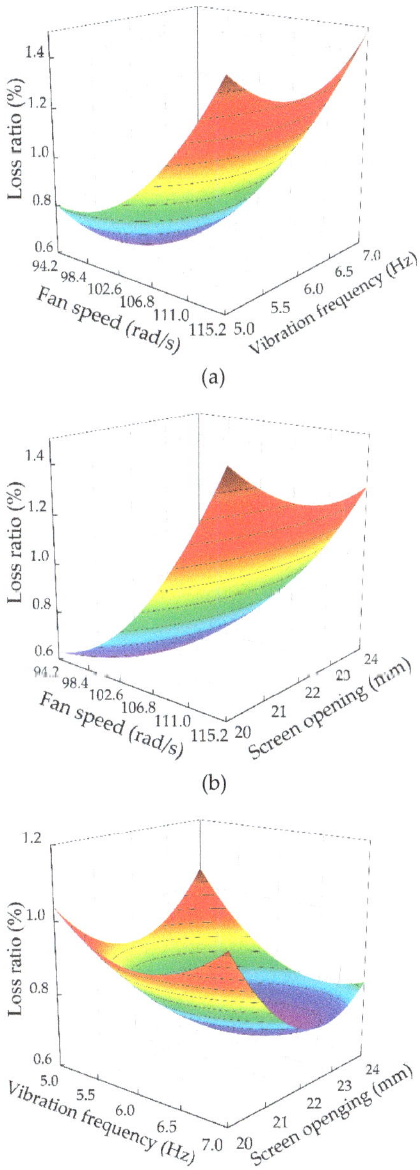

Figure 10. Response surface plots for interaction between the pairs (**a**) FS–VF, (**b**) FS–SO, and (**c**) VF–SO.

3.5. Optimization and Verification

The analysis above shows that the IR and the LR varied inversely with the variation in process parameters. Response surface optimization [35,36] was carried out. The aim of optimization was to obtain the optimal combination of FS, VF, and SO to simultaneously minimize the IR and LR in the process of frozen corn cleaning. Therefore, the response values of the IR and LR were taken as the minimum values. The ranges of FS, VF, and SO were constrained to 94.2–115.2 rad/s, 5–7 Hz, and 20–24 mm, respectively. The optimal combination was obtained using Design-Expert 2021 software. The IR and LR were both minimized when the FS was 102.7 rad/s, the VF was 6.42 Hz, the SO was 21.9 mm. The predicted values of IR and LR were 0.80% and 0.61%, respectively. Three validation experiments were then carried out under the optimal parameters. As shown in Table 11, the experimental results were highly consistent with the predicted values, with a relative error of less than 5%. These results indicate that the optimization model is reliable in the frozen corn cleaning operation.

Table 11. Results of validation experiments and predicted values.

	1	2	3	Mean	Prediction	Relative Error (%)
IR (%)	0.74	0.82	0.76	0.77	0.80	3.89
LR (%)	0.59	0.66	0.68	0.64	0.61	4.69

4. Discussion

4.1. Physical Properties

When frozen corn is harvested, the ice contained in the mixture causes the phenomenon of corn freeze-adhesion. According to statistical data, adhering kernels accounted for 1% of the corn-cleaning mixture. Amongst them, there were 74% cases with two-kernel adhesion, 24% cases with three-kernel adhesion, and 2% with four-kernel adhesion. Some physical properties of components were determined, including characteristic dimensions, coefficient of restitution, coefficient of static friction, and coefficient of rolling friction. Compared with literature reports [13,30,31], no notable differences in characteristic dimensions of nonadhesive corn kernels, corn cobs and corn stalks were observed. However, the proportion of corn cobs decreased by 1.2% because in the frozen state, corn cobs are difficult to break when corn ears are threshed. Most corn cobs with full or half lengths were discharged at the end of the threshing drum and did not enter the cleaning system. The coefficient of restitution, coefficient of static friction, and coefficient of rolling friction of frozen corn were slightly higher than those of nonfrozen corn, which is similar to the results reported in the literature [33].

4.2. Single-Factor Experiment

With the increase in the fan speed, the impurity ratio continuously decreased while the loss ratio continuously increased. When the fan speed was 73.2 rad/s, the minimum loss ratio occurred, which was 0.69%; however, the maximum impurity ratio of 1.45% was observed. When the fan speed was 125.6 rad/s, the impurity ratio of 0.78% was the lowest, while the loss ratio peaked at 1.42%. For cleaning performance, both the low impurity ratio and loss ratio should be taken into account. Therefore, as presented in Figure 6, the fan speed interval 94.2–115.2 rad/s was superior, ensuring both a lower impurity ratio and loss ratio. Similarly, as shown in Figures 7 and 8, the superior interval of vibration frequency and screen opening were 5–7 Hz and 20–24 mm, respectively. The results of the single-factor experiment provided the data support that selected the levels of each experimental factor in the Box–Behnken experiment.

4.3. Response Surface Analysis

The ANOVA results for response surface demonstrated that fan speed was the most important factor for affecting both the impurity ratio and loss ratio. The response surface plots also showed that the change of impurity ratio with fan speed (Figure 9a,b) was stronger than those with vibrational frequency (Figure 9a,c) and screen opening (Figure 9b,c). This changing tendency was the same for the loss ratio. For the interaction terms, the function was realized by synergism between FS and VF, which led to a decrease in impurity ratio and an increase in loss ratio. The optimal combination was fan speed 102.7 rad/s, vibration frequency 6.42 Hz, and screen opening 21.9 mm. Correspondingly, the impurity ratio was 0.80% and loss ratio was 0.61%. Results of validation tests were a 0.77% impurity ratio and a 0.64% loss ratio, which meet the requirements of Chinese National Standard GB/T 21962-2020 for the impurity ratio and loss ratio in corn combine harvesters (IR $\leq$ 2%, LR $\leq$ 2%).

4.4. Cleaning Process Parameters between Frozen and Nonfrozen Corn

At present, no studies have reported on cleaning process parameters for frozen corn. In contrast to harvesting at above-zero temperature, there were certain differences in the parameter values. In literature reports [12,13,35], the values of fan speed covered a range from 78.5 to 94.2 rad/s, and the values of vibrational frequency covered a range from 4.45 to 6 Hz. In this study, the optimal values of the fan speed and vibration frequency were 102.7 rad/s and 6.42 Hz, which were higher than optimum values for harvesting above zero. This is because of the physical properties of crop components and ice in the mixture. Firstly, the increase of static friction coefficient reduced the moving velocity of the mixture. Secondly, as the screen slices of chaffer screen were inclined upwards, this caused rebound forward of mixture to occur frequently when the screen was in reciprocating movement. The increase of coefficient of restitution enabled the backward motion velocity of the mixture increase, but it also meant that the forward rebound velocity of the mixture was increased. Thus, these physical properties affected the movement of the mixture. Thirdly, frozen kernel adhesion weakened the fluidity of the mixture as small pieces of ice were present. In particular, this effect was exacerbated by ice attaching to the oscillating plate and screen surface. Hence, the fan speed and vibrational frequency needed to be larger to promote the flow of the mixture. Additionally, there was not much difference in screen opening values.

5. Conclusions

In this study, the process parameters of an air-screen cleaning system for frozen corn were optimized by using the response surface method (RSM). The physical properties of materials were measured. The influences of fan speed, vibration frequency, and screen opening on evaluation indexes were analyzed. The single-factor experiment results indicate that the impurity ratio was negatively correlated with fan speed and vibration frequency, and it was positively correlated with screen opening. The loss ratio was positively correlated with fan speed and vibration frequency, and it was negatively correlated with screen opening. The optimal combination of process parameters was a fan speed of 102.7 rad/s, vibrational frequency of 6.42 Hz, and screen opening of 21.9 mm. Under this condition, the impurity ratio was 0.80% and the loss ratio was 0.61%. Compared to the results of validation experiments, the relative error of the predicted values was less than 5%, proving the reliability of the regression models that were determined. This study provided a theoretical basis for the process of optimizing air-screen cleaning systems for frozen corn.

It is worth noting that this study has some limitations. First, the influence of the variation in moisture content was not explored. The moisture content influences the physical properties of corn and then affects the parameter ranges. Second, in the study, we only investigated the cleaning of a single corn cultivar. The differences between corn varieties in frozen state merit further study.

Therefore, further research is needed to address these limitations. The effect of moisture content as a factor on cleaning performance will be investigated. Additional corn varieties require consideration. More types of screen and hole sizes in the screen will be selected. Based on the test results, a new cleaning system will be designed to improve higher cleaning performance for frozen corn harvesting.

Author Contributions: Conceptualization, J.F.; methodology, J.F.; validation, N.Z.; investigation, N.Z.; resources, J.F.; writing—original draft preparation, N.Z.; writing—review and editing, J.F.; supervision, Z.C., X.C., and L.R.; project administration, J.F.; funding acquisition, J.F. All authors have read and agreed to the published version of the manuscript.

Funding: This research was funded by the International Science and Technology Cooperation Project of Science and Technology Development Program of Jilin Province, grant number 20190701055GH.

Institutional Review Board Statement: Not applicable.

Informed Consent Statement: Not applicable.

Data Availability Statement: Not applicable.

Acknowledgments: The authors are grateful for the frozen corn provided by the Experimental Base of Agriculture of Jilin University.

Conflicts of Interest: The authors declare no conflict of interest.

References

1. Chen, S.; Gong, B. Response and adaptation of agriculture to climate change: Evidence from China. *J. Dev. Econ.* **2021**, *148*, 102557. [CrossRef]
2. Qian, F.; Yang, J.; Torres, D. Comparison of corn production costs in China, the US and Brazil and its implications. *Agric. Sci. Technol.* **2016**, *17*, 731–736.
3. Hou, L.; Wang, K.; Wang, Y.; Li, L.; Ming, B.; Xie, R.; Li, S. In-field harvest loss of mechanically-harvested maize grain and affecting factors in China. *Int. J. Agric. Biol. Eng.* **2021**, *14*, 29–37.
4. Wang, K.; Xie, R.; Ming, B.; Hou, P.; Xue, J.; Li, S. Review of combine harvester losses for maize and influencing factors. *Int. J. Agric. Biol. Eng.* **2021**, *14*, 1–10.
5. Fu, J.; Yuan, H.; Zhang, D.; Chen, Z.; Ren, L. Multi-Objective Optimization of Process Parameters of Longitudinal Axial Threshing Cylinder for Frozen Corn Using RSM and NSGA-II. *Appl. Sci.* **2020**, *10*, 1646. [CrossRef]
6. Yang, L.; Cui, T.; Qu, Z.; Li, K.; Yin, X.; Han, D.; Yan, B.X.; Zhao, D.; Zhang, D. Development and application of mechanized maize harvesters. *Int. J. Agric. Biol. Eng.* **2016**, *9*, 15–28.
7. Wang, L.; Yu, Y.; Ma, Y.; Feng, X.; Liu, T. Investigation of the Performance of Different Cleaning Devices in Maize Grain Harvesters Based on Field Tests. *Trans. ASABE* **2020**, *63*, 809–821. [CrossRef]
8. Li, H.; Li, Y.; Gao, F.; Zhao, Z.; Xu, L. CFD–DEM simulation of material motion in air-and-screen cleaning device. *Comput. Electron. Agric.* **2012**, *88*, 111–119. [CrossRef]
9. Badretdinov, I.; Mudarisov, S.; Lukmanov, R.; Permyakov, V.; Ibragimov, R.; Nasyrov, R. Mathematical modeling and research of the work of the grain combine harvester cleaning system. *Comput. Electron. Agric.* **2019**, *165*, 104966. [CrossRef]
10. Ueka, Y.; Matsui, M.; Inoue, E.; Mori, K.; Okayasu, T.; Mitsuoka, M. Turbulent flow characteristics of the cleaning wind in combine harvester. *Eng. Agric. Environ. Food* **2012**, *5*, 102–106. [CrossRef]
11. Gebrehiwot, M.G.; De Baerdemaeker, J.; Baelmans, M. Effect of a cross-flow opening on the performance of a centrifugal fan in a combine harvester: Computational and experimental study. *Biosyst. Eng.* **2010**, *105*, 247–256. [CrossRef]
12. Wang, L.; Wu, Z.; Feng, X.; Li, R.; Yu, Y. Design and experiment of curved screen for maize grain harvester. *Trans. Chin. Soc. Agric. Mach.* **2019**, *50*, 90–101.
13. Wang, L.; Feng, X.; Zheng, Z.; Yu, Y.; Liu, T.; Ma, Y. Design and test of combined sieve of maize screening. *Trans. Chin. Soc. Agric. Mach.* **2019**, *50*, 104–113.
14. Sabashkin, V.A.; Sukhoparov, A.A.; Sinitsyn, V.A.; Zakharov, S.E. Removing straw impurities from grain heaps by cylindrical sieve. *Sib. Her. Agric. Sci.* **2017**, *47*, 80–87. [CrossRef]
15. Ivanov, N.M.; Fedorenko, I.Y.; Zakharov, S.E.; Sukhoparov, A.A. Evaluating grain feed at separation by planetary cylindrical sieve with round holes. *Sib. Her. Agric. Sci.* **2017**, *47*, 72–79. [CrossRef]
16. Krzysiak, Z.; Samociuk, W.; Zarajczyk, J.; Kaliniewicz, Z.; Pieniak, D.; Bogucki, M. Analysis of the sieve unit inclination angle in the cleaning process of oat grain in a rotary cleaning device. *Processes* **2020**, *8*, 346. [CrossRef]
17. Cheng, C.; Fu, J.; Hao, F.; Chen, Z.; Zhou, D.; Ren, L. Effect of motion parameters of cleaning screen on corn cob blocking law. *J. Jilin Univ.* **2021**, *51*, 761–771.
18. Xu, L.; Wei, C.; Liang, Z.; Chai, X.; Li, Y.; Liu, Q. Development of rapeseed cleaning loss monitoring system and experiments in a combine harvester. *Biosyst. Eng.* **2019**, *178*, 118–130. [CrossRef]
19. Craessaerts, G.; Saeys, W.; Missotten, B.; De Baerdemaeker, J. A genetic input selection methodology for identification of the cleaning process on a combine harvester, Part I: Selection of relevant input variables for identification of the sieve losses. *Biosyst. Eng.* **2007**, *98*, 166–175. [CrossRef]
20. Craessaerts, G.; de Baerdemaeker, J.; Missotten, B.; Saeys, W. Fuzzy control of the cleaning process on a combine harvester. *Biosyst. Eng.* **2010**, *106*, 103–111. [CrossRef]

21. Li, Y.; Xu, L.; Zhou, Y.; Li, B.; Liang, Z.; Li, Y. Effects of throughput and operating parameters on cleaning performance in air-and-screen cleaning unit: A computational and experimental study. *Comput. Electron. Agric.* **2018**, *152*, 141–148. [CrossRef]
22. Yuan, J.; Li, H.; Qi, X.; Hu, T.; Bai, M.; Wang, Y. Optimization of airflow cylinder sieve for threshed rice separation usin1g CFD-DEM. *Eng. Appl. Comput. Fluid Mech.* **2020**, *14*, 871–881.
23. Wang, Z.; Liu, C.; Wu, J.; Jiang, H.; Zhao, Y. Impact of screening coals on screen surface and multi-index optimization for coal cleaning production. *J. Clean. Prod.* **2018**, *187*, 562–575. [CrossRef]
24. Zendehboudi, A.; Li, X. Desiccant-wheel optimization via response surface methodology and multi-objective genetic algorithm. *Energy Convers. Manag.* **2018**, *174*, 649–660. [CrossRef]
25. Sun, X.; Kim, S.; Yang, S.D.; Kim, H.S.; Yoon, J.Y. Multi-objective optimization of a Stairmand cyclone separator using response surface methodology and computational fluid dynamics. *Powder Technol.* **2017**, *320*, 51–65. [CrossRef]
26. Ren, L.Q. *Design of Experiment and Optimiz-Ation*; Higher Education Press: Beijing, China, 2009; pp. 246–257.
27. Kim, I.; Ha, J.-H.; Jeong, Y. Optimization of Extraction conditions for antioxidant activity of acer tegmentosum using response surface methodology. *Appl. Sci.* **2021**, *11*, 1134. [CrossRef]
28. Babić, L., Radojèin, M.; Pavkov, I., Babić, M., Turan, J., Zoranović, M., Stanišić, S. Physical properties and compression loading behaviour of corn seed. *Int. Agrophys.* **2013**, *27*, 119–126. [CrossRef]
29. Selvam, T.A.; Manikantan, M.R.; Chand, T.; Sharma, R.; Seerangurayar, T. Compression loading behaviour of sunflower seeds and kernels. *Int. Agrophys.* **2014**, *28*, 543–548. [CrossRef]
30. Chen, Z.; Yu, J.; Xue, D.; Wang, Y.; Zhang, Q.; Ren, L. An approach to and validation of maize-seed-assembly modelling based on the discrete element method. *Powder Technol.* **2018**, *328*, 167–183. [CrossRef]
31. Castiglioni, C.A.; Drei, A.; Carydis, P.; Mouzakis, H. Experimental assessment of static friction between pallet and beams in racking systems. *J. Build. Eng.* **2016**, *6*, 203–214. [CrossRef]
32. González-Montellano, C.; Fuentes, J.M.; Ayuga-Téllez, E.; Ayuga, F. Determination of the mechanical properties of maize grains and olives required for use in DEM simulations. *J. Food Eng.* **2012**, *111*, 553–562. [CrossRef]
33. Wang, L.; Wu, B.; Wu, Z.; Li, R.; Feng, X. Experimental determination of the coefficient of restitution of particle-particle collision for frozen maize grains. *Powder Technol.* **2018**, *338*, 263–273. [CrossRef]
34. Sharma, R.K.; Bilanski, W.K. Coefficient of restitution of grains. *Trans. ASAE* **1971**, *14*, 216–218.
35. Cheng, C.; Fu, J.; Chen, Z.; Hao, F.; Cui, S.; Ren, L. Optimization experiment on cleaning device parameters of corn kernel harvester. *Trans. Chin. Soc. Agric. Mach.* **2019**, *50*, 151–158.
36. Park, C.I.L. Multi-objective optimization of the tooth surface in helical gears using design of experiment and the response surface method. *J. Mech. Sci. Technol.* **2010**, *24*, 823–829. [CrossRef]

agriculture

Article

Optimization of Structure Parameters of the Grouser Shoes for Adhesion Reduction under Black Soil

Jun Fu [1,2,*], Jian Li [1,2], Xinlong Tang [3], Ruixue Wang [4] and Zhi Chen [1,4]

[1] Key Laboratory of Bionic Engineering, Ministry of Education, Jilin University, Changchun 130025, China; jianli20@mails.jlu.edu.cn (J.L.); chenzhi@caamm.org.cn (Z.C.)
[2] College of Biological and Agricultural Engineering, Jilin University, Changchun 130025, China
[3] Agricultural Experimental Base, Jilin University, Changchun 130062, China; tangxl@jlu.edu.cn
[4] Chinese Academy of Agricultural Mechanization Sciences, Beijing 100083, China; wangruixue@caams.org.cn
* Correspondence: fu_jun@jlu.edu.cn; Tel.: +86-043-1850-9576-0609

Abstract: The black soil of Northeast China has a strong adhesion ability because of its unique physical properties, and the soil often adheres to the surface of the grouser shoes of tracked vehicles during the operation. The adhesion performance depends largely on structure parameters of the grouser shoes. The grouser height, the grouser thickness, and the grouser splayed angle were selected as structure parameters. The adhesion force and adhesive soil mass were selected as indicators of adhesion performance. Therefore, the mathematical model between structure parameters and response indicators was established by the response surface method (RSM). The optimal parameters combination was that the grouser height was 20 mm, grouser thickness was 6.34 mm, and grouser splayed angle was 40.45°. The average data of verification experiments occurred when the adhesion force reached 1.11 kPa and adhesive soil mass reached 22.68 g. Compared with the average value of un-optimized experiment results, the adhesion force and adhesive soil mass reduced by 12.84% and 4.63%, respectively. The relative error of the predicted values and measured values was less than 5%, proving the reliability of the regression models. This study could provide a reference method for parameters optimization, and a new structure of the grouser shoes of tracked vehicles will be designed to reduce adhesion.

Keywords: soil adhesion; grouser shoe; structure parameters; optimization; black soil; RSM

Citation: Fu, J.; Li, J.; Tang, X.; Wang, R.; Chen, Z. Optimization of Structure Parameters of the Grouser Shoes for Adhesion Reduction under Black Soil. *Agriculture* **2021**, *11*, 795. https://doi.org/10.3390/agriculture11080795

Academic Editor: José Pérez-Alonso

Received: 8 August 2021
Accepted: 18 August 2021
Published: 20 August 2021

1. Introduction

Black soil is mainly distributed in the Northeast China, with high moisture content, low soil porosity, stickiness, and easy agglomeration [1,2]. After the rainy season, the moisture content often exceeds 20% which is close to rheological properties, therefore, the adhesion performance is very prominent [3,4]. Because of the strong adhesion, the phenomenon of soil adhesion frequently occurs between the soil and the grouser shoes of tracked vehicles, which significantly increases the energy consumption and the working resistance and has greatly affected the production efficiency [5–7].

In order to reduce the undesirable influence of the soil adhesion, various studies have been carried out in the past few years. Massah et al. [8] designed a bionic electro-osmotic soil-engaging plate to investigate the effect of several design and functional parameters on soil adhesion reduction during a simulated tillage operation. The results showed that using the bionic electro-osmosis technology, plates' soil adhesion reduced by 29.8%. Barzegar et al. [9] applied the ultra-high molecular weight polyethylene (UHMW-PE) to the surface of tine compared with the uncoated tine in the heavy clay soil to evaluate the soil-tool adhesion of the UHMW-PE-coated tine. The soil did not adhere to the UHMW-PE plastic tine while moving through the heavy clay soil. The results presented the self-scouring ability and low adhesion characteristics of UHMW-PE. Marani et al. [10] studied the effect of Nano-coating materials on reduction of soil adhesion and external friction and developed a precise measuring system to measure the adhesion and friction coefficient of soil and metal. Nano-coated plates exhibited a much lower friction angle and less adhesion than steel due to hydrophobic surfaces and roughness in the nanoscale. Ucgul et al. [11] accounted for the soil adhesion and cohesion by using the hysteretic spring contact model to establish the three-dimensional discrete element modelling (DEM) of tillage. Results of modelling the direct shear test showed that the DEM could predict tillage draft and vertical tillage forces in a range of soil and operating conditions. Zenkov et al. [12,13] analyzed the temperature–time characteristics of three self-regulating heating elements and presented the thermal effect as a method to combat ground-and-machine part adhesion. For reducing the adhesion of soils to working bodies of machines, an intermediate layer at the interface between the phases of the system was created, similar to a screen for intermolecular interaction forces of phases, which has a lubricating effect. Ren et al. [14–16], based on the principles of bionics, used the electro-osmosis method to reduce adhesion of soil against loading shovel and designed and tested the pole plates of the non-smooth surfaces with different dimensions. The results showed that the adhesion of soil to a shovel was substantially reduced at a low electro-osmosis voltage. Meanwhile, theoretical analysis was carried out to reveal the mechanism of resistance reduction/anti-adhesion. Tong et al. [17] studied the geometrical features and wettability of dung beetles and discovered that the head was similar to bulldozing blades, and the forelegs have a tooth-like structure with strong burrowing ability. Wang et al. [18] used the mild steel plate as a model to study on the use of vibration to reduce soil adhesion. The vibration was produced by blowing compressed air, and the frequencies between 60 Hz and 100 Hz were significantly effective in removing soil from the blade during operation. The study explored a new mechanism to prevent soil adhesion to machines. Jia et al. [19] analyzed the characteristic of earthworm motion and designed the anti-adhesion structure of a press roller, which revealed the anti-adhesion mechanism of the rubber bulge structure. Field experiments showed that the press roller with rubber bulge structure had a lower adhesive soil mass than the conventional press roller. Sun et al. [20] extracted and analyzed the features of the bear claws and designed a new soil plowing device for resistance reduction/anti-adhesion using the discrete element method (DEM) simulation of the bear-claw bionic ditcher to verify the accuracy of the model and obtained the motion characteristics. Zhang et al. [21] and Sun et al. [22] summarized the biological structures of some soil-burrowing animals by designing and developing the biomimetic disc furrow opener and other tillage implements, which presented the application of biomimetic for designing soil-engaging tillage implements in Northeast China. From the present research, it could be concluded that the methods of the biomimetic of bionic electro-osmotic, the UHMW-PE/Nano coating, the DEM simulation, the self-regulating heating effect, the mechanical vibration, and the biomimetic of soil-burrowing animals are commonly used to reduce the soil adhesion.

However, almost all researchers focused on the tractive performance of the grouser shoes of tracked vehicles [23–26]. Additionally, the soil types were mainly paddy soil, clay soil, and red soil that were distributed in the south. Therefore, there is a paucity of theory and method to estimate the effect of structure parameters of the grouser shoes of tracked vehicles on the adhesion performance under black soil in Northeast China. To address the above problem, the theoretical analysis based on the Bekker's sinkage model needs to be built to explore the impact factors of structure parameters of the grouser shoes on the adhesion force. Furthermore, the appropriate structure parameters (i.e., the grouser height, the grouser thickness, and the grouser splayed angle) and the response indicators of adhesion performance (i.e., the adhesion force and adhesive soil mass) also should be optimized synchronously.

The objectives of this study were to: (a) evaluate the adhesion performance of the grouser shoes with different structure parameters; (b) utilize the RSM to optimize the structure parameters of the grouser shoes of tracked vehicles for minimum soil adhesion; (c) establish the verification experiment in the soil bin to verify the performance of the optimal structure parameters on soil adhesion reduction.

2. Materials and Methods

2.1. Soil Sampling

The soil samples were collected from the Agricultural Experimental Base of Jilin University (N43°56′45.99″, E125°14′52.44″) in Changchun, Jilin province, Northeast China. In order to simulate the real subsidence amount of tracked vehicles working in the field, considering the previous crop of the test field was mainly wheat, the depth of the plow layer was between 250 mm and 300 mm. Therefore, we selected the characterization of the soil samples up to 300 mm from the surface for the adhesion test. The soil stress distribution was uniform under the compaction of the crawler walking mechanism [27]. The black soil physical properties of each sample are listed in Table 1. The test was repeated three times on the same condition to obtain the precise results. The average hardness of the black soil was 854 kPa, measured by a SC-900 soil hardness tester (Spectrum Technologies Inc., Aurora, IL, USA). The average bulk density of the black soil was a 1269 kg/m^3, measured by a TJSD-750-IV soil bulk density determination detector. The results showed that the bulk density was positively correlated with the hardness.

Table 1. Measured black soil physical properties.

Sample Depth (mm)	Bulk Density (kg·m^{-3})	Hardness (kPa)	Moisture Content (%)
0–100	1130	42	24.31
100–200	1352	972	21.28
200–300	1326	1165	19.87

The soil moisture content has an important effect on the adhesion force, and the mechanical properties of soil samples vary greatly under different moisture contents. The study mainly optimized different structure parameters of the grouser shoes for adhesion reduction. In order to minimize the influence of other variables on the study, we performed the experiments at only one soil moisture content. Additionally, we tested the plastic limit and liquid limit of soil samples by using the LP-100D digital display soil liquid-plastic limit combined tester (Kedaluqiao Electric Equipment Co., Ltd., Cangzhou, China). The plastic limit and the liquid limit of soil samples were 15.83% and 32.25%, respectively. The average moisture content of each layer of soil samples was dried in a DZF-6050 thermostatic drying oven (Rongshida Electric Equipment Co., Ltd., Kunshan, China) at 105 ± 5 °C for 6 h.

The soil moisture content for all the soil samples was in the range of 19.87–24.31%, and the average moisture content was 21.82% (d.b.) (i.e., dry basis). The results showed that the soil samples were higher than the plastic limit and lower than the liquid limit. Overall, the average moisture content of all the soil samples were within the plastic state, which had a strong adhesion to the grouser shoes of tracked vehicles. This critically restricted the efficiency of the tracked vehicles during the operation.

2.2. Vertical Stress Analysis of the Grouser Shoes

When the vertical load was applied on the grouser shoes, the entire grouser shoes sank in the black soil. The vertical stress analysis of grouser shoes is shown in Figure 1.

As is presented in Figure 1, the vertical forces of the grouser shoes were expressed by Equation (1):

$$W = P_1(L - d) + P_2 d \quad (d = \lambda L) \tag{1}$$

where W is vertical load, L is the length of grouser shoe, d is grouser thickness. P_1 is the contact stress of the top surface of the grouser, and P_2 is the stress at the bottom surface of the grouser shoe. According to the Beeker's sinkage formula [28,29], the stress was expressed as follows:

$$P_1 = \left(\frac{K_c}{b} + K_\varphi\right) \times Z^n \tag{2}$$

$$P_2 = \left(\frac{K_c}{b} + K_\varphi\right) \times (Z + h)^n \tag{3}$$

where k_c is soil cohesive modulus, k_φ is soil frictional modulus, Z is sinkage depth, n is sinkage exponent, h is the grouser height, and b is the width of grouser shoe. Apart from the fixed value b, these parameters are only depended on the soil properties, and tested by the penetration test.

Therefore, the vertical load equation could make the following Equation (4):

$$\begin{aligned} W &= \left(\tfrac{K_c}{b} + K_\varphi\right) bL\{(h+z)^n\lambda + (1-\lambda)Z^n\} \\ &= \left(K_c + bK_\varphi\right) L\{(h+z)^n\lambda + (1-\lambda)Z^n\} \end{aligned} \tag{4}$$

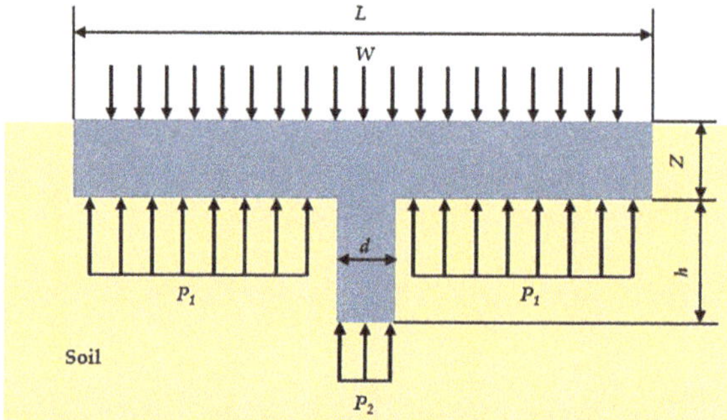

Figure 1. Vertical stress analysis of the grouser shoes. L is the length of grouser shoe, W is vertical load, Z is sinkage depth, h the grouser height, d is the grouser thickness. P_1 is the contact stress of the top surface of the grouser, and P_2 is the stress at the bottom surface of the grouser shoe.

2.3. Experimental Equipment

In this experiment, the soil bin was 500 mm (length) × 350 mm (width) × 200 mm (height) and filled with black soil. Electronic universal material testing machine was custom developed and used for the adhesion force tests, which was multi-functional and had high precision. The overall structure of the test platform is shown in Figure 2. The test platform is composed of computer system, TMC digital measurement, and control system. The ZLF-D 1093 spoke sensor (Zhenhai Electric Equipment Co., Ltd., Ningbo, China), which is equipped with 500 N force range, has a precision class of 0.02% and the output sensitivity range is 1.5–2 mV/V.

Figure 2. Overall structure of the test platform. (**a**) A soil bin test; (**b**) structure of the grouser shoes. L is the length of grouser shoe, b is the width of grouser shoe, h is the grouser height, d is the grouser thickness, and α is the grouser splayed angle.

2. Materials and Methods

2.1. Soil Sampling

The soil samples were collected from the Agricultural Experimental Base of Jilin University (N43°56′45.99″, E125°14′52.44″) in Changchun, Jilin province, Northeast China. In order to simulate the real subsidence amount of tracked vehicles working in the field, considering the previous crop of the test field was mainly wheat, the depth of the plow layer was between 250 mm and 300 mm. Therefore, we selected the characterization of the soil samples up to 300 mm from the surface for the adhesion test. The soil stress distribution was uniform under the compaction of the crawler walking mechanism [27]. The black soil physical properties of each sample are listed in Table 1. The test was repeated three times on the same condition to obtain the precise results. The average hardness of the black soil was 854 kPa, measured by a SC-900 soil hardness tester (Spectrum Technologies Inc., Aurora, IL, USA). The average bulk density of the black soil was a 1269 kg/m^3, measured by a TJSD-750-IV soil bulk density determination detector. The results showed that the bulk density was positively correlated with the hardness.

Table 1. Measured black soil physical properties.

Sample Depth (mm)	Bulk Density (kg·m^{-3})	Hardness (kPa)	Moisture Content (%)
0–100	1130	42	24.31
100–200	1352	972	21.28
200–300	1326	1165	19.87

The soil moisture content has an important effect on the adhesion force, and the mechanical properties of soil samples vary greatly under different moisture contents. The study mainly optimized different structure parameters of the grouser shoes for adhesion reduction. In order to minimize the influence of other variables on the study, we performed the experiments at only one soil moisture content. Additionally, we tested the plastic limit and liquid limit of soil samples by using the LP-100D digital display soil liquid-plastic limit combined tester (Kedaluqiao Electric Equipment Co., Ltd., Cangzhou, China). The plastic limit and the liquid limit of soil samples were 15.83% and 32.25%, respectively. The average moisture content of each layer of soil samples was dried in a DZF-6050 thermostatic drying oven (Rongshida Electric Equipment Co., Ltd., Kunshan, China) at 105 ± 5 °C for 6 h.

The soil moisture content for all the soil samples was in the range of 19.87–24.31%, and the average moisture content was 21.82% (d.b.) (i.e., dry basis). The results showed that the soil samples were higher than the plastic limit and lower than the liquid limit. Overall, the average moisture content of all the soil samples were within the plastic state, which had a strong adhesion to the grouser shoes of tracked vehicles. This critically restricted the efficiency of the tracked vehicles during the operation.

2.2. Vertical Stress Analysis of the Grouser Shoes

When the vertical load was applied on the grouser shoes, the entire grouser shoes sank in the black soil. The vertical stress analysis of grouser shoes is shown in Figure 1.

As is presented in Figure 1, the vertical forces of the grouser shoes were expressed by Equation (1):

$$W = P_1(L - d) + P_2 d \quad (d = \lambda L) \tag{1}$$

where W is vertical load, L is the length of grouser shoe, d is grouser thickness. P_1 is the contact stress of the top surface of the grouser, and P_2 is the stress at the bottom surface of the grouser shoe. According to the Beeker's sinkage formula [28,29], the stress was expressed as follows:

$$P_1 = \left(\frac{K_c}{b} + K_\varphi\right) \times Z^n \tag{2}$$

$$P_2 = \left(\frac{K_c}{b} + K_\varphi\right) \times (Z + h)^n \tag{3}$$

where k_c is soil cohesive modulus, k_φ is soil frictional modulus, Z is sinkage depth, n is sinkage exponent, h is the grouser height, and b is the width of grouser shoe. Apart from the fixed value b, these parameters are only depended on the soil properties, and tested by the penetration test.

Therefore, the vertical load equation could make the following Equation (4):

$$\begin{aligned} W &= \left(\tfrac{K_c}{b} + K_\varphi\right) bL\{(h+z)^n\lambda + (1-\lambda)Z^n\} \\ &= (K_c + bK_\varphi)L\{(h+z)^n\lambda + (1-\lambda)Z^n\} \end{aligned} \tag{4}$$

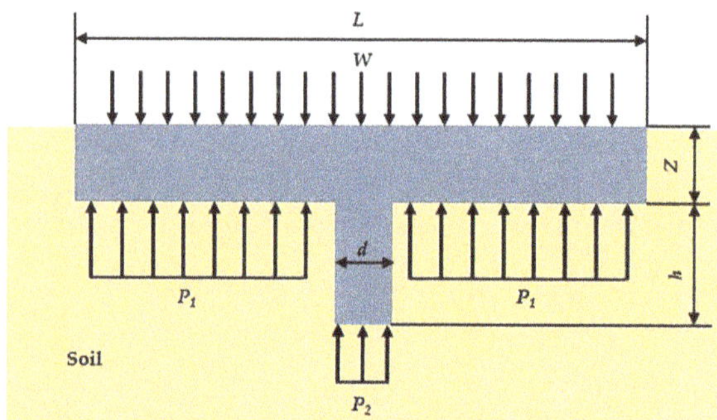

Figure 1. Vertical stress analysis of the grouser shoes. L is the length of grouser shoe, W is vertical load, Z is sinkage depth, h the grouser height, d is the grouser thickness. P_1 is the contact stress of the top surface of the grouser, and P_2 is the stress at the bottom surface of the grouser shoe.

2.3. Experimental Equipment

In this experiment, the soil bin was 500 mm (length) × 350 mm (width) × 200 mm (height) and filled with black soil. Electronic universal material testing machine was custom developed and used for the adhesion force tests, which was multi-functional and had high precision. The overall structure of the test platform is shown in Figure 2. The test platform is composed of computer system, TMC digital measurement, and control system. The ZLF-D 1093 spoke sensor (Zhenhai Electric Equipment Co., Ltd., Ningbo, China), which is equipped with 500 N force range, has a precision class of 0.02% and the output sensitivity range is 1.5–2 mV/V.

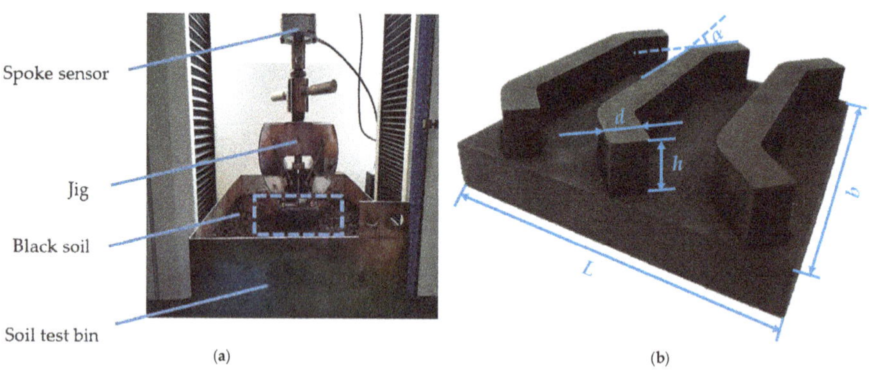

Figure 2. Overall structure of the test platform. (**a**) A soil bin test; (**b**) structure of the grouser shoes. L is the length of grouser shoe, b is the width of grouser shoe, h is the grouser height, d is the grouser thickness, and α is the grouser splayed angle.

This paper mainly considered the static normal adhesion performance of the grouser shoes. When the normal loading speed of the equipment was 5 mm/min, the relative displacement of soil samples along the tangential direction of the contact surface of the grouser shoes was less than 1 mm, and the effect of tractor traction forces on the adhesive force between the belt and the ground may be ignored in this study.

During the test, when the cross beam started moving upward at a speed of 5 mm/min, the pressure on the grouser shoes also started decreasing from the maximum until the moment of separation of black soil and the grouser shoes. At that moment, the adhesion force appeared and the test platform had to apply a force (*P*) more than the weight of the grouser shoes (*G*) to separate the grouser shoes from the black soil. The difference between the force applied by the test platform and the grouser shoes' weight divided by the contact area of the grouser shoes (*A*) was defined as the adhesion force (F_a). The methodology for measuring the adhesion force was consistent with the related papers studied by Massah, Ren, and Zhang et al. [8,14,30,31]. The adhesion force could be established by utilizing Equation (5). Additionally, the adhesive soil mass to the grouser shoes was measured by LQ-A 30002 electronic scales (Ante Weighing Electric Equipment Co., Ltd., Ruian, China), and the level of accuracy was 1% (i.e., reading of 0.01).

$$\begin{aligned} F_a &= \frac{P}{A} = \frac{W-G}{3[dh\cos\alpha + d(b-2h\cos\alpha)]} \\ &= \frac{(K_c+bK_\varphi)L\{(h+z)^n\lambda+(1-\lambda)Z^n\}-mg}{3d(b-h\cos\alpha)} \end{aligned} \tag{5}$$

where *m* is the mass of the grouser shoes, *g* is the acceleration of gravity (*g* = 9.8 kg/s), *d* is the grouser thickness, *h* is the grouser height, α is the grouser splayed angle, and the other variables are fixed values. Equation (5) showed that the adhesion force is mainly dependent on the three structure parameters of the grouser shoes, such as the grouser height (*h*), the grouser thickness (*d*), and the grouser splayed angle (α). Thus, these parameters were selected as the experimental factors to be optimized.

2.4. Experiment Scheme

In order to determine the optimal level interval of the factors, pre-experiments were carried out. The results presented the reasonable range of the level value of the factors: the grouser height was 20–30 mm, the grouser thickness was 5–7 mm, and the grouser splayed angle was 35°–45°. The grouser height (*h*), the grouser thickness (*d*), and the grouser splayed angle (α) were considered as the independent variables, the adhesion force (F_a) and adhesive soil mass (M_a) were determined as the response indicators. The functional relationship between the independent variables and the response indicators was established by the RSM, and a Box-Behnken Design (BBD) was adopted to optimize structure parameters of the grouser shoes [32,33]. According to the BBD, it took five center points per block and three levels of each factor and took rotatable alpha, which the factor level encodes (−1, 0, 1). There was a total of 17 sets of coded conditions for the modeling of the experiment results, and each group of tests was repeated at least 3 times to obtain the real average value. The coded levels of the independent variables are shown in Table 2. The experimental scheme and results are presented in Table 3.

Table 2. Coded levels of independent variables.

Coded Levels	Grouser Height (mm)	Grouser Thickness (mm)	Grouser Splayed Angle (°)
−1	20	5	35
0	25	6	40
1	30	7	45

Table 3. The experimental scheme and results.

Numbers	Factors			Evaluation Indexes	
	Grouser Height (mm)	Grouser Thickness (mm)	Grouser Splayed Angle (°)	Adhesion Force (kPa)	Adhesive Soil Mass (g)
1	1	0	1	1.48	23.55
2	1	1	0	1.54	23.62
3	−1	0	−1	0.97	24.18
4	−1	1	0	1.31	23.63
5	1	−1	0	1.47	23.41
6	−1	0	1	1.35	23.79
7	0	1	−1	1.15	24.12
8	0	1	1	1.60	23.65
9	−1	−1	0	1.02	24.18
10	0	−1	1	1.24	23.92
11	0	−1	−1	1.27	23.99
12	1	0	−1	1.42	23.51
13	0	0	0	1.27	23.71
14	0	0	0	1.09	23.74
15	0	0	0	1.18	23.78
16	0	0	0	1.15	23.75
17	0	0	0	1.14	23.73

2.5. Data Analysis

The data of the experiment were analyzed by using the analysis of variance (ANOVA) and least-significant difference (LSD) at the significance levels of 0.01 and 0.05 in Origin 8.0 software [34,35]. The Design-Expert software was employed to generate response surfaces and conduct multi-objective optimization. The quadratic polynomial equation of the independent variables and the response indicators could be established by Equation (6):

$$y_i = \beta_0 + \sum_{j=1}^{p} \beta_j z_j + \sum_{h<j} \beta_{hj} z_h z_j + \sum_{j=1}^{p} \beta_{jj} z_j^2 + \varepsilon_i \; j = 1,\ 2,\ \ldots \mathrm{N} \tag{6}$$

where y_i is the response indicators, z_j is the independent variables, β_0 is the fixed value, β_j is the first order term of coefficient, β_{hj} is the interaction term coefficient, β_{jj} is the second order term of coefficient, ε_i is the error value.

3. Results and Discussion

3.1. ANOVA Analysis

ANOVA analysis and the determination coefficient are usually used to evaluate the fitting quality of a model [36,37]. According to the experiment result of BBD, the results of analysis of variance for the response indicators are presented in Table 4.

Table 4. Analysis of variance for the response indicators.

Source	Adhesion Force (kPa)				Adhesive Soil Mass (g)			
	Sum of Squares	Degrees of Freedom	F-Value	*p* Value	Sum of Squares	Degrees of Freedom	F-Value	*p*-Value
Model	0.5150	9	19.24	0.0004 ***	0.8227	9	44.73	<0.0001 ***
h	0.1985	1	66.72	<0.0001 ***	0.3570	1	174.70	<0.0001 ***
d	0.0450	1	15.13	0.0060 **	0.0288	1	14.09	0.0071 *
α	0.0925	1	31.08	0.0008 ***	0.0990	1	48.45	0.0002 ***
h-d	0.0121	1	4.07	0.0835	0.1444	1	70.66	<0.0001 ***
h-α	0.0256	1	8.61	0.0219 *	0.0462	1	22.62	0.0021 **
d-α	0.0576	1	19.37	0.0032 **	0.0400	1	19.57	0.0031 **

Table 4. *Cont.*

Source	Adhesion Force (kPa)				Adhesive Soil Mass (g)			
	Sum of Squares	Degrees of Freedom	F-Value	*p* Value	Sum of Squares	Degrees of Freedom	F-Value	*p*-Value
h^2	0.0266	1	8.95	0.0202 *	0.0398	1	19.49	0.0031 *
d^2	0.0337	1	11.34	0.0120 *	0.0179	1	8.77	0.0211 *
α^2	0.0149	1	5.01	0.0602	0.0535	1	26.19	0.0014 **
Residual	0.0208	7			0.0143	7		
Lack of Fit	0.0031	3	0.2333	0.8692	0.0116	3	5.78	0.0615
Pure Error	0.0177	4			0.0027	4		
Cor Total	0.5358	16			0.8370	16		

F-value = Fischer's variance ratio; *p* value = probability value. *** Extremely significant level ($p < 0.001$); ** Very Significant level ($p < 0.01$); * Significant level ($p < 0.05$).

The p values of the two models were less than 0.001, indicating that the grouser height, the grouser splayed angle, and the grouser thickness were all very significant. The lack of fit was greater than 0.05, which was not significant. Hence, the model reflects the experimental data more realistically, which could predict and analyze the influence of various factors on the response indicators. Within the level range of the selected factors, according to the F-value of each factor, the weight of the factors affecting the adhesion force and the adhesive soil mass were both the grouser height > the grouser splayed angle > the grouser thickness. Based on the orthogonal test results, eliminating non-significant factors ($p > 0.05$), experimental data were analyzed and fitted with the quadratic polynomials that were obtained using the regression model. Equation (7) showed the regression models of the adhesion force, and Equation (8) represented adhesive soil mass regression response surface models.

$$F_a = 7.054 - 5.763 \times 10^{-3}h - 1.997d - 0.043\alpha - 3.2 \times 10^{-3}h * \alpha + 0.024d * \alpha + 3.305 \times 10^{-3}h^2 + 0.093d^2 \quad (7)$$

$$M_a = 38.382 - 0.248\,h - 0.993d - 0.371\alpha + 0.038h * d + 4.3 \times 10^{-3}h * \alpha - 0.02d * \alpha - 3.89 \times 10^{-3}h^2 + 0.065d^2 + 4.51 \times 10^{-3}\alpha^2 \quad (8)$$

To ensure the adaptability and accuracy of the model, it was necessary to evaluate the predictive ability by analyzing the determination coefficient. The response model fit summary output for indicators is illustrated in Table 5. The coefficients of determination (R^2) of the adhesion force and the adhesive soil mass were all larger than 0.95, suggesting that more than 95% of variations in the indices could be explained by the models. Moreover, both of the models' difference between predicted R^2 and adjusted R^2 less than 0.2 indicates that the response surfaces of the two models established had good consistency and predictability for the experimental results [20]. The predicted vs. actual values for the response indicators are shown in Figure 3. It could be seen from the scatter plot that the actual values were basically distributed on the predicted curve, consistent with the trend of the predicted values, and linearly distributed. The predicted precision of the models was 15.72 and 23.61, which was far larger than four, suggesting the models were precise [35].

Table 5. Fit statistics for the response indicators.

Response Indicators	R^2	Adjusted R^2	Predicted R^2	Predicted Precision
Adhesion force (kPa)	0.96	0.91	0.86	15.72
Adhesive soil mass (g)	0.98	0.96	0.77	23.61

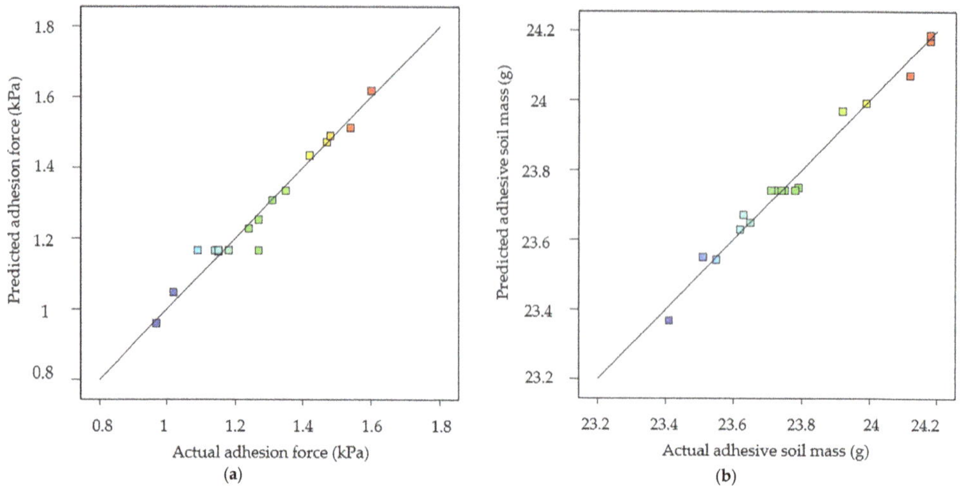

Figure 3. Predicted vs. actual values of the response indicators. (**a**) Predicted vs. actual values of adhesion force; (**b**) predicted vs. actual values of adhesive soil mass.

3.2. Analysis of Interactive Factors

According to the established quadratic Equations (7) and (8), a 3D response surface diagram was drawn to assess the interactive effects of the variables on the adhesion force and the adhesive soil mass. The interactive impact of two factors on the response indicators were analyzed by maintaining one factor as constant. The 3D response surface diagram of pairwise interaction of the grouser height, the grouser thickness, and the grouser splayed angle on adhesion force is shown in Figure 4. The 3D response surface diagram of pairwise interaction of the grouser height, the grouser thickness, and the grouser splayed angle on adhesive soil mass is shown in Figure 5.

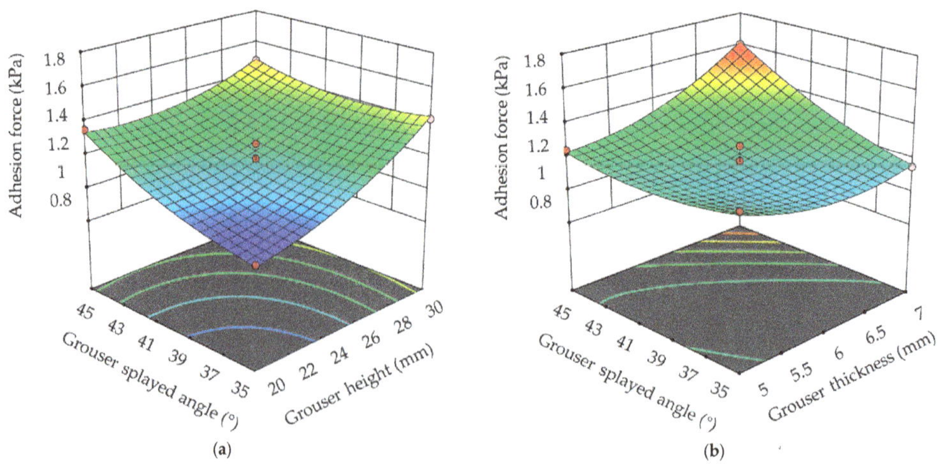

Figure 4. 3D response surface diagram of pairwise interaction of the grouser height, grouser thickness, and grouser splayed angle on adhesion force. (**a**) Interaction of grouser height and grouser splayed angle; (**b**) interaction of grouser thickness and grouser splayed angle.

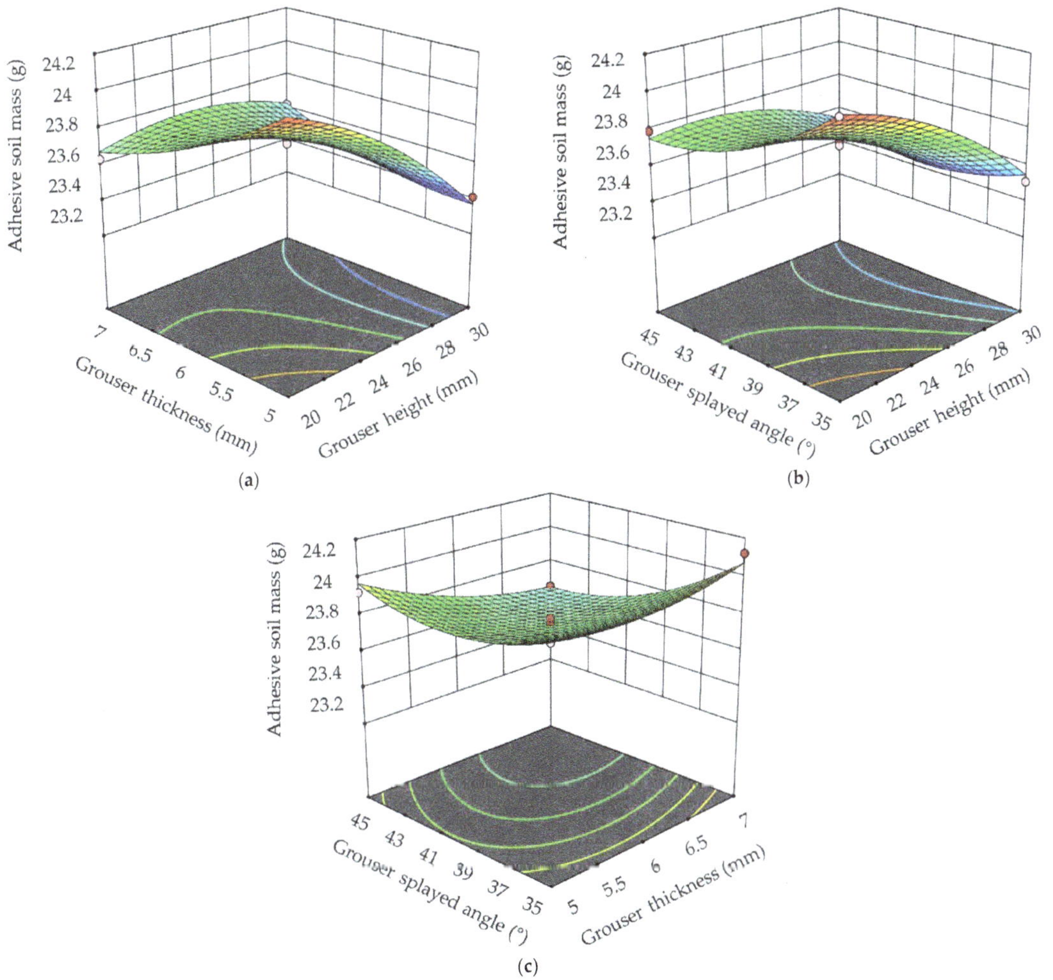

Figure 5. 3D response surface diagram of pairwise interaction of the grouser height, grouser thickness, and grouser splayed angle on adhesive soil mass. (**a**) Interaction of grouse height and grouser thickness; (**b**) interaction of grouser height and grouser splayed angle; (**c**) interaction of grouser thickness and grouser splayed angle.

3.2.1. Analysis of Interactive Factors on Adhesion Force

From the results of Table 4, the p-value of the grouser height and grouser thickness was 0.0835, which did not reach a significant level ($p < 0.05$). Therefore, the analysis of interaction of grouser height and grouser thickness on adhesion force does not need to be discussed. It could be seen from Figure 4a,b that when the one factor was fixed, another two factors were considered as interactive factors on adhesion force and the adhesion force exhibited an increasing trend, which ranged from 0.97 kPa to 1.6 kPa. Figure 4a showed that when the grouser thickness was maintained at 6 mm, the grouser splayed angle was fixed and the adhesion force increased with the growth of the grouser height. The grouser splayed angle was 35°; when the grouser height increased from 20 mm to 30 mm, the adhesion force increased from 0.97 kPa to 1.42 kPa, which was an increase of 46.39%. The grouser splayed angle was 45°; when the grouser height increased from 20 mm to 30 mm, the adhesion force increased from 1.35 kPa to 1.48 kPa, which was an increase of 9.63%. The results showed that the grouser splayed angle and the grouser height had an interactive effect on soil adhesion force, and the effect of the grouser splayed angle on soil adhesion force exceeded that of the grouser height. Figure 4b showed that when the grouser height was maintained at 25 mm, the grouser splayed angle was fixed and the adhesion force decreased with the growth of the grouser thickness. The grouser splayed angle was 35°; when the grouser thickness increased from 5 mm to 7 mm, the adhesion force decreased from 1.27 kPa to 1.15 kPa, which decreased by 9.45%. The grouser splayed angle was 45°; when the grouser height thickness increased from 5 mm to 7 mm, the adhesion force increased from 1.24 kPa to 1.6 kPa, which was an increase of 29.03%. The results showed that the grouser splayed angle and the grouser thickness had an interactive effect on soil adhesion force, and the effect of grouser splayed angle on soil adhesion force exceeded that of the grouser thickness. In summary, the interactive impact of combined factors on adhesion force was ranked as $d\alpha > h\alpha$.

3.2.2. Analysis of Interactive Factors on Adhesive Soil Mass

As was presented in Figure 5a–c, when one factor was fixed, another two factors were considered as interactive factors on adhesive soil mass. Figure 5a,b showed that the adhesive soil mass appeared to be a declining trend, which ranged from 24.18 g to 23.41 g, and Figure 5c showed that adhesive soil mass initially appeared as a declining trend and subsequently an increasing trend. Figure 5a showed that when the grouser splayed angle was maintained at 40°, the adhesive soil mass declined slowly with increases in the grouser height and the grouser thickness. The maximum adhesive soil mass was obtained when the grouser height was 20 mm and the grouser thickness was 5 mm. When the grouser height was 30 mm and the grouser thickness was 5 mm, the adhesive soil mass reached the minimum. The grouser thickness and the grouser height had an interactive effect on adhesive soil mass, and the effect of the grouser height on adhesive soil mass exceeded that of the grouser thickness. Figure 5b showed that when the grouser thickness was maintained at 6 mm, the adhesive soil mass declined with increases in the grouser height and the grouser splayed angle. The maximum adhesive soil mass was obtained when the grouser height was 20 mm and the grouser splayed angle was 35°. When the grouser height was 30 mm and the grouser splayed angle was 35°, the adhesive soil mass reached the minimum. The grouser height and the grouser splayed angle had an interactive effect on adhesive soil mass, and the effect of grouser height on adhesive soil mass exceeded that of the grouser splayed angle. Figure 5c showed that when the grouser height was maintained at 25 mm, the adhesive soil mass initially declined and subsequently increased with increases in the grouser height and the grouser splayed angle. The maximum adhesive soil mass was obtained when the grouser thickness was 7 mm and the grouser splayed angle was 35°. When the grouser thickness was 5 mm and the grouser splayed angle was 35°, the adhesive soil mass reached the minimum. The grouser thickness and the grouser splayed angle had an interactive effect on adhesive soil mass, and the effect of the grouser splayed angle on adhesive soil mass exceeded that of the grouser thickness. In a word, the interactive impact of combined factors on adhesive mass was ranked as $hd > h\alpha > d\alpha$.

3.3. *Multi-Objective Optimization Model*

We used the multi-objective optimization method of the Design-Expert 12.0 software (Stat-Ease Inc., Minneapolis, MN, USA) to predict the optimal parameters combination, the minimum value for the adhesion force, and the adhesive soil mass in this study. The grouser height (h), the grouser thickness (d), and the grouser splayed angle (α) were selected as parameters to be optimized. The constraint condition of the multi-objective optimization model was as followed:

$$\begin{cases} 20\text{ mm} \leq h \leq 30\text{ mm} \\ 5\text{ mm} \leq d \leq 7\text{ mm} \\ 35^{\circ} \leq \alpha \leq 45^{\circ} \end{cases} \tag{9}$$

Under the constraints given in Equation (9), the Design-Expert 12.0 software (Minneapolis, MN, USA) presented the optimal parameters combination when the grouser height was 20 mm, grouser thickness was 6.34 mm, and grouser splayed angle was 40.45°. The predicted value for the adhesion force and the adhesive soil mass were 1.16 kPa and 23.76 g, respectively.

3.4. Verification Experiment

Although the results from the ANOVA analysis indicated that the established mathematical models had a good accuracy of predicting the adhesion force and the adhesive soil mass in this study, the verification experiment was still needed to verify the model performance. Each experiment was repeated at least three times to obtain the average data. The results of the verification experiment are presented in Table 6. The average data of the measured adhesion force reached 1.11 kPa, and adhesive soil mass reached 22.68 g. The relative error of the predicted values and measured values was less than 5%, proving the reliability of the regression models. Compared with the average value of the un-optimized experiment results in Table 3, the adhesion force and adhesive soil mass reduced by 12.84% and 4.63%, respectively. The results showed that the mathematical models were reliable, and the optimized structure parameters of the grouser shoes by the RSM could reduce soil adhesion.

Table 6. Results of the verification experiment.

Indicators	Adhesion Force (kPa)	Adhesive Soil Mass (g)
Predicted value	1.16	23.76
Measured value	1.11	22.68
Relative error value	0.045	0.0476

4. Conclusions

Theoretical analysis was carried out to explore the impact factors of structure parameters of the grouser shoes on the adhesion performance. The structure parameters were optimized by the RSM. The optimal parameter combination occurred when the grouser height was 20 mm, grouser thickness was 6.34 mm, and grouser splayed angle was 40.45°. The average data of verification experiments occurred when the adhesion force reached 1.11 kPa and adhesive soil mass reached 22.68 g. Compared with the average value of un-optimized experiment results, the adhesion force and adhesive soil mass reduced by 12.84% and 4.63, respectively. The relative error of the predicted values and measured values was less than 5%, proving the reliability of the regression models. This study can provide a reference method for parameters optimization, and a new structure of the grouser shoes of tracked vehicles will be designed to reduce adhesion.

Even though the established mathematical model has a good accuracy in predicting structure parameters of the grouser shoes on the adhesion force and adhesive soil mass in this study, it is worth noting that there are some limitations. Firstly, the optimal structure parameters combination may reduce the adhesion but may also affect the traction, and the effect of traction on the adhesive force between the belt and the ground were not analyzed. Secondly, there are many parameters that could cause adhesion to vary, including not only the structure parameters, but other parameters (e.g., materials, geometry, or surface porosity). In this study, we mainly analyzed the effect of structure parameters of the grouser shoes on the adhesion, and the other parameters were not explored. Thirdly, the adhesion-related parameters (i.e., soil cohesive and friction modulus, soil cohesion, or internal friction angle) under different soil moisture content were not studied.

Therefore, further study needs to be carried out to address the limitations. We propose to establish a multi-objective evaluation system based on adhesion and traction, and the effect of traction on the adhesive force between the belt and the ground also need to be analyzed. The impact of other parameters (e.g., materials, geometry, or surface porosity) on the soil adhesion should be explored and compared with the structure parameters of the grouser shoes. Additionally, the adhesion-related parameters under different soil moisture contents by penetration test will be carried out in a following study.

Author Contributions: Conceptualization, J.F.; methodology, J.F.; validation, J.L.; investigation, J.L.; resources, X.T.; writing—original draft preparation J.L.; writing—review and editing, J.F. and J.L.; supervision, Z.C. and R.W.; project administration, J.F.; funding acquisition, J.F. All authors have read and agreed to the published version of the manuscript.

Funding: This research was funded by the International Science and Technology Cooperation Project of Science and Technology Development Program of Jilin Province (No. 20190701055GH).

Institutional Review Board Statement: Not applicable.

Informed Consent Statement: Not applicable.

Data Availability Statement: The datasets in this study are available within this article. The data that were derived from the original datasets but not included herein are available upon request.

Acknowledgments: The authors would like to thank the soil samples provided by the Experimental Base of Agriculture of Jilin University.

Conflicts of Interest: The authors declare no conflict of interest.

References

1. Zhang, Y.; Wu, Y.; Liu, B.; Zheng, Q.; Yin, J. Characteristics and factors controlling the development of ephemeral gullies in cultivated catchments of black soil region, Northeast China. *Soil Tillage Res.* **2007**, *96*, 28–41. [CrossRef]
2. Liu, D.; Wang, Z.; Zhang, B.; Song, K.; Duan, H. Spatial distribution of soil organic carbon and analysis of related factors in croplands of the black soil region, Northeast China. *Agric. Ecosyst. Environ.* **2006**, *113*, 73–81. [CrossRef]
3. Li, J.; Tong, J.; Hu, B. Calibration of parameters of interaction between clayey black soil with different moisture content and soil-engaging component in northeast China. *Trans. Chin. Soc. Agric. Eng.* **2019**, *35*, 130–140.
4. Jiang, Q.; Cao, M.; Wang, Y. Quantification of the soil stiffness constants using physical properties of paddy soils in Yangtze Delta Plain, China. *Biosyst. Eng.* **2020**, *200*, 89–100. [CrossRef]
5. Jia, X.; Ren, L. Theoretical Analysis of Adhesion Force of Soil to Solid Materials. *Trans. Chin. Soc. Agric. Eng.* **1995**, *87*, 489–493. [CrossRef]
6. Ren, L.; Tong, J.; Li, J.; Chen, B.C. Soil Adhesion and Biomimetics of Soil-engaging Components: A Review. *J. Agric. Eng. Res.* **2001**, *79*, 239–263. [CrossRef]
7. Baek, S.H.; Shin, G.B.; Lee, S.H. Evaluation of the Slip Sinkage and its Effect on the Compaction Resistance of an Off-Road Tracked Vehicle. *Appl. Sci.* **2020**, *10*, 3175. [CrossRef]
8. Massah, J.; Fard, M.R.; Aghel, H. An optimized bionic electro-osmotic soil-engaging implement for soil adhesion reduction. *J. Terramechanics* **2021**, *95*, 1–6. [CrossRef]
9. Barzegar, M.; Karimi, R.; Hashemi, S.J.; Nazokdast, H. Evaluating the draft force and soil-tool adhesion of a UHMW-PE coated furrower. *Soil Tillage Res.* **2016**, *163*, 160–167. [CrossRef]
10. Marani, S.M.; Shahgholi, G.; Moinfar, A. Effect of nano coating materials on reduction of soil adhesion and external friction. *Soil Tillage Res.* **2019**, *193*, 42–49. [CrossRef]
11. Ucgul, M.; Fielke, J.M.; Saunders, C. Three-dimensional discrete element modelling (DEM) of tillage: Accounting for soil cohesion and adhesion. *Biosyst. Eng.* **2015**, *129*, 298–306. [CrossRef]
12. Zenkov, S.A. Application of thermal effect as a means to combat ground-and-machine part adhesion. *Mater. Sci. Eng.* **2020**, *862*, 32–60. [CrossRef]
13. Zenkov, S.A.; Lkhanag, D. Use of repair-recovery compositions for reducing the adhesion of soils to working bodies of machines. *Mater. Sci. Eng.* **2020**, *971*, 52–55.
14. Ren, L.; Han, Z.; Li, J. Experimental investigation of bionic rough curved soil cutting blade surface to reduce soil adhesion and friction. *Soil Tillage Res.* **2006**, *85*, 1–12. [CrossRef]
15. Ren, L.; Deng, S.Q.; Wang, J.C. Design Principles of the Non-smooth Surface of Bionic Plow Moldboard. *J. Bionic Eng.* **2004**, *1*, 9–19. [CrossRef]
16. Ren, L. Progress in the bionic study on anti-adhesion and resistance reduction of terrain machines. *Sci. China* **2009**, *52*, 273–284. [CrossRef]
17. Tong, J.; Sun, J.; Chen, D. Geometrical features and wettability of dung beetles and potential biomimetic engineering appli cations in tillage implements. *Soil Tillage Res.* **2005**, *80*, 1–12. [CrossRef]
18. Wang, X. Study on use of vibration to reduce soil adhesion. *J. Terramech.* **1998**, *35*, 87–101. [CrossRef]
19. Jia, H.; Wang, W. Application of anti-adhesion structure based on earthworm motion characteristics. *Soil Tillage Res.* **2018**, *178*, 159–166. [CrossRef]

20. Sun, J.; Chen, H.; Wang, Z. Study on plowing performance of EDEM low-resistance animal bionic device based on red soil. *Soil Tillage Res.* **2020**, *196*, 104336. [CrossRef]
21. Zhang, Z.; Jia, H.; Sun, J. Review of application of biomimetics for designing soil-engaging tillage implements in Northeast China. *Int. J. Agric. Biol. Eng.* **2016**, *9*, 12–21.
22. Sun, J.; Wang, Y.; Zhang, S. The mechanism of resistance-reducing/anti-adhesion and its application on biomimetic disc furrow opener. *Math. Biosci. Eng.* **2020**, *17*, 4657–4677. [CrossRef] [PubMed]
23. Shaikh, S.A.; Li, Y.; Zheng, M. Effect of grouser height on the tractive performance of single grouser shoe under different soil moisture contents in clay loam terrain. *Sustainability* **2021**, *13*, 1156. [CrossRef]
24. Ge, J.; Wang, X.; Kito, K. Comparing tractive performance of steel and rubber single grouser shoe under different soil moisture contents. *Int. J. Agric. Biol. Eng.* **2016**, *9*, 11–20.
25. Grisso, R.; Perumpral, J.; Zoz, F. An empirical model for tractive performance of rubber-tracks in agricultural soils. *J. Terramechanics* **2006**, *43*, 225–236. [CrossRef]
26. Park, W.; Chang, Y.; Lee, S.S. Prediction of the tractive performance of a flexible tracked vehicle. *J. Terramech.* **2008**, *45*, 13–23. [CrossRef]
27. Ding, Z.; Li, Y.; Ren, L.; Tang, Z. Distribution uniformity of soil stress under compaction of tracked undercarriage. *Trans. Chin. Soc. Agric. Eng.* **2020**, *36*, 52–58.
28. Bekker, M. *Introduction of Terrain Vehicle Systems*; University of Michigan Press: Ann Arbor, MI, USA, 1969.
29. Wong, J. *Theory of Ground Vehicles*, 3rd ed.; John Wiley & Sons: New York, NY, USA, 2001.
30. Ren, L. *Soil Adhesion Mechanics*, 1st ed.; Machinery Industry Press: Beijing, China, 2011; pp. 4–7, 214–215.
31. Zhang, L.; Yang, Z.; Xu, S. Application of electronic universal testing machine in soil adhesion testing. *Test. Technol. Test. Mach.* **2005**, *45*, 27–29. (In Chinese)
32. Mosayebi, A. Methanol steam reforming over Co-Cu-Zn/γ-Al_2O_3 catalyst: Kinetic and RSM-BBD modeling approaches. *Int. J. Energy Res.* **2021**, *45*, 3288–3304. [CrossRef]
33. Alipour, M.; Vosoughi, M.; Mokhtari, S.A. Optimizing the basic violet 16 adsorption from aqueous solutions by magnetic graphene oxide using the response surface model based on the Box–Behnken design. *Int. J. Environ. Anal. Chem.* **2019**, *101*, 758–777. [CrossRef]
34. Fu, J.; Yuan, H.; Zhang, D. Multi-Objective Optimization of Process Parameters of Longitudinal Axial Threshing Cylinder for Frozen Corn Using RSM and NSGA-II. *Appl. Sci.* **2020**, *10*, 1646. [CrossRef]
35. Chen, T.; Jia, H.; Zhang, S. Optimization of cold pressing process parameters of chopped corn straws for fuel. *Energies* **2020**, *13*, 652. [CrossRef]
36. Armstrong, R.A.; Eperjesi, F.; Gilmartin, B. The application of analysis of variance (ANOVA) to different experimental designs in optometry. *Ophthalmic Physiol. Opt.* **2002**, *22*, 248–256. [CrossRef] [PubMed]
37. Nagelkerke, N. A note on a general definition of the coefficient of determination. *Biometrika* **1991**, *78*, 691–692. [CrossRef]

agriculture

Article

Optimization Design and Experiment of Ear-Picking and Threshing Devices of Corn Plot Kernel Harvester

Ranbing Yang [1], Dongquan Chen [2], Xiantao Zha [1], Zhiguo Pan [2] and Shuqi Shang [2,*]

1 College of Mechanical and Electrical Engineering, Hainan University, Haikou 570228, China; yangranbing@hainanu.edu.cn (R.Y.); zhaxt@hainanu.edu.cn (X.Z.)
2 College of Mechanical and Electrical Engineering, Qingdao Agricultural University, Qingdao 266109, China; 20192204153@stu.qau.edu.cn (D.C.); 200401043@qau.edu.cn (Z.P.)
* Correspondence: sqshang@qau.edu.cn

Abstract: In order to solve the problems of easy-to-break kernels and substantial harvest losses during kernel harvesting in breeding trials plot of corn, an ear-picking device and a threshing device of corn plot kernel harvester has been optimized. To automatically change the gap of the ear-picking plate, a self-elastic structure with compression spring and connecting rod is used. The ear-picking plate is glued, and an elastic rubber gasket is placed underneath it, which effectively improves the adaptability of the ear-picking device and reduces corn kernel collision damage during ear-picking. To ensure the self-purification of the ear-picking device, a combination of auger sieve hole cleaning device and lateral pneumatic auxiliary cleaning system is used. A dual-axial flow threshing device is designed, which uses a "U"-shaped conveying system to transport maize ears in the threshing chamber. The spacing of the concave sieve may be adjusted, and the residual kernels in the threshing chamber can be cleaned up after harvesting one plot by combining three cleanings, which meets the requirements of no mixing between plots. The force analysis of corn ears in the threshing chamber determines the best design plan for the forward speed, the speed of the second threshing drum, and the threshing gap. The breakage rate and non-threshing rate regression models were created using the quadratic regression orthogonal combination test, and the parameters were optimized using MATLAB. The verification test results showed that when the forward speed was 0.61 m/s, the second threshing drum speed was 500 r/min, and the threshing gap was 40 mm, the breakage rate was 1.47%, and the non-threshing rate was 0.89%, which met the kernel harvesting requirements in corn plots.

Keywords: breeding plots; corn; low-damage harvesting; optimization design; field experiment

Citation: Yang, R.; Chen, D.; Zha, X.; Pan, Z.; Shang, S. Optimization Design and Experiment of Ear-Picking and Threshing Devices of Corn Plot Kernel Harvester. *Agriculture* 2021, *11*, 904. https://doi.org/10.3390/agriculture11090904

Academic Editor: José Pérez-Alonso

Received: 8 August 2021
Accepted: 16 September 2021
Published: 21 September 2021

1. Introduction

The need for seeds continues to rise as the breeding business develops at a rapid pace. Seed integrity is important for seed transportation, storage, and breeding, etc. [1–4]. In our country, there are few studies on corn breeding machines, although foreign research on corn kernel plot harvesters is rather developed [5–7]. For example, the WINTERSTEIGER (WINTERSTEIGER AG, Ried im Innkreis, Austria) combine plot harvester can harvest rice, corn, soybeans, and other different crop plots at the same time, and by using an axial flow threshing device and a pneumatic seed conveying system, it can basically guarantee low damage and no mixed seeds when working in two consecutive plots. However, foreign machines have not been widely used in China due to their expensive cost and differences in cropping patterns.

In recent years, researchers have performed a study in order to overcome the problems of excessive harvest losses and high kernel damage in corn harvesters. Pickard [8] investigated the effect of cylinder and concave bar variations on the threshing of corn. It was discovered that the rasp-type cylinder bar outperformed the angle-type cylinder bar in terms of shelling efficiency and kernel damage, meanwhile covering the cylinder or concave bars with rubber had no impact on shelling efficiency or kernel damage. Srivastava et al. [9] investigated the mechanical properties of corn kernels and discovered certain impact features of corn kernels that are related to damage during harvesting and handling by measuring corn impact parameters. When maize kernels are subjected to impact forces, they discovered that longitudinal shear is weaker than transverse shear. Voicu et al. [10] investigated dimensional analysis theory in order to develop a mathematical model of the seeds separation process at the cleaning system level of grain harvester combines in order to anticipate seed losses. Fu et al. [11] analytically evaluated the collision force of corn ears during the ear-picking process and built a wheel-type rigid-flexible coupling loss-reducing head with a flexible surface and a buffer spring to efficiently decrease kernel loss. Geng et al. [12] developed a mathematical model of corn ear force by analyzing the influencing factors and trends of corn damage during the ear-picking process, which led to the discovery of damage mechanism and major influencing factors of mechanical ear-picking. Li et al. [13] developed a corn bionic threshing machine based on the discrete first and then threshing principle, which refers to the chicken beak cut into the kernel gap and the bare-hand low-damage threshing mechanism. Li et al. [14] used discrete element method solution (DEM) software to analyze the threshing process of corn ears in order to reduce the rate of broken and non-threshing during the corn threshing process. They discovered an effective combination of structural parameters that can be used to design a high moisture content kernel threshing device and combine harvester by combining the characteristics of high moisture content during harvesting. Di et al. [15] investigated the factors that influence the breakage rate and non-threshing rate of corn kernels and designed a combined axial flow corn threshing drum with rasp bar and nail tooth to meet low-loss rate and non-determinate rate field harvesting requirements. In particular, the above researches are all focused on the field, and the demand of breeding companies for breakage rate and non-threshing rate during plot harvesting process far outweighs field technical requirements. The needs of field breeding experiments cannot be met by domestic research in this area.

Therefore, aiming at the demand of Huanghuaihai region corn breeding harvesting machinery, a flexible gap self-adjusting ear-picking plate and a dual-axial flow threshing device were designed to achieve low-damage and self-purification harvesting in plots, and the field experiment was completed to verify the performance of ear-picking and threshing devices. This study achieves non-mixed kernel harvesting between plots and serves as a useful reference for future research into low-damage, self-purification combined harvesting in corn plots.

2. Structure Design

2.1. Structure Design of Ear-Picking Device

The corn plot kernel harvester requires to be equipped with an undamaged ear-picking device. This paper chose an ear-picking device combined with an ear-picking plate and snapping rolls in order to ensure a low breakage rate during the ear-picking process. If the gap between ear-picking plates is too large, corn ears will directly contact the snapping rolls, and if the gap is too small, thick stalk feeding and impurity discharge will be affected. Therefore, this paper develops a gap self-adjusting ear-picking plate, as shown in Figure 1.

This institution uses a self-elastic structure with a compression spring and connecting rod. Stalks give the ear-picking plate a lateral thrust when stalks are fed into the ear-picking device through the stalk chain, and the spring is compressed by a connecting rod. Due to the preload of spring, the ear-picking plate returns to its initial state after the ear-picking operation, allowing for automatic adjustment of the ear-picking plate gap. According to the previous measurement data of different corn varieties, the minimum diameter of corn ears at the big end is 46.1 mm, and the maximum diameter of stalks at a distance of 200 mm from the root is 31.9 mm. The adjustment range of this institution is 20–36 mm.

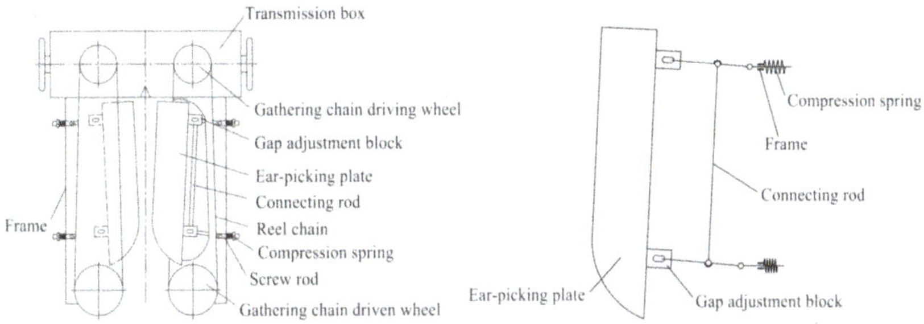

Figure 1. Installation and working diagram of gap self-adjusting ear-picking plate.

2.1.1. Structure Design of Picking Plate

According to the reference [16], increasing the diameter at the point where the ear-picking plate and corn ears come into contact can reduce the instantaneous impact force. However, the flat plate or rounded edge is usually used in the existing corn ear-picking device that includes an ear-picking plate and snapping rolls. It has a small equivalent contact radius and cannot buffer the impact force generated during the ear-picking process. Therefore, the inverted right angle ear-picking plate of this paper has an inclination angle between the edge and horizontal plane. According to ear size measurement data, the bending length of the ear-picking plate is 33 mm, and the angle between the plate and horizon is 17°. Because the force area of the ear in relation to the ear-picking plate is greater than that of the plane plate, the impact force is insufficient to collide kernels. At the same time, two flexible treatments on the ear-picking plate parts have been made to reduce the damage of corn ears during the working process. As shown in Figure 2, the surface layer of the ear-picking plate is treated with glue for 3~5 mm, and an elastic rubber gasket is added under the ear-picking plate to further reduce the impact force received by corn ears in the working process. When the ear-picking device receives the impact of corn ears, the buffering effect of the elastic rubber pad allows it to convert the impact force of corn ears into the potential energy of the elastic rubber pad. The elastic rubber pad returns to its original shape after corn ears are picked, reducing kernels damage.

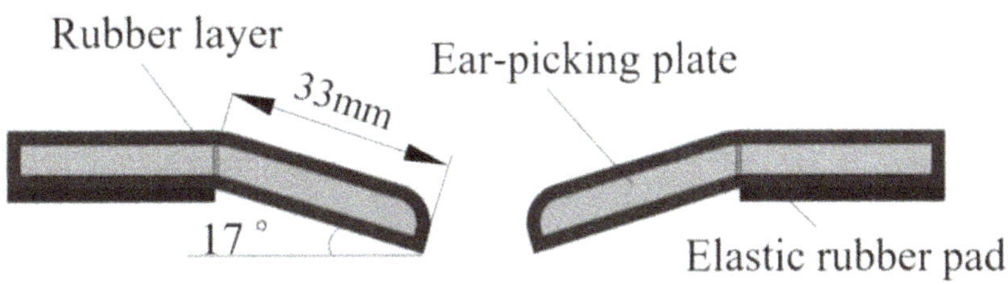

Figure 2. Structural scheme of V-shaped ear-picking plate.

2.1.2. Structure Design of Snapping Rolls

Snapping rolls are welded with ribs on the surface in combination with processing technology to ensure the continuity of the pulling stalk process and improve working efficiency. The snapping rolls have a six-ribbed structure that increases the diameter of the roller [17,18], which can improve the ability of the snapping rolls to grab stalk, and the working principle of the ear-picking device is shown in Figure 3.

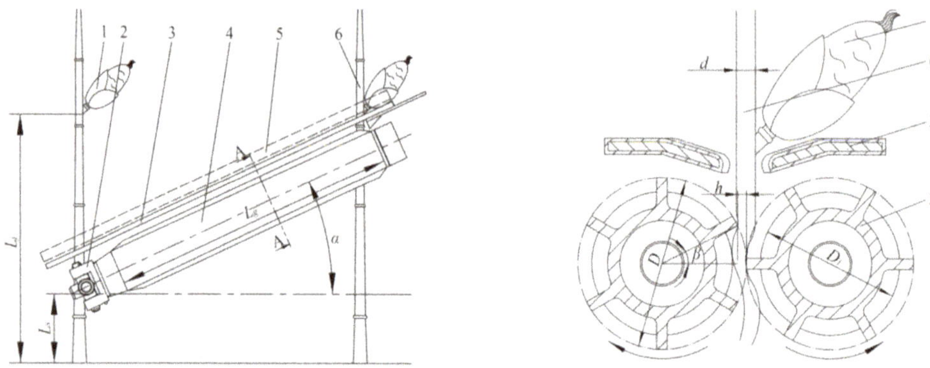

Figure 3. Working scheme of snapping rolls. 1. Corn ear 2. Snapping rolls 3. Ear-picking plate 4. Snapping rolls gap adjustment device 5. Gathering chain 6. Corn stalk. (**a**) Working scheme, (**b**) A-A Cross section scheme. **Note:** L_j is the ear height of corn plant, mm; L_s is the height of the center line in front of snapping rolls above ground, mm; L_g is the working length of snapping rolls, mm; α is the angle between snapping rolls and horizontal plane, (°); d is the diameter of the stalk at corn ear, mm; h is the working gap of snapping rolls, mm; D is the outer diameter of snapping rolls, mm; D_1 is the diameter of snapping rolls drum, mm; β is the grabbing angle of snapping rolls to the stalk, (°).

When calculating the actual working length of snapping rolls, it should be noted that it can adjust to different corn varieties' height at corn ear setting, avoiding the phenomenon of snapping rolls pulling two corn plants at the same time.

According to Figure 3, the length L_g is satisfied by the following equation:

$$Lg = (Lj - Ls) \cdot \cos\alpha \tag{1}$$

where L_j is the ear height of corn plant, mm; L_s is the height of the center line in front of snapping rolls above ground, mm; L_g is the working length of snapping rolls, mm; α is the angle between snapping rolls and horizontal plane, (°).

According to information collected and sorted from the mainly promoted breeding corn in the Huanghuaihai region of China, the height L_j is 700~1100 mm, the height L_s is 300~500 mm, and the angle α is 25~35° when corns are harvested [19]. It can be seen from Equation (1) that the length of snapping rolls is 328~725 mm. When snapping rolls are installed obliquely, and the length is less than 400 mm, the ear-picking effect is not sui. Combined with the overall structure of the machine, the length L_g is selected as 550 mm, which can satisfy the corn breeding harvest of various varieties and growth.

The compression ratio of stalks by the snapping rolls 'i' is described as the following equation:

$$i = \frac{d-h}{d} \tag{2}$$

where d is the diameter of the stalk at corn ear, mm; h is the working gap of snapping rolls, mm.

The gripping force produced by snapping rolls will not be able to pick off corn ears if the compression ratio is too small, and the stalk will be prone to breakage if the compression ratio is too high. According to the references [20,21], the best compression ratio is 0.5~0.7, and the corn plant has suitable trafficability when the gap between snapping rolls is 1/2~1/3 of the stalk diameter. Through measurement, the diameter d is 16.1~26.8 mm, and the average value is 21.45 mm. Therefore, the working gap h is set to 10 mm.

The outer diameter D and the diameter D_1 can be expressed as the following equation:

$$D = \frac{d-h}{1-\cos\beta} \tag{3}$$

$$D_1 = D - h \quad (4)$$

where D is the outer diameter of snapping rolls, mm; D_1 is the diameter of snapping rolls drum, mm; β is the grabbing angle of snapping rolls to the stalk, (°).

In the expected state, the rib on the one snapping roll moves to the lowest position when the rib on the other snapping roll just touches the stalk. On snapping rolls, six ribs are evenly distributed, and the angle of two adjacent ribs is 60°, then $\beta \approx 30°$. Substitute the working gap h and the average diameter d into Equations (3) and (4). The outer diameter D is 85 mm, and the diameter D_1 is 75 mm.

2.1.3. Structure Design of Cleaning System

A type of ear-conveying auger with kernel removal function is designed to meet the self-purification agronomic requirements of corn plot harvesters. The structure is shown in Figure 4, including a pneumatically assisted seed cleaning system, auger discharge port, auger blades, ear-conveying auger, scrapers, sieve hole cleaning device, and auger-type kernel recovery devices. The auger rotational speed is 267 r/min, which can fit the input amount of ear-picking device. A centrifugal fan with a rotational speed of 860 r/min and a maximum wind force of 0.36 m^3/s is used in the system. The fan's air outlet is fitted with a pneumatic assisted system that allows the cleaning wind size to be adjusted in real-time according to the harvesting operation.

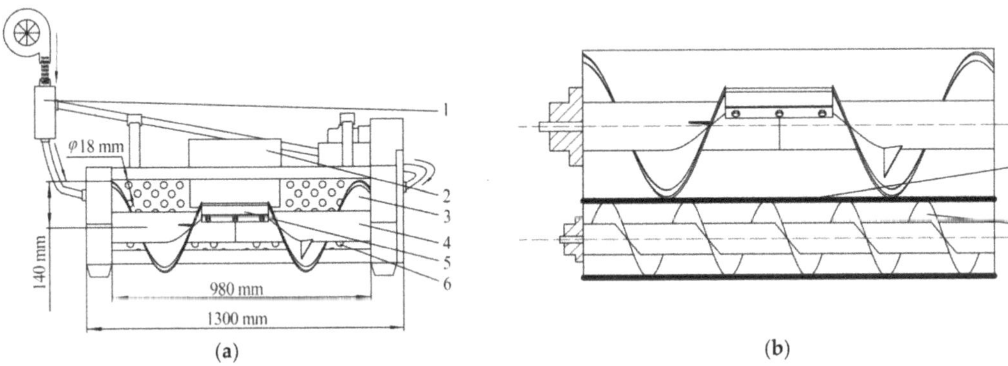

Figure 4. Cleaning system scheme of ear-picking device. 1. Pneumatic assisted seed cleaning system 2. Auger discharge port 3. Auger blades 4. Ear-conveying auger 5. Scraper 6. Sieve hole cleaning device 7. Auger-type kernel recovery device. (**a**) Top view, (**b**) Main view.

The ear-conveying auger and the auger-type kernel recovery device are driven by the hydraulic motor to rotate, and the auger transports ears and bracts upwards to the ear lifter. The rubber is wrapped around the edge of the auger blade, which could not only reduce the collision force of corn ears but also eliminate the gap between the auger blade and the bottom of the auger. The inner wall of the auger channel is designed as a streamlined diversion surface with a sieve hole at the bottom. When a plot is harvested, a small number of kernels left in the conveying process will be removed by a sieve hole and lateral pneumatic assisted kernel cleaning system, and the kernels will fall into the kernel recovery device through the sieve hole. A kernel storage box is attached to one end of the auger-type kernel recovery device. After cleaning, the extra kernels will be poured out to prevent mixing between different plots and ensure self-purification of corn harvest in the plot.

2.2. Structure Design of Threshing Device

As shown in Figure 5, the threshing device is composed of a dual-axial flow threshing drum, concave sieve, clutch adjusting device, spiral cleaning device, fan, vibration sieve, etc.

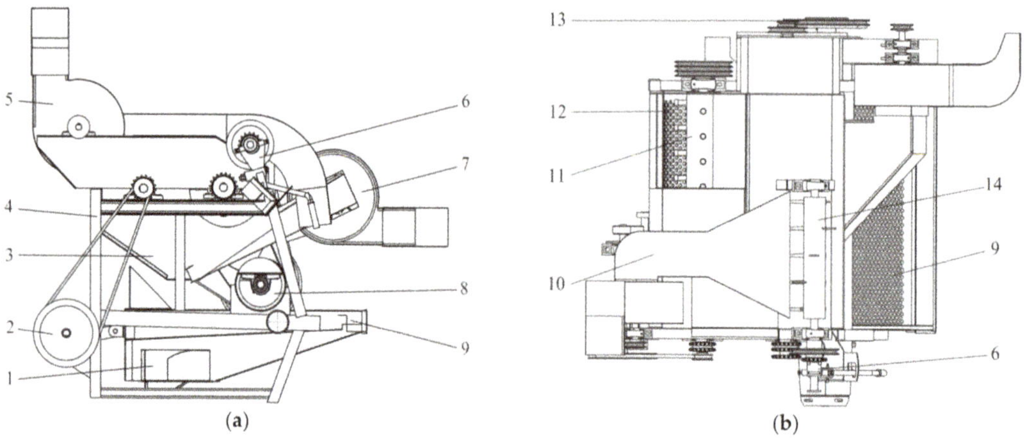

Figure 5. Structural scheme of threshing device. 1. kernel discharge port 2. Crank drive wheel 3. Inclined guide plate 4. Frame 5. Front fan 6. Clutch adjustment device 7. Rear fan 8. Core discharging device 9. Vibrating sieve 10. Front fan air supply channel 11. Threshing roller 12. Concave plate screen 13. Drive wheel 14. Spiral impurity removal device. (**a**) Main view, (**b**) Top view.

In the threshing chamber, corn ears are threshed by the blow, friction, and rubbing of the dual-axial flow threshing drum, concave sieve, and cover plate. The removed kernels, bracts, and part of broken cores pass through the holes of the concave sieve and fall into a vibrating sieve via an inclined guide plate, while the rear fan blows impurities out of the machine via wind force, completing the first cleaning. A spiral cleaning device at the feeding end of the second threshing chamber is designed to lead a small number of impurities and entrained kernels out of the threshing chamber and fall into a vibrating sieve via the wind force of the front fan to complete the second cleaning. Vibrating separation of kernels and impurities is realized by shaking the sieve under the rotation of the crank, and kernels fall into the unloading port for pneumatic conveying. The kernels and remaining impurities are transported to the zone-bagging device, where the lighter impurities are blown out of the upper part of the zone-bagging device, completing the third cleaning. After harvesting a plot, the machine stops moving forward and continues the wind cleaning operation until the entire threshing device achieves self-purification.

2.2.1. Structure Design of Threshing Drum

The function of the threshing drum is to separate the kernels by beating them, and its structure directly affects the extraction effect of kernels [22–25]. This paper uses a dual-axial flow threshing drum with a rubber sleeve is provided on the top of each threshing element to reduce the non-threshing and breakage rates in the breeding harvest. The drum speed is low in the front and high in the back, with a speed ratio of 1:2. Figure 6a shows the model and expansion scheme for the threshing element, which uses a hemispherical spike-tooth and spiral arrangement on the threshing shaft. A single drum consists primarily of a drum body and four rows of spike-teeth, with each row of spike-teeth consisting of nine threshing units evenly spaced on each line. So as to lessen the impact on corn.

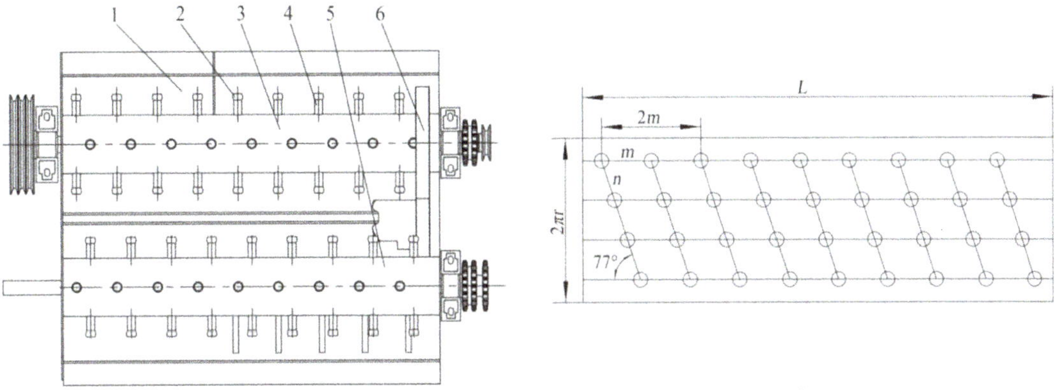

Figure 6. Model and expansion scheme of threshing drum. 1. Feeding inlet 2. Rubber 3. First drum body 4. Threshing tooth 5. Ear guide plate 6. Second drum body. (**a**) Structure scheme, (**b**) Cylinder unfolded view. **Note:** m is the distance between two adjacent teeth on the same axis, mm; n is the arc length of two adjacent teeth on the same spiral line, mm; L is the drum length, mm.

The feed quantity should be the main design basis of the threshing drum. Because this threshing device is installed on a 2-row corn kernel plot harvester, it should meet the initial setting parameter of 2.5~4 kg/s of the harvester's feed quantity. Then the drum length L is described as the following equation:

$$L \geq \frac{q}{q_0} \tag{5}$$

where q is the feed quantity, kg/s; q_0 is the allowable feed quantity per unit length, which is 3~4 (kg/s·m) for the combined harvester.

Substituting the data into Equation (5) will get the drum length L as 0.625~1.33 m, according to the overall design requirements of the whole machine and drum, take L = 960 mm. Figure 6b shows an expanded view of the threshing drum. In the figure, m is the distance between two adjacent teeth on the same axis, and n is the arc length of two adjacent teeth on the same spiral line. In order to increase the threshing time and beat frequency in the threshing chamber, the threshing teeth on the same axis should satisfy that m is less than the length of corn ears and n is greater than the length of corn ears. Equation (6) exists and is described as follows:

$$\begin{cases} m < l \\ n > l \end{cases} \tag{6}$$

where l is the length of corn ears, mm.

In order to satisfy the axial pushing effect of the threshing drum on corn ears, it is necessary to analyze the distance between two adjacent teeth on the same spiral line. The mathematical model diagram is shown in Figure 7.

Assuming that the shortest length of threshing teeth is x, it should satisfy Equation (7) and is described as follows:

$$\begin{cases} l_1{}^2 + (\frac{m}{4})^2 = l_2{}^2 \\ l_2 \leq l \\ l_1 = \sqrt{2}(r + x) \end{cases} \tag{7}$$

where l_1 is the shortest distance between two teeth tangent to drum body on the same section, mm; l_2 is the distance between two teeth tangent to the drum body on the same spiral line, mm; r is the radius of the drum, mm.

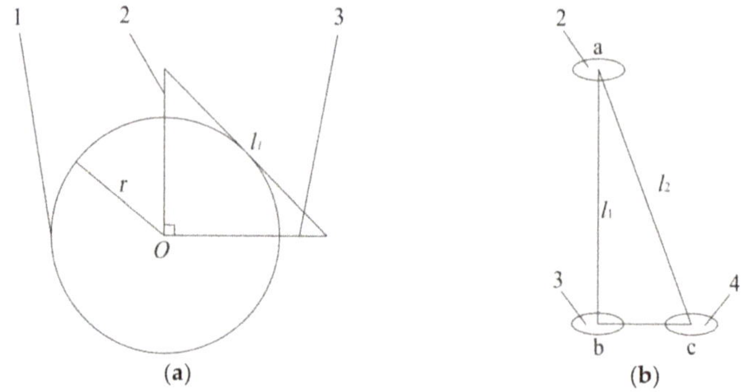

Figure 7. Mathematical model diagram. 1. Threshing drum 2. Threshing gear *a* 3. Virtual threshing gear *b* 4. Threshing gear *c*. (**a**) Cylinder section view, (**b**) Distribution of threshing teeth. **Note:** l_1 is the shortest distance between two teeth tangent to drum body on the same section, mm; l_2 is the distance between two teeth tangent to the drum body on the same spiral line, mm; r is the radius of the drum, mm.

According to the corn ears length of the plot (l is 175 mm) and the actual working conditions of the harvester, the radius r is selected as 80 mm and m is 100 mm, and the shortest length x of threshing teeth is 42.47 mm by Equation (7). The length of the threshing teeth is 45 mm.

2.2.2. Structure Design of Concave Sieve

The concave sieve is the main threshing component [26,27]. As shown in Figure 8, the concave sieve is grid-type in order to meet the requirements of self-purification during the harvesting of corn kernels in the plot. In order to improve the adaptability of different varieties of harvest, the upper part of the concave plate screen is equipped with a cylindrical pin to enable rotation, and the lower part is drilled with a slot-shaped hole, which is adjusted and fixed by a stud. The bars are 8 mm wide, which have a 30 mm gap between them, and each radial bar extends 20 mm. When the gap between the corncob and vibrating screen is adjusted, the corncob will not fall into the vibrating screen below. Increase the gap after harvesting a plot so that the corn kernels, corn cobs, and impurities in the threshing chamber are cleaned up, and the seeds do not mix when harvesting the next plot.

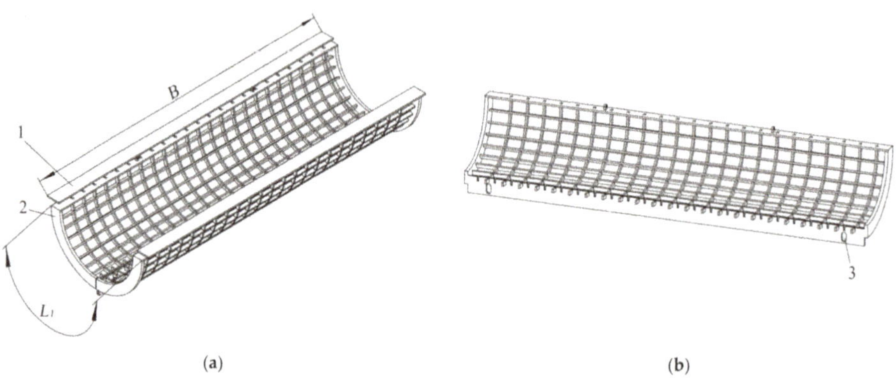

Figure 8. Concave sieve model diagram. 1. Fixed plate 2. Movable sieve 3. Slot hole. (**a**) Concave Screen Combination Diagram, (**b**) Concave sieve split diagram. **Note:** B is the width of concave, mm; L_1 is the arc length of concave, mm.

The concave sieve is in direct contact with the ears of corn during the threshing process, which aids in the shedding of the kernels. The equation of concave sieve area A with respect to arc length L_1 is as follows:

$$A = BL_1 \geq (1 - \varepsilon)q/0.6q_a \tag{8}$$

where B is the width of concave sieve, mm; L_1 is the arc length of concave sieve, mm; q is the feed quantity, kg/s; ε is the ratio of kernels fed into crops; q_a is the allowable feed quantity per unit concave sieve area when ε was 0.4, the combine harvester takes 5~8.

The wrap angle δ of concave sieve should satisfy Equation (9) and is described as follows:

$$\delta = \frac{180L_1}{\pi D} \tag{9}$$

Among them, the width B is equal to the length L, which is 960 mm, and the range of arc length L_1 can be calculated as 360 mm $\leq L_1 \leq$ 830 mm. The higher the arc length, the stronger the threshing ability, but the power consumption and the arc length wrap angle will also increase. The wrap angle is mostly 90°~120°, under the premise of satisfying the work quality, take L_1 = 500 mm, and put the data into Equation (9) to obtain the wrap angle δ = 120°.

3. Materials and Methods

As shown in Figure 9, firstly, select evaluation indicators and factors according to the analysis and related requirements, and then use the quadratic regression orthogonal combination design experiment to carry out the field test so as to obtain the regression equation corresponding to evaluation indicators. The regression equation is used to analyze the influence of factors and optimize the index, and finally, the optimized value is verified and compared with the corresponding field test.

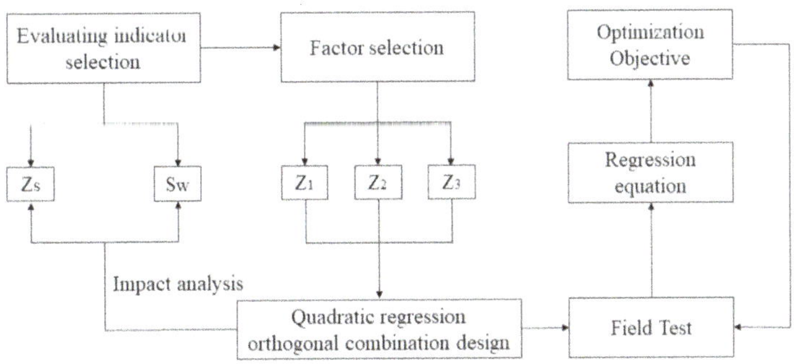

Figure 9. Flowchart of materials and methods.

3.1. Force Analysis of Threshing Device

The force involved in threshing corn ears is more complicated. The single corn ear threshing process is taken as the research object under the ideal state of low feed rate, the force on the ear is translated to the barycenter, and the force analysis is shown in Figure 10a without considering the torque. Using the barycenter position O of corn ears as the origin, the connecting line between the barycenter of corn ears and the center of the threshing drum is the Y axis, the movement direction of corn is the positive direction of the X axis, and A is the contact point where the ear meets the concave sieve.

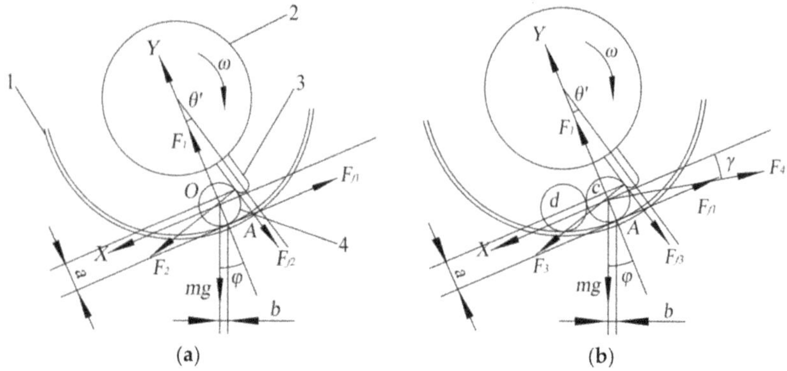

Figure 10. Stress analysis of corn. 1. Concave sieve 2. Threshing drum 3. Threshing teeth 4. Corn. (**a**) Low-feeding ear force diagram, (**b**) High-feeding ear force diagram. **Note:** F_1 is the supporting force of concave sieve on the ear, N; F_2 is the impact of threshing teeth on the ear, N; F_3 is the impact force of threshing tooth on ear c, N; F_4 is the pressure of ear *d* on ear *c*, N; μ_1 is the friction coefficient between the ear and concave sieve; μ_2 is the friction coefficient between the ear and threshing teeth; F_{f1} is the friction between the ear and concave sieve, N; F_{f2} is the friction between the ear and threshing teeth, N; F_{f3} is the friction force between ear *c* and threshing tooth, N; θ' is the included angle between *Y* axis and the axis of the threshing tooth, °; φ is the angle between *Y* axis and the direction of gravity, °; γ is the angle between the pressure F_4 and the friction force F_{f1}, °; ω is the angular speed of the drum, rad/s; *r* is ear radius, mm; *a* is the distance from the intersection of ear and the threshing tooth to the concave sieve, mm; *b* is the distance from the barycenter of the ear to the concave sieve, mm.

Then the force acting on the ear with low feeding should satisfy the following equation:

$$F_2 \cos\theta' + mg\sin\phi - F_{f1} - F_{f2}\sin\theta' = 0 \tag{10}$$

$$F_1 - F_2\sin\theta' - mg\cos\phi - F_{f2}\cos\theta' + mr\omega^2 = 0 \tag{11}$$

where F_1 is the supporting force of concave sieve on the ear, N; F_2 is the impact of threshing teeth on the ear, N; μ_1 is the friction coefficient between the ear and concave sieve; usually, μ_1 is 0.059; μ_2 is the friction coefficient between the ear and threshing teeth; usually, μ_2 is 0.102 [28]; F_{f1} is the friction between the ear and concave sieve, $F_{f1} = \mu_1 F_1$, N; F_{f2} is the friction between the ear and threshing teeth, $F_{f2} = \mu_2 F_2$, N; θ' is the included angle between *Y* axis and the axis of the threshing tooth, °; usually, θ' is 17 °; φ is the angle between *Y* axis and the direction of gravity, °; *r* is ear radius, mm; ω is the angular speed of the drum, rad/s.

According to Equations (10) and (11), the force F_2 is simplified by Equation (12):

$$F_2 = \frac{mg(\mu_1\cos\phi - \sin\phi) - \mu_1 mr\omega^2}{\cos\theta'(1-\mu_1\mu_2) - \sin\theta'(\mu_1+\mu_2)} \tag{12}$$

The equation of the moment of the force acting on the ear to point *A* is as follows:

$$\begin{aligned}\sum M_A = F_2 a\cos\theta' + mgb\sin\phi - F_2 r\sin^2\theta' \\ -F_{f2}a\sin\theta' - F_{f2}r\sin\theta'\cos\theta'\end{aligned} \tag{13}$$

where *a* is the distance from the intersection of the ear and the threshing tooth to the concave sieve, mm; *b* is the distance from the barycenter of the ear to the concave sieve, mm.

From the above mechanical analysis, it can be seen that the ear force with low feeding is mainly related to the impact force F_2, the included angle θ', and the length *a*, etc. Therefore, the drum speed and threshing gap have an effect on the damage of kernels.

Under the ideal state of high-feeding, take ear c as the research object, and its force analysis is shown in Figure 10b, then the Equations (14) and (15) exists and is described as follow:

$$F_3 \cos\theta' + mg\sin\phi - F_{f1} - F_{f3}\sin\theta' - F_4\cos\gamma = 0 \tag{14}$$

$$F_1 - F_3\sin\theta' - mg\cos\phi - F_{f3}\cos\theta' - F_4\sin\gamma + mr\omega^2 = 0 \tag{15}$$

where F_3 is the impact force of threshing tooth on the ear c, N; F_4 is the pressure of ear d on ear c, N; F_{f3} is the friction force between ear c and threshing tooth, $F_{f3} = \mu_2 F_3$, N; γ is the angle between the pressure F_4 and the friction force F_{f1}, °.

According to Equations (10)~(15), the force F_3 is simplified by Equation (16):

$$F_3 = \frac{mg(\mu_1\cos\phi - \sin\phi) + F_4(\cos\gamma + \mu_1\sin\gamma) - \mu_1 mr\omega^2}{\cos\theta'(1-\mu_1\mu_2) - \sin\theta'(\mu_1+\mu_2)} \tag{16}$$

According to Equations (12) and (16), the impact of threshing teeth on the ear can be express by Equation (17):

$$F_3 - F_2 = \frac{F_4(\cos\gamma + \mu_1\sin\gamma)}{\cos\theta'(1-\mu_1\mu_2) - \sin\theta'(\mu_1+\mu_2)} \tag{17}$$

According to Equation (17), when $0° < \gamma < 90°$, $F_3 - F_2 > 0$. Therefore, the feed quantity increases, then the impact force of threshing teeth on the ear increases under the condition of a certain drum speed.

3.2. *Key Performance Parameters of Corn Plot Kernel Harvester*

3.2.1. Breakage Rate

In the measurement area, extract kernels from kernel outlet at least 2000 g after threshing and cleaning. Pick out the machine-damaged, cracked, and broken kernels and weigh the quality of damaged kernels and the total mass of sample kernels, respectively. Their test method in this research followed the National Standards of China, *Corn combine harvester* (GB/T 21962-2020). The breakage rate is calculated according to Equation (18), and it is described as follows:

$$Z_s = \frac{W_s}{W_i} \times 100\% \tag{18}$$

where Z_s is the breaking rate, %; W_s is the mass of damaged kernels, g; W_i is the total mass of sample kernels, g.

3.2.2. Non-Threshing Rate

The quality of non-threshing kernels accounts for a percentage of the total quality of kernels after one plot has been harvested and threshing has been completed. The non-threshing rate is calculated according to Equation (19), which is described as follows:

$$S_w = \frac{W_w}{W} \times 100\% \tag{19}$$

where S_w is the non-threshing rate, %; W_w is the quality of kernels that are not decontaminated, g; W is the total mass of kernels, g.

3.3. *Key Experimental Factors of Corn Plot Kernel Harvester*

From the above analysis, the factors that affect kernels breakage during the harvesting process of corn plot are drum speed, threshing gap, and feed quantity. In order to facilitate calculation, the forward speed of the machine is used instead of feed quantity.

3.3.1. Speed of Threshing Drum

The speed of the threshing drum is closely related to the breakage rate, which directly affects the impact, friction, and rubbing force of ears during threshing [29]. The front drum is faster than the back drum in this article. Therefore, the second threshing drum speed is selected as the test factor for the convenience of measurement. If the speed of the threshing drum is too fast, the impact force increased, resulting in an increase in the rate of kernel breakage and broken straw impurities. The burden of cleaning is greatly increased, and the power consumption will increase as well. If the speed is too low, the impact force will be insufficient, and kernels will not be removed. Meanwhile, the material staying and rubbing repeatedly in the threshing chamber for a long time will also increase the breakage rate. Therefore, the adjustment range of the second drum speed is set to 500~800 r/min.

3.3.2. Threshing Gap

The threshing gap is the minimum distance between the top of the drum threshing tooth and the concave sieve, and its size is related to the diameter of the ears. The gap is narrower than the average diameter of ears generally so that the ear can be squeezed and rubbed during the feeding process [30]. The basic characteristic parameters of Zhengdan 958 ears are shown in Table 1. Based on the statistical results in Table 1 and the overall structure of the machine, the adjustment range of the threshing gap is set to 20~40 mm.

Table 1. Corn basic characteristic parameters in the plot.

Parameters	Value
Average ear length/mm	174.36
Average weight per ear/kg	0.348
Average diameter of large end/mm	54.2
Average diameter of small end/mm	35.8
Kernel moisture content/%	21~26

3.3.3. Forward Speed

While increasing the feed rate can achieve higher harvest efficiency and shorten the working time, the quality of the operation will suffer, resulting in an increase in the rate of corn breakage during harvest, which affects the quality of the harvest. The relational function equation between the feeding amount and the forward speed of the harvester is described as follow:

$$q = n \times m \times \frac{v}{s} \tag{20}$$

where q is the feed quantity, Kg/s; n is the number of corn rows; m is the quality of single corn, Kg; v is the forward speed of the harvester, m/s; s is the distance between corn plants, m.

According to the planting mode of the test base, the feed quantity is proposed to be 1~2.5 kg/s, and the adjustment range of forward speed is 0.5~1.26 m/s based on Equation (20).

3.4. *Materials of Field Test*

The test site was conducted in Zhangpan Town, Xuchang County, Henan Province. The test plot was 100 × 70 m in size, with corn rows 800 mm apart, plants 350 mm apart, the plant lodging rate was less than 5%, and the minimum scion height was 1100 mm. The corn variety used in the field test was Zhengdan 958, and its basic characteristic parameters are listed in Table 1.

As shown in Figure 11, the flexible gap self-adjusting ear-picking plate and the dual-axial flow threshing device were installed to a 4YZL-2 corn plot kernel harvester (Henan Haofeng Machinery Manufacturing Co., Ltd., Henan, China), which can complete ear-picking lifting, threshing, and collection in one operation. The lossless harvest of corn plots and the realization of no mixed kernels between plots are the technical features. The harvester has extremely high technical requirements, and the working parameters are shown in Table 2.

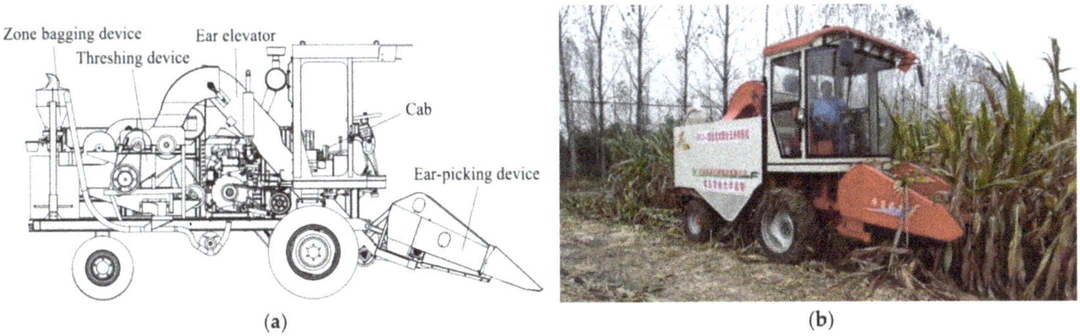

Figure 11. Ear-picking device and threshing device with 4YZL-2 corn plot kernel harvester. (**a**) Structural diagram, (**b**) Real machine diagram.

Table 2. Main technical parameters of plot corn kernel harvester.

Parameters	Value
Structure form	Self-propelled wheel
Type of diesel engine	JDM490
Rated power of engine/kW	37
Rated speed of engine/($r \cdot min^{-1}$)	2400
Vehicle weight/kg	2600
Harvest rows/row	2
Minimum ground clearance/mm	250
Applicable line spacing/mm	550~650
Size/(mm × mm × mm)	5200 × 1300 × 2650
Working width/mm	1200

3.5. Methods of Field Test

Combined with the above analysis and the actual working conditions of the prototype, taking the forward speed z_1, the second threshing drum speed z_2 and the threshing gap z_3 as the experimental factors. The experimental factors and levels are shown in Table 3.

Table 3. Experimental factors and levels.

Levels \ Factors	Forward Speed z_1/($m \cdot s^{-1}$)	Speed of Second Threshing Drum z_2/ ($r \cdot min^{-1}$)	Threshing Gap z_3/mm
1	0.5	500	20
2	0.88	650	30
3	1.26	800	40

This test adopts a three-factor quadratic regression orthogonal combination design and sets three zero-level tests (m_0 = 3), which can be obtained r^2 = 1.831 [31]. The experimental factors need to be coded before the test, and the coding method is shown in Table 4.

The formula for centralizing the quadratic term is described as follows:

$$x'_{ij} = x_{ij}^2 - 0.686\ i = 1,2,\cdots,17; j = 1,2,3 \tag{21}$$

The breakage rate y_1 and the non-threshing rate y_2 has been used as the test evaluating indexes. The total number of tests is 17, and each group of tests is repeated 3 times to obtain the average value, and the test scheme and results are shown in Table 5.

Table 4. Coding method of experimental factors.

Factors \ Code	Forward Speed Z_1/(m· s^{-1})	Speed of Second Threshing Drum Z_2/ (r·min^{-1})	Threshing Gap Z_3/mm
r (Z_{2j})	1.26	800	40
$1(Z_{0j} + \Delta_j)$	1.16	760.86	37.39
$0(Z_{0j})$	0.88	650	30
$1(Z_{0j} - \Delta_j)$	0.6	539.14	22.61
$-r$ (Z_{1j})	0.50	500	20
$\Delta_j = (Z_{2j} - Z_{1j})/2r$	0.28	110.86	7.39
$x_j = (Z_j - Z_{0j})/\Delta_j$	$x_1 = 3.571(z_1 - 0.52)$	$x_2 = 0.009(z_2 - 650)$	$x_3 = 0.135(z_3 - 30)$

Note: Δ_j is the minimum change interval of the j-th factor, j = 1, 2, 3; x_j is the encoded factor; r is the asterisk arm.

Table 5. Test program and results by quadratic regression orthogonal combination design.

Test \ Factors	x_0	x_1	x_2	x_3	x_1x_2	x_1x_3	x_2x_3	$x_1{}^2$	$x_2{}^2$	$x_3{}^2$	y_1	y_2
1	1	1	1	1	1	1	1	1	1	1	1.54	0.83
2	1	1	1	−1	1	−1	−1	1	1	1	1.66	1.12
3	1	1	−1	1	−1	1	−1	1	1	1	1.48	0.78
4	1	1	−1	−1	−1	−1	1	1	1	1	1.55	0.95
5	1	−1	1	1	−1	−1	1	1	1	1	1.62	1.07
6	1	−1	1	−1	−1	1	−1	1	1	1	1.87	1.19
7	1	−1	−1	1	1	−1	−1	1	1	1	1.53	0.98
8	1	−1	−1	−1	1	1	1	1	1	1	1.76	1.05
9	1	$-r$	0	0	0	0	0	r^2	0	0	1.69	1.13
10	1	r	0	0	0	0	0	r^2	0	0	1.59	0.81
11	1	0	$-r$	0	0	0	0	0	r^2	0	1.49	0.92
12	1	0	r	0	0	0	0	0	r^2	0	1.61	1.09
13	1	0	0	$-r$	0	0	0	0	0	r^2	1.67	0.96
14	1	0	0	r	0	0	0	0	0	r^2	1.51	0.84
15	1	0	0	0	0	0	0	0	0	0	1.56	0.93
16	1	0	0	0	0	0	0	0	0	0	1.57	0.91
17	1	0	0	0	0	0	0	0	0	0	1.54	0.95

Note: x_j is the encoded factor, j = 1, 2, 3; r is the asterisk arm.

4. Results and Discussion

4.1. Results of Filed Test

The test project and field test results by quadratic regression orthogonal combination design are shown in Table 5, while the regression analysis results are shown in Table 6. Figure 12 shows that the kernels were undamaged and that threshing was completed by the optimized and threshing device.

(**a**)

(**b**)

Figure 12. Presentation of harvest result. (**a**) Harvested corn grain, (**b**) threshed corn.

Table 6. Variance analysis for breakage rate.

Source	Sum of Squares	Degree of Freedom	Mean Square	*F* Value	*p*-Value
x_1	0.040	1	0.0402	58.98	0.00012 **
x_2	0.024	1	0.0243	35.59	0.00056 **
x_3	0.067	1	0.0674	98.70	0.00002 **
x_1x_2	0.0001	1	0.00011	0.165	0.69693
x_1x_3	0.011	1	0.0105	15.40	0.00572 **
x_2x_3	0.0006	1	0.00061	0.897	0.37511
${x_1}^2$	0.016	1	0.0161	23.65	0.00183 **
${x_2}^2$	0.00000004	1	0.00000004	0.00006	0.99391
${X_3}^2$	0.0032	1	0.0032	4.653	0.06791
Regression	0.163	9	0.0181	26.45	0.00014 **
Residuals	0.0048	7	0.00068	/	/
Lack of fit	0.0043	5	0.00086	1.58	0.2691
Error	0.00047	2	0.00023	/	/
Total	0.1673	16	/	/	/

Note: ** indicates highly significant ($p \leq 0.01$).

4.2. Regression Analysis of Filed Test

As shown in Table 6, the p-value of lack of fit P_l is 0.2691, the p-value of residuals model P_R is 0.00014, indicating that the regression equation of the experimental index fits well. Taking $\alpha = 0.05$ as the significance level, the primary term x_1, x_2, x_3, the square term ${x_1}^2$, and the interaction term x_1x_3 have extremely significant effects, while other factors have no significant effects. After removing the insignificant terms, the regression equation of the dimensionless code value of the breakage rate is obtained as:

$$\hat{y}_1 = 1.554 - 0.059x_1 + 0.046x_2 - 0.076x_3 + 0.036x_1x_3 + 0.049x_1^2 \quad (22)$$

From Equation (22), it can be seen that the influence order of experimental factors on breakage rate is that threshing clearance, forward speed, and second threshing cylinder speed. According to Table 4, the actual regression equation of breakage rate was obtained after sorting:

$$y_1 = 2.737 - 1.828x_1 + 0.0004x_2 - 0.026x_3 + 0.017x_1x_3 + 0.622x_1^2 \quad (23)$$

As shown in Table 7, the p-value of lack of fit P_l is 0.1894, the p-value of residuals model P_R is 0.00052, indicating that the regression equation of the experimental index fits well. Taking $\alpha = 0.05$ as the significance level, the primary term x_1, x_2, and x_3 have highly significant effects, the square term ${x_1}^2$ and the interaction term x_1x_3 have significant effects, while other factors have no significant effects. After removing the insignificant terms, the regression equation of the dimensionless code value of the non-threshing rate is obtained as:

$$\hat{y}_2 = 0.924 - 0.089x_1 + 0.058x_2 - 0.070x_3 - 0.034x_1x_3 + 0.048x_2^2 \quad (24)$$

Table 7. Variance analysis for non-threshing rate.

Source	Sum of Squares	Degree of Freedom	Mean Square	*F* Value	*p*-Value
x_1	0.09328	1	0.0933	65.77	0.000084 **
x_2	0.03965	1	0.0397	27.96	0.001138 **
x_3	0.05659	1	0.0566	39.901	0.000398 **
x_1x_2	0.000013	1	0.000013	0.0088	0.9278
x_1x_3	0.00911	1	0.0091	6.425	0.03896 *
x_2x_3	0.00361	1	0.0036	2.547	0.1545

Table 7. Cont.

Source	Sum of Squares	Degree of Freedom	Mean Square	F Value	p-Value
x_1^2	0.00575	1	0.0058	4.051	0.08401
x_2^2	0.0157	1	0.0157	11.07	0.01264 *
X_3^2	0.00054	1	0.00054	0.3793	0.5575
Regression	0.22425	9	0.0249	17.568	0.00052 **
Residuals	0.00993	7	0.00141	/	/
Lack of fit	0.00913	5	0.00183	4.564	0.1894
Error	0.0008	2	0.0004	/	/
Total	0.2342	16	/	/	/

Note: ** indicates highly significant ($p \leq 0.01$); * indicates significant ($0.01 \leq p \leq 0.05$).

From Equation (24), it can be seen that the influence order of experimental factors on the non-threshing rate is that forward speed, threshing clearance, and second threshing cylinder speed. According to Table 4, the actual regression equation of non-threshing rate was obtained after sorting:

$$y_2 = 2.394 + 0.169x_1 - 0.0046x_2 + 0.005x_3 - 0.016x_1x_3 + 0.000004x_2^2 \tag{25}$$

The regression analysis results are shown in Table 8. The correlation coefficient R of the breakage rate y_1 and the non-threshing rate y_2 reached 0.9856 and 0.9786, which indicating that there is a strong linear relationship between experimental factors and test indicators in the regression model. The coefficient of determination R^2 of the breakage rate y_1 and the non-threshing rate y_2 reached 0.9714 and 0.9576, which indicating that the regression model has a high degree of fit. In order to avoid the limitation of coefficient of determination R^2, and adjusted coefficient of determination Adj R^2 is obtained. It shows that the adjusted coefficient of determination of breakage rate y_1 is 0.9668, which means the independent variables in the model can explain the 96.68% variation of breakage rate. Similarly, the adjusted coefficient of determination of non-threshing rate y_2 is 0.9503, which means the independent variables in the model can explain the 95.03% variation of the non-threshing rate. Durbin-Watson statistic satisfies the conditions of $1 < d < 3$ at the same time, so the test data have strong reliability.

Table 8. Significance analysis of regression coefficient on the test results.

Inspection Items	y_1	y_2
R	0.9856	0.9786
R^2	0.9714	0.9576
Adj R^2	0.9668	0.9503
S	0.0261	0.0377
Durbin-Watson statistic	2.7176 ($1 < d < 3$)	2.1973 ($1 < d < 3$)

The distribution of test measured and fitted values for the two regression models are shown in Figure 13. The test measured values are represented by the blue line, while the sample fitted values are represented by the orange line. It can be seen that the two values are very well matched.

4.3. Interaction Analysis of Filed Test

In order to obtain the influence law of the experimental factors on each test index more intuitively, study the interaction effect of the other two factors by fixing a certain factor at zero levels, and the regression equation is transformed into a three-dimensional contour map through MATLAB, which are shown in Figures 14–16.

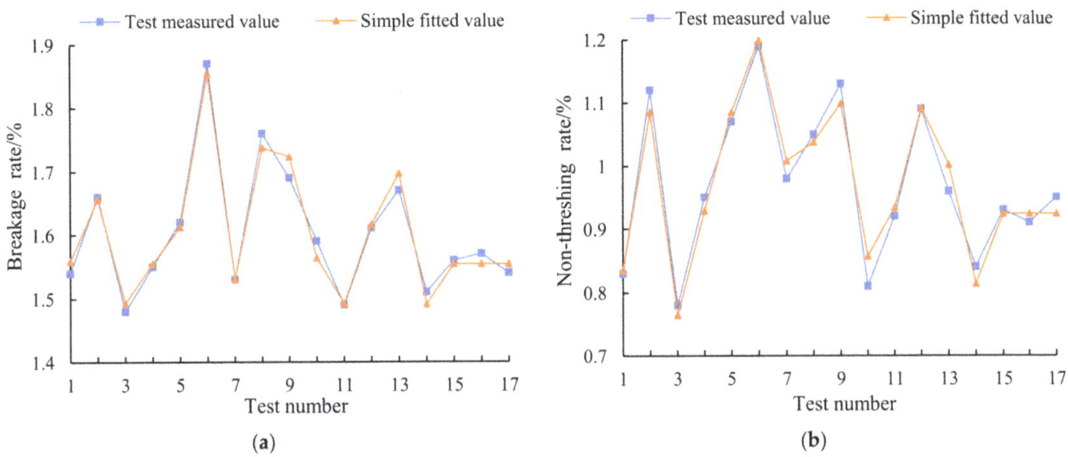

Figure 13. Distribution of test measured and fitted values. (**a**) Fitting curve of breakage rate, (**b**) fitting curve of non-threshing rate.

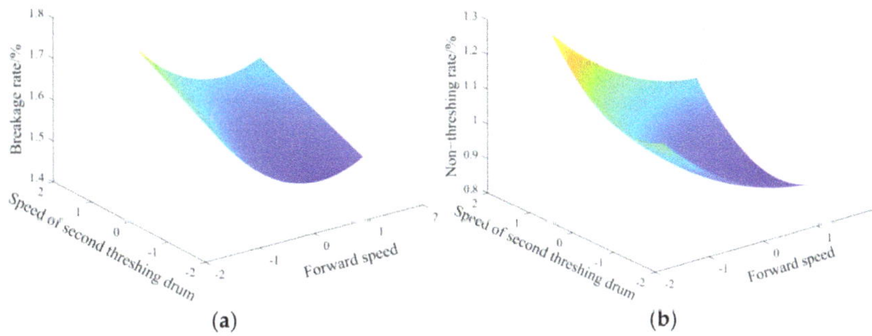

Figure 14. Interaction between the forward speed and the speed of the second threshing drum. (**a**) Effect on breakage rate, (**b**) effect on the non-threshing rate.

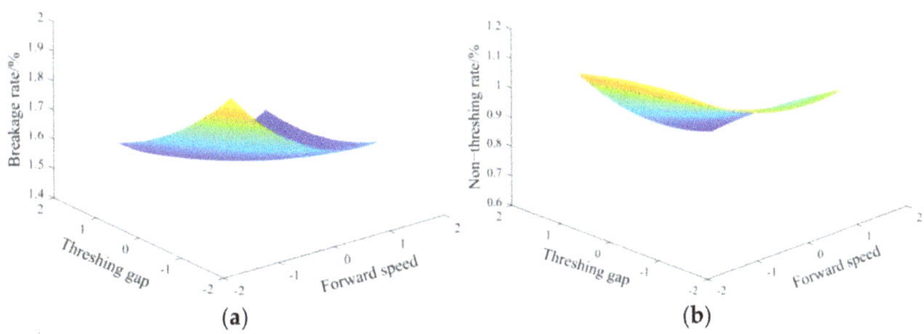

Figure 15. Interaction between the forward speed and the threshing gap. (**a**) Effect on breakage rate, (**b**) effect on the non-threshing rate.

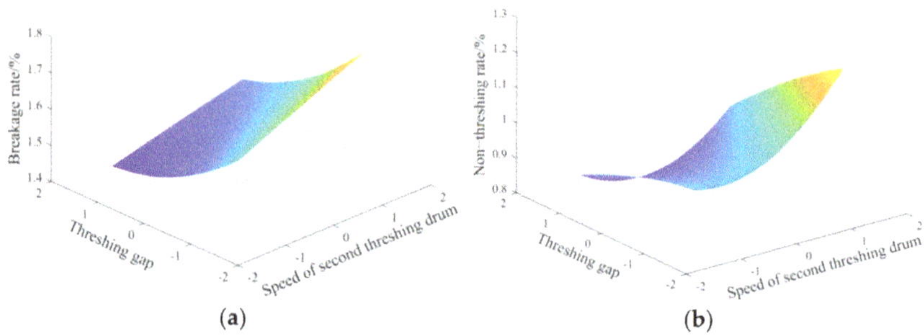

Figure 16. Interaction between the speed of the second threshing drum and the threshing gap. (**a**) Effect on breakage rate, (**b**) effect on the non-threshing rate.

4.3.1. Interaction Analysis between Forward Speed and Second Threshing Speed Drum

Figure 14a shows that the breakage rate decreases at first and then increases as the forward speed increase. This is because when the feed amount is low, the strike force and strikes number of nail teeth are limited. After that, according to Equation (17), when $0° < \gamma < 90°$, the impact force of threshing gear under high feed amount F_3 is greater than the impact force of the threshing gear under low feed amount F_2. The feeding amount increases with the forward speed increase, and the impact force of threshing teeth will also increase, resulting in a higher breakage rate. Therefore, there will be a double change phenomenon. In order to ensure a small and reasonable breakage rate, the forward speed can be controlled within the range of 0.5~0.88 $m \cdot s^{-1}$.

Duane et al. [32] showed that the kernel damage increased with increased kernel velocity and was related more to this than any of the other factors tested. With the increase in threshing drum speed, the overall breakage rate shows an upward trend. Increasing the speed of the threshing drum will increase the hitting frequency of threshing teeth and the number of impacts and rubbing, which increases the breakage rate. The drum speed should be controlled at a minimum of 500 $r \cdot min^{-1}$. These findings are consistent with the findings of Di et al. [15]. It can be concluded that the effect of forward speed on the breakage rate is significantly higher than that of the threshing drum speed from the response cloud map.

Figure 14b shows that the non-threshing rate presents a downward trend as the forward speed increases. This is due to the increase in feed quantity, which increases the chances of contact between drum and ear. As the speed of the threshing drum increases, the non-threshing rate decreases first and then increases. This is because the drum speed increases, allowing for more contact opportunities. It is known from Equation (12) that the impact of threshing teeth on ear F_2 decreases when the angular speed of drum ω increases, thereby lowering the non-threshing rate. It can be concluded that the forward speed has a higher impact on the non-threshing rate than the speed of the threshing drum from the response cloud map.

4.3.2. Interaction Analysis between Forward Speed and Threshing Gap

Figure 15a shows that the increase in the threshing gap causes the breakage rate to decrease significantly. This is because as the threshing gap widens, the effect of threshing elements on the ear was weakened, the grain crushing rate was reduced, the mutual squeezing and rubbing effect between ear was weakened, resulting in the non-threshing rate was increased. This is similar to the findings of Chen et al. [29]. The breakage rate decreases at first then increases as the forward speed increases. This is because, with the increase in the feeding amount, the ears cannot move in the axial direction in time, which will cause accumulation. It can be concluded that the effect of the threshing gap on the breakage rate is slightly higher than the forward speed from the response cloud map.

Figure 15b shows that the non-threshing rate increases slightly as the threshing gap widens and decreases significantly as the forward speed increases, and the response speed of forward speed in the map is faster than the threshing gap. When the feeding amount is large, there are more ears in the gap between the threshing drum and concave plate, and the threshing elements are in full contact with the corn ears to improve the non-threshing rate [33]. It can be concluded that the effect of the threshing gap on the non-threshing rate is lower than the forward speed from the response cloud map.

4.3.3. Interaction Analysis between Second Threshing Drum Speed and Threshing Gap

Figure 16a shows that the breakage rate decreases significantly as the threshing gap widens and increases as the threshing drum speed increases. This is because when the feeding amount is constant, the threshing gap increases and the beating frequency of corn will not increase rapidly with the increase in drum speed. It can be concluded that the effect of the threshing gap on the breakage rate is slightly higher than the speed of the threshing drum from the response cloud map.

Figure 16b shows that when the forward speed is consistent, the non-threshing rate decreases at first and then increases as the threshing drum speed increases and decreases slightly as the threshing gap widens, but the response speed of the threshing gap in the map is much slower than the threshing drum speed. This is because with the increase in drum speed, the threshing efficiency of the drum was also improved, and the retention time of corn in the threshing chamber was reduced, resulting in the increase in non-threshing rate [30]. It can be concluded that the effect of threshing drum speed on the non-threshing rate is significantly higher than the threshing gap from the response cloud map.

4.4. Performance Optimization of Filed Test

According to the requirements of corn plot kernel harvesting, the minimum breakage rate and non-threshing rate of kernels should be met during harvest. Create the objective function for the two indicators above, and substitute the obtained regression model into the function. The objective function and constraint conditions are described as follows:

$$\left\{\begin{array}{c} F_{\min}(z)=\left\{\begin{array}{c} y_1(z_1,z_2,z_3) \\ y_2(z_1,z_2,z_3) \end{array}\right\} \\ s.t.\left\{\begin{array}{c} 0.5\ \mathrm{m/s}\le z_1\le 1.26\ \mathrm{m/s} \\ 500\ \mathrm{r/min}\le z_2\le 800\ \mathrm{r/min} \\ 20\ \mathrm{mm}\le z_3\le 40\ \mathrm{mm} \end{array}\right. \end{array}\right. \quad (26)$$

Using MATLAB to optimize the parameters, the optimal value is obtained as the forward speed is 0.61 m/s, the speed of the threshing drum is 500 r/min, and the threshing gap is 40 mm. The verification test uses Zhengdan 958, which is the same variety as the regression orthogonal combination design test. The number of tests is three times, and the average value is taken. The results are shown in Table 9. The test results show that in the optimal parameters test, the field test validation value is close to the theoretical value.

Table 9. Optimum parameters test results.

	Breaking Rate y_1/%	Non-Threshing Rate y_2/%	F_{min}/%
Theoretical value	1.45	0.82	2.27
Field test validation value	1.47	0.89	2.36

5. Conclusions

(1) Based on the existing ear-picking mechanism, the basic parameters of the ear-picking device are theoretically studied by sorting out and collecting information on the main popularized breeding corn to ensure the reasonable and reliability of the design.

Combining with the structural characteristics of corn plot kernel harvester, a flexible gap self-adjusting ear-picking device and a dual-axial flow corn threshing device were optimized through a pneumatically assisted seed cleaning system and three times of cleaning to realize the non-mixed harvest of kernels between plots.

(2) Through the mechanical analysis of corn ears, it is concluded that the factors affecting kernel damage were drum speed, threshing gap, and feed quantity. Combining the actual working conditions of the prototype, it is determined that the forward speed of the unit varies from 0.4 to 1.0 m/s, the speed of the threshing drum varies from 300 to 500 r/min, and the threshing gap is arbitrarily adjustable within the range of 20~40 mm.

(3) The field test carried out the quadratic regression orthogonal combination design by taking the breakage rate and the non-threshing rate as the test evaluating indexes. The test data were analyzed to produce a quadratic regression model between the breakage rate and the non-threshing rate. Through the significance analysis of regression coefficients, it is known that the regression model has a high degree of fit. Meanwhile, the distribution of test measured and fitted values for the two regression models are very well matched.

(4) Furthermore, through the interactive analysis of three factors, it can be concluded that the forward speed has a significant impact on the breakage rate and the non-threshing rate. The threshing gap has a greater impact on the breakage rate than the speed of the threshing drum. The effect of the threshing drum speed on the non-threshing rate is greater than the effect of the threshing gap. Then the optimal operating parameters of the machine are calculated by optimization, which showed that the forward speed was 0.61 m/s, the speed of the threshing drum was 500 r/min, and the threshing gap was 40 mm. More importantly, the minimum breakage rate and non-threshing rate were 1.47% and 0.89%, respectively in the field validation test.

6. Patents

The flexible gap self-adjusting ear-picking device and dual-axial flow corn threshing device reported in this manuscript have been applied for patents in China (Application No. CN108293421A).

Author Contributions: Conceptualization, R.Y. and D.C.; Data curation, D.C. and X.Z.; Formal analysis, R.Y. and D.C.; Funding acquisition, S.S.; Investigation, R.Y., X.Z. and Z.P.; Methodology, R.Y. and D.C.; Project administration, S.S.; Resources, R.Y. and X.Z.; Software, R.Y. and D.C.; Supervision, S.S.; Validation, R.Y., D.C., X.Z. and Z.P.; Visualization, R.Y. and X.Z.; Writing—original draft, R.Y. and D.C.; Writing—review and editing, R.Y. and D.C. All authors have read and agreed to the published version of the manuscript.

Funding: This work was financially supported by the research and development of seed breeding technology and equipment (Grant No. 2017YFD0701200), which is the National Key Research and Development Project of China.

Institutional Review Board Statement: Not applicable.

Informed Consent Statement: Not applicable.

Data Availability Statement: The data presented in this study are available on-demand from the second author at (20192204153@stu.qau.edu.cn).

Acknowledgments: This work was financially supported by the National Key Research and Development Project of China (research and development of seed breeding technology and equipment, 2017YFD0701200). We thank Guoying Li for its fund and investigation assistance during the preparation of this manuscript. All supports and assistance are sincerely appreciated.

Conflicts of Interest: The authors declare no conflict of interest.

References

1. Zhu, M.; Chen, H.; Li, Y. Investigation and development analysis of seed industry mechanization in China. *Int. J. Agr. Biol. Eng.* **2015**, *31*, 1–7.
2. Qin, X.; Feng, F.; Li, Y.; Xu, S.; Siddique, K.H.M.; Liao, Y. Maize yield improvements in China: Past trends and future directions. *Plant. Breed.* **2016**, *135*, 166–176. [CrossRef]
3. Huang, M.; Zou, Y. Integrating mechanization with agronomy and breeding to ensure food security in China. *Field Crops Res.* **2018**, *224*, 22–27. [CrossRef]
4. Hou, L.; Wang, K.; Wang, Y.; Li, L.; Ming, B.; Xie, R. In-field harvest loss of mechanically-harvested maize kernel and affecting factors in China. *Int. J. Agric. Biol. Eng.* **2021**, *14*, 29–37. [CrossRef]
5. Babić, L.; Radojèin, M.; Pavkov, I.; Babić, M.; Turan, J.; Zoranović, M.; Stanišić, S. Physical properties and compression loading behaviour of corn seed. *Int. Agrophysics* **2013**, *27*, 119–126. [CrossRef]
6. Volkovas, V.; Petkevičius, S.; Špokas, L. Establishment of maize kernel elasticity on the basis of impact load. *Mechanika* **2006**, *6*, 64–67.
7. Petre, I.M. *Combine Harvesters Theory, Modeling and Design*; Taylor & Francis Group: Boca Raton, FL, USA, 2016.
8. Pickard, G.E. Laboratory studies in corn combining. *Agric. Eng.* **1955**, *36*, 792–794.
9. Srivastava, A.K.; Herum, F.L.; Stevens, K.K. Impact parameters related to physical damage to corn kernel. *Trans. ASABE* **1976**, *19*, 1147–1151. [CrossRef]
10. Voicu, G.; Căsăndroiu, T.; Stan, G. Using the dimensional analysis for a mathematical model to predict the seeds losses at the cleaning system of the cereals harvesting combines. *U.P.B. Sci.* **2007**, *69*, 29–39.
11. Fu, Q.; Fu, J.; Chen, Z.; Han, L.; Ren, L. Effect of impact parameters and moisture content on kernel loss during corn snapping. *Int. Agrophys.* **2019**, *33*, 493–502. [CrossRef]
12. Geng, A.; Yang, J.; Zhang, J.; Zhang, Z.; Yang, Q.; Li, R. Influence factor analysis of mechanical damage on corn ear picking. *Trans. Chin. Soc. Agric. Eng.* **2016**, *32*, 56–62.
13. Li, X.; Wu, K. Design and Experiment of Bionic Discrete Devices Based on Corn Threshing System. *Chem. Eng. Trans.* **2016**, *51*, 127–132. [CrossRef]
14. Li, X.; Du, Y.; Guo, J.; Mao, E. Design, Simulation, and Test of a New Threshing Cylinder for High Moisture Content Corn. *Appl. Sci.* **2020**, *10*, 4925. [CrossRef]
15. Di, Z.; Cui, Z.; Zhang, H.; Zhou, J.; Zhang, M.; Bu, L. Design and experiment of rasp bar and nail tooth combined axial flow corn threshing cylinder. *Int. J. Agric. Biol. Eng.* **2018**, *34*, 28–34.
16. Li, K. *Design and Experimental Study on Gap-Adjustable Combined Ear-Picking Mechanism of Corn*; China Agricultural University: Beijing, China, 2018.
17. Du, Y.; Mao, E.; Zhu, Z.; Wang, X.; Yue, X.; Li, X. Design and Experiment of Two row Corn Harvester Header. *Trans. Chin. Soc. Agric. Mach.* **2013**, *44 (Supp2)*, 22–26.
18. He, J. *Biomimetic Surfaces of Snapping Rolls with Lower Damage Action and Simulation of No-Row Feed-in Mechanism of Maiz Harvesters*; Jilin University: Jilin, China, 2007.
19. Zhang, Z.; Chi, R.; Du, Y.; Pan, X.; Dong, N.; Xie, B. Experiments and modeling of mechanism analysis of maize picking loss. *Int. J. Agric. Biol. Eng.* **2021**, *14*, 11–19. [CrossRef]
20. Zhang, Z. *Study on a New Corn Combine Harvester Header for Reaping both Corn Stalk and Spike*; China Agricultural University: Beijing, China, 2018.
21. Xin, S. *Study on the Mechanism and Key Technology of Corn Ear Picking with Vertical Roller*; Gansu Agricultural University: Lanzhou, China, 2020.
22. Li, X.; Xiong, S.; Du, Z.; Geng, L.; Ji, J. Design and experiment on floating corn single panicle threshing device. *Int. J. Agric. Biol. Eng.* **2017**, *48*, 104–111. [CrossRef]
23. Yi, S.; Tao, G.; Mao, X. Comparative experiment on the distribution regularities of threshed mixtures for two types of axial flow threshing and separating installation. *Int. J. Agr. Biol. Eng.* **2008**, *24*, 154–156.
24. Wan, X.; Liao, Q.; Xu, Y.; Yuan, J.; Li, H. Design and evaluation of cyclone separation cleaning devices usding a conical sieve for rape combine harvesters. *Appl. Eng. Agric.* **2018**, *34*, 677–686. [CrossRef]
25. Yang, L.; Wang, W.; Wang, M.; Zhang, H.; Hou, M. Structural Dynamics of Corn Threshing Drum Based on Computer Simulation Technology. *Wireless Pers Commun.* **2018**, *102*, 701–711. [CrossRef]
26. Gao, L.; Zhang, S.; Chen, R.; Yang, D. Design and experiment on soybean breeding thresher of double feeding roller and combined threshing cylinder. *Trans. Chin. Soc. Agric. Mach.* **2015**, *46*, 112–118. [CrossRef]
27. Yang, D.; Jiang, D.; Shen, Y.; Gao, L.; Wan, L.; Wang, J. Design and Test on Soybean Seed Thresher with Tangential-axial Flow Double-roller. *Trans. Chin. Soc. Agric. Mach.* **2017**, *48*, 102–110. [CrossRef]
28. Tang, H. *Design and Mechanism Analysis of Ripple Surface Pickup Finger Maize Precision Seed Metering Device*; Dongbei Agricultural University: Shenyang, China, 2018.
29. Chen, M.; Xu, G.; Wang, C.; Diao, P.; Zhang, Y.; Niu, G. Design and Experiment of Roller-type Combined Longitudinal Axial Flow Flexible Threshing and Separating Device for Corn. *Trans. Chin. Soc. Agric. Mach.* **2020**, *51*, 123–131. [CrossRef]

30. Yang, L.; Wang, W.; Zhang, H.; Li, L.; Wang, M.; Hou, M. Improved design and bench test based on tangential flow-transverse axial flow maize threshing system. *Int. J. Agric. Biol. Eng.* **2018**, *34*, 35–43. [CrossRef]
31. He, Y. *Experimental Design and Analysis*; Chemical Industry Press: Beijing, China, 2013.
32. Keller, D.; Converse, H.; Hodges, T.; Chung, D. Corn kernel damage due to high velocity impact. *Trans. ASAE* **1972**, *12*, 330–331. [CrossRef]
33. Wang, Z.; Wang, Z.; Zhang, Y.; Yan, W.; Chi, Y.; Liu, C. Design and test of longitudinal axial flexible hammer-claw corn thresher. *Trans. Chin. Soc. Agric. Mach.* **2020**, *51*, 109–117.

agriculture

Article

Effect of Planting Distance on the Mechanical Harvesting of Hot Pepper

Seokho Kang, Yeongsu Kim, Hyunggyu Park, Seungmin Woo, Daniel Dooyum Uyeh and Yushin Ha *

Department of Bio-Industrial Machinery Engineering, College of Agriculture and Life Sciences, Kyungpook National University, Daehak-ro 80, Buk-gu, Daegu 41566, Korea; deshshk@knu.ac.kr (S.K.); mvio9256@knu.ac.kr (Y.K.); pyd4796@knu.ac.kr (H.P.); kooger7571@naver.com (S.W.); Uyehdooyum@knu.ac.kr (D.D.U.)

* Correspondence: yushin72@knu.ac.kr; Tel.: +82-53-950-5792

Abstract: Hot peppers are well known for being spicy and also have a high nutrient content. Human resources have formerly been used to harvest hot peppers; however, a high level of musculoskeletal risk to the human workforce has been reported. Therefore, to reduce the risk to farmers and replace the human workforce, the mechanical harvesting of hot pepper and steps to improve the harvesting efficiency of farmers were conducted. To achieve this, the effect of planting distance on the mechanical harvesting of hot peppers was analyzed at three planting distances (30, 40, and 50 cm) with several cultivars. Subsequently, machine-harvested hot peppers were classified into five groups (marketable, damaged, lost, unharvested, and twigged hot pepper), depending on their postharvest status. The average weight ratio of each group was then calculated, after which statistical analyses were conducted. The effect of planting distance on mechanical harvesting was then analyzed by comparing the differences between each group's average weight ratio and the total weight of hot pepper, which was simultaneously harvested mechanically at each planting distance. Results showed that the average weight ratio of marketable, unharvested, and twigged hot pepper improved as the planting distance increased. However, no effect on the average weight ratio of damaged and lost hot pepper was observed. The highest yield of marketable hot pepper was found at a planting distance of 40 cm, and the average weight ratio to the whole was lower than at 50 cm of planting distance. Thus, the most suitable planting distance for mechanical harvesting was 40 cm.

Keywords: *Capsicum annuum* L.; hot pepper harvester; machinery harvest efficiency; planting distance

Citation: Kang, S.; Kim, Y.; Park, H.; Woo, S.; Uyeh, D.D.; Ha, Y. Effect of Planting Distance on the Mechanical Harvesting of Hot Pepper. *Agriculture* **2021**, *11*, 945. https://doi.org/10.3390/agriculture11100945

Academic Editor: Jose Pérez-Alonso

Received: 26 August 2021
Accepted: 27 September 2021
Published: 29 September 2021

1. Introduction

Hot pepper (*Capsicum annuum* L.) is among the most popular and demanded vegetables globally because of its intense flavor, attractive color, and high nutrient content [1,2]. Its nutrient content, such as L-ascorbic acid, soluble sugars, and carotenoids, are beneficial for human nutrition. Therefore, breeding methods have been studied previously to enhance its chemical quality [3–5].

Hot pepper is harvested manually, and it's cultivated area has been increasing globally [6]. It is well known that harvesting hot pepper requires long periods of manual labor; this results in the manifestation of many musculoskeletal problems among farmers. Studies have shown that farmers are exposed to higher risks of hazard when they manually harvest the crop. These includes musculoskeletal symptoms on their backs, knees, necks, and shoulders after daily work [7,8]. These long periods of field work also results in high labor costs, and continuous increase in base wage rates are annually [9]. Hence, the mechanical hot pepper harvester was developed and used for convenience to hazards faced by growers and replace human workforce. Unlike the manual harvesting process conducted several times, mechanical harvesting of hot pepper is undertaken only once. The quantity of mechanically harvested hot pepper, in addition to its harvesting efficiency, is better than handwork. Also, hot pepper cultivars should be highly resistant against anthrax which occurs during late harvesting of the fruit. Field performance tests were conducted on double-opened helix-type harvesters by calculating their harvesting efficiency, successful harvesting rates, pepper with twig rates, pepper left on plant rates, and ground fall loss rates at different rotational speeds of the helix. As reported, the harvesting efficiency increased with an increase in the speed of the helix; however, pepper with twigs and damage rates also increased. Thus, successfully harvesting rates require improvements in all rotational speeds of the helix [10].

Furthermore, comparisons of mechanical harvesting performance among five harvest mechanisms, including comparisons of harvest efficiency among four hot pepper cultivars, were conducted. From the study, all cultivars' harvest efficiencies were <70%, and the double-open helix mechanism had the best machine-harvest performance among those five, with a harvest efficiency of 88.3%. However, this mechanism also had the highest ratio of the whole to damaged fruit, and the low yield of marketable hot pepper was unsatisfactory [11]. Likewise, the mechanical harvest efficiency of six new Mexican hot pepper cultivars showed an efficiency below 60%, and the average yield of the six cultivars under study was 27.3 ton/ha. Nevertheless, considering that the yield of pepper in studies conducted in the United States of America, and Europe was 31.8 and 32.02 ton/ha, respectively, in 2019, his yield value desperately required improvement for sustainability. Another factor for yield differences among cultivars that were also analyzed was the architecture of the plant [6,12]. Moreover, a study reported that commercialized harvesters had a severe problem with the low harvest efficiency of commercial hot pepper that reduced the yield of these hot peppers. Thus, the hot pepper harvester needed improvement by reducing its damage to the crop and minimizing the ground-fallen peppers [13].

A previous study indicated that mechanical harvest efficiency was related to plant structure [14], proposing that the growth of the pepper plant is related to the planting distance. Therefore, planting distance affected the growth of pepper plants, including yield, number per plant, and weight of plant [15,16]. Furthermore, the growth of the pepper plant, such as plant height, fruit diameter, fruit thickness, and fruit length, increased with distance [17,18]. The yield of hot peppers per plant and stem girth also increased with distance [19–21]. These results, wherein planting distance influenced the number of fruits and weight of fruits per plant were confirmed in other plants, such as cocoyam, mango, onion, rice, and tomato [22–26].

Based on these findings, it was evident that the planting distances of hot pepper plants were highly related to the quantity of the pepper obtained at harvest and affected mechanical harvesting efficiency. However, planting distance was used at a variety of levels: 35 cm in Canada [27], 38–45 cm in the United Kingdom [28], 30–45 cm in America [29,30], 30 cm in Brazil [31], and 30–40 cm in the Republic of Korea [32]. Based on the total yield of hot peppers per plant and marketable fruit yield per plant, a previous study indicated that a low planting density resulted in a high total yield of hot pepper [33–35]. There are suggested planting distance for manual work in each country, but there is no recommended planting distance for mechanical harvesting to the best of our knowledge. Consequently, a comparison of different planting distances to improve mechanical harvesting efficiency was conducted in this study.

2. Materials and Methods

2.1. Hot Pepper Harvester

Hot pepper harvesters (CH301, TYM CO., LTD., Daegu, Republic of Korea) comprised of three main components: threshing, transporting, and loading bay. The threshing component was placed in front of the machine, where the rotating double-helix system threshes the whole plant and pulls out the fruit. Then, the harvested fruit is transported to the loading bay, after which a motor-created airflow was used to separate leaves. The specifications of the hot pepper harvesting machine used in this study are presented in Figure 1 and Table 1 [36].

Figure 1. Hot pepper harvester used for the field test.

Table 1. Specifications of the hot pepper harvester.

Hot Pepper Harvester	Specifications
Model	CH301 (TYM CO., LTD./Republic of Korea)
Company/country	TYM CO., LTD./Republic of Korea
Length/width/height (mm)	4380/1810/1800
Ground clearance (mm)	600
Engine power (kW)	20.1
Revolution speed of helix (RPM)	700
Weight (kg)	2065
Work speed (km/h)	1.6
Maximum capacity of loading bay (kg)	200
Diameter of the helix (mm)	100
Length of helix (mm)	1700

2.2. Cultivars

Three hot pepper cultivars were selected: AR Legend (Pepper & Breeding Institute, Gimje, Republic of Korea), Maeuntan (Syngenta Korea, Daegu, Republic of Korea), and Jeockyoung (Rural Development Administration, Jinju, Republic of Korea).

2.3. Planting Distance

Hot peppers were harvested using the double-helix harvester with a threshing component, where the whole pepper plant goes through the gap between the double-helix. Consequently, controlling the planting distance is proposed to decrease the impact between pepper plants, thereby earning enough time for the harvester to thresh the whole pepper plant before a new inflow.

Therefore, to obtain a low planting density, a narrow planting distance was required. For planting distance, 30–50 cm is used globally. Thus, in this study, three levels of planting distances were selected for the mechanical harvesting experiment (Figure 2).

(a) (b) (c)

Figure 2. Three levels of planting distance selected for harvesting experiment (**a**) 30 cm planting distance, (**b**) 40 cm planting distance, (**c**) 50 cm planting distance.

2.4. Field Conditions

The experiment was conducted at the upland-field machinery research center at Gunwi-gu, Gyeongsangbuk-do, Republic of Korea (latitude 36°06′55.1″-N, longitude 128°38′28.1″-E). The field area was 1487 $m^{2,}$ and the soil was clay loam with a pH of 6.2. The experimental region had an average annual air temperature between 30.2 °C and 19.2 °C, and a 460 mm average annual rainfall during summer. The distance between rows was 130 cm, with three rows in each cultivar, and the length of each row was 90 m. All cultivars were sown on 3 March 2020, raised for 75 days, and transplanted on 18 May 2020. The pepper plant reached 80 cm in height in all cultivars.

2.5. Mechanical Harvest

The mechanical harvest was conducted at 10 m for each planting distance. The constant forward speed of the harvester was 1.6 km/h, and the rotational speed of the helix was 700 rpm. After 10 m mechanical harvesting, harvested peppers were manually classified into five groups depending on their status, and fresh weight was used for measurement (Figure 3). The groups included marketable, damaged, lost, unharvested, and twigged hot peppers. Marketable hot pepper was defined as well-harvested pepper, which is whole and safely transported into the loading bay; damaged hot pepper was defined as that which was ripped and chopped; lost hot pepper was defined as ground-fallen peppers during mechanical harvesting; unharvested hot pepper was the fruit left after mechanical harvesting; and twigged hot pepper was defined as those peppers still having twigs and leaves attached postharvest.

(a) (b) (c)

(d) (e)

Figure 3. Image of classified machinery-based harvested peppers (**a**) marketable hot pepper, (**b**) damaged hot pepper, (**c**) lost hot pepper, (**d**) unharvested hot pepper, and (**e**) twigged hot pepper.

Following this, each group was weighed using an electric scale (PAG4102, OHAUS, Parsippany, NJ, USA). The mechanically harvested hot pepper in the loading bay consisted of five groups of hot pepper. For analysis of the percentage of each group in the loading bay after 10 m of harvesting, the weight ratio of the whole weight of mechanically harvested pepper was calculated using Equation (1) below with the parameters shown in Table 2.

$$R_n = \frac{S_n\ (kg)}{\sum_{n=1}^{5} S_n\ (kg)} \times 100\ (\%) \tag{1}$$

Table 2. Symbols and definitions for equation one.

Symbols	Definitions	Symbols	Definitions
R_1	The weight ratio of marketable hot pepper	S_1	Weight of marketable hot pepper
R_2	The weight ratio of damaged hot pepper	S_2	Weight of damaged hot pepper
R_3	The weight ratio of lost hot pepper	S_3	Weight of lost hot pepper
R_4	The weight ratio of unharvested hot pepper	S_4	Weight of unharvested hot pepper
R_5	The weight ratio of twigged hot pepper	S_5	Weight of twigged hot pepper

2.6. Statistical Analysis

Statistical analysis was conducted using SPSS (IBM SPSS Statistics 26, IBM, Armonk, NY, USA) at a 5% significance level. Results are presented for each group regarding the mean weight ratio, conducted nine times during mechanical harvesting experiments. Furthermore, the difference in weight ratios between planting distances was elaborated using one-way analysis of variance (ANOVA) with Duncan's test. Pearson's correlation test was conducted to elaborate on the relationship between the planting distance and the weight ratio of each classified group. Furthermore, multiple ways ANOVA was conducted to analyze the effect of planting distance on the variables and to compare the three cultivars statistically.

3. Results

3.1. Marketable Hot Pepper

Mechanical harvesting efficiency at different planting distances was evaluated using the weighted ratio of marketable, damaged, lost, unharvested, and twigged hot peppers. Table 3 presents the mean statistical analysis of the weight ratio obtained from marketable harvested hot peppers in all cultivars. Comparing the planting distances, the average weight ratios were 71.1 ± 4.0% (SD), 77.1 ± 3.5% (SD), and 81.5 ± 2.2% (SD) for AR legend; 74.3 ± 3.0% (SD), 80.0 ± 1.4% (SD), and 82.1 ± 1.9% (SD) for Maeuntan; and 71.0 ± 1.5% (SD), 74.5 ± 2.1% (SD), and 81.1 ± 2.2% (SD) for Jeockyoung. Results showed that the average weight ratio of marketable hot pepper increased with planting distance. Moreover, the highest weight ratio of marketable hot pepper was indicated at a planting distance of 50 cm for all cultivars. Furthermore, results of the ANOVA on the effects of planting distance and cultivar on the weight ratio of marketable hot pepper showed a significantly different effect as all p-values were <0.05. However, there was no interaction effect between planting distance and cultivar as the p-value was >0.05.

Furthermore, previous studies have reported that yield per plant was significantly influenced by planting distance. The yield per plant was also increased with an increase in planting distances [19,33,35]. However, further planting distances caused a leak in pepper yield because increased planting distances meant fewer plants per unit area of hot pepper fields. Therefore, the highest weight of marketable hot pepper was obtained at a 40 cm planting distance (Table 3), with 7.0 ± 0.2 kg (SD) for AR Legend, 9.5 ± 0.7 kg (SD) for Maeuntan, and 6.2 ± 0.5 kg (SD) for Jeockyoung obtained for all cultivars. Hence, we propose that the most suitable planting distance for mechanical harvesting of hot pepper is 40 cm.

Table 3. Statistical analysis results on marketable hot pepper.

Cultivar	PD (cm)	M ± SD (%)	Weight (kg)	PCC	F-Value	*p*-Value
AR Legend	30	71.1 ± 4.0 a [(1)]	5.9 ± 0.25	0.780 **	11.928	0.001
	40	77.1 ± 3.5 b	7.0 ± 0.2			
	50	81.5 ± 2.2 b	6.3 ± 0.1			
Maeuntan	30	74.3 ± 3.0 a	8.0 ± 0.3	0.802 **	16.331	0.000
	40	80.0 ± 1.4 b	9.5 ± 0.7			
	50	82.1 ± 1.9 b	8.2 ± 0.4			
Jeockyoung	30	71.0 ± 1.5 a	5.3 ± 0.4	0.889 **	33.191	0.000
	40	74.5 ± 2.1 b	6.2 ± 0.5			
	50	81.1 ± 2.2 c	5.1 ± 0.3			
Group	**Source**	**DF**	**ANOVA SS**	**Mean Square**	**F-Value**	***p*-Value**
Marketable hot pepper	PD	2	804.111	402.056	49.478	0.000 **
	Cultivar	2	100.778	50.389	6.201	0.004 **
	PD*Cultivar	4	35.444	8.861	1.090	0.373
	Error	45	365.667	8.126		
	Total	53	1306.000			

PD: Planting Distance; M: Mean; SD: Standard Deviation; PCC: Pearson's Correlation Coefficient; SS: Sum of Squares. [(1)]: Mean separation within columns by Duncan's Multiple Range test at $p = 0.05$. ** $p < 0.01$.

3.2. Damaged Hot Pepper

The physical quality of hot pepper, which was pre-harvested manually, and hot pepper, which was post-mechanically harvested are shown in Figure 4. This method of quality assessment was adopted because it was difficult to carry out the standard wet chemistry methods. These wet chemistry methods are also destructive and time-consuming. The assessment showed that the physical quality of the two hot peppers (Figure 4a,b) was defined with no chop or rip on their pericarp. The statistical analysis of the average weight ratio of damaged hot pepper is shown in Table 4. Comparing between the planting distances, the average weight ratios were 4.2 ± 1.4% (SD), 2.5 ± 0.7% (SD), and 2.2 ± 0.3% (SD) for AR legend; 3.8 ± 1.6% (SD), 3.7 ± 0.9% (SD), and 3.5 ± 0.9% (SD) for Maeuntan; and 4.8 ± 1.8% (SD), 4.2 ± 1.0% (SD), and 3.0 ± 1.5% (SD) for Jeockyoung. The average weight ratio of damaged hot pepper was only related to the planting distance; no significantly different interaction effect was found between planting distance and cultivar. Furthermore, most damaged hot peppers were recorded during transportation to the loading bay; thus, we suspect that too much force was applied to the hot pepper, resulting in the peppers being chopped and ripped out. Compared with a previous study on mechanical harvesting experiments, the damaged ratio observed in this study was much lower (8.37%) because the broader and blockier structure of the hot pepper ridges, characterized by the different planting distances, was more acceptable in managing the mechanical damage [10]. Furthermore, additional studies with fruit damage measurement in each harvester unit are required and would be done in the future.

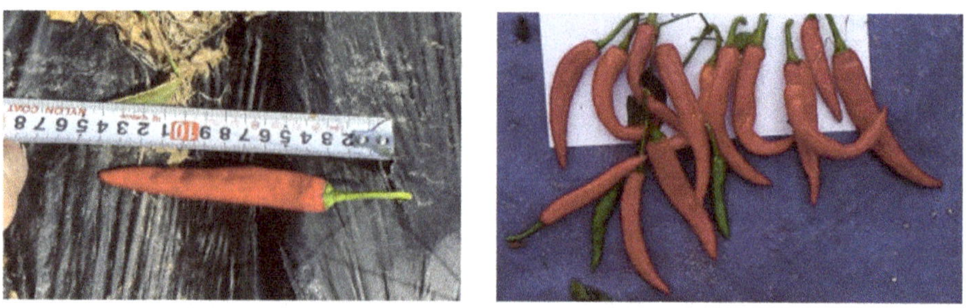

(a) (b)

Figure 4. Physical quality of hot pepper (**a**) manually pre-harvested and (**b**) post-mechanically harvested.

Table 4. Statistical analysis results on damaged hot pepper.

Cultivar	PD (cm)	M ± SD (%)	Weight (kg)	PCC	F-Value	*p*-Value
AR Legend	30	4.2 ± 1.4 b [(1)]	0.4 ± 0.2	−0.623 **	6.019	0.012
	40	2.5 ± 0.7 a	0.2 ± 0.1			
	50	2.2 ± 0.3 a	0.2 ± 0.1			
Maeuntan	30	3.8 ± 1.6 a	0.4 ± 0.2	−0.109	0.090	0.914
	40	3.7 ± 0.9 a	0.4 ± 0.1			
	50	3.5 ± 0.9 a	0.3 ± 0.1			
Jeockyoung	30	4.8 ± 1.8 a	0.2 ± 0.1	−0.440	1.860	0.190
	40	4.2 ± 1.0 a	0.4 ± 0.1			
	50	3.0 ± 1.5 a	0.2 ± 0.1			
Group	**Source**	**DF**	**ANOVA SS**	**Mean Square**	**F-Value**	***p*-Value**
Damaged hot pepper	PD	2	17.593	8.796	4.576	0.016 **
	Cultivar	2	10.481	5.241	2.726	0.076
	PD*Cultivar	4	6.852	1.713	0.891	0.477
	Error	45	86.500	1.922		
	Total	53	121.426			

PD: Planting Distance; M: Mean; SD: Standard Deviation; PCC: Pearson's Correlation Coefficient; SS: Sum of Squares. [(1)]: Mean separation within columns by Duncan's Multiple Range test at $p = 0.05$. **: $p < 0.01$.

3.3. Lost Hot Pepper

The statistical analysis of the average weight ratio of lost hot pepper are shown in Table 5. Comparing planting distances, the average weight ratios were 4.2 ± 1.3% (SD), 3.8 ± 1.3% (SD), and 4.8 ± 2.6% (SD) for AR legend; 3.0 ± 1.0% (SD), 2.8 ± 0.9% (SD), and 5.1 ± 1.2% (SD) for Maeuntan; and 4.0 ± 1.0% (SD), 4.1 ± 2.3% (SD), and 2.1 ± 1.3% (SD) for Jeockyoung. Results also showed that the average weight ratios of lost hot pepper cultivars did not relate to the planting distance or cultivar; therefore, no significantly different effect was observed. The result of two-way ANOVA indicated that there was an interaction effect between planting distance and cultivar. Therefore, the weight ratio of lost hot pepper was significantly different when planting distance and cultivar were considered. Most lost hot peppers were observed during transportation and probably caused by airflow, which was created in the harvester for separating the fruits from the leaves.

3.4. Unharvested Hot Pepper

Table 6 presents the statistical analysis results of average weight ratios of unharvested hot peppers in all cultivars. Comparing the planting distances, the average weight ratio was 10.0 ± 1.2% (SD), 8.8 ± 2.6% (SD), and 5.5 ± 1.3% (SD) for AR Legend; 9.3 ± 2.4% (SD), 7.5 ± 1.9% (SD), and 5.0 ± 1.0% (SD) for Maeuntan; and 10.5 ± 0.5% (SD), 8.6 ± 0.7% (SD), and 7.5 ± 0.7% (SD) for Jeockyoung. From the results, the weight ratio of unharvested hot pepper decreased with the planting distance and cultivar. The lowest weight ratio of unharvested hot pepper was at a planting distance of 50 cm for all cultivars. Nevertheless, results of the ANOVA regarding the effect of planting distance on the average weight ratio of unharvested hot pepper showed a significantly different effect as all *p*-values were <0.05, and but no interaction effect between planting distance and cultivar was found as the *p*-value was >0.05.

3.5. Twigged Hot Pepper

Statistical analysis results of the average weight ratio of twigged hot pepper are shown in Table 7. Comparing the planting distances, the average weight ratios were 10.5 ± 1.3% (SD), 7.5 ± 1.2% (SD), and 6.1 ± 2.6% (SD) for AR legend; 9.8 ± 0.6% (SD), 5.3 ± 0.4% (SD), and 4.1 ± 0.9% (SD) for Maeuntan, and 10.3 ± 1.9% (SD), 8.3 ± 1.8% (SD), and 6.1 ± 0.9% (SD) for Jeockyoung. As shown, the weighed ratios of twigged hot pepper cultivars decreased with the planting distance, and the lowest weight ratio of twigged hot peppers was at a planting distance of 50 cm for all cultivars. Moreover, the results of the ANOVA on the effect of planting distance and cultivar on average weight ratios of twigged hot peppers showed significantly different effects as all p-values were <0.05, but no interaction effect between planting distance and cultivar was found. Likewise, in a previous study, the whole to twigged ratios of AR Legend and Jeockyoung were 44.3% and 23.1%, respectively, respectively [9].

Table 5. Statistical analysis results for lost hot pepper.

Cultivar	PD (cm)	M ± SD (%)	Weight (kg)	PCC	F-Value	p-Value
AR Legend	30	4.3 ± 1.3 a [(1)]	0.4 ± 0.1	0.105	0.346	0.713
	40	3.8 ± 1.3 a	0.4 ± 0.1			
	50	4.8 ± 2.6 a	0.2 ± 0.1			
Maeuntan	30	3.0 ± 1.0 a	0.3 ± 0.1	0.593 **	7.754	0.005
	40	2.8 ± 0.9 a	0.4 ± 0.1			
	50	5.1 ± 1.2 b	0.6 ± 0.1			
Jeockyoung	30	4.0 ± 1.0 a	0.2 ± 0.1	−0.396	2.232	0.142
	40	4.1 ± 2.3 a	0.3 ± 0.1			
	50	2.1 ± 1.3 a	0.2 ± 0.1			
Group	**Source**	**DF**	**ANOVA SS**	**Mean Square**	**F-Value**	***p*-Value**
Lost hot pepper	PD	2	1.815	0.907	0.304	0.739
	Cultivar	2	7.704	3.852	1.290	0.285
	PD*Cultivar	4	36.296	9.074	3.040	0.027 **
	Error	45	134.333	2.985		
	Total	53	180.148			

PD: Planting Distance; M: Mean; SD: Standard Deviation; PCC: Pearson's Correlation Coefficient; SS: Sum of Squares. [(1)]: Mean separation within columns by Duncan's Multiple Range test at $p = 0.05$. ** $p < 0.01$.

Table 6. Statistical analysis results on unharvested hot pepper.

Cultivar	PD (cm)	M ± SD (%)	Weight (kg)	PCC	F-Value	p-Value
AR Legend	30	10.0 ± 1.2 b [(1)]	0.9 ± 0.1	−0.690 **	7.874	0.005
	40	8.8 ± 2.6 b	1.1 ± 0.1			
	50	5.5 ± 1.3 a	0.5 ± 0.1			
Maeuntan	30	9.3 ± 2.4 b	1.3 ± 0.2	−0.681 **	6.568	0.009
	40	7.5 ± 1.9 ab	1.1 ± 0.2			
	50	5.0 ± 1.0 a	0.5 ± 0.1			
Jeockyoung	30	10.5 ± 0.5 c	0.8 ± 0.1	−0.869 **	24.700	0.000
	40	8.6 ± 0.7 b	0.7 ± 0.1			
	50	7.5 ± 0.7 a	0.5 ± 0.1			
Group	**Source**	**DF**	**ANOVA SS**	**Mean Square**	**F-Value**	***p*-Value**
Unharvested hot pepper	PD	2	141.593	70.796	23.512	0.000 **
	Cultivar	2	23.370	11.685	3.881	0.028 **
	PD*Cultivar	4	8.074	2.019	0.670	0.616
	Error	45	135.500	3.011		
	Total	53	308.537			

PD: Planting Distance; M: Mean; SD: Standard Deviation; PCC: Pearson's Correlation Coefficient; SS: Sum of Squares. (1): Mean separation within columns by Duncan's Multiple Range test at $p = 0.05$. **: $p < 0.01$.

3.6. Twigged Hot Pepper

Statistical analysis results of the average weight ratio of twigged hot pepper are shown in Table 7. Comparing the planting distances, the average weight ratios were 10.5 ± 1.3% (SD), 7.5 ± 1.2% (SD), and 6.1 ± 2.6% (SD) for AR legend; 9.8 ± 0.6% (SD), 5.3 ± 0.4% (SD), and 4.1 ± 0.9% (SD) for Maeuntan, and 10.3 ± 1.9% (SD), 8.3 ± 1.8% (SD), and 6.1 ± 0.9% (SD) for Jeockyoung. As shown, the weighed ratios of twigged hot pepper cultivars decreased with the planting distance, and the lowest weight ratio of twigged hot peppers was at a planting distance of 50 cm for all cultivars. Moreover, the results of the ANOVA on the effect of planting distance and cultivar on average weight ratios of twigged hot peppers showed significantly different effects as all p-values were <0.05, but no interaction effect between planting distance and cultivar was found. Likewise, in a previous study, the whole to twigged ratios of AR Legend and Jeockyoung were 44.3% and 23.1%, respectively, respectively [9].

Table 7. Statistical analysis results for twigged hot pepper.

Cultivar	PD (cm)	M ± SD (%)	Weight (kg)	PCC	F-Value	p-Value
AR Legend	30	10.5 ± 1.3 b [(1)]	1.0 ± 0.1			
	40	7.5 ± 1.2 b	0.8 ± 0.1	−0.682 **	7.170	0.007
	50	6.1 ± 2.6 a	0.7 ± 0.1			
Maeuntan	30	9.8 ± 0.6 c	1.1 ± 0.1			
	40	5.3 ± 0.4 b	0.6 ± 0.1	−0.910 **	89.537	0.000
	50	4.1 ± 0.9 a	0.4 ± 0.1			
Jeockyoung	30	10.3 ± 1.9 b	0.9 ± 0.1			
	40	8.3 ± 1.8 ab	0.8 ± 0.1	−0.723 **	8.228	0.004
	50	6.1 ± 0.9 a	0.4 ± 0.1			
Group	**Source**	**DF**	**ANOVA SS**	**Mean Square**	**F-Value**	**p-Value**
Twigged hot pepper	PD	2	208.481	104.241	39.641	0.000 **
	Cultivar	2	36.037	18.019	6.852	0.000 **
	PD*Cultivar	4	10.185	2.546	0.968	0.434
	Error	45	118.333	2.630		
	Total	53	373.037			

PD: Planting Distance; M: Mean; SD: Standard Deviation; PCC: Pearson's Correlation Coefficient; SS: Sum of Squares. [(1)]: Mean separation within columns by Duncan's Multiple Range test at $p = 0.05$. ** $p < 0.01$.

4. Discussion

Previous studies have reported that yield per plant was significantly influenced by planting distance. The yield per plant was also increased with an increase in planting distance [19,33,35]. However, increased planting distances caused a decrease in pepper yield because further planting distances meant fewer plants per unit area of hot pepper field. Therefore, the highest weight of marketable hot pepper was indicated at the 40 cm planting distance (Table 3), with 7.0 ± 0.2 kg (SD) for AR Legend, 9.5 ± 0.7 kg (SD) for Maeuntan, and 6.2 ± 0.5 kg (SD) for Jeockyoung obtained for all cultivars. Hence, we propose that the most suitable planting distance for mechanical harvesting of hot pepper is 40 cm.

The average weight ratio of damaged hot pepper was only related to the planting distance. The average weight ratio of lost hot pepper was significantly different when planting distance and cultivar were considered together. However, most damaged hot peppers were found during transportation to the loading bay causing too much force to be applied to the hot pepper. To improve the mechanical harvesting efficiency, the average weight ratio of damage should be reduced. The harvester consisted of three main components (threshing, transporting, and loading bay). A study to determine the specific harvester unit where the damage occurred needs to be conducted in the future with measurement of fruit damage. It was observed that the rotational speed of the helix had an effect on unharvested hot pepper rate in a previous study [9]. However, another factor was found in this study; according to Table 6, shortening the planting distance reduced the average weight ratio of unharvested hot peppers. Through the results of this study, the planting distance can also be said to affect the unharvested hot pepper rate. By controlling the planting distance and rotational speed of the helix, a more consistent performance for mechanically harvesting hot pepper can be conducted.

The average weight ratio of twigged hot pepper was decreased with further planting distance. The minimum weight ratio of twigged hot pepper was 4.1% in this study. Considering the twigged ratios of AR Legend and Jeockyoung were 44.3% and 23.1%, respectively, in a previous study [9], controlled planting distance seems to be a controlling factor for twigged hot pepper.

5. Conclusions

This study classified mechanically harvested hot peppers into five groups—marketable, damaged, lost, unharvested, and twigged. Evaluation of mechanical harvesting performances was conducted at three levels of planting distance with three different cultivars. We verified the effect of planting distance and cultivars on mechanical harvesting of hot pepper. The weight ratio of marketable hot pepper was highest at a planting distance of 50 cm for Maeuntan, followed by AR Legend and Jeockyoung. However, the yield of marketable hot pepper was highest at a planting distance of 40 cm. Moreover, the weight ratios of unharvested and twigged hot peppers were lowest at a planting distance of 50 cm for Maeuntan, followed by AR Legend and Jeockyoung. Therefore, we propose that increasing the planting distance provided enough time for the double helix to thresh the whole pepper plant, thereby reducing the intersection between the pepper plants and facilitating proper harvest. The proposed distance also resulted in increased weight ratios of marketable hot pepper and decreased weight ratios of damaged, unharvested, and twigged hot peppers. An interaction effect between planting distance and cultivar was found with the weight ratio of lost hot pepper. However, the damaged hot peppers were more related to the mechanical structure of the transporting component of the harvester; additional study is required with measurements of damage in each component of the machineTherefore, considering the yield of marketable hot pepper and the observed decrease in the weight ratio of unharvested and twigged hot peppers with the further planting distances used in this study, we propose the most suitable planting distance for mechanical harvesting of hot pepper is 40 cm.

Author Contributions: Authors have contributed to this work in the following roles: Conceptualization, S.K. and Y.K.; Methodology, S.K. and S.W.; Software, S.K. and Y.K.; Formal analysis, S.K. and H.P.; Investigation, S.K. and H.P.; Resources, S.K., Y.K. and H.P.; Writing—Original Draft Preparation, S.K.; Writing—Review and Editing, D.D.U. and Y.H.; Supervision, Y.H. All authors have read and agreed to the published version of the manuscript.

Funding: This work was supported by the Technology Innovation Program (2003975, Development of intelligent 30 kW crawler-based traveling platform for multi-purpose farming) funded by the Ministry of Trade, Industry& Energy (MOTIE, Korea).

Institutional Review Board Statement: Not applicable.

Informed Consent Statement: Not applicable.

Data Availability Statement: The data presented in this study are available in the article.

Conflicts of Interest: The authors declare no conflict of interest.

References

1. Crapnell, R.D.; Banks, C.E. Electroanalytical overview: The pungency of chile and chilli products determined via the sensing of capsaicinoids. *Analyst* **2021**, *146*, 2769–2783. [CrossRef]
2. Nair, K.P. *The Geography of Black Pepper (Piper nigrum): The "King" of Spices*; Springer Nature: London, UK, 2020; Volume 1.
3. Cebula, S.; Jakubas, A.; Sekara, A.; Kalisz, A.; Pohl, A. The relationship between morphological features and nutritive value of sweet pepper fruits. *Acta Sci. Pol. Hortorum Cultus* **2015**, *14*, 79–91.
4. Conforti, F.; Statti, G.A.; Menichini, F. Chemical and biological variability of hot pepper fruits (*Capsicum annuum* var. *acuminatum* L.) in relation to maturity stage. *Food Chem.* **2007**, *102*, 1096–1104. [CrossRef]
5. Sharma, V.K.; Srivastava, A.; Mangal, M. *Accelerated Plant Breeding Volume 2 Vegetable Crops*; Gosal, S.S., Wani, S.H., Eds.; Springer: Berlin/Heidelberg, Germany, 2020; pp. 417–444.
6. FAO. Database for World Area Harvested of Hot Pepper in the World 2019. Available online: https://fao.org/faostat/en/#data/QC (accessed on 22 July 2021).
7. Jain, R.; Meena, M.L.; Dangayach, G.S.; Bhardwaj, A.K. Association of risk factors with musculoskeletal disorders in manual-working farmers. *Arch. Environ. Occup. Health* **2018**, *73*, 19–28. [CrossRef] [PubMed]
8. Momeni, Z.; Choobineh, A.; Razeghi, M.; Ghaem, H.; Azadian, F.; Daneshmandi, H. Work-related musculoskeletal symptoms among agricultural workers: A cross-sectional study in Iran. *J. Agromed.* **2020**, *25*, 339–348. [CrossRef]
9. National Agricultural Statistics Service. Farm Labor. Available online: https://downloads.usda.library.cornell.edu/usda-esmis/files/x920fw89s/q237jp329/1g05g852h/fmla0521.pdf (accessed on 17 September 2021).

10. Kim, T.H.; Kim, D.C.; Cho, Y.J. Performance comparison and evaluation of two small chili pepper harvester prototypes that attach to walking cultivators. *Appl. Sci.* **2020**, *10*, 2570–2580. [CrossRef]
11. Funk, P.A.; Walker, S.J. Evaluation of five green chile cultivars utilizing five different harvest mechanisms. *Appl. Eng. Agric.* **2010**, *26*, 955–964. [CrossRef]
12. Joukhadar, I.S.; Walker, S.J.; Funk, P.A. Comparative mechanical harvest efficiency of six New Mexico pod-type green chile pepper cultivars. *Hort. Technol.* **2018**, *28*, 310–318. [CrossRef]
13. Funk, P.A.; Walker, S.J.; Herbon, R.P. A systems approach to chile harvest mechanization. *Int. J. Veg. Sci.* **2011**, *17*, 296–309. [CrossRef]
14. Akay, O.E.; Guzel, E.; Ozcan, M.T. Development of a harvest mechanism for the Kahramanmaras chili pepper. *Int. J. Agric. Biosyst. Eng.* **2016**, *10*, 666–670.
15. Decoteau, D.R.; Graham, H.A.H. Plant spatial arrangement affects growth, yield, and pod distribution of cayenne peppers. *Hortic. Sci.* **1994**, *29*, 149–151. [CrossRef]
16. Russo, V.M. Yield in nonpungent jalapeño pepper established at different in-row spacings. *Hort. Sci.* **2008**, *43*, 2018–2021. [CrossRef]
17. Edgar, O.N.; Gweyi-Onyango, J.P.; Korir, N.K. Plant row spacing effect on growth and yield of green pepper (*Capsicum annuum* L.) in Western Kenya. *Arch. Curr. Res. Int.* **2017**, *7*, 1–9. [CrossRef]
18. El-Gazzar, T.M.; Tartoura, E.A.; Nada, M.M.; Ismail, M.E. Effect of some treatments to reduce the injury of high temperature on sweet pepper grown in late summer season. *J. Plant Prod.* **2020**, *11*, 855–860. [CrossRef]
19. Islam, M.; Satyaranjan, S.A.H.A.; Akand, H.; Rahim, A. Effect of spacing on the growth and yield of sweet pepper (*Capsicum annuum* L.). *J. Cent. Eur. Agric.* **2001**, *12*, 328–335. [CrossRef]
20. Jadama, L.; Jammeh, P.T.; Cham, A.K.; Diedhiou, I. Effect of different spacing on the growth and yield of California wonder bell pepper (*Capsicum annuum*) on sandy loam soil in the Gambia. *Asian J. Biol.* **2021**, *12*, 1–9. [CrossRef]
21. Santos, B.M.; Salame-Donoso, T.P.; Obregon-Olivas, H.A.; Inestroza, J.E.; Galeano, R.; Saenz, M.V.; Monge-Perez, J.E.; Cuevas, M.G.; Torres-Quezada, E.A.; Mendez-Urbaez, C.J. Evaluation of planting densities and shoot pruning practices for indeterminate bell pepper production in Dominican Republic, Honduras, and Costa Rica. In Proceedings of the Florida State Horticultural Society, Crystal River, FL, USA, 6–8 June 2010; Volume 124, pp. 191–193.
22. Ansari, A.M.; Ahmad, E.; Bhagat, B.K.; Singh, D.N. Effect of planting space and pruning intensity in mango (*Mangiferaindica* L.) cv. *Amrapali. J. Pharm. Phytochem.* **2018**, *1*, 198–201.
23. Ginoya, A.V.; Patel, J.B.; Delvadiya, I.R.; Jethva, A.S. Effect of bulb size and plant spacing on seed quality parameters of onion (*Allium cepa* L.) cv. GJWO 3. *J. Pharm. Phytochem.* **2018**, *7*, 2511–2515.
24. Mamnoie, E.; Dolatkhahi, A. Plant spacing and cultivar affects yield components, qualitative traits and early ripening of tomato (*Lycopersicon esculentum*). *Not. Sci. Biol.* **2013**, *5*, 494–498. [CrossRef]
25. Ogbonna, P.E.; Orji, K.O.; Nweze, N.J.; Opata, P. Effect of planting space on plant population at harvest and tuber yield in taro (*Colocasia esculenta* L.). *Afr. J. Agric. Res.* **2015**, *10*, 308–316. [CrossRef]
26. Samanhudi, A.Y.; Dinana, A. Liquid organic fertilizer and planting space influencing the growth and yield of rice (*Oryza sativa* L.) in system of rice intensification (SRI) methods. *J. Agric. Sci. Technol.* **2011**, *5*, 232–238
27. Growveg. Chili Pepper Growing Guide. 2021. Available online: https://growveg.com/plants/us-and-canada/how-to-grow-chili-peppers/ (accessed on 22 July 2021).
28. RHS. Peppers and Chillies. 2021. Available online: https://rhs.org.uk/advice/profile?pid=664 (accessed on 22 July 2021).
29. HGIC. Pepper. 2020. Available online: https://hgic.clemson.edu/factsheet/pepper/ (accessed on 22 July 2021).
30. UI. Growing Peppers in Short-Season, High-Altitude Idaho Gardens. 2020. Available online: https://www.extension.uidaho.edu/publishing/html/BUL974-Growing-Peppers-in-Short-Season-High-Altitude-Gardens.aspx (accessed on 22 July 2021).
31. Ragassi, C.F.; Zucolotto, J.; Gomes, L.M.; Ribeiro, C.S.; Madeira, N.R.; Reifschneider, F.J.B. Productivity, quality of fruits and architecture of jalapeño pepper at different planting densities. *Hortic. Bras.* **2019**, *37*, 331–337. [CrossRef]
32. NIHHS. Technical Information for Horticultural Special Crops No. 89. National Institute of Horticultural and Herbal Science. 2020. Available online: https://www.nihhs.go.kr/usr/nihhs/news_Public_view.do?idx=27927 (accessed on 22 July 2021).
33. Aminifard, M.H.; Aroiee, H.; Karimpour, S.; Nemati, H. Growth and yield characteristics of paprika pepper (*Capsicum annum* L.) in response to plant density. *Asian J. Plant Sci.* **2010**, *9*, 276–280. [CrossRef]
34. Cavero, J.; Ortega, R.G.; Gutierrez, M. Plant Density Affects Yield, Yield Components, of Color of Direct-seeded Paprika Pepper. *Hort. Sci.* **2001**, *36*, 76–79. [CrossRef]
35. Nasto, T.; Balliu, A.; Zeka, N. The influence of planting density on growth characteristics and fruit yield of peppers (*Capsicum annuum* L.). *Acta Hortic.* **2009**, *28*, 609–612. [CrossRef]
36. Kang, S.H. Effect of Planting Distance and NaCl Treatment on Machinery Harvest by Hot Pepper Cultivar. Master's Thesis, Kyungpook National University, Daegu, Korea, June 2021.

agriculture

MDPI

Article

Research on Loading Method of Tractor PTO Based on Dynamic Load Spectrum

Yu Wang, Ling Wang *, Jianhua Zong, Dongxiao Lv and Shumao Wang

Beijing Key Laboratory of Optimized Design for Modern Agricultural Equipment, College of Engineering, China Agricultural University (East Campus), Beijing 100083, China; wysummer@cau.edu.cn (Y.W.); zongjianhua@cau.edu.cn (J.Z.); lvdongxiao@cau.edu.cn (D.L.); wangshumao@cau.edu.cn (S.W.)
* Correspondence: wangling2017068@cau.edu.cn; Tel.: +86-138-1123-6360

Abstract: The torque load spectrum is an important basis for the strength design and durability test verification of tractor power take-off (PTO), and the performance and reliability of tractor PTO directly affect the quality and efficiency of agricultural operations. In this paper, taking the PTO torque load as the object, a PTO loading method based on the dynamic load spectrum acquired in the actual field work was proposed in this paper. Based on the Peak Over Threshold model, the extrapolation of the PTO load spectrum was realized, and the load spectrum throughout the whole life cycle was obtained. On the basis of this, the mobile tractor PTO loading test bench and Fuzzy-Proportional-Integral-Derivative (Fuzzy-PID) controller were developed to achieve the dynamic loading of the PTO load spectrum, and the dynamic characteristics were analyzed and verified by the simulation and laboratory test. The results showed that with the time domain extrapolation method, the load extreme value was expanded from (63.24, 469.50) to (60.88, 475.18), and the coverage was expanded by 1.98%. By comparing with the fitting results, statistical characteristics and rain flow counting results, the load spectrum extrapolation method was effective. In addition, the response time of simulation and laboratory test were 0.05s and 0.75s, respectively; the maximum error was 1.77% and 4.03%, respectively; and the goodness of fit was 16.78 N·m, which indicated that the PTO loading test bench, can accurately restore the dynamic loading of the tractor and the Fuzzy-PID controller had better accuracy and stability. It would provide a reference for the practical application of PTO load spectrum of the tractors.

Keywords: dynamic load spectrum; power take-off; tractor PTO loading test bench; Fuzzy-PID controller

Citation: Wang, Y.; Wang, L.; Zong, J.; Lv, D.; Wang, S. Research on Loading Method of Tractor PTO Based on Dynamic Load Spectrum. *Agriculture* **2021**, *11*, 982. https://doi.org/10.3390/agriculture11100982

Academic Editor: José Pérez-Alonso

Received: 6 September 2021
Accepted: 4 October 2021
Published: 9 October 2021

1. Introduction

Tractor power take-off (PTO) is an important component for driving agricultural implement such as rotary tillage and harrowing [1,2], and its performance and reliability directly affect the quality and efficiency of agricultural operations [3,4]. Nowadays, the tractor PTO loading test bench is a commonly used piece of equipment for PTO performance test, sensor device calibration, and reliability test. Wang Jiankang [5] and Sumer et al. [6] built a PTO loading test-bed with alternating current (AC) dynamometer and eddy current dynamometer, respectively, which was mainly used to test the performance of tractor PTO and determine its load characteristics. Siddique et al. [7] used PTO dynamometer to analyze the fuel consumption level of tractor under different loads in order to optimize the load characteristics. Roeber et al. [8] used a PTO loading test bench dynameter of Nebraska tracker test lab to calibrate the tractor torque measurement system. In addition, some companies such as HORIBA group (Kyoto, Japan) and Powerlink Company (Bangkok, Thailand) have also developed PTO loading systems, which could realize the detection of speed and torque performance of tractors. Until recently, these loading test benches are performed in a fixed load mode, which is different from the load borne by the tractor in actual field operation, resulting in errors in the detection results [9]. Moreover, there are few studies on dynamic loading methods for tractor PTO.

The PTO load spectrum can reflect the time history of the torque load borne by tractor PTO in actual operation [9], which is an important way to explore the relationship between the operating load and the performance and reliability of tractor PTO. However, due to the limitation of time and cost, it is impossible to obtain the load time-history during its whole life cycle in the load spectrum collection process. Therefore, it is necessary to extend the load coverage by extrapolating the load spectrum. The commonly used load spectrum extrapolation methods can be divided into frequency domain extrapolation method (FDE), rain-flow domain extrapolation method (RDE) and time domain extrapolation method (TDE) [10]. The main approach of the FDE method is to use normal distribution and Weibull distribution to fit the mean value and the amplitude value of the load spectrum, and then the extrapolation can be realized based on the load joint distribution function [11]. Jixin Wang et al. achieved the extrapolation of transmission load of wheel loader based on FDE method [10]. For the RDE method, it uses nonparametric estimation to achieve data fitting based on the distribution characteristics of rain flow matrix [12,13]. However, there is a common problem with both approaches, that is, the load signal in the time domain should be converted to the frequency domain in the process of extrapolation, which will break the time sequence of the original load and bring inconvenience to the application of the load spectrum. The TDE method was proposed by Johannesson and Thomas [14] and Johannesson [15], and the extrapolation can be actualized on the basis of the extreme value distribution characteristics of the original load based on the extreme value theory [11]. This approach offers a way to retain the time-domain order and reduce the conversion error. Yang Zihan et al. [9] extrapolated the load of tractor drive shaft based on the Peak Over Threshold model. The optimal thresholds of the upper and lower limits in the time-domain load data were determined by comparing with the values of grey correlation degree corresponding to different threshold, and the extrapolation model was built to realize the extrapolation of the load spectrum of the tractor drive shaft. Yan Xianghai et al. [16] studied the compilation method of the PTO dynamic torque load spectrum in time domain under multiple working conditions and the durability test was carried out on PTO torque loading test bench by using the load spectrum. The test results showed that the actual failure time of tractor PTO was basically consistent with the theoretical, which verified the effectiveness of the load spectrum.

At present, although there are many studies on the extrapolation and compilation methods of the load spectrum, there are relatively few studies on the application in actual field load test. Therefore, in this paper, taking the tractor PTO torque load as the object, the load spectrum compilation and loading method of tractor were studied. Firstly, the extrapolating method of the torque load spectrum of tractor PTO was studied based on the Peak Over Threshold model. The threshold was selected by using the ForwardStop sequential test method, and the extrapolation of tractor PTO torque load spectrum was realized. Secondly, a movable tractor PTO loading test bench was developed. The torque control of the eddy current dynamometer was realized by using Fuzzy-Proportional-Integral-Derivative (Fuzzy-PID) controller and the dynamic response characteristics of the system were simulated and analyzed. On this basis, the load spectrum loading test was carried out. The test results showed that the tractor PTO loading test bench can realize the dynamic control of torque. It can provide reference for the compilation and practical application of PTO torque load spectrum of tractors.

This paper is organized as follows: Section 2 outlines the method applied and presents the data used. The experiments and results are presented in Section 3; Sections 4 and 5 discuss and summarize the article, respectively. The paper ends with the conclusions.

2. Materials and Methods

2.1. Extrapolation Method of Tractor PTO Torque Load Spectrum

In this paper, the TDE method of the tractor PTO torque load based on Peak Over Threshold model is realized. Under such approach, the extreme value obtained from the actual operation conditions can further extrapolate the possible theoretical extreme load [17], as well as the bidirectional extrapolation of the maximum and minimum values, expand the load coverage range, and improve the effectiveness of load spectrum loading. The extrapolation process is shown in Figure 1.

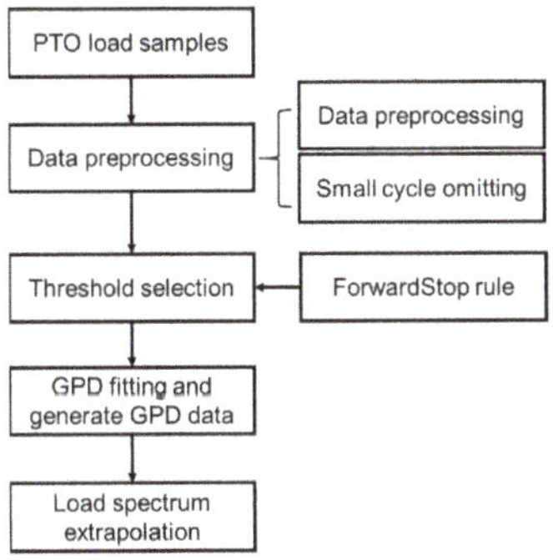

Figure 1. Extrapolation process based on Peak Over Threshold theory.

The generalized pareto distribution (GPD) is the basis for the extrapolation of excess. The cumulative distribution function (CDF) of GPD is as follows:

$$G(x;u,\sigma,\xi)=\begin{cases}1-\left(1+\xi\frac{x-u}{\sigma}\right)^{-\frac{1}{\xi}},\xi\neq 0,x\geq u\\ 1-\exp\left(-\frac{x-u}{\sigma}\right),\xi=0,x\geq u\end{cases} \tag{1}$$

The probability density function (PDF) of GPD is as follows:

$$g(x;u,\sigma,\xi)=\begin{cases}\frac{1}{\sigma}\left(1+\xi\frac{x-u}{\sigma}\right)^{-\frac{1}{\xi-1}},\xi\neq 0,x\geq u\\ \frac{1}{\sigma}\exp\left(-\frac{x-u}{\sigma}\right),\xi=0,x\geq u\end{cases} \tag{2}$$

where, u means the threshold, $x-u$ means the excess, ξ means shape parameter, and σ means scale parameter.

The GPD shape parameters and scale parameters can be calculated by the maximum likelihood estimation [18], and it is necessary to select an appropriate threshold value to establish an accurate Peak Over Threshold model [19]. In this paper, the ForwardStop rule is used to calculate the threshold value. Based on the multiple test method, the p-value of the hypothesis test of the threshold fitting results is corrected. The accuracy of the threshold selection can be improved by controlling the false discovery rate (FDR) of the fitting results [20]. The specific process is as follows:

Considering the torque load data $X=(x_1,x_2,\cdots,x_n)$ and the candidate threshold sequence $U=(u_{\min}<u_2<\cdots<u_{\max})$, for each candidate threshold u_i, the excess sequence is $y_i=x|_{xi>ui}-u_i$. For each peak over sequence y_i, the corresponding null hypothesisH0i is proposed as:

H_0^i: The distribution of the excess sequence y_i follows the GPD.

Therefore, for the candidate threshold sequence $U=(u_{\min}<u_2<\cdots<u_{\max})$, there are l null hypotheses in total: $H_0=(H_0^1,H_0^2,\cdots,H_0^i)$. To perform the hypothesis testing on H_0^i sequentially, the corresponding sequence of p-values: $p_1,p_2,\cdots,p_i\in[0,1]$ are obtained.

The p-value of a single hypothesis test is calculated by using Anderson–Darling test (AD test) in this paper. The weight coefficient is introduced in the AD test, which makes it possible to pay more attention to the tail-distributed data in the calculation process and is more sensitive for the detecting data that deviates from the sample [21,22].

For the threshold u_i, when the corresponding AD test statistic does not meet the expected threshold value, the null hypothesis H_0^i is rejected. Additionally, according to the distribution properties of GPD, for $u_{\min}\leq u_k<u_i$, H_0^i H_0^k will be rejected because the

hypothesis test is performed in a monotonic order of the threshold value. On this basis, the ForwardStop rule is applied. Firstly, in the monotonically increasing order, the null hypothesis H_0^i with each threshold value is inspected by the AD test, and the corresponding AD test statistic can be obtained to construct the p-value sequence; then, the Equation (3) is used to correct the p-value sequentially, and the new p-value sequence is constructed at the same time; finally, the minimum $\hat{k}$ that meets the expected threshold α is found, and the threshold value corresponding to the sequence $\hat{k}$ is the optimal threshold value.

$$\hat{k} = max\left\{ k \in \{1, \cdots, l\} : -\frac{1}{k}\sum_{i=1}^{k} \log(1 - p_i) \leq \alpha \right\} \tag{3}$$

When the threshold value is selected, the maximum likelihood estimation is used to fit the GPD of the excess. The extrapolated data can be generated randomly based on this GPD function. Then, the 1-time extrapolated load spectrum can be obtained by replacing the extreme value data of the original load.

2.2. Data Acquisition and Preprocessing of Tractor PTO Torque Load Spectrum

On 7 November 2018, the load spectrum collection test was carried out at Shangzhuang Experimental Station of China Agricultural University (Haidian District, Beijing, China). The detailed parameters are described in Table 1. The tractor PTO torque is measured in real time by using the torque wireless monitoring system, and the trend term and abnormal data are removed by preprocessing. The load cycle, being less than 10 % of the maximum amplitude of the load cycle, is omitted to simplify the load time history. The time history of tractor PTO torque load after preprocessing is showed in Figure 2.

Table 1. Parameters of load spectrum acquisition test.

Item	Parameter
Tractor type	TS404
Manufacturer	Shandong Wuzheng Group Co., Ltd. (Rizhao, Shandong, China)
Test condition	Rotary tillage
Implement type	1GLN-0145
Tillage depth	200 mm
Tillage width	1800 mm
Average operating speed	7.3 km/h
Torque sampling frequency	20 Hz
Torque sampling duration	900 s

2.3. Extrapolation Model Establishment of Tractor PTO Torque Load Spectrum

The ForwardStop rule is used to select the threshold value of the processed data; afterwards, the load spectrum extrapolation is realized. For the maximum value in the load sample, the candidate threshold interval is selected as (63.00, 469.00), increasing by 0.5. Similarly, for the minimum value, the candidate threshold interval is (469.00, 63.00), decreasing by 0.5. Finally, the calculated upper threshold is 419, and the number of excess data is 125; the lower threshold is 97.50, and the number of excess data is 170.

According to the threshold selection results, the GPD parameters is estimated by using maximum likelihood estimation method in Matlab. The GPD fitting result is shown in Table 2.

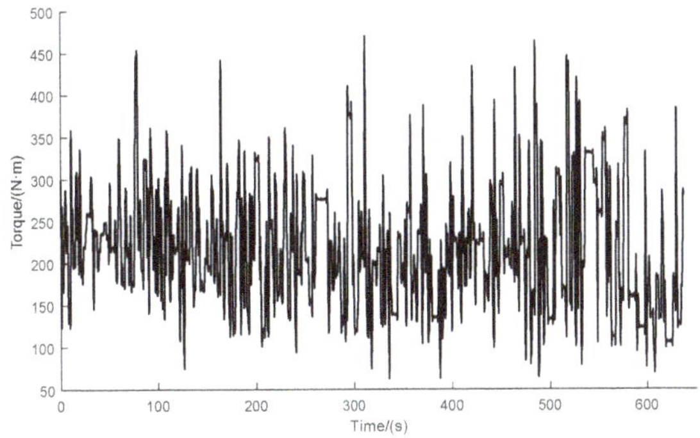

Figure 2. The time history of tractor power take-off (PTO) torque load after preprocessing.

Table 2. Generalized Pareto Distribution (GPD) fitting result.

Threshold	Shape Parameters	Scale Parameters
419.00	−0.4737	25.858
97.50	−0.6776	23.486

Therefore, the GPD functions of the PTO load sample for the maximum and minimum value are shown in Equations (4) and (5).

$$G(x;u,\sigma,\xi) = 1 - \left(1 - 0.4737 \times \frac{x - 419}{25.858}\right)^{2.1107} \quad (4)$$

$$G(x;u,\sigma,\xi) = 1 - \left(1 - 0.6776 \times \frac{x - 97.5}{23.486}\right)^{1.4756} \quad (5)$$

Furthermore, the load spectrum with extrapolation factor of 1 is obtained by replacing the excess of the original load with the random data from Equations (4) and (5). The load spectrum with the extrapolation factor of 10 and 131 can be obtained by repeating the above steps. When the extrapolation factor is 131, the cumulative frequency of the load spectrum is extrapolated to 106, and it represents that the load spectrum contains the load data throughout the whole life cycle of the PTO torque. The comparison of original load time history and 10 times extrapolated load time history is shown in Figure 3. The sinusoidal curve is used to be a smooth transition at the starting point of the PTO torque in order to avoid the sudden change of load.

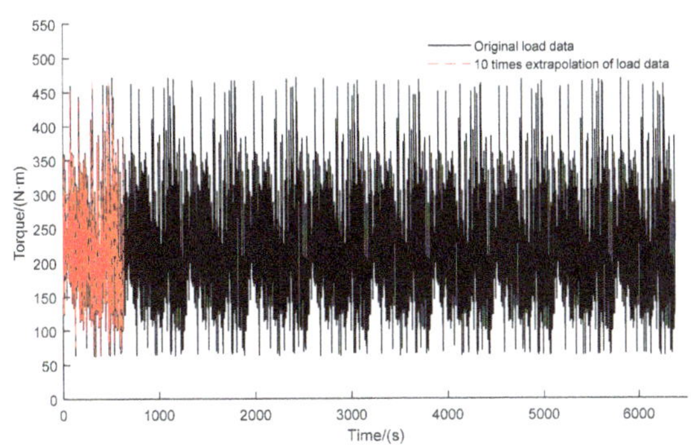

Figure 3. The comparison of original load time history and 10-times-extrapolated load time history.

2.4. Development of Tractor PTO Loading Test Bench

The structure principal diagram of the tractor PTO loading test bench is shown in Figure 4, which is developed for the load spectrum loading under indoor conditions to simulate the loads in actual field operations. The loading test bench is mainly composed of the movable chassis, PTO loading system, data acquisition system (DAQ) and personal computer (PC) software.

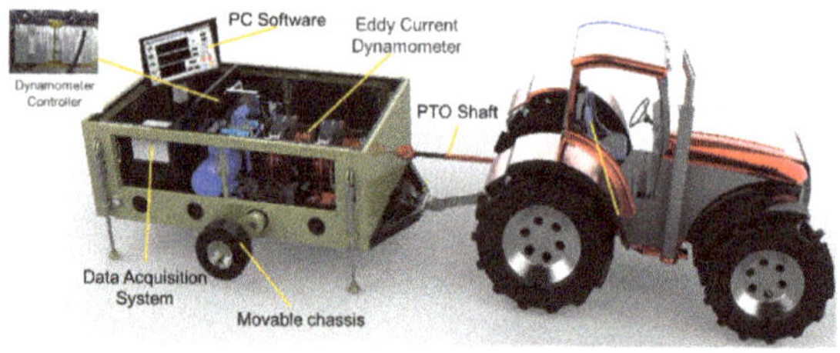

Figure 4. The structure principal diagram of tractor PTO loading test bench.

The PTO loading system consists of two eddy current dynamometers, eddy current dynamometer controller, torque sensor, coupling, and torque output shaft. The PTO torque loading is implemented by the two eddy current dynamometers in series, and the maximum torque loading range is up to 3000 N·m. The torque sensor is used to measure the real-time loading torque and speed, and through the feedback adjustment of the dynamometer controller, the dynamic control of PTO torque is achieved. The measurement and control signals are obtained and output by the DAQ, which adopts NI-CompactRIO embedded system based on NI-FPGA (National Instrument Co., Ltd, Austin, Texas, USA). The structure of PTO loading system is shown in Figure 5.

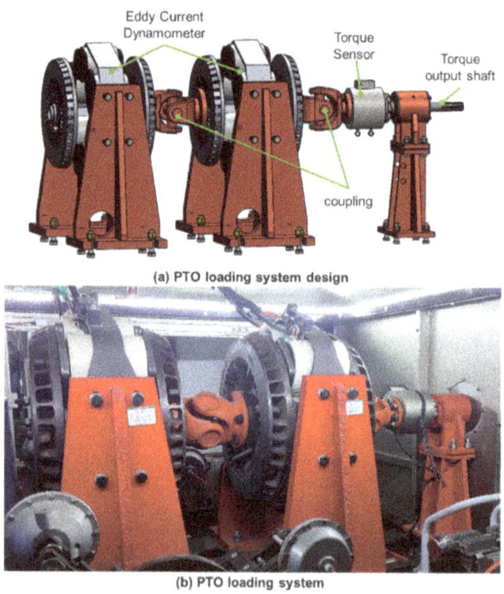

Figure 5. The structure of PTO loading system.

2.5. Development of Fuzzy-PID Controller

The dynamic loading response characteristics and accuracy of the eddy current dynamometer determine the reproducibility of the load spectrum. The loading torque of the eddy current dynamometer is related to its structure, material, speed, temperature, and other factors, and it is dynamic in the running state of the eddy current dynamometer. The Fuzzy-PID controller has better robustness and dynamic characteristics than PID controller by the online correction of the PID parameters according to the error changes and deviations on the basis of fuzzy rules [23].

2.5.1. Determination of Input and Output Variables

The Fuzzy-PID control includes two parts: the fuzzy controller and PID controller. The input of fuzzy controller is torque deviation (E) and deviation change rate (EC), and the output are the correction values of PID three parameters ΔKp, ΔKi, and ΔKd. The input of PID controller are torque deviation (E) and parameter correction value, and the output is the eddy current dynamometer control signal. The principle of Fuzzy-PID controller is shown in Figure 6.

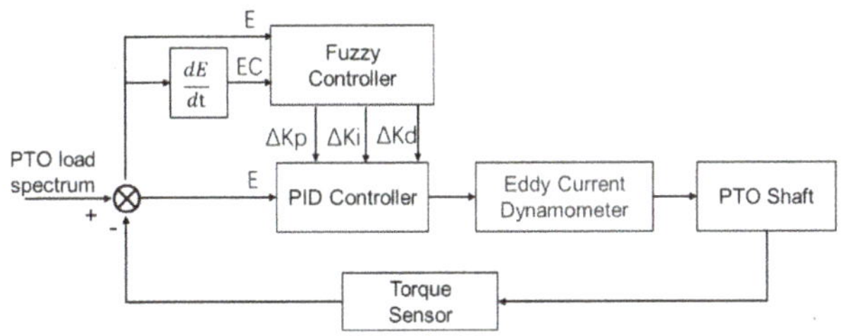

Figure 6. The principle of Fuzzy-Proportional-Integral-Derivative (Fuzzy-PID) controller.

2.5.2. Determination of the Membership Function and Fuzzy Subset

According to the torque loading range, the universes of input variables E and EC of the fuzzy controller are determined as (−600 600), and the universes of output variables ΔKp, ΔKi, and ΔKd are determined as (−2 2), (−0.5 0.5), and (−0.1 0.1), respectively. The quantization levels of input and output variables are divided into 7 levels, and the corresponding fuzzy subset is {NB, NM, NS, ZO, PS, PM, PB}, which represents negative big, negative middle, negative small, zero, positive small, positive middle, positive big, respectively. The Gaussian function is selected as the membership function of each fuzzy subset.

2.5.3. Determination of the Fuzzy Rules

The fuzzy control rule is determined based on the influence of PID parameters on the torque control of eddy current dynamometer and the actual operation experience. The fuzzy control rules are shown in Table 3. The Fuzzy controller adopts two-dimensional Mamdani reasoning method, in which, the max–min method is used for control decision, and the defuzzification is realized by the Centriod method.

Table 3. The fuzzy rules of ΔKp, ΔKi, and ΔKd.

ΔKp/ΔKi/ΔKd		**EC**						
		NB	**NM**	**NS**	**ZO**	**PS**	**PM**	**PB**
E	**NB**	PB/NB/PS	PB/NB/NS	PM/NM/NB	PM/NM/NB	PS/NS/NB	ZO/ZO/ZO	ZO/ZO/PS
	NM	PB/NB/PS	PB/NB/NS	PM/NM/NB	PS/NS/NM	PS/NS/NM	ZO/ZO/NS	NS/ZO/ZO
	NS	PM/NB/ZO	PM/NM/NS	PM/NS/NM	PS/NS/NM	ZO/ZO/NS	NS/PS/PS	NS/PS/ZO
	ZO	PM/NM/ZO	PM/NM/NS	PS/NS/NS	ZO/ZO/NS	NS/PS/NS	NM/PM/NS	NM/PM/ZO
	PS	PS/NM/ZO	PS/NS/ZO	ZO/ZO/ZO	NS/PS/ZO	NS/PS/ZO	NM/PM/ZO	NM/PS/ZO
	PM	PS/ZO/PB	ZO/ZO/NS	NS/PS/PS	NM/PS/PS	NM/PM/PS	NM/PB/PS	NB/PB/PB
	PB	ZO/ZO/PB	ZO/ZO/PM	NM/PS/PM	NM/PM/PM	NM/PM/PS	NB/PB/PS	NB/PB/PB

2.5.4. Simulation Model of Fuzzy-PID Controller

The mathematical model of the eddy current torque loading system is established by using the step response identification method. Based on that, the simulation model is built in Simulink, which is shown in Figure 7. The initial parameters of the PID controller are 0.785, 0.2117, and 0.0148.

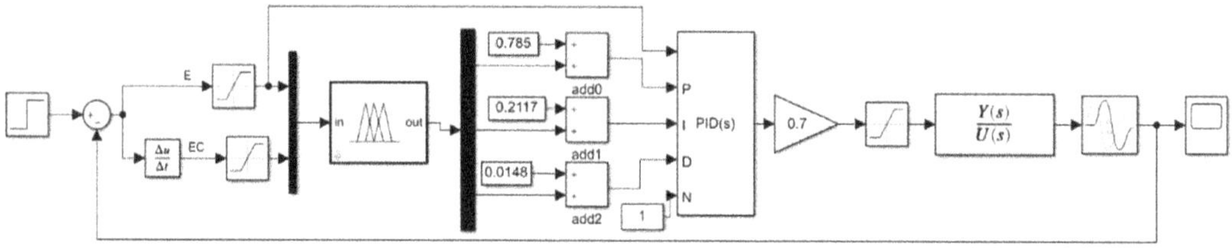

Figure 7. System simulation model.

3. Results

3.1. The Results of Load Spectrum Extrapolation

3.1.1. Validation of GPD Fitting Results

In order to verify the GPD fitting results, the fitting test, standard deviation of parameter estimation, correlation coefficient, and goodness of fit are analyzed. The results are shown in Table 4. It can be seen that the GPD fitting results of the maximum and minimum values both pass the Kolmogorov–Smirnov test (K-S test) and Chi-square test at the confidence level of 95%; that is, the fitting results are acceptable. The correlation coefficient between the fitted GPD and the original data is greater than 0.99, and the fitting results have a high coincidence with the excess data; the goodness of fit is 2.8676 N·m and 1.9804 N·m, respectively, indicating that the overall error of the fitting results is small. Therefore, the GPD fitting results are effective and accurate.

Table 4. Analysis of GPD fitting results.

Item	Upper Threshold	Lower Threshold
Threshold	419.00	97.50
Shape parameter	−0.4737	−0.6776
Scale parameter	25.858	23.486
Standard deviation of Shape parameter	0.0781	0.0626
Standard deviation of Scale parameter	2.8676	1.9804
K-S statistic value	0.9762	0.7082
Chi-square statistic	0.4427	0.6818
Correlation coefficient	0.9913	0.9971
Goodness of fit	2.8676	1.9804

3.1.2. Comparison of Statistical Characteristics of Extrapolated Load Spectrum

By comparing the statistical characteristics of the original load spectrum and the extrapolated load spectrum, the effectiveness of the extrapolated load spectrum is verified. It can be seen from Table 5 that with the increase of the data volume, the range of load extreme value expanded from (63.24, 469.50) to (60.88, 475.18), and the load coverage expanded by 1.98%; in addition, the mean value, standard deviation, and variance of the extrapolated load spectrum are basically consistent with the statistical characteristics of the original load data, and the extrapolated load spectrum is effective.

Table 5. Statistical characteristics of power take-off (PTO) load spectrum after extrapolation.

Extrapolation Factor	Data Volume	Maximum Value/(N·m)	Minimum Value/(N·m)	Mean Value/(N·m)	Standard Deviation/(N·m)	Variance /(N·m)^2
Load sample	17892	469.50	63.24	211.78	70.73	5050.22
After pre-processing	12743	469.50	63.24	222.18	70.74	5003.93
1 time	12743	464.26	63.02	215.79	70.76	5006.91
10 times	127430	471.78	61.89	215.81	70.74	5004.35
131 times	1669333	475.18	60.88	216.54	70.73	5003.33

3.1.3. Comparison of Rain Flow Counting

The rain flow matrixes of the original load data and 1-time extrapolation load data are shown in Figure 8. From the overall distribution of the rain flow matrix, the extrapolated load spectrum has the same distribution law as the original load data, which proves that the extrapolation method is feasible and reasonable. Moreover, the essential function of the time domain extrapolation method is to reconstruct the data above the threshold value, while the data below the threshold value remains unchanged. It can be seen from Figure 8b that the extrapolated load spectrum has two peaks marked by red circles in the distribution area of the maximum and minimum values, and they are distributed at both ends of the graph, which mean that the extreme value has been successfully extrapolated. The frequency in Figure 8 represents the concentration degree of the load cycle; the higher it is, the higher the cumulative frequency is. Compared with the original data, the frequency corresponding to the extrapolated extreme value is higher. Therefore, it can be considered that the extrapolation effect with this approach is obvious.

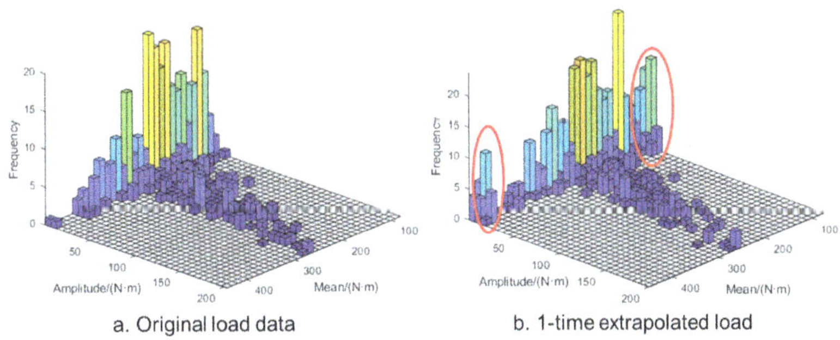

Figure 8. Comparison of rain flow matrixes: (**a**) original load; (**b**) 1-time extrapolated load.

3.2. Simulation Test of Fuzzy-PID Controller

3.2.1. Simulation of Step Response

The step response is used to simulate the Fuzzy-PID controller and PID controller, respectively, in Simulink; the result is shown in Figure 9.

It can be seen that the response characteristics of both controllers are fast and accurate. The signal tends to be stable after 2.15 s of the Fuzzy-PID controller, while it is 3.25 s of the PID controller, which has a maximum overshoot of 358.7 at 2.35 s. The steady-state errors are 0 of both controllers.

3.2.2. Simulation of the Load Spectrum

The load spectrum is used to verify the dynamic response characteristics of the two controllers. The simulation results in the first 35 s are shown in Figure 10. The response time of Fuzzy-PID controller and PID controller are 0.05 s and 0.02 s, respectively, and the maximum error are 1.77% and 7.27%, respectively. The simulation results show that the dynamic response characteristics are fast, and the PID controller has a larger overshoot.

The response time is long for the Fuzzy-PID controller; however, the stability and accuracy are better to simulate the load spectrum loading.

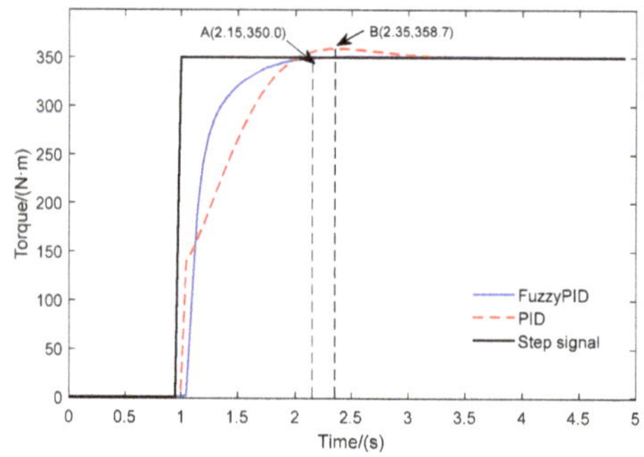

Figure 9. Simulation results of step response.

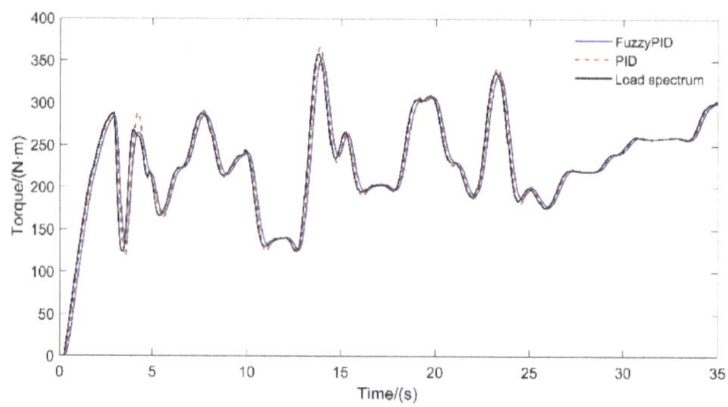

Figure 10. Simulation results of load spectrum loading.

3.2.3. Test Verification

The load spectrum loading test is carried out to verify the actual performance of the PTO loading test bench and the Fuzzy-PID controller. The loading test site is shown in Figure 11, and the parameters of the test equipment are shown in Table 6. The three-phase motor is used to simulate the tractor PTO for the loading test, and the motor rotational speed is 540 r·min^{-1}. The actual loading load spectrum is consistent with the load spectrum used in the simulation, and the loading frequency of the load spectrum is 20 Hz. The test results of the two controllers are shown in Figure 12.

It can be seen that the dynamic loading of the load spectrum can be realized based on both the Fuzzy-PID controller and the PID controller. The change trend and results are similar with the simulation. The response time of the two controllers is 0.75 s and 1.05 s, respectively, and due to the mechanical lag of the PTO loading test bench, the response time is a little slower compared to the load change at the initial stage. In terms of the loading accuracy, the maximum overshoot is 15 N·m, the maximum error is 4.03%, and the goodness of fit is 16.78 N·m for the Fuzzy-PID controller. In comparison, there is a large overshoot when the load changes greatly for the PID controller, the maximum overshoot is 52 N·m, the maximum error is 14.53%, and the goodness of fit is 29.46 N·m.

The results show that it has high accuracy and stability for the dynamic loading control of the load spectrum of the Fuzzy-PID controller proposed in this paper, and the PTO loading test bench is available for the loading of the dynamic load spectrum, which can provide the reference for the tractor performance test.

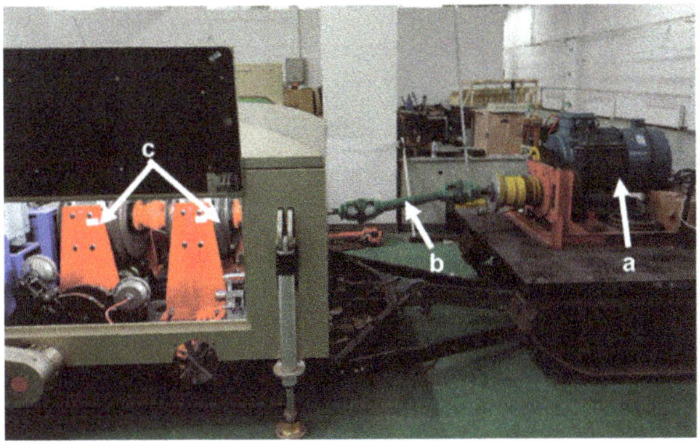

Figure 11. The loading test site. (**a**) 37.5 kW motor; (**b**) universal joint connection shafts; (**c**) eddy current dynamometer.

Table 6. The parameters of the test equipment.

Item	Parameter
Manufacturer	Lu'an Yisheng Motor Co., Ltd. (Wenling, Zhejiang, China)
Model	YVP-225S-8
Power	37.5 kW
Rated voltage	380 V
Rated speed	1000 r·min^{-1}
Rated torque	356 N·m
Maximum torque	523 N·m
Transmission ratio	1:2.78

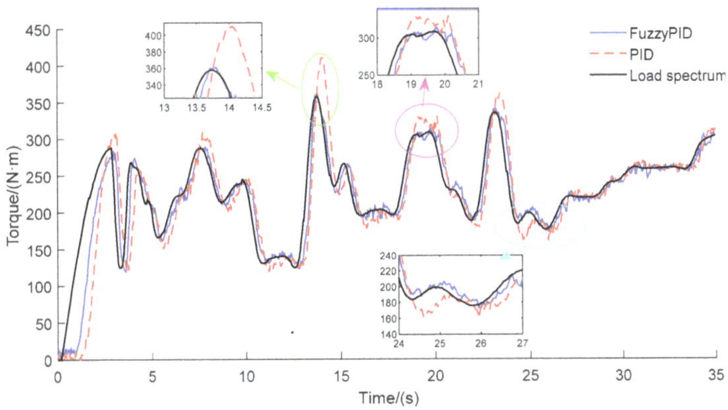

Figure 12. Test results of the first 35s loading data.

4. Discussion

The tractor PTO is the key power transmission component when used in the rotary tillage and driven harrow operations, and the main loading part of the bench test for the tractor drive train. In the current tractor performance test, the torque loading of tractor PTO is mainly based on static loading or static step-by-step loading, and the dynamic characteristics of PTO load output by this mode is different from that of PTO load in the actual field operation, which leads to the difference in test results between the bench test and the actual field operation. To solve this problem, in this paper, the dynamic loading method of the tractor PTO torque based on load spectrum is studied.

First, an accurate PTO torque load spectrum using the extrapolating method of tractor PTO is studied based on the Peak Over Threshold model, and in order to obtain a more accurate Peak Over Threshold model, an automatic threshold selection method based on FDR control is proposed. Compared with the traditional method [9,16], the method proposed in this paper has a simple process and higher GPD fitting accuracy which is more suitable for the extrapolation of tractor PTO torque load. In addition, a novel movable tractor PTO loading test bench is developed which has a more compact structure and in which it is convenient to carry out various PTO loading tests compared with the test bench proposed in references [5–7]. More importantly, on this basis, the loading test method of the tractor PTO load spectrum is studied by using the PTO torque load spectrum as the input signal; the load change in actual operations can be fully reflected and the load characteristics are closer to the reality. Theoretically, the test results based on the load spectrum will be more real and reliable, which can provide new ideas and reference for the tractor PTO test.

5. Conclusions

There is relatively little research on the loading test of tractor PTO based on the load spectrum acquired in the actual field work; thus, the performance and reliability of tractor PTO is difficult to evaluate and test. In this study, the time domain extrapolation method based on a Peak Over Threshold model is proposed for extrapolating the tractor PTO torque load spectrum throughout the whole life cycle. The tractor PTO loading test bench with two eddy current dynamometers and control system are developed based on the Fuzzy-PID control algorithm, and the test bench and the control algorithm are evaluated for the step signal and dynamic load spectrum. The results are as follows:

1. With the approach of the time domain extrapolation method based on a Peak Over Threshold model, the extrapolation of the tractor PTO torque load spectrum is realized. The results show that the load extreme value is expanded from (63.24, 469.50) to (60.88, 475.18), and the coverage is expanded by 1.98%. In the GPD fitting part, the correlation coefficient between the fitting model obtained by the ForwardStop rule and the original data is greater than 0.99, and the goodness of fit is 2.8676 N·m and 1.9804 N·m, respectively, indicating that the fitting results are highly accurate.

2. The tractor PTO loading test bench and control system is developed, which adopts Fuzzy-PID controller to realize the torque loading control. The simulation results show that the response time of the Fuzzy-PID controller is 0.05 s, and the maximum error is 1.77%. Compared with the PID controller, it has a better follow effect on the loading control of the load spectrum of the Fuzzy-PID controller.

3. The loading test method of the tractor PTO load spectrum is explored. The actual test results show that the output torque of the loading system has a good response to the input load spectrum. The response time is 0.75 s, the maximum overshoot is 15 N·m, the maximum error is 4.03%, and the goodness of fit was 16.78 N·m, which indicate that the PTO loading test bench can meet the dynamic loading test demand of the tractor.

The limitation of this study is that only the preliminary research on the laboratory test of the loading method and effect based on the PTO load spectrum has been carried out and has not been applied in the field operation for the tractor PTO test. For future directions, we intend to take the actual tractor PTO as the object to explore the impact of PTO load spectrum on the fatigue damage and reliability of PTO, and compare it with the traditional test method. It is hoped that the research results can be used to guide the actual tractor test and provide a reference for the optimization method of agricultural machinery and equipment. We look forward to using the research results to conduct the actual tractor test and provide reference for the optimization method of agricultural machinery and equipment.

6. Patents

A patent has been applied for in China for the PTO test bench reported in this manuscript (Patent No. CN201910661680 and Application No. CN201921117795).

Author Contributions: Conceptualization, L.W.; data curation, Y.W.; funding acquisition, L.W.; methodology, Y.W.; software, J.Z. and D.L.; supervision, L.W. and S.W.; writing—original draft, Y.W.; Writing—review and editing, L.W. and S.W. All authors have read and agreed to the published version of the manuscript.

Funding: This research was funded by the National Natural Science Foundation of China (Grant No. 31801783).

Institutional Review Board Statement: Not applicable.

Informed Consent Statement: Not applicable.

Acknowledgments: The authors thank the National Science Foundation of China (Grant No. 31801783).

Conflicts of Interest: The authors declare no conflict of interest.

References

1. Lips, M. Length of Operational Life and Its Impact on Life-Cycle Costs of a Tractor in Switzerland. *Agriculture* **2017**, *7*, 68. [CrossRef]
2. Bulgakov, V.; Aboltins, A.; Beloev, H.; Nadykto, V.; Kyurchev, V.; Adamchuk, V.; Kaminskiy, V. Experimental Investigation of Plow-Chopping Unit. *Agriculture* **2021**, *11*, 30. [CrossRef]
3. Vigneault, C.; Amour, G..S.; Buckley, D.J.; Masse, D.I.; Savoie, P.; Tremblay, D. A trailer-mounted PTO torquemeter system. *Can. Agric. Eng.* **1989**, *31*, 89–91.
4. Kornecki, T.S.; Reyes, M.R. Equipment Development for Small and Urban Conservation Farming Systems. *Agriculture* **2020**, *10*, 595. [CrossRef]
5. Jiankang, W.; Mingliang, W.; Ping, J.; Songlin, S.; Fangping, X. Application of DC Electric Dynamometer in the PTO Power Test of Tractors. *Int. Conf. Intell. Comput. Technol. Automation. IEEE* **2011**, 89–91. [CrossRef]
6. Sumer, S.K.; Kocabiyik, H.; Say, S.M.; Cicek, G. Comparisons of 540 and 540E PTO operations in tractors through laboratory tests. *lgarian J. Agric. Sci.* **2010**, *16*, 526–533.
7. Siddique, M.A.A.; Baek, S.; Baek, S.; Kim, W.; Kim, Y.; Kim, Y.; Lee, D.; Lee, K.; Hwang, J. Simulation of Fuel Consumption Based on Engine Load Level of a 95 kW Partial Power-Shift Transmission Tractor. *Agriculture* **2021**, *11*, 276. [CrossRef]
8. Roeber, J.B.; Pitla, S.K.; Hoy, R.M.; Luck, J.D.; Kocher, M.F. Tractor Power Take-Off Torque Measurement and Data Acquisition System. *Appl. Eng. Agric.* **2017**, *33*, 679–686. [CrossRef]
9. Yang, Z.H.; Song, Z.H.; Yin, Y.Y.; Zhao, X.Y.; Liu, J.H.; Han, J.G. Time domain extrapolation method for load of drive shaft of high-power tractor based on POT model. *Trans. Chin. Soc. Agric. Eng.* **2019**, *35*, 40–47. (In Chinese) [CrossRef]
10. Wang, J.; Wang, N.; Wang, Z.; Zhang, Y.; Liu, L. Determination of the minimum sample size for the transmission load of a wheel loader based on multi-criteria decision-making technology. *J. Terramech.* **2012**, *49*, 147–160. [CrossRef]
11. Wang, J.; Chen, H.; Li, Y.; Wu, Y.; Zhang, Y. A Review of the Extrapolation Method in Load Spectrum Compiling. *Strojniški Vestnik-J. Mech. Eng.* **2016**, *62*, 60–75. [CrossRef]
12. Downing, S.D.; Socie, D.F. Simple rainflow counting algorithms. *Int. J. Fatigue* **1982**, *4*, 31–40. [CrossRef]
13. Hong, N. A modified rainflow counting method. *Int. J. Fatigue* **1991**, *13*, 465–469. [CrossRef]
14. Johannesson, P.; Thomas, J. Extrapolation of Rainflow Matrices. *Extremes* **2001**, *4*, 241–262. [CrossRef]
15. Johannesson, P. Extrapolation of load histories and spectra. *Fatigue Fract. Eng. Mater. Structures* **2006**, *29*, 209–217. [CrossRef]
16. Yan, X.H.; Zhou, Z.L.; Jia, F. Compilation and verification of dynamic torque load spectrum of tractor power take-off. *Trans. Chin. Soc. Agric. Eng.* **2019**, *35*, 74–81. (In Chinese) [CrossRef]
17. Kiriliouk, A.; Rootzén, H.; Segers, J.; Wadsworth, J.L. Peaks over Thresholds Modeling with Multivariate Generalized Pareto Distributions. *Technometrics* **2019**, *61*, 123–135. [CrossRef]
18. Zhao, X.; Zhang, Z.; Cheng, W.; Zhang, P. A New Parameter Estimator for the Generalized Pareto Distribution under the Peaks over Threshold Framework. *Mathematics* **2019**, *7*, 406. [CrossRef]
19. Thompson, P.; Cai, Y.; Reeve, D.; Stander, J. Automated threshold selection methods for extreme wave analysis. *Coast Eng.* **2009**, *56*, 1013–1021. [CrossRef]
20. G'Sell, M.G.; Wager, S.; Chouldechova, A.; Tibshirani, R. Sequential selection procedures and false discovery rate control. *J. R. Stat. Soc.* **2016**, *78*, 423–444. [CrossRef]
21. Choulakian, V.; Stephens, M.A. Goodness-of-Fit for the Generalized Pareto Distribution. *Technometrics* **2001**, *43*, 478–484. [CrossRef]

22. Langousis, A.; Mamalakis, A.; Puliga, M.; Deidda, R. Threshold detection for the generalized Pareto distribution: Review of representative methods and application to the NOAA NCDC daily rainfall database. *Water Resour. Res.* **2016**, *52*, 2659–2681. [CrossRef]
23. Siddique, M.A.A.; Kim, W.; Kim, Y.; Kim, T.; Choi, C.; Lee, H.; Chung, S.; Kim, Y. Effects of Temperatures and Viscosity of the Hydraulic Oils on the Proportional Valve for a Rice Transplanter Based on PID Control Algorithm. *Agriculture* **2020**, *10*, 73. [CrossRef]

agriculture

Article

Optimization and Experiment of Straw Back-Throwing Device of No-Tillage Drill Using Multi-Objective QPSO Algorithm

Hongbo Xu [1,2], Zhichao Hu [1], Peng Zhang [1,*], Fengwei Gu [3,4,*], Feng Wu [1], Wanli Song [5] and Chunci Wang [5]

1 Nanjing Institute of Agricultural Mechanization, Ministry of Agriculture and Rural Affairs, Nanjing 210014, China; xuhongbo@caas.cn (H.X.); huzhichao@caas.cn (Z.H.); wufeng@caas.cn (F.W.)
2 College of Engineering, China Agricultural University, Beijing 100083, China
3 Key Laboratory of Modern Agricultural Equipment, Ministry of Agriculture and Rural Affairs, Nanjing 210014, China
4 College of Energy and Power Engineering, Nanjing University of Aeronautics and Astronautics, Nanjing 210016, China
5 College of Mechanical Engineering and Automation, Northeastern University, Shenyang 110167, China; songwl@me.neu.edu.cn (W.S.); penggggg@163.com (C.W.)
* Correspondence: zhangpeng01@caas.cn (P.Z.); gufengwei@caas.cn (F.G.)

Abstract: Reducing operation energy consumption is the development demand of conservation tillage equipment. In order to solve the problems of high power consumption and the easy blockage of the no-tillage drill under full straw retention, the key parameters of the straw back-throwing device were optimized in this study. The Box–Behnken central combination test method was used to analyze the influence of the impeller rotating speed, feed quantity and cross-sectional area of the throwing pipe on the specific power consumption and throwing speed, the mathematical models of which were built with the aid of the least square method. In addition, the mathematical models were optimized by using a multi-objective quantum behaved particle swarm optimization (QPSO) with an improved target weighting coefficient. The optimization results indicated that, when the impeller rotating speed was 2287 r/min, the feed quantity was 1.1 kg/s and the cross-sectional area of the throwing pipe was 506.997 cm^2, the specific power consumption and throwing speed by the models were 7528 m^2/s^2 and 11.73 m/s, respectively. The models were verified by comparing the optimization results with the measured data in the simulation filed tests, which proved that the multi objective QPSO algorithm was feasible to optimize the working and structural parameters of the straw back-throwing device of the no-tillage drill under full straw retention. The results provide references for the parameter optimization of similar no-tillage drills under straw retention.

Keywords: no-tillage drill; straw back-throwing device; quantum behaved particle swarm optimization (QPSO); multi-objective optimization; verification test

Citation: Xu, H.; Hu, Z.; Zhang, P.; Gu, F.; Wu, F.; Song, W.; Wang, C. Optimization and Experiment of Straw Back-Throwing Device of No-Tillage Drill Using Multi-Objective QPSO Algorithm. *Agriculture* **2021**, *11*, 986. https://doi.org/10.3390/agriculture11100986

Academic Editor: José Pérez-Alonso

Received: 6 September 2021
Accepted: 7 October 2021
Published: 10 October 2021

1. Introduction

The sowing method of the no-tillage drill has the advantages of racing against time, high replanting index, protecting the soil ecological environment, low cost and so on [1]. Accordingly, it is adopted all over the world, especially in North America, Oceania, Europe and Asia, and is used in many crops, including corn, wheat, soybean and so on [2]. With the development of agricultural machinery, the efficiency of the no-tillage drill has become the focus of research [3]. The no-tillage drill under full straw retention can help to achieve the processes of straw chopping, straw transport, sowing, fertilization and straw mulching once and for all, which are widely used in many areas of China [4]. As an important working part of the no-tillage drill under full straw retention, the back-throwing device is of great significance to performance improvement [5]. Therefore, many scholars have studied the back-throwing device of the no-tillage drill under full straw retention [6–8].

Shinners et al. [9] studied the rotation direction of the throwing impeller and reduced the power consumption by changing the direction of the cutter. The experimental results showed that the specific power consumption could be reduced by 30%–34%, which greatly improved the work efficiency, but the throwing distance of the harvester was shortened by 27%. Shinners et al. [10] then went further, increasing the opening in the side of the cutting chamber to increase the air flow in the chamber and thus increase the throw distance. Chattopadhyay et al. [11] studied the effect of flail tip speed, knife rake angle and bevel angle on the conveying of chopped forage sorghum (S. bicolor) through a 90° deflector elbow under laboratory conditions, which showed that the static pressure head created by the rotating flails at the blower outlet (chute inlet) increased exponentially as the flail tip speed was increased and decreased linearly when the knife rake angle was increased. Jia et al. [12] studied the motion of the throwing device of a maize no-tillage drill, established the motion micro-equation aiming at power consumption and optimized the relevant working parameters by the orthogonal polynomial regression method and variance analysis. Zhai and Wang [13] simulated the internal airflow field of the throwing device through FLUENT. The test results showed that the simulated airflow velocity at the nozzle of the straight pipe was consistent with the test, proving the accuracy of the software numerical simulation. By comparing and simulating the shell shape and blade inclination angle of the different parameters of the throwing device, it was concluded that, when the structure of the throwing device is a round shell and the blade inclination is a rearward angle, it is more beneficial to the throwing of discarded straw, and the arc radius at the exit ought not to be too large. Then, the simulation and parameter optimization of the airflow motion of the blade throwing device were carried out, and they found that the blade throwing effect was the best when tilting back 10° [14]. In 2012, Zhai et al. [15] simulated the motion of discarded straw along the throwing blade, analyzed the throwing efficiency, power consumption and the movement of discarded straw in the throwing device and obtained that the optimal angle range for the movement of discarded straw in the volute was about 60°–130°. However, the problems of high power consumption and congestion affect the promotion of the straw back-throwing device of the no-tillage drill under full straw retention.

To improve the power consumption and smoothness of the straw back-throwing device, it is necessary to calculate the optimal parameters that affect the working process. However, power consumption and smoothness are often mutually restrictive optimization indexes, so an appropriate multi-objective optimization algorithm is needed to solve them. Particle swarm optimization (PSO) is an evolutionary computation that simulates the predation behavior of birds, which is widely used in the engineering field [16,17]. On this basis, the PSO was continuously improved in order to improve the search scope and convergence speed [18].

In this paper, the equation of specific power consumption and the throwing efficiency of the back-throwing device are established to determine the reasonable structure and motion parameters. Since the relationship between the working parameters of the back-throwing device are a non-linear correlation, the least square method is used to establish the equations of the specific power consumption and the throwing speed of the back-throwing device on the relationship between the rotating speed of the throwing impeller, the feeding quantity and the cross-sectional area of the throwing pipeline. After that, an improved multi-objective quantum particle swarm optimization (IMQPSO) algorithm and improved weight coefficient are introduced to optimize the multiple parameter equations. The goal of this study is to provide references for the parameter optimization of similar no-tillage drills under straw retention.

2. Materials and Methods

2.1. Back-Throwing Device of No-Tillage Drill under Full Straw Retention

The no-tillage drill under full straw retention is mainly composed of a straw cleaning device and a no-tillage drill. The straw cleaning device, including a straw crushing device, a straw back-throwing device and a straw dispersing device, is set in front of the drill [19]. As shown in Figure 1, during the operation, the tractor provides power to drive the operation of the straw crushing device and the straw back-throwing device. After the residual straw in the field is picked and crushed by the straw crushing device, the straw is transported and lifted by the straw back-throwing device, and it is evenly thrown back under the action of the dispersing device. Thus, the no-tillage drill is planting in the area without any residual straw, which can significantly improve the working smoothness and sowing quality.

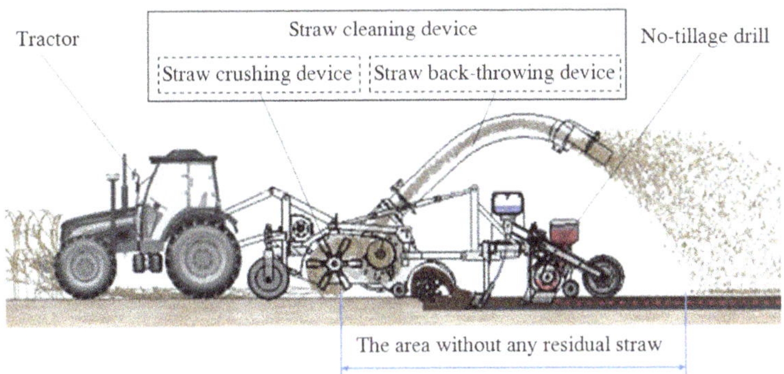

Figure 1. Operation principle of no-tillage drill under full straw retention.

The straw back-throwing device is mainly composed of throwing impeller and throwing pipeline (Figure 2). The crushed straw is thrown along the pipeline under the action of rotation of the impeller. The parameters of the back-throwing device affect the throwing smoothness and the power consumption of the device. Realizing smooth throwing with the lowest power consumption is the design goal of the back-throwing device.

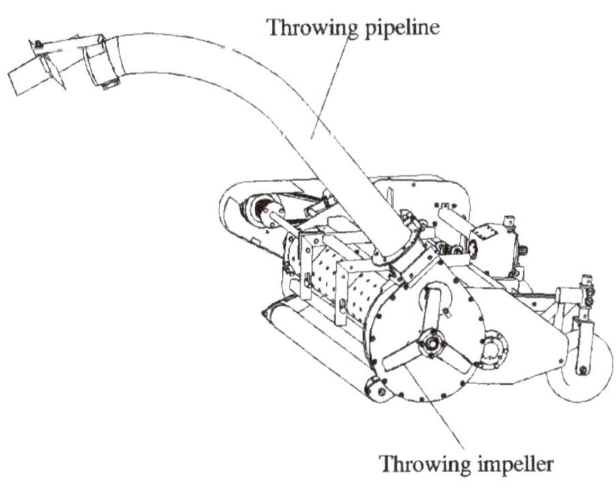

Figure 2. The structure of straw back-throwing device.

2.2. Performance Testing of Throwing Device

Specific power consumption reflects the power consumption per unit mass of straw, while throwing speed reflects the operation smoothness of straw back-throwing device. Thus, the specific power consumption and throwing speed under different working parameters of the straw back-throwing device were determined as evaluation indexes of the straw back-throwing device in the study. The installation mode of the test device is shown in Figure 3. A rotational speed and torque sensor was installed between the power output shaft of the tractor and the power input shaft of the transmission box of the straw cleaning device. A high-speed camera acquisition system was placed on the side of the forward path of the straw back-throwing device and focused on the outlet of the throwing pipeline.

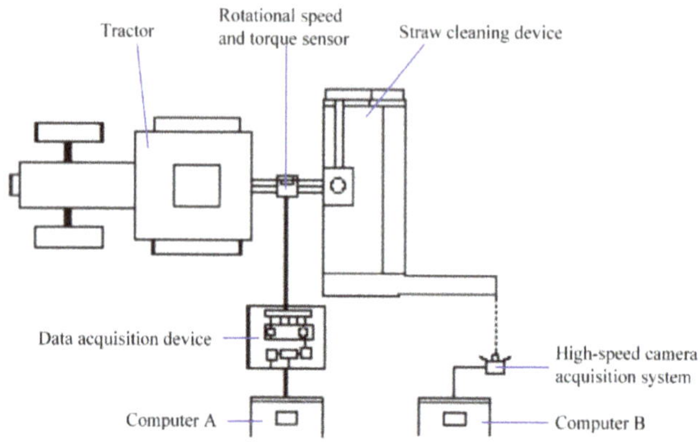

Figure 3. The installation method of testing device.

The experimental site was at the east area of Nanjing Institute of Agricultural Mechanization, Ministry of Agriculture and Rural Affairs (Nanjing, China). The straw was the chopped part of rice harvested by Jiangsu Academy of Agricultural Sciences (Jiangsu, China). The length of discarded straw after crushing was less than 120 mm, and the moisture content of straw was 65%. The simulated field was manually laid with a laying density of 2 kg/m^2 and a width of 2.2 m. The main instruments used in the test: powered tractor, measuring tape, electronic balance, the SL06 rotational speed and torque sensor (Beijing Sanjing United Technology Co., LTD., Beijing, China) and HiSpec5 high-speed camera acquisition system (FASTEC IMAGING Co., LTD., San Diego, CA, USA) with a resolution of 1376 × 1132 pixels, sampling frame frequency of 200 FPS, exposure time of 4998 μs) were used to measure specific power consumption and throwing speed.

Orthogonal rotational quadratic combination test with three factors and three levels was made to evaluate the combined influence of the factors on the test index value. Based on previous single-factor test results and practical operability, the throwing effect is well within the rotating speed of throwing impeller of 1800–2700 r·min^{-1}. According to the conventional operation efficiency of machines and tools and the amount of straw in the field, the variation range of feeding amount is controlled at 1.1–1.7 kg·s^{-1}. When the cross-sectional area of the throwing pipeline is 207–507 cm^2, the structure of throwing pipeline is reasonable and not easy to jam. The factors and levels of the test are shown in Table 1.

Table 1. Factors and levels of orthogonal rotational quadratic combination test.

Level	Rotate Speed of Throwing Impeller (r·min^{-1})	Feeding Quantity (kg·s^{-1})	Sectional Area of Throwing Pipeline (cm^2)
−1	1800	1.1	207
0	2250	1.4	357
1	2700	1.7	507

During the test, the power consumption P_j of the straw cleaning device under different working parameters with the rotational speed and torque sensor was measured, and the straw throwing image at the outlet of the throwing pipeline with the high-speed camera acquisition system was recorded. Then, the belt drive between the straw crushing device and the straw back-throwing device was removed, and then the above tests were repeated.

The power consumption P_q was recorded in the meantime. The specific power consumption of the straw back-throwing device is

$$P_s = \frac{P_j - P_q}{Q} \tag{1}$$

while P_s is the specific power consumption of the straw back-throwing device, m^2/s^2; P_j is the power consumption of the straw cleaning device, W; P_q is the power consumption of the straw cleaning device without back-throwing device, W; Q is the feeding quantity, kg/s. Straw throwing images under different working conditions were recorded by the high-speed camera acquisition system, and then the straw throwing speed V_p was obtained by the video processing software (Proanalyst Professional 2D, Xcitex Inc., Woburn, MA, USA).

2.3. Improved Multi-Objective Particle Swarm Optimization Algorithm

The multi-objective quantum particle swarm optimization (MQPSO) algorithm is an improvement of ordinary quantum particle swarm optimization (QPSO) algorithm [20]. In this paper, the dynamic weighting method (DWM) is used to improve the MQPSO algorithm. The DWM can be used to solve the problems of low efficiency, multi-objective optimization with convex shape and difficulty in ensuring the accuracy of the optimal value. By dynamically changing the value of the weight coefficient, the optimization results can be prevented from being incorrect due to the constant weight coefficient. When the problem of multi-objective optimization is two-objective, its weight coefficient can usually be changed according to Equation (2)

$$\omega_1 = m - (m - n) \times \frac{t}{MaxTimes} \omega_2 = 1 - \omega_1 \tag{2}$$

where t is the number of optimized iterations; m and n are the adjustment coefficients. In general, they are constants, and the specific values are determined according to the actual situation. $MaxTimes$ is the maximum number of iterations; ω_1, ω_2 are the inertia weights of the two targets, respectively.

The scholar Shi.Y [21] found that a small value of inertia weight was beneficial to local search, while a large value of inertia weight was beneficial to global search. Based on the previous conclusion, Shi.Y proposed the linear decreasing inertia weight, as shown in Equation (2), with the purpose of balancing the local search capability and the global search capability so as to better solve the multi-objective optimization problem. According to the traditional experience, the performance of the algorithm is best when the inertia weight of m is 0.9 and n is 0.4. The inertia weight decreases linearly from 0.9 to 0.4. In the early stage, the higher the inertia weight is, the better the global search ability of the algorithm will be. With the progress of iteration, the lower the inertia weight is, the stronger the local search capability of the algorithm will be.

The inertia weight is improved to change linear decline into parabolic decline so that the global search in the early stage is more comprehensive and the local search in the later stage is faster. The inertia weight after modification is shown as follows:

$$\omega_1 = m - (m - n) \times \left(\frac{t}{MaxTimes}\right)^3 \omega_2 = 1 - \omega_1 \tag{3}$$

Figure 4 shows the dynamic change of ω before and after the improvement.

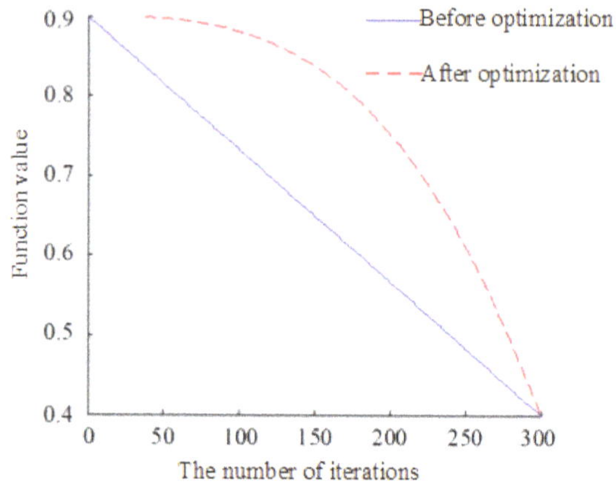

Figure 4. The change of two inertia weight.

The following is an analysis of the performance of the two ω changing algorithms, and the following functions are selected for testing:

$$f(x) = \frac{\sin\sqrt{(x^2+y^2)}}{\sqrt{(x^2+y^2)}} + e^{\frac{\cos 2\pi x + \cos 2\pi y}{2}} - 2.71289 \tag{4}$$

Figure 5 shows a graph of the function to test the performance of the algorithm. It shows that the function has a lot of local maximum points. This function is used to test the performance of inertia weight to check whether the function falls into local maximum. The extremum gets the maximum near (0,0) coordinate.

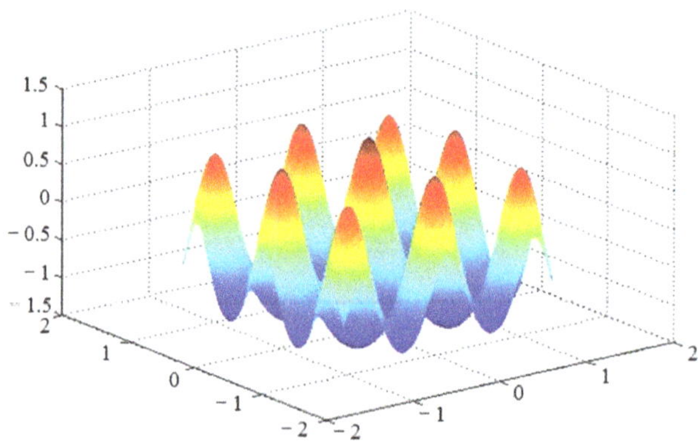

Figure 5. Function to test the algorithm performance.

The algorithm parameter is set to population size of 20 and evolution of 300 generations. The experiment is set up to run for 100 times, and the average of 100 times is taken as the final result. The two inertia weights are solved, and the average value, failure times and near-optimal times of the solutions are compared to analyze the convergence accuracy, convergence speed and other performances.

Table 2 shows that the improved inertia weight searches the same optimal solution, but the average value is closer to the optimal solution, and the times fall into suboptimal solution significantly less than before the improvement. The improved inertia weight function keeps a large value of the inertia weight ω at the beginning of the iteration, which can increase the global search time in the early stage and quickly enter the local search in the later stage so as to improve the search efficiency (Figure 6). After the iteration to 150 generations, the global optimum is achieved, the convergence rate is faster than before the improvement and the solution accuracy is higher.

Table 2. Performance comparison of different inertia weights.

	The Optimal Solution	The Average Value	The Number of Times to Get into a Suboptimal Solution	The Number of Times to Approach the Optimal Solution
$\omega_1 = m - (m-n) \times \frac{t}{MaxTimes}$	1.0054	0.9801	16	84
$\omega_1 = m - (m-n) \times \left(\frac{t}{MaxTimes}\right)^3$	1.0054	1.0052	0	100

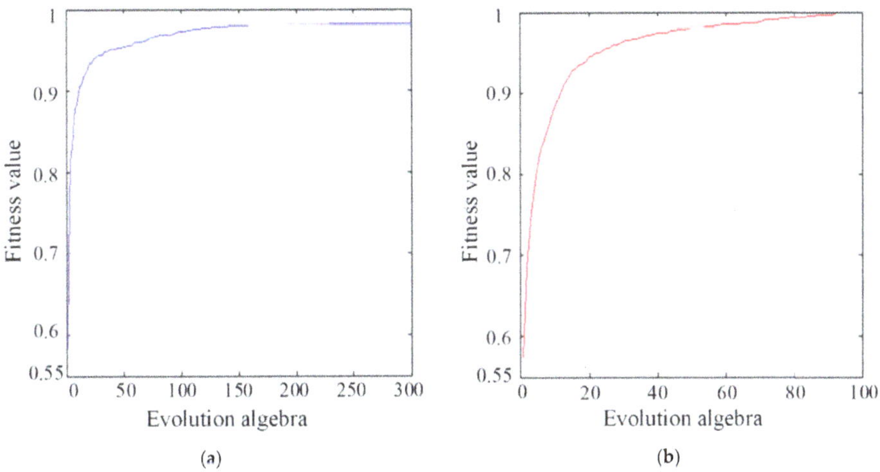

Figure 6. The convergence curve of the average value of the function before and after the improvement of inertia weight: (**a**) before improving; (**b**) after improving.

Assuming that the optimization function of a two-objective is $F(X) = (f_1(X), f_2(X))^T$, the purpose is to find the minimum value of F(X), then the main steps of optimization using IMQPSO algorithm are as follows:

(1) Initialize the particle swarm in which n particles are generated at random and are expressed in vector form. The position of the least adaptable of all particles is set as the initial position.
(2) The average optimal position value of the whole particle swarm can be obtained by using the Equation (5).

$$m_{best} = sum(p_{best})/n \tag{5}$$

(3) According to Equation (3), the weight coefficient of each objective function is obtained;
(4) The random points PP_{ij} and position of each particle are obtained according to the position update formula of QPSO algorithm.
(5) Calculate the p_{best} value of the new position of each particle;
(6) Find the global optimal position g_{best};
(7) Repeat steps Equations (2)–(6) above until the algorithm can meet the initial termination conditions. Finally, a set of Pareto optimal solution sets is given.

3. Results

3.1. The Equation of Response Surface

Before using the least square method to fit the function, it is necessary to process the experimental data with the help of Design-Expert software and make the equation fit closer to the actual value by analyzing the correlation of the parameters [22–24].

The variance analysis results of the throwing velocity [25] showed that the influencing factors were mainly the impeller speed and the cross-sectional area. The impeller speed and pipe cross-sectional area were extracted from the Box–Behnken test design table, and zero level was selected for the feeding quantity to obtain the relation table of the throwing speed on the impeller speed and the pipe cross-sectional area, as shown in Table 3.

Table 3. The relationship among the throwing speed, the rotate speed of throwing impeller and the sectional area of throwing pipeline.

Impeller Speed (r/min)	Cross-Sectional Area (cm^2)	Throwing Speed (m/s)
2700	207	13.1
2700	507	14.8
1800	507	9.8
2250	357	11.2
1800	207	8.2

Similarly, the relation table of the specific power consumption with the impeller speed, feeding quantity and cross-sectional area of the throwing pipeline is shown in Table 4.

Table 4. The relationship among the specific power consumption, the rotate speed of throwing impeller, feeding quantity and sectional area of throwing pipeline.

Impeller Speed (r/min)	Feeding Quantity (kg/s)	Cross-Sectional Area (cm^2)	Specific Power Consumption (m^2/s^2)
2700	1.4	207	16,930
2250	1.4	357	6440
2700	1.4	507	20,540
1800	1.4	507	5340
2250	1.1	207	7120
2700	1.1	357	18,090
2700	1.7	357	19,210
1800	1.7	357	6060
1800	1.1	357	5300
2250	1.7	507	9900
1800	1.4	207	5140
2250	1.1	507	7160
2250	1.7	207	7250

The variance analysis was used to eliminate the factors that have less influence on the specific power consumption and throwing speed. The response surface equation of the specific power consumption and throwing speed is as follows:

$$\begin{aligned} Y_1 = 100.927 - 0.0897X_1 - 2.776X_2 - 0.0582X_3 + 1.263 \times 10^{-5}X_1X_3 \\ +0.0145X_2X_3 + 2.2269 \times 10^{-5}X_1^2 + 8.941 \times 10^{-3}X_2^2 + 1.98 \times 10^{-5}X_3^2 \end{aligned} \quad (6)$$

$$Y_2 = -2.9185 + 5.28 \times 10^{-3}X_1 + 5.28 \times 10^{-3}X_3 \quad (7)$$

where Y_1 is the specific power consumption and Y_2 is the throwing speed.

3.2. The Detection of Regression Model

The reliability of the model was verified by the analysis of the predicted and actual test values and the residual analysis. Figure 7 shows the analysis diagram of the predicted value and the actual test value using the software of Design-Expert V8.0.6.1 (Stat-Ease Inc., Minneapolis, MN, USA). The proximity of the predicted value and the actual test value reflected the reliability of the equation.

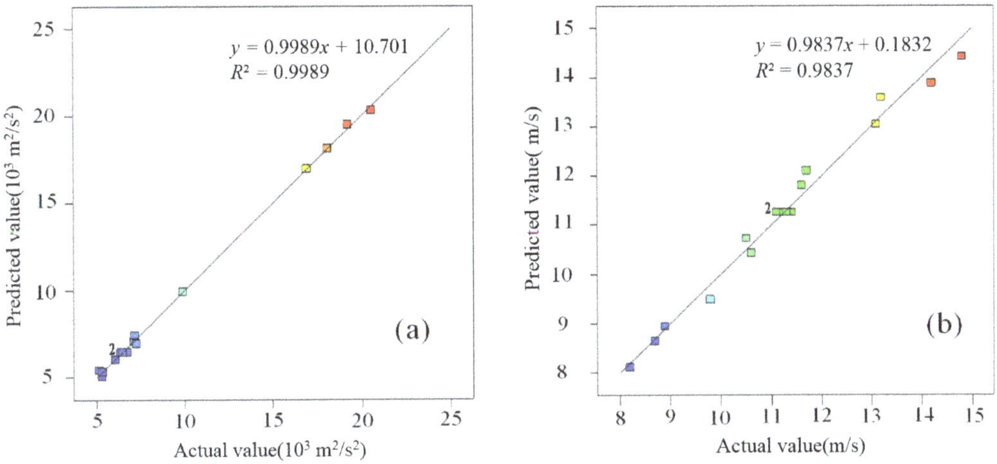

Figure 7. Analysis of predicted and actual values of specific power consumption and throwing speed: (**a**) Specific power consumption; (**b**) Throwing speed.

Figure 8 is the residual analysis diagram of the regression equation for the specific power consumption and throwing speed; the ordinate was the normal distribution probability of the residual, and the abscissa was the value of the residual after standardization. It can be seen that the residual is basically on a straight line, which demonstrates that the error of the regression equation of specific power consumption and throwing speed was normally distributed; that is, the fitted equation satisfied the requirements of the least square regression analysis method.

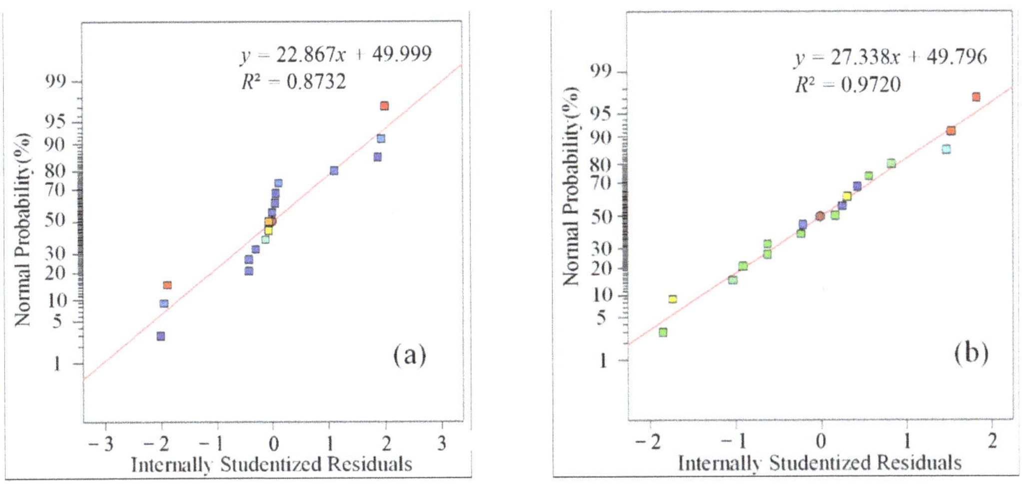

Figure 8. The residual analysis of specific power consumption and throwing speed: (**a**) Specific power consumption; (**b**) Throwing speed.

3.3. Parameters Optimization of Back-Throwing Device Based on IMQPSO Algorithm

The improved algorithm is used to optimize the function so that both objective functions are minimized. According to Equations (6) and (7), two objective optimization problems are expressed as:

$$\begin{aligned} f_1 = 100.927 - 0.0897A - 2.776B - 0.582C + 1.263 \times 10^{-5}AC + 0.0145BC \\ +2.2269 \times 10^{-5}A^2 + 8.941 \times 10^{-3}B^2 + 1.98 \times 10^{-5}C^2 \end{aligned} \quad (8)$$

$$f_2 = \frac{10}{Y_2} = \frac{10}{2.9185 + 5.28 \times 10^{-3} \times A + 5.28 \times 10^{-3}C} \quad (9)$$

Constraint conditions:

$$1800 \leq A \leq 2700$$

$$1.1 \leq B \leq 1.7$$

$$207 \leq C \leq 507$$

where A is the speed of the throwing impeller (r/min); B is the feed quantity (kg/s); C is the sectional area of the throw pipe (cm^2).

Function f_1 is the equation of the specific power consumption of the back-throwing device, and function f_2 is the equation of the throwing speed. The goal is to find one or more sets of solutions that minimize functions f_1 and f_2 at the same time. It is almost impossible for f_1 and f_2 to meet the minimum value at the same time, so our goal is to find a set of Pareto optimal solutions, making one target value as small as possible while not making another target value increase. The IMQPSO is used to optimize the above two objective functions, and the specific steps are carried out according to Figure 9.

When the flow chart of the algorithm is determined, it is necessary to program the MATLAB algorithm, of which the specific set values of the main parameters are shown in Table 5.

Table 5. The main parameters of IMQPSO.

Parameter	Parameter Value
Number of Particle Swarm n	50
Maximum Number of Iterations $MaxDT$	100
Maximum Weight Coefficient ω_{max}	1
Dimension of the Independent Variable	3

Figure 10 is the Pareto curve of all the optimal solutions obtained through the calculation results drawn by the software of MATLAB 2016b (MathWorks Inc., Natick, MA, USA). The figure is basically a curve, very smooth, and all the particles involved in the operation are at the edge of the Pareto or at least close to the edge, so all the solutions obtained are the solutions that meet the optimization requirements. It shows that the improved QPSO algorithm can solve the optimization problem of the working parameters of the back-throwing device of a no-tillage drill under full straw retention.

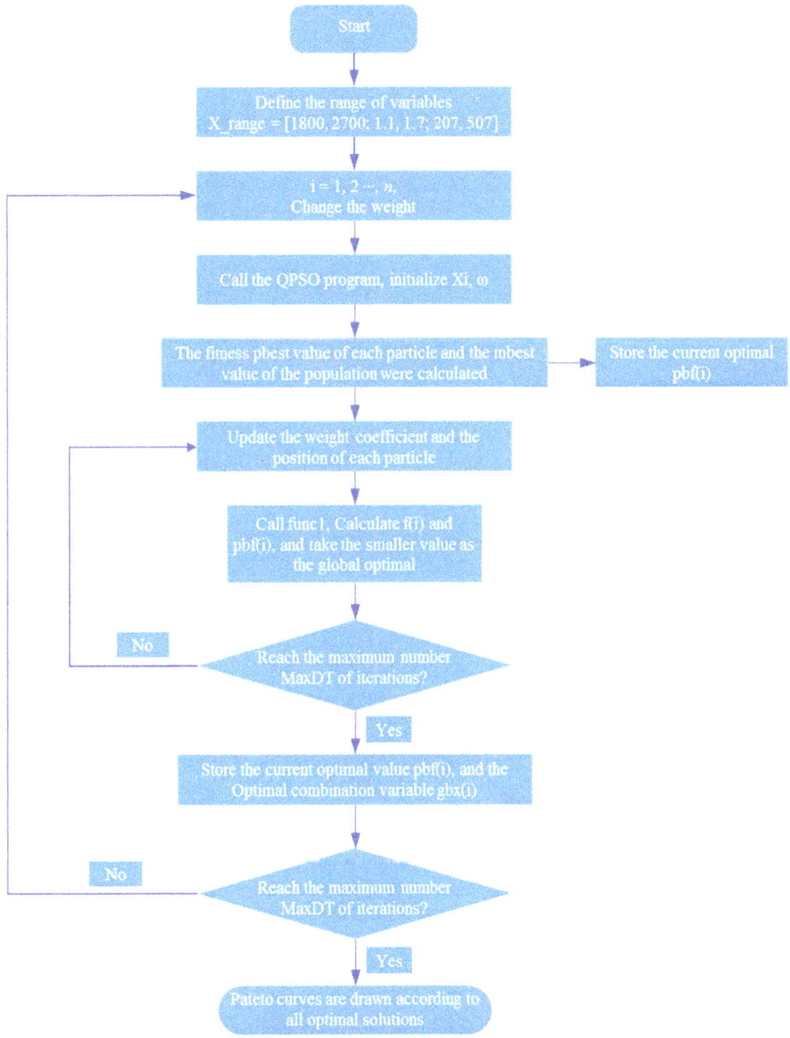

Figure 9. Flow chart of IMQPSO algorithm.

In Figure 10, all the points on the Pareto curve are the optimal solutions of the specific power consumption and the throwing speed parameters of the back-throwing device, indicating that they are mutually restricted. In practice, it is necessary to decide whether to choose a small, specific power consumption or a large throwing speed according to the demand. In this paper, it is required to achieve as little power consumption as possible and as much throwing speed as possible, so the center point on the Pareto frontier is selected as the best solution, which is (7528, 11.73). The values of the operating parameters of the back-throwing device corresponding to the optimal solution, and the results of the optimal solution are shown in Table 6.

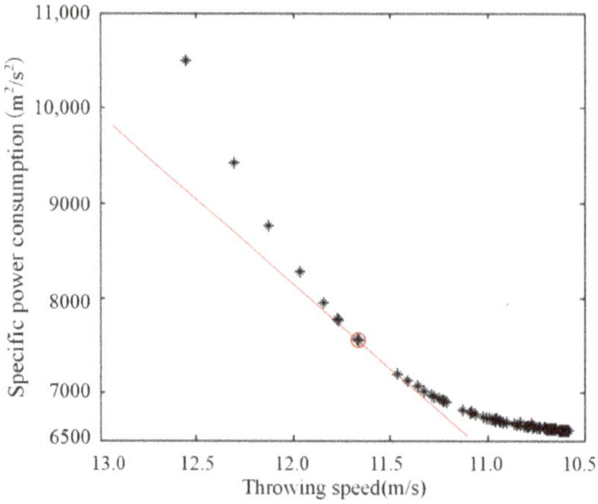

Figure 10. The Pareto graph of specific power consumption and throwing speed.

Table 6. Optimization of process parameters and corresponding optimal results.

Impeller Speed (r/min)	Feeding Quantity (kg/s)	Cross-Sectional Area (cm^2)	Specific Power Consumption (m^2/s^2)	Throwing Speed (m/s)
2287	1.1	506.997	7528	11.73

3.4. Verification Test

In order to verify the accuracy of the prediction model, the above parameters were used to conduct three repeated tests in the east area of Nanjing Institute of Agricultural Mechanization, Ministry of Agriculture and Rural Affairs (considering the feasibility of the test, the throwing impeller speed was set to 2270 r/min, the feeding quantity was 1.1 kg/s and the cross-sectional area of the throwing pipe was 507 cm^2), and the average value of the three tests was taken as the test verification value. The test scenario is shown in Figure 11.

(a) (b)

Figure 11. The scenario of verification test: (**a**) the simulative residual straw; (**b**) the operation process of back-throwing device.

Through the above method, the specific power consumption and throwing speed of the back-throwing device were measured, and the experimental results were compared with the optimized results, as shown in the Table 7.

Table 7. The relationship among the specific power consumption, the rotate speed of throwing impeller, feeding quantity and sectional area of throwing pipeline.

	The Optimization Results	The Experimental Results	Error (%)
Specific Power Consumption (m^2/s^2)	7528	7180	4.8
Throwing Speed (m/s)	11.73	11.60	1.1

Compared with the results in Table 7, the optimal solution of the specific power consumption and throwing speed of the back-throwing device obtained by the IMQPSO algorithm was very close to the experimental results, and the error of the specific power consumption and throwing speed was less than 5%. It could also be proved that the IMQPSO algorithm was effective, indicating that the improved optimization algorithm was feasible to optimize the back-throwing device.

4. Discussion

The results suggested that the specific power consumption increased with the increase of the speed of the throwing impeller and the cross-sectional area of the throwing pipe. The main reasons are as follows: when the speed of the throwing impeller increased, the energy consumed to provide its own operation and promote the movement of air flow and straw increased exponentially; therefore, the specific power consumption also increased exponentially. This result coincided with what Zhai et al. found in 2013 [26]. When the cross-sectional area of the throwing pipe is larger, the static pressure of the throwing impeller is smaller and the dynamic pressure of the throwing impeller is higher. Therefore, the air flow velocity increased the specific power consumption.

Considering the controllability of the index levels and test costs, we only conducted an artificial simulation test. Nevertheless, the existing experimental data were adequate to guide the parameter optimization of the straw back-throwing device. In the future, we will conduct a field test to further study the factors affecting the power consumption of the no-tillage drill.

By improving the particle swarm optimization algorithm, this paper theoretically optimized the structure and working parameters of the straw back-throwing device and also carried out some experimental verification. The power consumption of the improved device was reduced with the optimization of the working parameters. However, from the theoretical aspect, the optimization analysis of large agricultural machinery is still in an initial stage [27]. The improved particle swarm optimization algorithm in this paper only added quantum mechanics on the traditional basis so that it can accelerate convergence and avoid falling into local optimization, and dynamic weighting was added to make it meet the two objective functions. Furthermore, other algorithms can also be added for fusion to make the optimization method universal.

5. Conclusions

This article mainly studied the straw back-throwing device of a no-tillage drill under full straw retention through performance testing and established the equation of the operating parameters of the back-throwing device by using the test data. The quantum mechanics and weighting function were added to the traditional particle swarm optimization algorithm in order to find the combination of operating parameters with the minimum specific power consumption and the maximum throwing speed. By calculating, the most suitable solution was selected: the rotating speed of the throwing impeller was 2287 rad/min, the feeding quantity was 1.1 kg/s and the cross-sectional area of the throwing pipe was

506.997 cm^2. The specific power consumption of 7528 m^2/s^2 and the throwing speed of 11.73 m/s were obtained, which proves that the IMQPSO algorithm is feasible to optimize the back-throwing device. After verification, the test results were compared with the theoretical values, which showed that the IMQPSO algorithm was effective, indicating that the improved optimization algorithm was feasible to optimize the back-throwing device. This article provides a solution for the improvement of large agricultural machinery and also applies to other parts of the machine.

Author Contributions: Conceptualization, H.X. and Z.H.; methodology, W.S. and Z.H.; software, C.W.; validation, P.Z., H.X. and F.G.; formal analysis, W.S.; investigation, C.W.; data curation, H.X.; F.W.; writing—original draft preparation, H.X.; writing—review and editing, P.Z.; supervision, P.Z. and F.G. All authors have read and agreed to the published version of the manuscript.

Funding: This research was funded by Jiangsu Agricultural Science and Technology Innovation Fund (Grant No. CX (20)3066) and the Natural Science Foundation of Jiangsu Province (Grant No.BK20190140).

Institutional Review Board Statement: Not applicable.

Informed Consent Statement: Not applicable.

Data Availability Statement: The datasets used and/or analyzed during the current study are available from the corresponding author on reasonable request.

Conflicts of Interest: The authors declare that they have no competing interests.

References

1. Zhang, X.R.; He, J.; Li, H.W.; Li, W.Y.; Li, H. Design and experiment on the driving disc of anti-blocking unit for no-tillage planter. *Trans. CSAM* **2009**, *25*, 117–121.
2. Barr, J.; Desbiolles, J.M.; Fielke, J.; Ucgul, M. Development and field evaluation of a high-speed no–till seeding system. *Soil Tillage Res.* **2019**, *194*, 1–11. [CrossRef]
3. Shi, Y.; Xin, S.; Wang, X.; Hu, Z.; Newman, D.; Ding, W. Numerical simulation and field tests of minimum-tillage planter with straw smashing and strip laying based on EDEM software. *Comput. Electron. Agric.* **2019**, *166*, 1–9. [CrossRef]
4. Gu, F.; Hu, Z.; Chen, Y.; Wu, F. Development and experiment of peanut no-till planter under full wheat straw mulching based on "clean area planting". *Trans. CSAE* **2016**, *32*, 15–23.
5. Gu, F.W.; Gao, X.M.; Wu, F.; Hu, Z.C.; Chen, Y.Q.; Zhang, C. Improvement of uniform scattering device for straw-smashing, back-throwing, no-tillage planter under complete straw mulching condition. *Int. J. Agric. Biol. Eng.* **2018**, *11*, 49–57.
6. Zhao, J.; Wang, X.; Zhuang, J.; Liu, H.; Wang, Y.; Yu, Y. Coupled Bionic Design Based on Primnoa Mouthpart to Improve the Performance of a Straw Returning Machine. *Agriculture* **2021**, *11*, 775. [CrossRef]
7. Li, A.; Fan, X.; Wu, C.; Li, H. Situation and Development Trends of Conservation Tillage in the World. *Trans. CSAM* **2006**, *37*, 177–180.
8. Liu, X.; Wang, J. Brief analysis on the development status and existing problems of no-tillage planter. *Agric. Technol.* **2014**, *6*, 203–204.
9. Shinners, K.J.; Koegel, R.G.; Pritzl, P.J. An upward cutting cut-and-throw forage harvester to reduce machine energy requirements. *Trans. ASAE* **1991**, *34*, 2287–2290. [CrossRef]
10. Shinners, K.J.; Stelzle, M.; Koegel, R.G. Improving the throwing effectiveness of an upward cutting cut-and-throw forage harvester. *Trans. ASAE* **1994**, *37*, 1059–1067. [CrossRef]
11. Chattopadhyay, P.S.; Pandey, K.P. PM—Power and Machinery: Influence of Knife Configuration and tip speed on conveyance in flail forage harvesting. *J. Agric. Eng. Res.* **2001**, *78*, 245–252. [CrossRef]
12. Jia, H.; Ma, C.; Wang, Z.; Chen, Z. Chopping and Throwing Mechanism of Corn Straw. *Trans. CSAM* **2003**, *34*, 96–99.
13. Zhai, Z.; Wang, C. Numerical simulation and optimization for air flow in an impeller blower. *Trans. CSAM* **2008**, *39*, 84–87.
14. Zhai, Z.; Wang, C.; Yang, Z. Simulation of the Movement of Air Flow and Optimization of the Parameters of an Impeller Blower. *Mech. Sci. Technol. Aerosp. Eng.* **2010**, *29*, 1352–1356.
15. Zhai, Z.; Wu, Y.; Wang, C. Dynamic simulation and high-speed camera analysis on materials moving along throwing impellers. *Trans. CSAM* **2012**, *28*, 23–28.
16. Kennedy, J.; Eberhart, R.C. Particle Swarm Optimization. In Proceedings of the ICNN'95—International Conference on Neural Networks, Perth, Australia, 27 November–1 December 1995; pp. 1942–1948.
17. Huang, W.; Xu, J.; Zhu, D.-Y.; Wu, Y.-L.; Lu, J.-W.; Lu, K.-L. Semi-active Vibration Control Using a Magneto Rheological (MR) Damper with Particle Swarm Optimization. *Arab. J. Sci. Eng.* **2015**, *40*, 747–762. [CrossRef]
18. Shi, Z.; Chen, Q. Multi-objective optimization algorithm based on quantum-behaved particle swarm and adaptive grid. *Inf. Control* **2011**, *38*, 945–951.

19. Wei, Y.; Nu, W.; Fengwei, G.; Dezhi, L.; Xinxing, Z. Parameter optimization and experiment for the power consumption of impeller-blower. *J. China Agric. Univ.* **2017**, *22*, 99–106.
20. Omkar, S.N.; Khandelwal, R.; Ananth TV, S.; Naik, G.N.; Gopalakrishnan, S. Quantum behaved particle swarm optimization (QPSO) for multi-objective design optimization of composite structures. *Expert Syst. Appl. Int. J.* **2009**, *36*, 11312–11322. [CrossRef]
21. Shi, Y.; Eberhart, R.C. *Parameter Selection in Particle Swarm Optimization*; Springer: New York, NY, USA, 1998; pp. 591–600.
22. Wenfeng, L.; Wei, H. Optimization of polysaccharides extraction from Flammulina velutipes by box-behnken design and response surface methodology. *J. Nanjing Tech Univ.* **2016**, *38*, 95–100.
23. Montgomery, D.C. *Experimental Design and Analysis*, 3rd ed.; China Statistics Press: Beijing, China, 2009.
24. Cheng, G.R.; Wang, S.J.; Zou, Y.; Zhang, L. Optimization of the dacarbazine liposones preparation with Box-Behnken central composite design. *J. Chang. Univ. Technol. Nat. Sci. Ed.* **2011**, *32*, 348–353.
25. Wu, F.; Xu, H.; Gu, F.; Chen, Y.; Shi, L.; Hu, Z. Improvement of straw transport device for straw-smashing back-throwing type multi-function no-tillage planter. *Trans. CSAE* **2017**, *33*, 18–26.
26. Zhai, Z.; Gao, B.; Yang, Z.; Wu, Y. Power consumption and parameter optimization of stalk impeller blowers. *Trans. CSAE* **2013**, *29*, 26–33.
27. Jiang, Q.; Wang, Y.; Chen, J.; Wang, J.; Wei, Z.; He, Z. Optimizing the working performance of a pollination machine for hybrid rice. *Comput. Electron. Agric.* **2021**, *187*, 1–10. [CrossRef]

agriculture

MDPI

Article

An Octopus-Inspired Bionic Flexible Gripper for Apple Grasping

Jie Pi [1], Jun Liu [1,*], Kehong Zhou [2] and Mingyan Qian [1]

[1] Institute of Agricultural Facilities and Equipment, Jiangsu Academy of Agricultural Sciences, Key Laboratory of Protected Agriculture Engineering in the Middle and Lower Reaches of Yangtze River Ministry of Agriculture and Rural Affairs, Nanjing 210014, China; pijie@jaas.ac.cn (J.P.); 20190977@jaas.ac.cn (M.Q.)
[2] Institute of Agricultural Engineering, Jiangsu University, Zhenjiang 212001, China; 2221916050@stmail.ujs.edu.cn
* Correspondence: liujun@jaas.ac.cn; Tel.: +025-8439-9439

Abstract: When an octopus grasps something, the rigidity of its tentacle can change greatly, which allowing for unlimited freedom, agility, and precision. Inspired by this, a three-finger flexible bionic robot gripper was designed for apple picking. First, a flexible chamber finger was designed to drive the gripper finger to elongate, shorten, and bend, which works through a process of inflating and deflating. Further, we proposed a three-finger mode to achieve two kinds of motion states: grasping and relaxing, by simulating the movement of an octopus grasping at something. In this paper, we evaluated the bending property of the designed flexible bionic gripper through an apple grasping experiment. The experimental results show that the 100.0 g bionic gripper can load an apple with a weight of 246.5~350.0 g and a diameter of 69.0~99.0 mm, and the grasping success rate is 100%. It has a good grasping performance. Compared to other soft grippers, the proposed bionic flexible gripper has the advantages of being lightweight, and having good cushioning, low driving air pressure, and a strong grasping force.

Keywords: octopus tentacles; bionic flexible gripper; pneumatic drive; agriculture

Citation: Pi, J.; Liu, J.; Zhou, K.; Qian, M. An Octopus-Inspired Bionic Flexible Gripper for Apple Grasping. *Agriculture* **2021**, *11*, 1014. https://doi.org/10.3390/agriculture11101014

Academic Editor: José Pérez-Alonso

Received: 24 August 2021
Accepted: 14 October 2021
Published: 17 October 2021

1. Introduction

The apple is one of the most common fruits, which possesses huge consumer market, and is available in various kinds, and different weights and sizes. The traditional artificial approaches for picking and sorting imply a shape evaluation process, and the fruit pickers simply sort fruits by judging their shape, weight, and rough skin conditions, thus causing a huge workload in this process. In addition, the cost of apple picking reaches 50% to 70% of the total cost [1]. With the development of mechanization and automation technology, apple picking and sorting will gradually rely on mechanical operation, especially in the case of large-scale planting; since apple production is huge, the efficiency and scale of manual picking and sorting simply cannot satisfy the production needs [1]. In order to solve the objective problems, including huge workload, low efficiency, and high cost involved in manual picking and sorting operations, agricultural picking and sorting robots have emerged, becoming one of the research hotspots in the field of artificial intelligence agriculture.

The robot gripper structure is the last part which directly touches the apple, and its excellent performance effectively improves the working efficiency. Since the growing environment of apples is complex, their skin is fragile, and their shape is complex and changeable, it is easy to cause damage to the fruit in the process of picking and sorting, thus affecting the quality, storage, and transportation of apples, and ultimately harming the market price and economic benefits. Traditional industrial robots mainly have a rigid structure, with the characteristics of the "three highs": high stiffness, high strength, and high speed [2]. Thus, it is easy to damage the surface of apples during operation of these robots, which requires high precision but results in poor universality. Ma Lvzhong et al. [3] (2009) and Ji Changying et al. [4] (2011) have designed their own apple picking devices, but the structure is complex, which can cause varying degrees of damage to the apple epidermis. Therefore, owing to the particularity and delicacy of apples, soft robots with flexible grippers are more appropriate for apple sorting.

Currently, many scholars and engineers both at home and abroad have started to study the structure of the soft robot. Based on the bionics concept with intelligent materials such as silicone rubber, shape-memory alloys (SMAs) [5], electroactive polymers (EAPs) [6], etc. and new driving technologies such as SMA drivers [7], pneumatic drivers [8,9], magnetorheological drivers [10], EAP drivers, etc., they developed a series of novel robots without adopting the traditional rigid structure, by imitating the structure and behavior of invertebrates in nature, such as earthworms, starfish, octopuses, etc.. This kind of soft robot possesses a strong adaptive ability, and theoretically has infinite freedom, which can be applied to unstructured working environments. In terms of flexible robot manufacturing, many new processing technologies have been introduced, such as shape deposition manufacturing [11,12], 3D printing [13], flexible photolithography technology [14], and composite materials integrated through a variety of manufacturing methods [15]. Currently, led by Harvard University in the United States [13], the research on super-elastic silica gel material as the ontology material, combined with 3D printing technology, is the most popular branch of soft robot research. The GEORGE [16] team at Harvard University (2011) developed a multi-cavity pneumatic software gripper, which has a strong adaptive ability, but it is not good at end bending, thus it is difficult for the gripper to grasp objects. Further, due to the small end contact, instability appears during the working process. The PneuNet (pneumatic mesh) structure proposed by Shepherd et al. [17] from Harvard University is a flexible, multi-cavity pneumatic robot, which is capable of bending. Belgian company Octinion [18] developed a small strawberry picking robot, which is composed of two soft fingers, produced through 3D printing. After grasping the object, the grip will passively deform. Compared with human fingers, it has a larger contact area, and better pressure distribution. Hemming's team [19] designed a sweet pepper-picking claw with a fin-bar structure, which adopts fin-bar fingers and has a fin-bar effect, able bend according to the shape of the object's surface. Dimeas [20] (2015) et al. proposed a fuzzy controller-based hierarchical control gripper, which is combined with a pressure profile sensor, and can adjust the grasping force, and detect the uneven distribution of dislocation or force on the gripper. Yaguchi et al. [21] (2016) designed a tomato-picking robot with a rotating gripper, which can pick tomatoes within 23.0 s, with a success rate of approximately 62.6%. Paul Glick et al. [22] invented a flexible robot gripper with a gecko-like adhesive, which adopts Van der Waals forces to achieve adhesion, and can grasp large curvature objects with high grasping strength and fast execution speed, but this gripper is not suitable for grasping fragile objects. Tomokazu Takahashi et al. [23,24] designed an octopus-inspired vacuum clamper with miniature raised sections, which can grasp rigid objects with flat surfaces, curved surfaces, uneven surfaces, or grooves. Brown et al. [25,26] developed a bag-like gripper containing ground coffee, which is highly compliant and can conform to the shape of an object, thus wrapping the object safely. It is suitable for grasping all kinds of hard, small objects. The "vacuum cleaner" picking machine developed by Abundant Robotics Company [27] (2017) sucks mature apples from fruit trees through a large suction nozzle, but the machine can easily suck excess branches and leaves into the trachea, causing obstruction. Ma Shaochun et al. [28] (2018) developed a device for apple picking based on negative pressure suction, but the proportion of apples without is only 85%. In addition, the relevant apple-picking devices at home and abroad generally have a large body and a complex structure, so it is difficult to realize the non-destructive harvesting of apples. In addition, the cost is high, which is difficult for Chinese fruit farmers to bear.

In conclusion, despite the soft robot becoming a popular research topic in the field of robotics at home and abroad, it is still in the basic research stage in the field of agriculture, especially concerning the picking and sorting of fruits and vegetables, and is not widely used. Due to the good qualities of soft robots, developing a flexible gripper with adaptive abilities is the inevitable trend of agricultural intelligent robot development. In China, a bionic soft robots laboratory has been established in the Beijing University of Aeronautics and Astronautics, which has produced several kinds of soft robot for the purpose of grasping [29,30].

The flexible clamping device for apple gathering presented in this paper has the advantages of having a simple structure, convenient fabrication, and low cost. It is verified through experiments that the apple surface is not damaged. Based on an analysis of the muscles of octopus tentacles, we propose a novel bionic flexible agriculture grip structure for apple grasping, according to its structure and motion characteristics. The proposed flexible grip consists of three flexible fingers with the same structures, which are pneumatically controlled. Under certain pressures, the grip can practice two kinds of motion: grasping, and relaxing. In this paper, some assumptions are made on this flexible gripper. Through finite element analysis and experimental analysis, the experimental study of the proposed flexible gripper is conducted, and the conclusion is drawn. It is expected to provide reference for the development of non-destructive apple picking devices, imitating human hands.

2. Materials and Methods

Bionics' aim is to imitate the function and behavior of biological systems, that is, to improve technology based on inspiration from nature, and achieve effective applications in engineering. In this paper, a novel bionic flexible gripper was designed based on the principle of bionics, inspired by the structure and movement characteristics of octopus tentacles. The proposed gripper is mainly used for picking and sorting apples.

2.1. Analysis of the Tentacle Structure of Octopus

The octopus is an invertebrate marine animal with high agility and intelligence [31], as well as strong ability to learn. Most of the neurons in an octopus (about 60%) are distributed throughout its eight tentacles, forming an independent nerve center with an infinite degree of freedom, which can provide flexible changes of direction [32], so as to achieve free bending, stretching, winding, and other complex actions.

The movement pattern of octopus tentacles is from the roots to the ends, which is a kind of mobile bending. In any position of the tentacles, a combination of four basic movements may occur: elongation, shortening, twisting, and bending [33]. An octopus catches its prey using the method of enveloping grasp, that is, the tentacles stretch from its roots to contact the object, further cover the object until the tips of the tentacles touch the object, and completely wrap it up. In this way, an octopus can grasp an object without knowing its shape. This complex movement pattern of octopus tentacles is enabled by their unique muscle tissue structure. Its movement pattern and muscle tissue structure support are both based on the muscular hydrostatic skeleton, the interior of which is completely composed of arrays of muscle fibers and connective tissues of collagen fibers arranged in different directions [34]. This dense array of muscle fibers cannot be compressed, and it is difficult to compress its volume with external pressure, i.e., it has volume invariance. Volume invariance is one of the key biological properties of octopus tentacles, enabling them to achieve complex movements. The muscle tissues of octopus tentacles can be divided into radial, longitudinal, and oblique muscles, as shown in Figure 1. When one side of the longitudinal muscle is shortened, the diameter of the tentacle will increase, while the other muscle groups will work together to maintain the diameter of the tentacle. When one side of the longitudinal muscle is shortened by resistance, it causes the other side of the longitudinal muscle to stretch passively, which results in the bending motion along the direction of contraction force, and making the tentacle bend flexibly in the space form. With the cooperation of various muscle tissues, the diameter of octopus tentacles remains basically unchanged, and the contraction of the longitudinal muscle on one side can only cause slight irregular bending [35].

The muscular system of an octopus can act as a dynamic skeleton, which allows smooth changes in shape, thus resulting in potentially large movements of the tentacles [36].

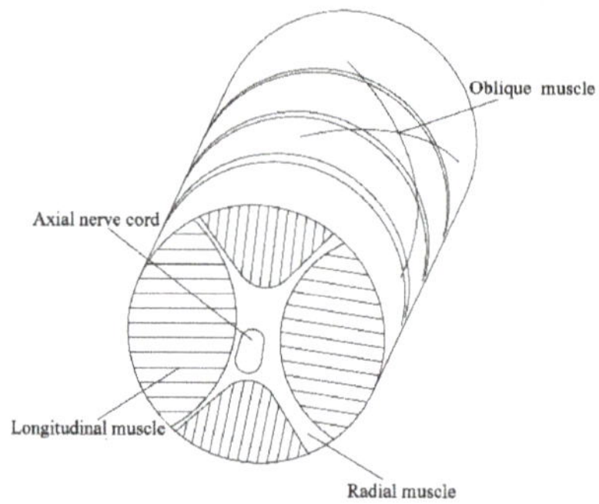

Figure 1. Muscle structure of octopus tentacles.

In order to simulate the muscle characteristics and movement of octopus tentacles, we designed a flexible fluid actuator, which can achieve the two motions of grasping and relaxing via gas driving of the extension, shortening, and bending actions of the grippers.

2.2. Analysis of Apple Morphology

The shape of the flexible gripper is designed according to the shape characteristics of apples, based on the principle to minimize the size of gripper. In the process of grasping the apple, the flexible gripper implements wrapped grasping by deformation of the flexible fingers through bending, and the gripper requires a certain deformation space. In order to better fit the shape of grasped target, it is necessary to pre-analyze the shape characteristics of the target, pre-determine the size of the required space, and estimate the deformation volume of the gripper, so as to set the range of clamping deformation and the angle of gripper.

Since the size and shape of apples vary in different stages of growth, the characteristic parameters considered in the design of the clamping claws are of apples in the ripening stage, including geometric parameters, such as mass, transverse diameter, longitudinal diameter, fruit shape index, and minimum enclosing rectangle, as shown in Figure 2.

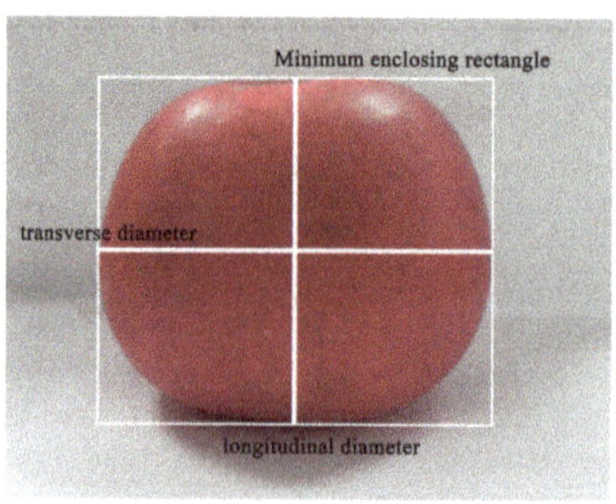

Figure 2. Apple shape measurement.

With the Fuji apple (producing area: Yantai, Shandong Province, China) as a representative sample, according to the shape selection rules, 100 apples with no mechanical damage, plant diseases, or insect pests were selected as the test samples, and were then numbered from 1 to 100. The disk scales (range of 500 g and precision of 0.1 g) and digital caliper (range of 150 mm and precision of 0.02 mm) were adopted for precision measurement and calculation of the sample quality and the shape parameters. Table 1 shows the related characteristic parameters of apple.

According to Table 1, the variation coefficients of apple mass, vertical axis diameter, horizontal axis diameter, and fruit shape indexes are 0.09, 0.05, 0.04, and 0.05, respectively, all of which are small, indicating that the samples have high quality on average, and the dispersion degree is minor.

Table 1. Characteristic parameters of sample apples.

Indicators	Quality m/g	Vertical Axis Diameter D_{va}/mm	Horizontal Axis Diameter D_{ha}/mm	Fruit Shape Index D_{va}/D_{ha}	Minimum Enclosing Rectangle A * b (mm)
Maximum	350.0	75.0	99.0	0.76	75 * 99
Minimum	246.5	69.0	82.0	0.94	69 * 82
Average	288.2	74.5	86.9	0.86	—
Standard deviation	24.8	3.7	3.2	0.05	—
Coefficient of variation	0.09	0.05	0.04	0.05	—

According to the above analysis, the basic indexes of the flexible clamping claw are put forward: the flexible clamping claw adopts the form of three fingers; the initial opening diameter of the clamping claw is 100 mm; the effective deformation length of the finger is more than 90 mm; the load capacity is higher than 0.5 kg; and, within the range of operating temperatures (−25~60 °C), it presents good adaptive deformation and flexible bending deformation capacity.

2.3. Design and Manufacture of a Single Flexible Finger

In order to simulate the particular texture and structure of octopus tentacles, this paper adopts the silicone rubber Dragonskin 30—a viscoelastic material—to create the flexible fingers, and a pneumatic driver is employed in the simulation.

A single flexible finger is composed of a top layer with a balloon structure, and a bottom layer with poor deformation capacity, as shown in Figure 3a. The top layer has the structure of a hollow elastic chamber, connected to the vent hole at the tip of finger, which is then connected to the external air source through an air pipe. A single flexible finger has 7 chambers in total, each of which is 7 mm wide and 13 mm high, with a wall thickness of 1.5 mm. Without deformation, the finger is 104 mm long, and the underside of the finger is 4 mm thick. A fixed structure is situated at the root of the finger, as shown in Figure 3b. In the initial state of the finger, the chamber pressure is consistent with the external atmospheric pressure, and the finger remains upright. As the finger is driven pneumatically, it is filled with gas, the pressure in the chamber becomes greater than the external pressure, and the chamber on the top layer of the finger expands, thus elongating the top part of the finger. However, due to the thicker bottom layer, the deformation capacity of the finger is limited. After removal of the gas, the pressures in and out of the chamber are once again equal, and the finger is restored to its original shape. Similarly, if the finger is pumped and decompressed, the chamber contracts, both layers of the finger are shortened, forcing the finger to bend outward. The deformation process of the chamber is shown in Figure 4.

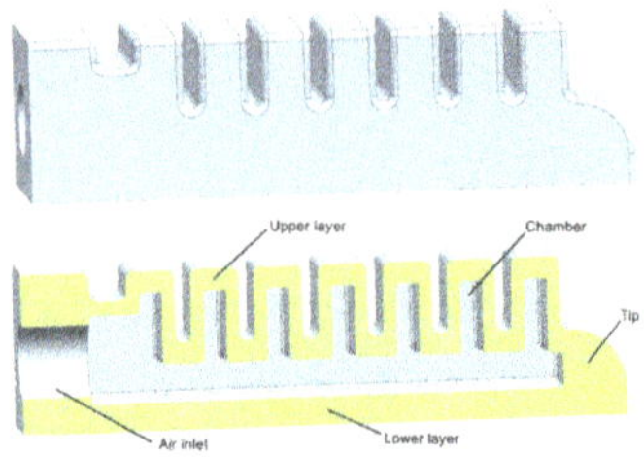

(**a**)

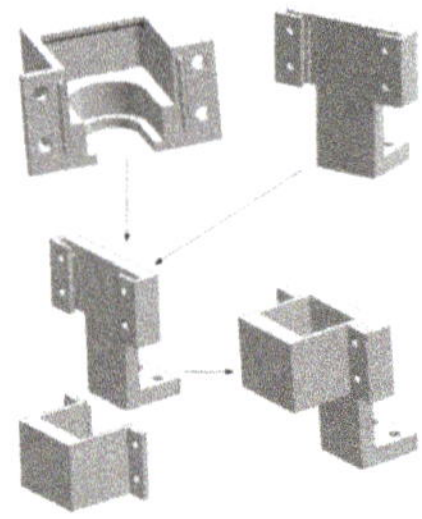

(**b**)

Figure 3. Schematic diagram of flexible finger: (**a**) schematic diagram of finger; (**b**) schematic diagram of fixed structure.

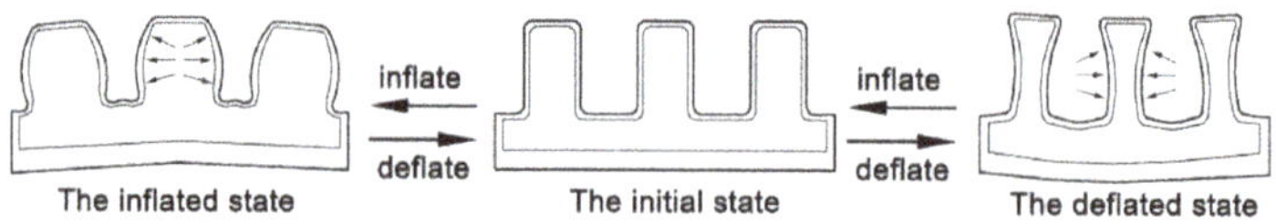

Figure 4. Schematic diagram of chamber deformation process.

The flexible pneumatic finger can be produced through 3D printing, which is more efficient and faster than other production methods. It has good integrity, and includes features of high strength and high tolerance to expansion pressure, meaning that the gripper can obtain greater grasping force. Therefore, the flexible gripper designed in this paper was also created through 3D printing technology.

2.4. Finite-Element Analysis and Experimental Analysis of Flexible Finger

The finite-element analysis software ANSYS (ANSYSV17.0) was used for the finite-element simulation and analysis. The material parameters for each part of the flexible finger are shown in Table 2. Firstly, the model of a single flexible finger was established by using the UG (Unigraphics NX) software, as shown in Figure 5a, and then the model was imported into ANSYS for simulation. The static structural module was created in ANSYS Workbench; subsequently, the material properties were integrated to obtain the ANSYS model, as shown in Figure 5b. By refining the grid and simplifying the model, the precision of the grid was determined, and 160,935 nodes and 62,146 elements were divided, as shown in Figure 5c. The load of 65 kPa and −65 kPa were added to each surface of the flexible finger, the fixed structure was set as rigid material, and the deformation was neglected. After fixing the end of the finger, the deformation data was obtained, based on the displacement of flexible finger. The simulation results are shown in Figure 6. The simulation results show that: under the pressure of 65 kpa, the bending angle of the flexible finger was 27.19° (α) and the displacement of the fingertip was 31.871 mm (Δs1). Under the pressure of −65 kpa, the bending angle of the flexible finger was 16.82° (β), and the displacement of the fingertip was 21.679 mm (Δs2).

Table 2. Material parameters.

Name	Material Type	Density/(kg·m^{-3})	Other Parameters		
Flexible fingers	Flexible	1.08×10^9	μ_1 = 0.264 MPa	α_1 = 3.0158	D_1 = 0
Fixed structure	Fixed	7850	Modulus of elasticity: 2×10^5 MPa		Poisson's ratio: 0.3

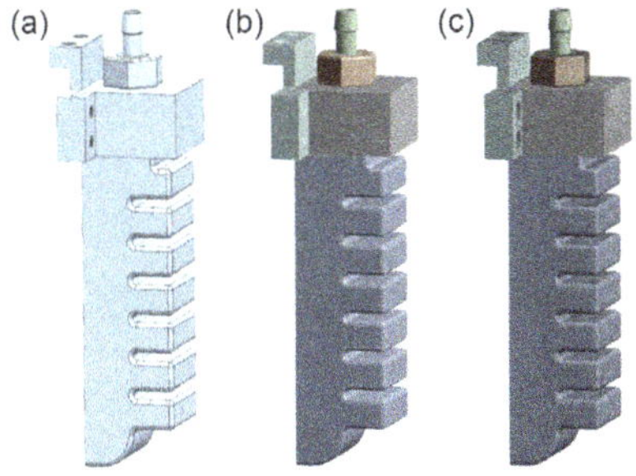

Figure 5. Model flexible finger: (**a**) UG model; (**b**) ANSYS model; (**c**) grid chart.

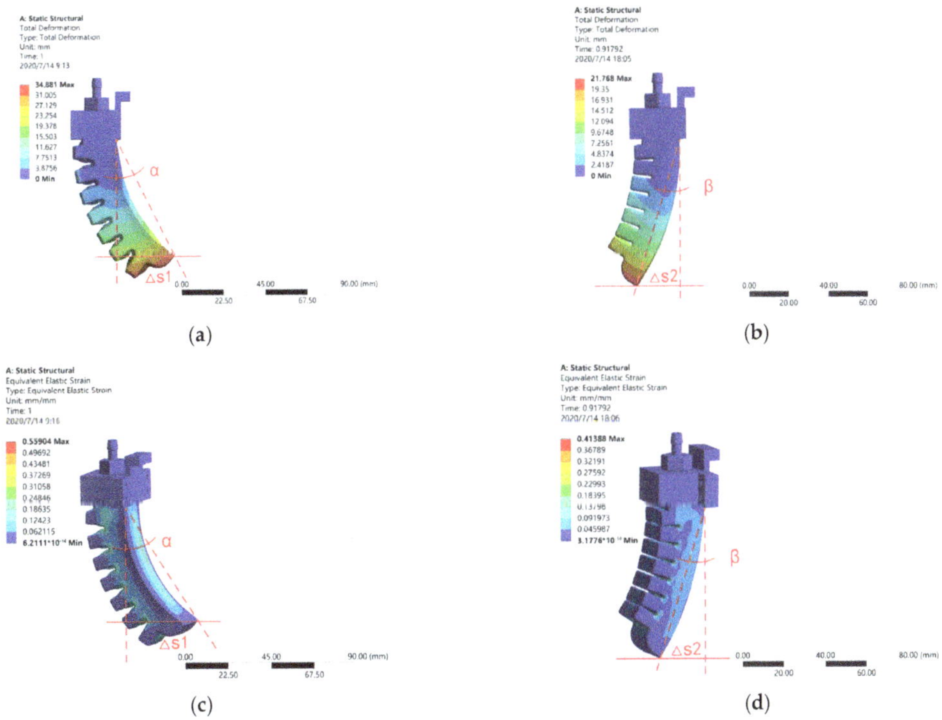

Figure 6. Simulation results. (**a**,**c**) show the deformation diagram and the stress diagram after inflation; (**b**,**d**) show the deformation diagram and the stress diagram after deflation.

The air path of the flexible finger was connected to the circuit to conduct the inflation and deflation experiments. The results are shown in Figure 7. The results show that: under the pressure of 65 kpa, the bending angle of the flexible finger was 28.05° (α), and the displacement of the fingertip was 32.948 mm (Δs1). Under the pressure of −65 kpa, the bending angle of the flexible finger was 17.35° (β), and the displacement of the fingertip was 23.029 mm (Δs2).

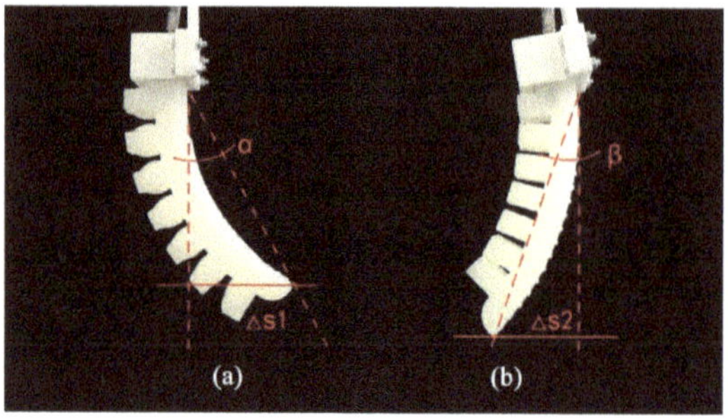

Figure 7. Experimental results: (**a**) inflation status; (**b**) deflation state.

It can be seen from Figures 6 and 7 that the simulation results are highly consistent with the experimental results, which proves the feasibility of the flexible finger, and the accuracy of the simulation model.

Since the bending capacity of flexible fingers has a greater influence on the structural design, in order to give the flexible gripper better bending deformation and fetching performance, the optimization structural design for flexible fingers is conducted based on the simulation model in Section 2.3. The main parameters of the original finger model are as follows: chamber spacing 2 mm; 7 chambers; chamber thickness 7 mm; chamber wall thickness 1.5 mm; finger side thickness 4 mm; finger underside thickness 4 mm. The control variable method was employed to change the factors which impact the length of flexible finger (spacing of chambers, number of chambers, thickness of chambers), the wall thickness of chambers, the thickness of lower layer, and other factors, and to compare the horizontal displacement at the end of the fingers. The larger the horizontal displacement, the greater the bending degree of the finger. The finger variable factors are shown in Figure 8.

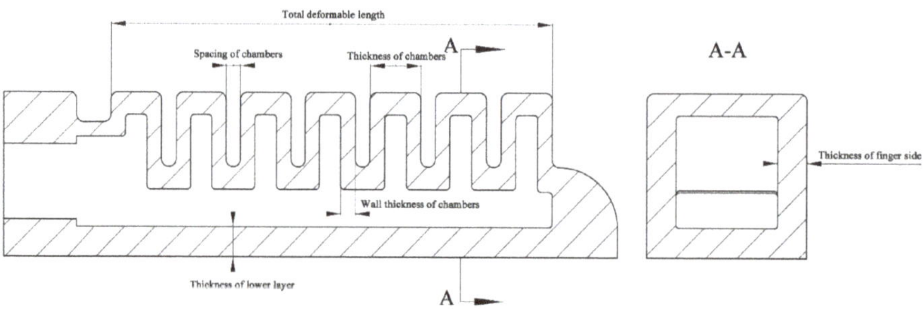

Figure 8. Finger variable factors.

2.5. Experiment of Flexible Gripper

The bionic flexible gripper designed in this work is a three-finger gripper, in order to present contact forces evenly. A flange structure is incorporated to fix the three flexible fingers. The quality of the flexible gripper is 100 g, excluding the flange structure. The structure diagram of the flexible gripper is shown in Figure 9. Both the flange structure and the flexible fingers are produced through 3D printing.

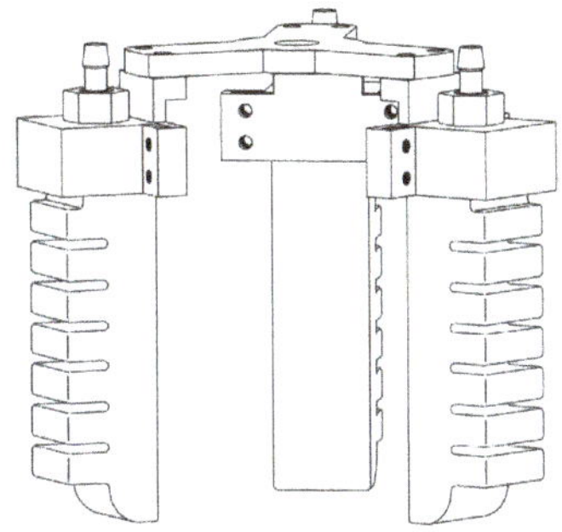

Figure 9. Structure diagram of flexible gripper.

As shown in Figure 10, the upper limit of the object mass that can be grasped by the gripper in different air pressure ranges is tested by adding a weight in the circular sphere.

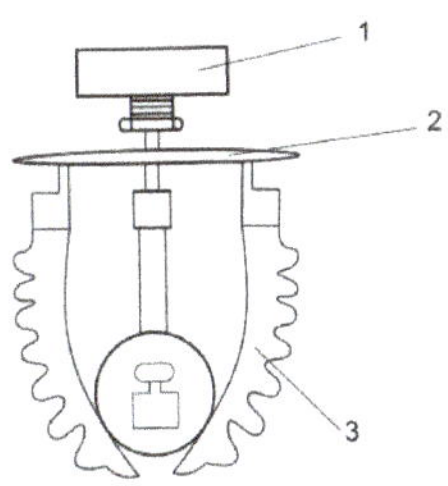

Figure 10. Maximum gripping quality test device of flexible gripper. 1: manipulator; 2: flange plate; 3: flexible finger.

2.5.1. Experiment of Apple Grabbing

In order to test the grasping ability of the gripper, an apple grasping experiment was carried out in a laboratory environment to evaluate it. The flexible gripper was fixed on the six-axis manipulator. The finger movement was accomplished with a pneumatic actuator. We connected the gas circuit and electric circuit to carry out the grasping experiments; the grasped object is the apple mentioned above.

We determined the initial position point A and the terminal position point B on the experimental platform, as shown in Figure 11a. The flexible gripper opened the finger to grab the apple from point A, and then the manipulator moved up 100 mm. The manipulator rotated horizontally above point B, and the manipulator moved down to point B to relax the flexible gripper, successfully completing the experiment. The motion path of the manipulator is shown in Figure 11b.

Apples numbered from 1 to 90 were selected for grab test, and apples numbered from 91 to 100 were taken as untreated comparison items, and kept in their original state. The working pressure was set to 50 kpa, and the working speed of the manipulator was set to 480 mm/s. The continuous loading time from the touch of the finger with the apple to the stable grasping of the apple was 3 s. Upon completion of each grab, the contact area was marked on the apple. The number of successful experiments and the number of failures were recorded, and the grasping success rate was calculated.

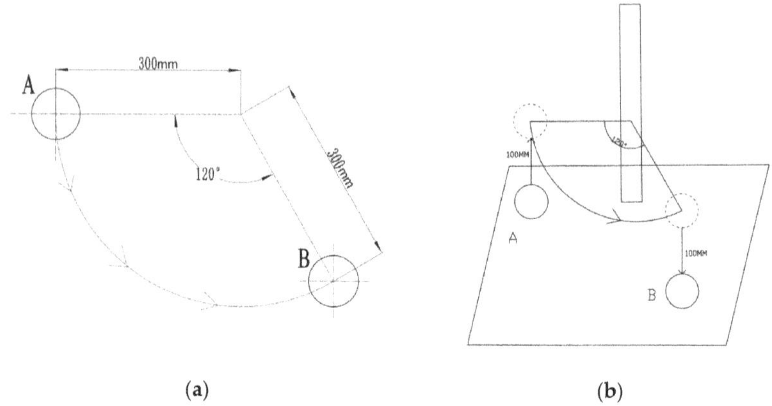

Figure 11. Motion path of the manipulator. (**a**) schematic diagram of rotation angle of manipulator; (**b**) schematic diagram of manipulator motion path.

2.5.2. Experiment of Apple Damage

In order to observe the damage induced on apples grabbed by the flexible gripper, we carried out an apple damage analysis.

(1) After completing the apple grab experiment, a 5× magnifying glass was used to observe whether there was any skin damage or juice overflow in the contact area between the apple and finger;
(2) The apple numbered 91 was selected, and grasped with a rigid claw. Subsequently, 10 apples were randomly selected from the successful grasping experiment. The surfaces which were touched by the finger were sliced respectively. The thickness was 2 mm, and placed under 5× microscope.
(3) 20 apples were randomly selected from the successful grasping experiment, and 9 untested apples numbered 92–100 were stored in the same environment. The temperature was maintained at 25 °C, and the humidity at 80%. After seven days, the surface condition of the apples was observed. The contact area between the apple and finger was peeled to observe the internal pulp.

3. Results and Discussion

3.1. Simulation Results

As shown in Table 3 and Figure 12a, three factors (spacing of chambers, number of chambers, and thickness of chambers) that affect the length of the finger were compared, and the main parameters of the original finger model were kept as follows: chamber wall thickness 1.5 mm; finger side thickness 4 mm; and finger lower thickness 4 mm. With the same amount of increase in the length of the finger, the horizontal displacement of the finger end presented a similar change, and the finger bending also changed with the same degree. Therefore, it is inferred that, when the wall thickness of finger is constant, the increase in finger length will lead to the increase in the degree of finger bending, and the degree of change is not significantly affected by the change of finger length.

We changed the thickness of the lower layer, and kept the main parameters of the original finger model as follows: chamber spacing 2 mm; 7 chambers; chamber thickness 7 mm; chamber wall thickness 1.5 mm; and finger lower thickness 4 mm. As shown in Table 4 and Figure 12b, the horizontal displacement of the finger end and the thickness of the lower layer present approximately linear change. The horizontal displacement decreases by approximately 2 mm for every 1 mm of increase in the thickness of lower layer. As the thickness of the lower layer increases, the horizontal displacement of the finger end also declines, and such decrease gradually becomes gentler. When the lower layer is 2 mm thick, the bending deformation of the finger is significant, and the underside of the finger shows transverse deformation, protruding outwards.

Table 3. Analysis of simulation results (spacing of chambers, number of chambers, thickness of chambers). The bold part indicates the variable factors in the experiment.

Number of Chambers/Units	Spacing of Chambers/mm	Thickness of Chambers/mm	Total Length/mm	Horizontal Displacement of Finger End/mm
7	2	7	76	15.791
8	2	7	85	20.434
9	2	7	94	25.719
10	2	7	103	31.468
11	2	7	112	37.841
7	**1**	7	86	20.824
7	**2**	7	94	25.719
7	**3**	7	102	30.286
7	**4**	7	110	35.053
7	**5**	7	118	40.131
7	2	**6**	85	20.824
7	2	**7**	94	25.719
7	2	**8**	103	30.286
7	2	**9**	112	35.053
7	2	**10**	121	40.131

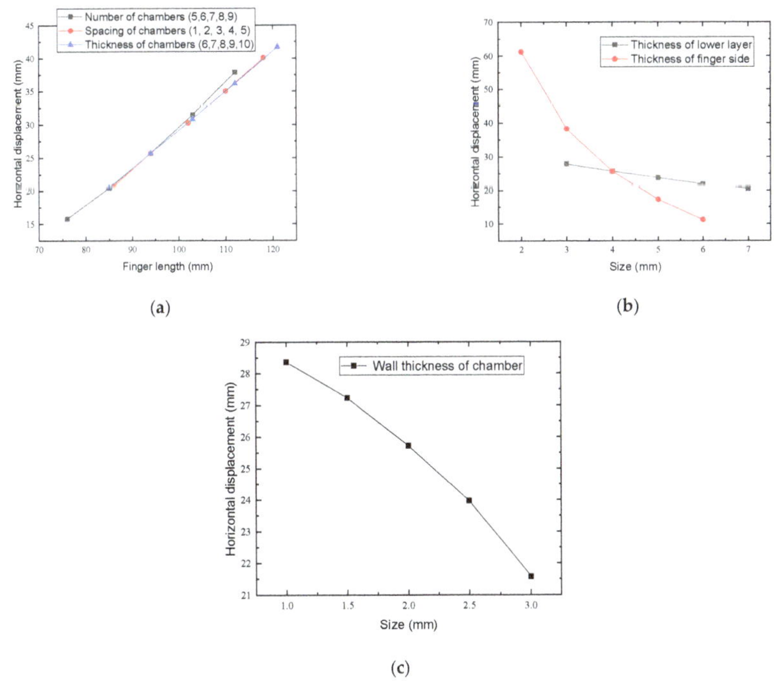

Figure 12. Simulation results: (**a**) shows the relationship between the spacing of chambers, the number of chambers, the thickness of chambers, and the horizontal displacement of finger ends; (**b**) shows the relationship between the thickness of finger bottom and the horizontal displacement of finger ends; (**c**) shows the relationship between the wall thickness of chambers and the horizontal displacement of finger ends.

Table 4. Analysis of simulation results (thickness of lower layer, the thickness of finger side, wall thickness of chambers). The bold part indicates the variable factors in the experiment.

Thickness of Lower Layer/ mm	Thickness of Finger Side/ mm	Wall Thickness of Chambers/ mm	Horizontal Displacement of Finger End/ mm
3	4	1.5	27.882
4	4	1.5	25.719
5	4	1.5	23.814
6	4	1.5	21.977
7	4	1.5	20.467
4	**2**	1.5	20.824
4	**3**	1.5	25.719
4	**4**	1.5	30.286
4	**5**	1.5	35.053
4	**6**	1.5	40.131
4	4	**1.0**	20.824
4	4	**1.5**	25.719
4	4	**2.0**	30.286
4	4	**2.5**	35.053
4	4	**3.0**	40.131

We changed the thickness of finger side, and kept the main parameters of the original finger model as follows: chamber spacing 2 mm; 7 chambers; chamber thickness 7 mm; chamber wall thickness 1.5 mm; and finger side thickness 4 mm. As shown in Table 4 and Figure 12b, the horizontal displacement of the finger end and the thickness of the finger underside present approximately linear change. The horizontal displacement decreases by approximately 2 mm for every 1 mm of increase in the thickness of finger underside. As the thickness of the finger underside decreases, the horizontal displacement of the finger end also declines, and such decrease gradually becomes gentler. When the finger underside is 2 mm thick, the bending deformation of the finger is significant, and the underside of the finger shows transverse deformation, protruding outwards.

We changed the wall thickness of the chambers and the main parameters of the original finger model as follows: chamber spacing 2 mm; 7 chambers; chamber thickness 7 mm; finger side thickness 4 mm; and finger lower thickness 4 mm. As shown Table 4 and in Figure 12c, with the increase of the wall thickness of the flexible finger chambers, the horizontal displacement gradually decreases at increasingly higher rate. This is because when the balloon wall thickness increases, the thin wall can bear the same pressure, and the volume of deformation will decrease. The deformation of this kind of pneumatic chamber finger is mainly caused by the deformation of the thin wall of the balloon. Therefore, the deformation degree of the flexible finger will also decrease.

This section may be divided by subheadings. It should provide a concise and precise description of the experimental results and their interpretation, as well as the experimental conclusions that can be drawn.

According to the simulation results above, by comparing the influencing factors on the bending degree of flexible fingers, and based on the minimum size principle for the clip parts of the flexible clamp claw, the structure of the flexible fingers is finally determined. In other words, the structure is a single flexible finger with 7 chambers in total, each measuring 10 mm wide and 15 mm high with a wall thickness of 2 mm, with chamber spacing of 4 mm; the thickness of lower layer is 4 mm, the total length of the finger is 116 mm with no deformation, and the thickness of finger bottom is 4 mm. In order to increase friction, the mesh stripe structure of finger was designed.

3.2. Experiment Results and Analysis of Flexible Gripper

The experimental results are shown in Figure 13. The pressure in the finger cavity of the flexible gripper is directly proportional to the mass of the grasping target. The maximum mass of the apple used in this paper is 350.0 g, thus the driving air pressure is 50.0 kPa, which can meet the requirements.

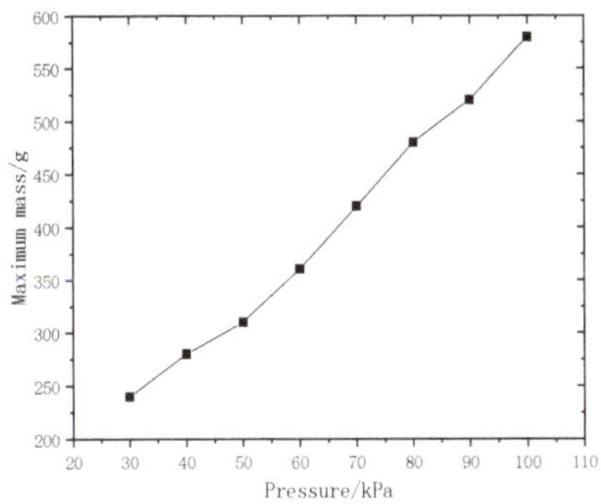

Figure 13. Maximum mass of the flexible gripper can grasp under different air pressures.

The gripper successfully completed the grasping experiment 90 times, failed 0 times, thus the success rate was 100%. It has been proven that the gripper can meet the grasping requirements of conventional apples, and that the grasping effect is good.

3.3. Experimental Results and Analysis for Apple Damage

Using a 5× magnifying glass to observe the apple grabbed by the flexible clamping claw, it can be found that there is no damage to the epidermis at the contact point between the finger and the apple, thus the damage rate is 0%.

The microscopic observation experiment shows that the apple surface cells grasped by the flexible gripper are round and full, and the microstructure is not damaged, as shown in Figure 14a. The apple surface grabbed by the rigid clamping claw produces dark marks, cells are damaged, and the microstructure is damaged, as shown in Figure 14b.

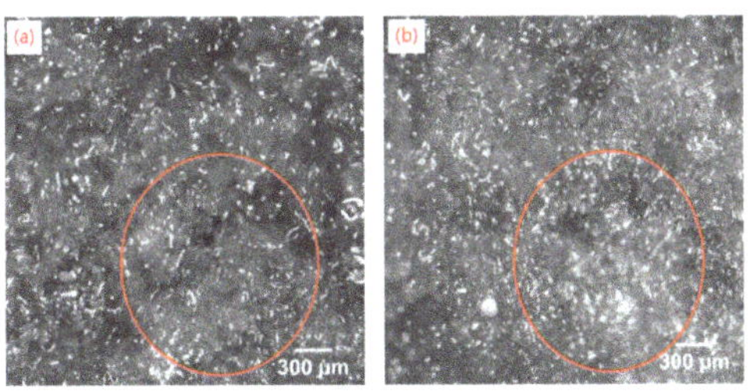

Figure 14. Apple surface under 5 times microscope. (**a**) apple held by flexible gripper. (**b**) apple held by rigid gripper.

The apples which had been kept under constant temperature and humidity conditions were removed after seven days. Upon observation of the surface of apple, there was no visible damage on the skin on the contact point between finger and apple, as shown in Figure 15a. After peeling, the internal pulp was observed, and there was no visible, as shown in Figure 15b. It is thus proven that the flexible picking claw does not mechanically damage the apple.

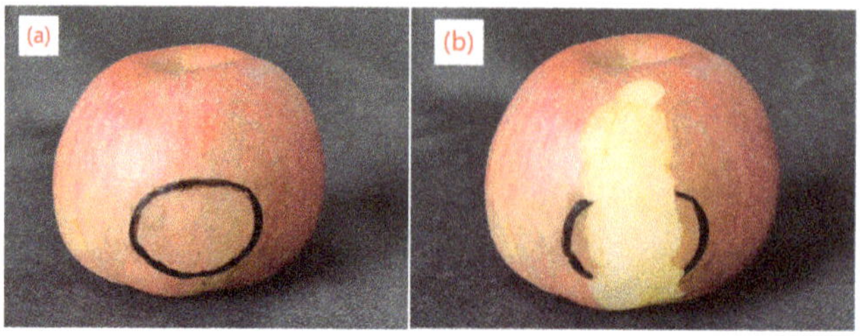

Figure 15. Apples stored for seven days. (**a**) apple skin condition; (**b**) apple pulp condition.

At present, there are many soft grippers that can be used for robot harvesting. For different capture targets, each study puts forward different strategies [37]. In many cases, it is difficult to make a comparative evaluation. The current harvesting robots in China, such as the apple-picking robot of Jiangsu University, the elevated strawberry-picking robot of China Agricultural University and the cucumber-picking robot in greenhouses [38], are still in their infancy. At present, SDM, software lithography, and wax loss manufacturing are the most commonly used processing technologies for software robots [11,39]. However, there are many processing procedures, such as long cycle and single material, so it is difficult to process a more complex three-dimensional cavity soft robot. Compared with their manufacturing process, we successfully used silica gel material and the FDM-based 3D printing technique to fabricate directly bionic flexible grippers with complex inner geometry. In addition, this effectively shortens the production cycle, and reduces cost. Moreover, we have proven through experiments that the proposed bionic flexible gripper has the advantages of being lightweight, and having good cushioning, low driving air pressure, and a strong grasping force.

4. Conclusions

This paper presents a bionic flexible gripper inspired by octopus tentacles, used to grasp apples, which is driven pneumatically. Firstly, the muscle structure and movement characteristics of octopus tentacles and the apple shape were analyzed theoretically, and the basic design indexes of flexible finger were put forward. Subsequently, by using the ANSYS finite-element analysis software, a single flexible finger model was established for simulation and optimization of structural design, and the structure of a single flexible finger was determined. Finally, based on the structure of a single flexible finger, the structure of a three-finger flexible gripper was proposed. The gripper is driven by pneumatic force, and the grasping and relaxing motions are realized based on the process of inflation and deflation, so as to complete the grasping and relaxing of the apple.

In order to test the performance of flexible gripper, we conducted a test using the apple gripper. We set the moving path of the flexible gripper after grasping the apple by using the six-axis manipulator, and completed 90 grasping experiments. The apple did not fall during the grasping process, and the success rate was 100%. We conclude that the flexible gripper can successfully grasp apples with a diameter of 69.0~99.0 mm, and a mass of 246.5~350.0 g when the driving air pressure is 50.0 kPa. The experiment was completed with great performance.

In order to observe the damage caused to apples grabbed by the flexible gripper, we carried out an apple damage analysis. The experimental results show that the bionic flexible gripper can not only effectively grasp apples, but also cause no damage the apple skin and inner pulp tissues.

The flexible gripper is driven pneumatically, and both the flexible fingers and fixed structure were all produced with 3D printing technology, which has the advantages of strong deformation capacity, durability, convenient manufacturing, and low cost. It is not only suitable for grasping apples, but also for grasping other fruits and vegetables.

However, our current research work has some limitations. On the one hand, the structural design of the flexible gripper needs further improvement. Since the material, size, and shape of the flexible gripper are important factors that need to be considered during the design of gripper, we will try to make it more universal in our future work, and achieve faster and more stable working conditions. On the other hand, this study only carries out the structural design and experimental verification of the flexible gripper for these two operations, and proves that the bionic gripper we designed can successfully complete the two operations of grasping and loosening an apple without damaging the skin of apple. This is only the first step of our research work. As a future work, we will study the whole flexible picking and sorting robot, including the vision system, positioning system, obstacle avoidance navigation system, and so on.

Author Contributions: Conceptualization, J.P. and J.L.; data curation, K.Z. and M.Q.; writing—original draft preparation, J.P.; All authors have read and agreed to the published version of the manuscript.

Funding: This work was funded by Jiangsu Agriculture Science and Technology Innovation Fund [Grant No.CX(21)1007], and the Key Laboratory for Protected Agricultural Engineering in the Middle and Lower Reaches of Yangtze River, Ministry of Agriculture and Rural Affairs.

Institutional Review Board Statement: Not applicable.

Informed Consent Statement: Not applicable.

Data Availability Statement: The data presented in this study are available on request from the corresponding author.

Conflicts of Interest: The authors declare no conflict of interest.

References

1. Li, J.J. *Research on Design Methods of Flexible Grasping Devices with Variable Stiffness*; University of Chinese Academy of Sciences: Beijing, China, 2019.
2. Deng, X.L.; Luo, Z.W.; Pang, J.Q.; Zhang, Y.M.; Yang, C.J.; Li, R.Q. Design and experiment of bionic nondestructive handheld suction apple picker. *J. China Agric. Univ.* **2019**, *24*, 106–114.
3. Ma, L.Z.; Yang, W.L.; Wang, C.J.; Chen, X.X.; Xue, C.; Lin, Y.Z.; Liu, A.R. Structure design and experiment of the end-effector for apple-harvesting robot. *J. Agric. Mech. Res.* **2009**, *31*, 65–67.
4. Zhang, Q.L.; Ji, C.Y.; Gao, F.; Zhao, W.Y. Effect of apple picking manipulator on fruits damage. *Sci. Technol. Food Ind.* **2011**, *32*, 404–405.
5. Laschi, C.; Cianchetti, M.; Mazzolai, B.; Margheri, L.; Follador, M.; Dario, P. Soft Robot Arm Inspired by the Octopus. *Adv. Robot.* **2012**, *26*, 709–727. [CrossRef]
6. Shen, Q.; Trabia, S.; Stalbaum, T.; Palmre, V.; Kim, K.; Oh, I. A multiple-shape memory polymer-metal composite actuator capable of programmable control, creating complex 3D motion of bending, twisting, and oscillation. *Sci. Rep.* **2016**, *6*, 24462. [CrossRef] [PubMed]
7. Cianchetti, M.; Calisti, M.; Margheri, L.; Kuba, M.; Laschi, C. Bioinspired locomotion and grasping in water: The soft eight-arm OCTOPUS robot. *Bioinspir. Biomim.* **2015**, *10*, 035003. [CrossRef]
8. Paek, J.; Cho, I.; Kim, J. Microrobotic tentacles with spiral bending capability based on shape-engineered elastomeric microtubes. *Sci. Rep.* **2015**, *5*, 10768. [CrossRef]
9. Sun, Y.; Yap, H.K.; Liang, X.; Guo, J.; Qi, P.; Ang, M.H.; Yeow, C.H. Stiffness customization and ratterning for property Modulation of Silicone-Based Soft Pneumatic Actuators. *Soft Robot.* **2017**, *4*, 251–260. [CrossRef]
10. Kashima, S.; Miyasaka, F.; Hirata, K. Novel soft actuator using magnetorheological elastomer. *IEEE Trans. Magn.* **2012**, *48*, 1649–1652. [CrossRef]
11. Gafford, J.; Ding, Y.; Harris, A.; McKenna, T.; Polygerinos, P.; Holland, D.; Walsh, C.; Moser, A. Shape Deposition Manufacturing of a Soft, Atraumatic, and Deployable Surgical Grasper. *J. Mech. Robot.* **2015**, *7*, 021006. [CrossRef]
12. Cho, K.-J.; Koh, J.-S.; Kim, S.; Chu, W.-S.; Hong, Y.; Ahn, S.-H. Review of manufacturing processes for soft biomimetic robots. *Int. J. Precis. Eng. Manuf.* **2009**, *10*, 171–181. [CrossRef]
13. Bartlett, N.W.; Tolley, M.T.; Overvelde, J.T.B.; Weaver, J.C.; Mosadegh, B.; Bertoldi, K.; Whitesides, G.M.; Wood, R.J. A 3D-printed, functionally graded soft robot powered by combustion. *Science* **2015**, *349*, 161–165. [CrossRef]
14. Morin, S.A.; Shepherd, R.F.; Kwok, S.W.; Stokes, A.A.; Nemiroski, A.; Whitesides, G.M. Camouflage and Display for Soft Machines. *Science* **2012**, *337*, 828–832. [CrossRef]

15. Wehner, M.; Truby, R.L.; Fitzgerald, D.J.; Mosadegh, B.; Whitesides, G.M.; Lewis, J.A.; Wood, R.J. Anintegrated design and fabrication strategy for entirelysoft autonomous robots. *Nature* **2016**, *536*, 451–455. [CrossRef]
16. Ilievski, F.; Mazzeo, A.D.; Shepherd, R.; Chen, X.; Whitesides, G.M. Soft Robotics for Chemists. *Angew. Chem. Int. Ed.* **2011**, *50*, 1890–1895. [CrossRef] [PubMed]
17. Shepherda, R.F.; Ilievskia, F.; Choia, W.; Morina, S.A.; Stokesa, A.A.; Mazzeoa, A.D.; Chen, X.; Wanga, M.; Whitesides, G.M. Multigait soft robot. *Proc. Natl. Acad. Sci. USA* **2011**, *108*, 20400–20403. [CrossRef] [PubMed]
18. Preter, A.D.; Anthonis, J.; Baerdemaeker, J.D. Development of a robot for harvesting strawberries. *ScienceDirect* **2018**, *51*, 14–19. [CrossRef]
19. Hemming, J.; Van Tuijl, B.; Gauchel, W.; Wais, E. Field test of different end-effectors for robotic harvesting of sweet-pepper. *Acta Hortic.* **2016**, *1130*, 567–574. [CrossRef]
20. Dimeas, F.; Sako, D.V.; Moulianitis, V.C.; Aspragathos, N.A. Design and fuzzy control of a robotic gripper for efficient strawberry harvesting. *Robotica* **2015**, *33*, 1085–1098. [CrossRef]
21. Yaguchi, H.; Nagahama, K.; Hasegawa, T.; Inaba, M. Development of an autonomous tomato harvesting robot with rotational plucking gripper. In Proceedings of the 2016 IEEE/RSJ International Conference on Intelligent Robots and Systems (IROS), Daejeon, Korea, 9–14 October 2016; Volume 9, pp. 652–657. [CrossRef]
22. Glick, P.; Suresh, S.A.; Ruffatto, D.; Cutkosky, M.; Tolley, M.T.; Parness, A. A Soft Robotic Gripper with Gecko-Inspired Adhesive. *IEEE Robot. Autom. Lett.* **2018**, *3*, 903–910. [CrossRef]
23. Takahashi, T.; Kikuchi, S.; Suzuki, M.; Aoyagi, S. Octopus-bioinspired Vacuum Gripper with Micro Bumps. *J. Robot. Soc. Jpn.* **2017**, *35*, 62–69. [CrossRef]
24. Tomokazu, T.; Kikuchi, S.; Suzuki, M.; Aoyagi, S. Vacuum gripper imitated octopus sucker-effect of liquid membrane for absorption. In Proceedings of the 2015 IEEE/RSJ International Conference on Intelligent Robots and Systems (IROS), Hamburg, Germany, 28 September–3 October 2015; pp. 2929–2936.
25. Brown, E.; Rodenberg, N.; Amend, J.; Mozeika, A.; Steltz, E.; Zakin, M.R.; Lipson, H.; Jaeger, H.M. Universal robotic gripper based on the jamming of granular material. *Proc. Natl. Acad. Sci. USA* **2010**, *107*, 18809–18814. [CrossRef]
26. Shintake, J.; Cacucciolo, V.; Floreano, D.; Shea, H. Soft Robotic Grippers. *Adv. Mater.* **2018**, *30*, e1707035. [CrossRef]
27. Abundant Robotics. Apple-Picking Robot Prepares to Competer for Farm Jobs. *[EB/OL]* **2017**, *5*, 8. Available online: https://www.sohu.com/a/139050700_355045 (accessed on 23 August 2021).
28. Ma, S.C.; Xie, R.P.; Huang, Q.; Liu, M. Experimental study on the interaction force between ground-wheel and soil of dibble-type transplanter. *J. Agric. Mech. Res.* **2018**, *40*, 158–161, 166.
29. Hao, Y.; Gong, Z.; Xie, Z.; Guan, S.; Yang, X.; Wang, T.; Wen, L. A Soft Bionic Gripper with Variable Effective Length. *J. Bionic Eng.* **2018**, *15*, 220–235. [CrossRef]
30. Hao, Y.; Wang, T.; Ren, Z.; Gong, Z.; Wang, H.; Yang, X.; Guan, S.; Wen, L. Modeling and experiments of a soft robotic gripper in amphibious environments. *Int. J. Adv. Robot. Syst.* **2017**, *14*, 14. [CrossRef]
31. Mazzolai, B.; Laschi, C.; Cianchetti, M.; Patane, F.; Bassi-Luciani, L.; Izzo, I.; Dario, P. Biorobotic Investigation on the Muscle Structure of an Octopus Tentacle. In Proceedings of the 2007 29th Annual International Conference of the IEEE Engineering in Medicine and Biology Society, Montréal, QC, Canada, 20–24 July 2007; Volume 2007, pp. 1471–1474.
32. Tramacere, F.; Beccai, L.; Sinibaldi, E.; Laschi, C.; Mazzolai, B. Adhesion Mechanisms Inspired by Octopus Suckers. *Procedia Comput. Sci.* **2011**, *7*, 192–193. [CrossRef]
33. Yekutieli, Y.; Sagiv-Zohar, R.; Aharonov, R.; Engel, Y.; Hochner, B.; Flash, T. Dynamic Model of the Octopus Arm. I. Biomechanics of the Octopus Reaching Movement. *J. Neurophysiol.* **2005**, *94*, 1443–1458. [CrossRef] [PubMed]
34. Gutfreund, Y.; Flash, T.; Fiorito, G.; Hochner, B. Patterns of Arm Muscle Activation Involved in Octopus Reaching Movements. *J. Neurosci.* **1998**, *18*, 5976–5987. [CrossRef]
35. Kier, W.M.; Smith, K.K. Tongues, tentacles and trunks: The biomechanics of movement in muscular-hydrostats. *Zool. J. Linn. Soc.* **1985**, *83*, 307–324. [CrossRef]
36. Yekutieli, Y.; Sagiv-Zohar, R.; Hochner, B.; Flash, T. Dynamic Model of the Octopus Arm. II. Control of Reaching Movements. *J. Neurophysiol.* **2005**, *94*, 1459–1468. [CrossRef] [PubMed]
37. Navas, E.; Fernández, R.; Sepúlveda, D.; Armada, M.; Gonzalez-De-Santos, P. Soft Grippers for Automatic Crop Harvesting: A Review. *Sensors* **2021**, *21*, 2689. [CrossRef] [PubMed]
38. Chang, L.; Na, L.; Ja, L. Dynamic simulation analysis of ADAMS on pneumatic flexible strawberry picking manipulator. *Chin. Sci. Pap.* **2017**, *12*, 390–393.
39. Marchese, A.D.; Katzschmann, R.K.; Rus, D. A recipe for soft fluidic elastomer robots. *Soft Robot.* **2015**, *2*, 7–25. [CrossRef] [PubMed]

agriculture

Review

Variable Rate Seeding in Precision Agriculture: Recent Advances and Future Perspectives

Egidijus Šarauskis [1], Marius Kazlauskas [1], Vilma Naujokienė [1,*], Indrė Bručienė [1], Dainius Steponavičius [1], Kęstutis Romaneckas [2] and Algirdas Jasinskas [1]

[1] Institute of Agricultural Engineering and Safety, Agriculture Academy, Vytautas Magnus University, Studentu Str. 15A, Kaunas Distr., LT-53362 Akademija, Lithuania; egidijus.sarauskis@vdu.lt (E.Š.); marius.kazlauskas@vdu.lt (M.K.); indre.bruciene@vdu.lt (I.B.); dainius.steponavicius@vdu.lt (D.S.); algirdas.jasinskas@vdu.lt (A.J.)

[2] Institute of Agroecosystems and Soil Science, Agriculture Academy, Vytautas Magnus University, Studentu Str. 11, Kaunas Distr., LT-53361 Akademija, Lithuania; kestutis.romaneckas@vdu.lt

* Correspondence: vilma.naujokiene@vdu.lt

Abstract: The main objective of this study was to analyze variable rate seeding (VRS) methods and critically evaluate their suitability and effectiveness for the challenges under field conditions. A search was performed using scientific databases and portals by identifying for analysis and evaluation 92 VRS methodologies, their impact and economic benefits depending on the main parameters of the soil and environment. The results of the review identified that VRS could adapt the appropriate seeding rate for each field zone, which was based on site-specific data layers of soil texture, ECa, pH and yield maps. Then, remotely detected images or other data which identify yield-limiting factors were identified. The site-specific sowing method (with a variable sowing rate for each field area) allows the optimization of crop density to obtain the best agronomic and economic results. Various proximal and remote sensor systems, contact and contactless equipment, mapping and VRS modeling technologies are currently used to determine soil and crop variability. VRS depends on the field characteristics' sowing equipment capabilities, the planned harvest, soil productivity and machine technology interactions with the environment. When forecasting the effective payback of a VRS over the desired period, the farm size should on average be at least 150 ha. In future studies, to achieve the best solutions and optimal methods, it is important to test, evaluate and put into practice the latest methodologies on farms, to perform complex assessments of changes in sensor, soil, plant and environmental parameters.

Keywords: precision farming; site-specific seeding; prescription maps; apparent electrical conductivity; proximal sensing

Citation: Šarauskis, E.; Kazlauskas, M.; Naujokienė, V.; Bručienė, I.; Steponavičius, D.; Romaneckas, K.; Jasinskas, A. Variable Rate Seeding in Precision Agriculture: Recent Advances and Future Perspectives. *Agriculture* **2022**, *12*, 305. https://doi.org/10.3390/agriculture12020305

Academic Editor: Yongchao Tian

Received: 26 January 2022
Accepted: 18 February 2022
Published: 21 February 2022

1. Introduction

Variable rate seeding (VRS) is a precise agricultural technology that can properly and accurately adjust the seeding rate according to the variability of soil properties, terrain, meteorological conditions and other factors. VRS not only provides better opportunities for the use of variable soil nutrient and water storage capacity characteristics, it can also increase crop yields by reducing seed consumption. Seed germination, crop development and yield potential may vary in different areas of a field, and thus VRS is a method of linking seed quantities to a specific area, thereby increasing crop yields and production profits. By implementing VRS practices, farmers can better manage farm risk and focus more on investing in areas with higher return potential. In most regions of the world, the implementation of VRS in agriculture has been relatively low [1,2]. A prescription map is an electronic data file containing specific information about input rates to be applied in every zone of a field. One of the main reasons for an increased interest in VRS is that VRS technological solutions have been introduced into agricultural machinery and have become easier to implement. In addition, VRS innovations combined with high-precision global navigation satellite systems allow farmers to create and implement VRS prescriptions to optimize seed placement and yield [2,3].

Seeding and planting at a variable rate are particularly useful in very heterogeneous fields, i.e., in fields with very different water retention capacities or soil organic matter levels. Measurements of soil properties, plant condition and yield are performed using a precision farming system. Using the data obtained, the technological parameters of the drill are adjusted, the number of seeds to be inserted is optimized, the yield potential is increased, and the quality of the plants is improved [4]. VRS is likely to work well in changing crop protection and crop nutrition strategies [2].

When developing a VRS application program at the level of a specific farm, a correct choice of precision agricultural technologies for the whole complex and a good understanding of the growth environment of different plants for each field is required. Forming this understanding requires not only good intuition by the farmer himself but also spatial layers of field data that allow the field to be divided into separate soil management zones (MZ) in which each MZs is subject to a unique VRS. General spatial layers of field data should include maps of soil properties, height difference data and yield maps of previously grown crops [5]. Precision agriculture (PA) technologies, variable rate fertilization and liming and seeding operations are applied to individual field MZs [6]. Other authors emphasize that one of the most important tasks for the successful application of PA technology is the assignment of optimal rates of fertilizers, limes or seeding in individual field MZs [7,8]. Modern farmers are well-aware of the differences in their soil productivity and recognize the potential of using variable rate technologies compared to uniform rates. Images depicting highly variable crop growth in the fields are often used to promote the attractiveness of intuitive variable rate farming. However, to recoup the costs of applying a variable rate, it is necessary to use only well-managed and accurately predictable field changes. An element of physical soil inspection will also be required because the relationship with crop establishment is related to stone content and soil texture and so it will not be reliable to use EC maps alone in predicting seed bed quality and establishment. Yield maps have been used to identify zones of different yield potential and planting rates [7].

VRS is a very important, but still emerging, PA technological operation, which has a particularly important impact on the further stages of plant development and the production efficiency of the whole farm. Therefore, it is crucial to have sufficient information for the application of VRS to be successful. Unfortunately, the resources of the scientific literature on the application of VRS to different plants are still quite limited. In particular, few research results have been published on seedings under VRS with one of the most popular plants—winter wheat. Winter wheat is one of the most popular plants in the world and is the most popular plant in Lithuania. In addition, the number of studies conducted with wheat VRS remains limited. The main aim of this study was to review and provide a synthesis of the recent advances in VRS methods and to critically analyze their suitability in view of the challenges posed under field conditions.

2. Search Methods

The review discusses scientific, technical and other documents from a particular time period (the search time duration was from 1998 to 2021). The review was written as a report on the subject of the research area and as an introduction before the planned experimental research. The survey drafting method was a piecemeal and systematic method, rather than impulsive. The scientific literature was analyzed and the review plan was prepared based on information that was clearly divided into sections. These methods were used as they were most convenient for the reader.

The literature search was performed using databases such as ScienceDirect, Wiley Online Library, Springer Link, Scopus and Google Scholar. We refined publications by domain (engineering, life sciences, agriculture, environmental science, earth and planetary sciences, energy, etc.) and publication type (journals, books, etc.). The subjects in the ScienceDirect database were physical sciences and engineering, earth and planetary sciences, energy, engineering, life sciences, agricultural and biological sciences and environmental science. The subjects in the Scopus database were agricultural and biological sciences, environmental science and engineering. The subjects in the Wiley Online Library were life sciences, earth space and environmental sciences, agriculture, aquaculture and food science and technology. The subjects in Springer Link were engineering, environment and life sciences (Figure 1).

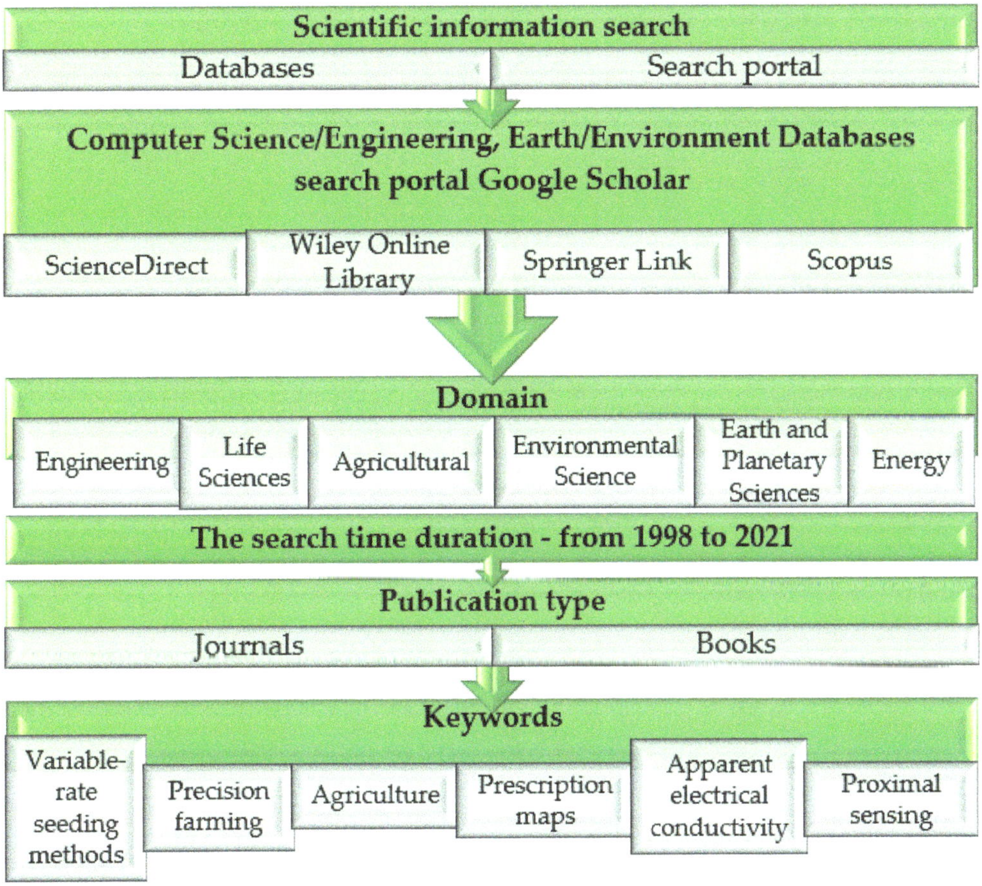

Figure 1. Scientific review information search strategy and enforcement selection process.

The time duration for articles and books was set from 1998 to 2021. Our in-depth analysis and evaluation largely focused on the most recent literature, the most relevant and useful of which was published between 2000 and 2021. Filtering by journal or book title was performed using the query string. We searched for the keywords "variable rate seeding methods", "precision farming", "agriculture", "prescription maps", "apparent electrical conductivity" and "proximal sensing", and a large volume of work from diverse professional fields was retrieved.

The scientific analysis first involved identifying the different techniques and methodologies used and then discussing and abstracting the results obtained by applying different methods depending on the variation of the main parameters of the soil and the environment. The conclusions provide suggestions for practitioners by highlighting the most optimal options that yielded the best results for the improvement of the environment and production (Figure 2).

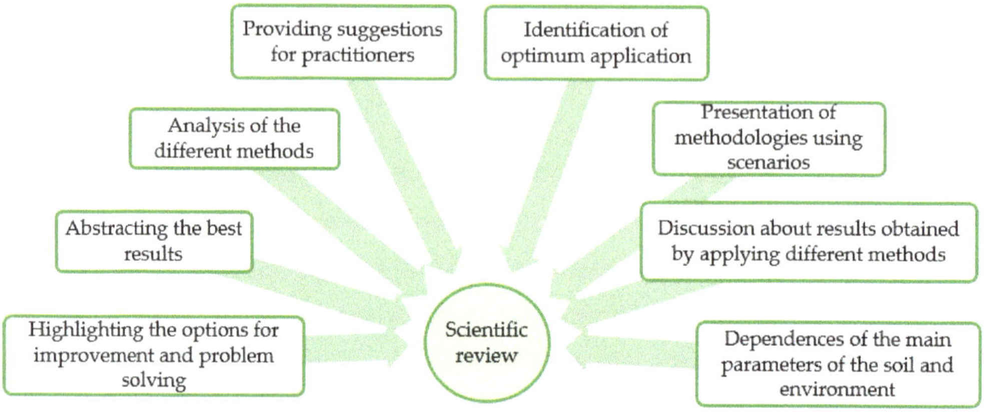

Figure 2. The components of the scientific information selection process of the review. These parts formed our review structure.

Most of the articles focused on general information on seeding and thus the collected information related to soil property maps, soil sensing methods and equipment used for VRS was purposefully concentrated. First, we analyzed the soil sensing methods and equipment used for VRS, then the influence of VRS on wheat growth characteristics and finally, what economic benefits could be obtained. Each document was reviewed and questions, including what soil sensing methods and equipment were used for VRS, how this affected wheat growth characteristics and what economic benefits were obtained, were answered (Figure 3).

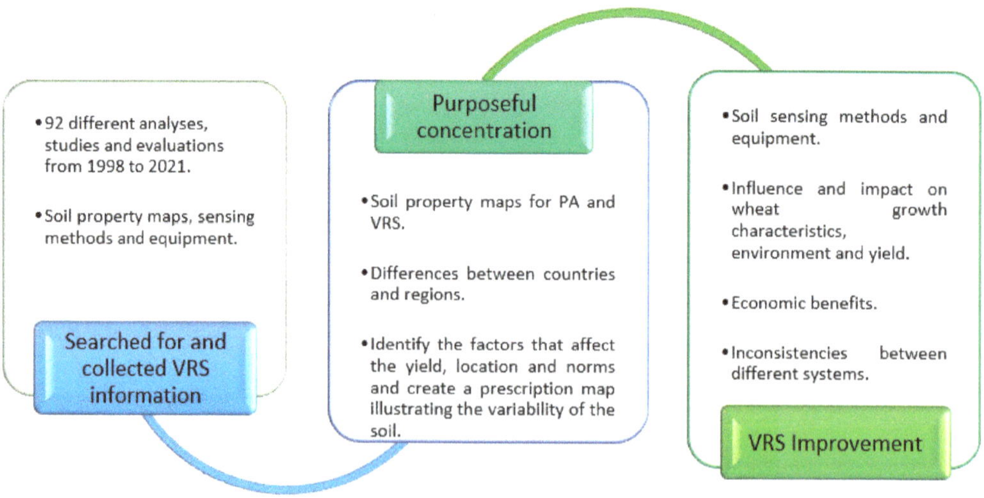

Figure 3. Method of evaluation of variable rate seeding methods and their applications in practice.

In summary, a significant beneficial effect has been identified as a result of the application of certain processes as a multifunctional filter for farmers. The main aim of this analysis was to provide a synthesis of recent advances in VRS methods and to analyze their suitability in view of the challenges posed under field conditions.

3. Findings

3.1. The Assessment of Soil Property Maps for PA and VRS

Depending on the country and the region, different methods are used to develop VRS. An essential element of VRS implementation is to identify the factors influencing the yield and the specific location in order to classify the sowing areas and to assign the necessary norms to them, thus creating a prescription map illustrating the variability of the soil [2]. In order to increase the value of VRS, it is necessary to define appropriate crop management zones or decision zones, in which soil types, topography, irrigation, long-term yield history, apparent electrical conductivity, etc. can be described [9]. A crucial element of VRS implementation is the establishment of site-specific factors impacting yield in order to create seeding zones and assign rates, thus generating a seeding prescription (RX) map. Fundamentally, two main methods are distinguished for the application of variable rate seeding—VRS based on a map and VRS based on sensor data [10,11]. Using the mapping method, soil and crop properties are determined, samples are taken, modeling and mapping are performed and variable rate seeding recommendations are prepared. These steps are performed in advance of the actual use of precision seeding in the field. Meanwhile, in a sensor data-based approach, these different steps mentioned are performed in real time using advanced algorithms, hardware, and software [11]. Recent VRS innovations increase our ability to insert two different plant varieties into the field. This corresponds to a multi-layer seeding method where seeds or two different varieties are distributed differently in the same field at the same time [2].

With the development of PA technologies, variable rate application technologies are increasingly used. They are often based not only on previous crop performance but also on soil productivity, soil structure, organic matter, landscape position, topography, or some combination of these [12]. Kaspar et al. [13] found that the yield potential of higher landscapes and steep slopes was lower than that of lower landscape positions in years when precipitation was below average. Griffin and Hollis [14] used landscape position elevation maps to identify areas of different yields in the field. Another common way to produce a variable rate seed plan is to start with a soil electrical conductivity survey that, along with other soil properties, identifies changes in soil texture. Figure 4 shows a map of the field terrain highlighting the difference in altitude at different locations in the field.

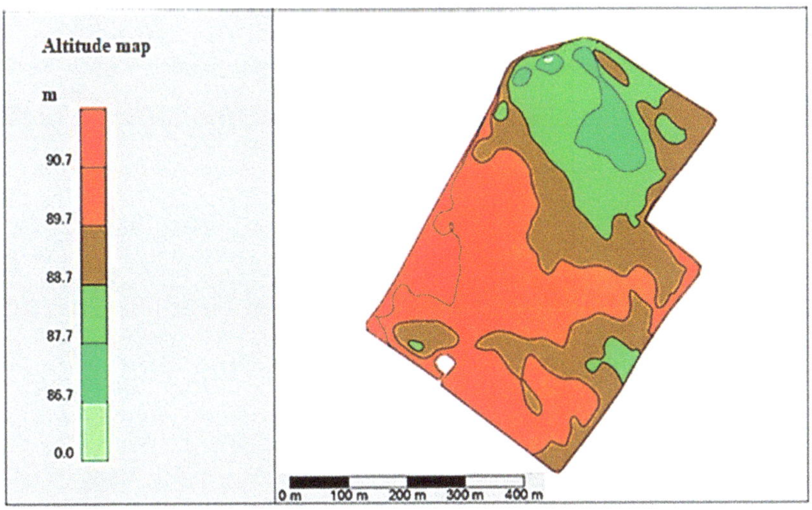

Figure 4. Elevation map representing relief differences in the field. The colored parameter variation internal maps were created by the authors using a Veris MSP 3150.

Soil property maps are probably the most important datasets used for the implementation of PA technologies. Table 1 provides a summary of the analysis of scientific sources showing the main soil properties applied to PA technologies by other authors.

Table 1. The main soil properties applicable to PA technologies.

Soil Properties	Application of PA Technologies	Reference
Organic matter, pH	Variable-rate seeding application on farms	[15]
pH	An on-the-go sensor used for mapping	[16]
SOC, soil texture	Exploring the driving forces and digital mapping using remote sensing	[17]
Soil texture	Electromagnetic induction for regional data coordination	[18]
	Radar and optical data from Sentinel-1 and Sentinel-2	[19]
Apparent electrical conductivity (ECa)	Applications of soil electrical conductivity mapping	[20]
	Scientific equipment and measurements in agriculture	[21]
	Effect of variable rate seeding on seedbed and germination parameters	[22]
Bulk density	Novel electromechanical system application	[23]

Soil productivity is defined as the ability of soil to provide plants with the necessary nutrients [24] and sufficient water. The ability of soil to perform various functions of biological productivity, such as ensuring the function of ecosystems, maintaining environmental sustainability and promoting plant and animal habitats, is often understood as soil quality [25]. The soil quality indicator is sensitive to any changes in the soil [11].

Soil pH affects the availability of nutrients important to plants, such as phosphorus and trace elements and herbicide activity. With the field area divided according to soil pH, lime and other sources of calcium can be used more accurately. This increases the likelihood that the amount required by the right product will be accurately distributed in the right places, which can increase the utility and efficiency of the product used as well as the crop yield [15]. Soil fertility and pH influence crop yield potential, and with precision soil management, can provide information that can be used for generating VRS RX maps. This soil information could reveal opportunities around determining high and low yield potential areas in a field depending on water availability; water represents the most important driver of crop yield.

Soil organic matter is another very important soil property, which has a significant effect on crop productivity. Organic matter (OM) is one of the main components of soil structure and porosity, influencing soil water retention capacity, biodiversity, the activity of soil organisms and the availability of plant nutrients, especially nitrogen [15]. The amounts of OM can vary significantly in different field areas, which is perfectly demonstrated by the field study performed in Lithuania with Veris MSP 3150 and the OM map presented in Figure 5.

PA technologies are increasingly being implemented using the variability of apparent electrical conductivity. Apparent electrical conductivity (ECa) is a property of soil that indicates its ability to conduct electricity. ECa field measurements started with soil salinity measurements, which were linked to the irrigation of arid agricultural areas [21]. This electrical soil property is influenced by a combination of physical and chemical properties, including soluble salts, clay content, soil water content, soil temperature, organic matter content and bulk density [21]. ECa varies depending on the amount of moisture retained by the soil particles. Thus, ECa strongly correlates with the size and the structure of soil particles. Saline soils and clay have high conductivity, silt has medium conductivity and sand has low conductivity [2,26].

ECa measurements, together with other soil properties, determine the changes in soil structure. Another common way to produce a variable rate seed plan is to start with a soil electrical conductivity survey that, along with other soil properties, identifies changes in soil texture [20]. Apparent electrical conductivity is related to soil structure and previous studies are mainly based on higher ECa values associated with higher yield areas, so these areas can grow a larger plant population. This method has been used mainly for corn crops in North America, but it is not necessarily the correct method for winter wheat or other crops that, due to their ability to bush, do not show a linear relationship between number plants and yield [14]. Experimental studies were carried out in Lithuania when an ECa map (Figure 6) was prepared after estimating the differences in soil granulometric composition, according to which the winter wheat seeding map of the study field was created. For winter wheat sowing, an average seeding rate of 180 kg ha^{-1}, typical for this region [27], was chosen. Then, variable rate seeding was applied to each of the five soil management zones (MZ), varying between the zones by about 20%—from a minimum of 146 kg ha^{-1} to a maximum of 214 kg ha^{-1} [22].

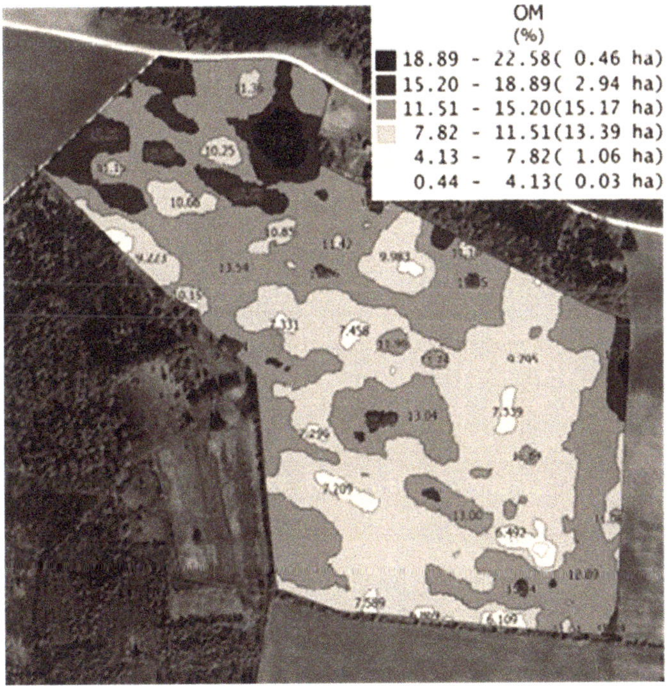

Figure 5. Map of soil organic matter differences. Colored parameter variation internal maps were created by the authors using a Veris MSP 3150; as a background layer, we used an ORT10LT digital raster orthophoto map of a territory in the Republic of Lithuania, M 1:10,000 (compiled 2012–2013) in the precision farming software AgLeader SMS Advanced.

When analyzing the influence of VRS on winter wheat production indicators, it was found that the unevenness of the field soil had a significant impact on the quality of seed placement, germination and plant tillering [22]. Using the same winter wheat seeding rate of 180 kg ha^{-1}, the germination of winter wheat seeds was significantly lower (MZ1—62.7% and MZ2—72.6%) in the two MZs with the highest soil ECa than in the remaining three soil MZs. There were no significant differences between MZs in the application of VRS. Changes in seed germination may have been influenced by soil structure and seedbed quality [14].

Soil ECa maps can be created for a variety of soil depths, from 5 to 150 cm. Figure 7 shows two ECa field maps drawn at the depths of 30 and 90 cm using a Veris MSP 3150 machine. These images illustrate that the apparent electrical conductivity was higher over the entire field as the measurement depth was increased but the differences between the zones remained similar.

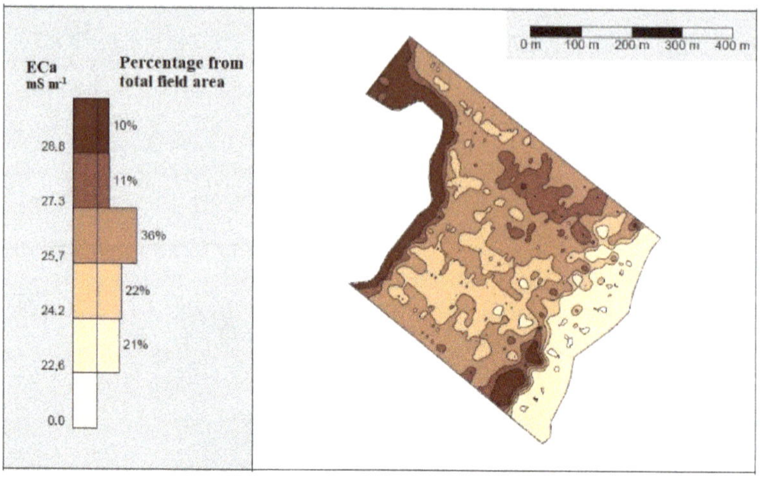

Figure 6. ECa map representing differences in soil properties in the field. The colored parameter variations were created by the authors using an EM38-MK2 device [22].

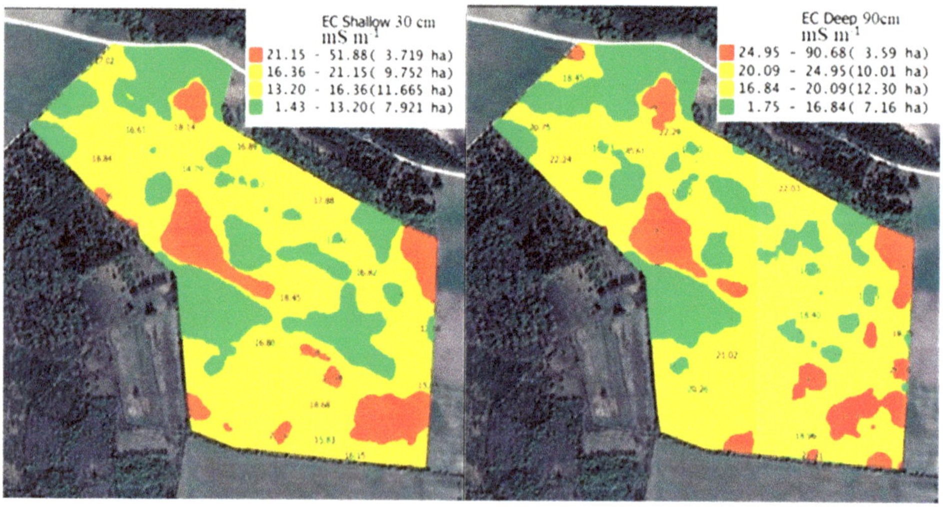

Figure 7. ECa maps of soil surface structure differences at the depth of 30 cm (**left**) and 90 cm (**right**). The colored parameter variation internal maps were created by the authors using a Veris MSP 3150; as a background layer, we used an ORT10LT digital raster orthophoto map of a territory in the Republic of Lithuania, M 1:10,000 (compiled 2012–2013) in the precision farming software AgLeader SMS Advanced.

Chemical and physical soil properties and topographical features are highly interrelated. As one soil characteristic changes, it can cause changes to other characteristics [28]. Soil organic carbon, pH, phosphorus, bulk density, water accumulation and other physical, chemical and biological properties are the most often reflected soil quality indicators [11,29]. It is not possible to assess soil functionality based on the individual soil property responsible for regulating crop yields [30], as changes in crop yields are influenced by a number of biotic and abiotic factors [31–34].

Studies by Reining et al. [35] with winter wheat showed that factors such as soil quality and precipitation had the greatest influence on the change in sowing rate. In order to optimize the seeding rate in VRS technologies, it is necessary to have as much information as possible about the properties of the soil and the crop. The seeding rate is optimized based on measurements such as soil ECa, soil pH, crop nitrogen content, previous yield and grain protein indices [36]. Water availability and accumulation are important parameters for determining the yield potential and optimal seeding rate in specific field areas, especially in dry weather conditions [12]. It is generally accepted that the specific optimal sowing rate depends on soil conditions, such as texture. However, optimal seeding rates can almost never be accurately known [37].

The mapping of soil properties such as pH and organic matter also holds great potential for delineating spatial variation in fields. Soil organic matter maps can be used to develop management zones for variable nitrogen and seeding rates. It is generally thought that areas of higher organic matter can support higher seeding rates. Soil sensing tools can better characterize the within-field variability into MZ and perhaps serves as a better source of information than soil surveys for deriving variable rate prescriptions of crop inputs.

3.2. *The Assessment of Soil Sensing Methods and Equipment Used for VRS*

The use of actual small-area data from a number of locations can provide an approximation of a variable rate seeding plan in order to increase the economic return on agricultural production. However, determining the optimal seeding rate for each field remains a challenge. The definition of specific control zones may be possible using different soil properties [12], but it may take time to establish consistent models under different crop rotations or tillage regimes [38–40].

Based on soil ECa, Taylor et al. [41] created VRS maps and evaluated them using GIS global information systems. In their study, soil ECa was considered an indicator of soil quality, which showed a different (positive and negative) correlation between the crop yield and the seeding rate in different years. ECa data collected using Veris or electromagnetic EM devices show soil productivity zones. The elevation and the slope of the terrain, collected using real-time kinematics (RTK), is also valuable information. All these data help to determine the MZs of the soil and allow us to establish the appropriate plant density and yield potential for each of these zones. This should result in the identification of the maximum yield for each zone. Topography and landscape position often have a significant impact on soil properties and, consequently, crop productivity, and can therefore enhance proximal and remote soil sensing. VRS has been proven to be profitable for fields with highly variable levels of productivity [42].

Munnaf et al. [43] argue that uniform rate seeding (URS), where field characteristics differ, is an inappropriate approach to the sustainable management of farm resources. They performed a study using field scanning online visible and near-infrared (vis-NIR) spectroscopy and an electromagnetic induction (EMI) sensor. The uniform rate seeding method was compared with two VRS methods, the first based on a soil MZs map generated using EMI data and the second treatment based on the fusion of vis-NIR measured soil data with the Sentinel-2-derived normalized difference vegetation index (vis-NIRsen). The latter method makes it possible to assess the interrelationships between the main characteristics of the soil and the yield using different seeding intervals. The research results obtained showed that both VRS methods resulted in higher yields of potato tubers and higher economic returns compared to URS [43].

A soil permeability map is a simple, inexpensive tool that farmers can use to describe soil differences in farm fields quickly and accurately. Soil ECa is the ability of a soil to transfer or transmit an electric current in units of millisiemens per meter [26,44–46].

In the agricultural industry, crop yield and seeding maps are becoming increasingly important as these two-parameter data layers help us to better understand the feasibility of VRS. The variable rate seeding strategy for each soil MZs is based on the following data: long-term yield history, field productivity, soil dryness and moisture, apparent electrical conductivity, environmental response units, soil type, topography, landscape, slope, drainage and color, crop, soil and plant vegetation index [9]. Site-specific field data, such as digital soil maps, aerial photographs, apparent electrical conductivity maps, or previous-year yield maps, are the basis for calculating the optimal site-specific seeding rate. In this way, VRS application maps are created, showing differences in the seeding rate due to different conditions related to location, cultivation system and short-term parameters (Figure 8). This means that the generated VRS maps describe the situation only at a certain point in time [47]. Prepared VRS prescription maps are recognized by most seed drills [8].

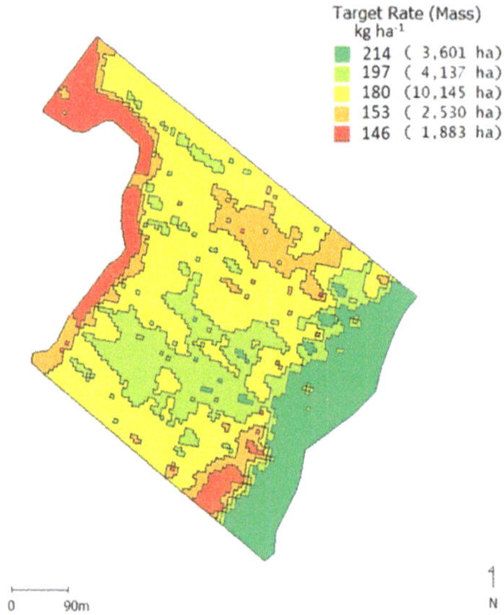

Figure 8. Winter wheat VRS map. The colored parameter variation was created by the authors using SMS software with the data received from an ECa map.

In heterogeneous fields, it is useful to change the number of seeds according to the soil potential. In precision agriculture, machinery makes it possible to adapt the optimal number of seeds, thus saving farm resources and achieving more environmentally friendly agricultural production [47]. The main process of VRS is the creation of a pre-scription map for individual fields. Research has shown that the development of such prognostic maps needs to be managed separately due to differences in the unique variability and historical management of each field. When designing zones, it is necessary to consider around four to five variables (yield, soil, terrain, etc.), paying attention to those that limit the yield [2].

When creating a prescription map, the number and diversity of plant populations in each area must be determined before loading the map onto the tractor cab screen. In most cases, there can be around four to five variables to consider in designing zones (yield, soil, terrain, organic matter, etc.), with attention being paid to those that are yield-limiting factors. When working with a population (seeds per hectare), a different zoning rate of at least 15–20% must be maintained in order to maximize the yields in response to VRS. In general, higher seed populations will be found in soil areas with higher yield potential and in lower yield areas seed population rates are lower [48]. Methods for zoning may include assessments of aerial imagery [49], soil type, farmers' knowledge of fields [50], soil characteristics [51], crop history, water availability, terrain, yield maps or remote sensing [52].

Mapping systems are not suitable for large variations in soil conditions that are highly dependent on weather conditions. Therefore, future advanced farming requires systems that overcome the limitations of a map-based approach. Sensor-based, real-time, site-specific seeding (SSS) requires high-resolution data collected using advanced sensor technology. This continuous flow of sensor data is consistently translated into information and recommendations to be implemented by the right controller, all in real time [10].

Seeding by sensor makes it possible to overcome the limitations of the map method. The design of the sensing and control device is very important, as the incorrect design of the system can sometimes cause problems [11]. SSS is an accurate application method that takes into account inherent soil heterogeneity to maximize crop yields and to minimize seed yields. Most SSS program practices are based on management zone (MZ) maps generated using different inputs [11], which include total field productivity, land and moisture, soil and crop vegetation, environmental response index [9], terrain attributes [2], soil fertility status, structure, [53], color [9], apparent electrical conductivity [53,54] and historical yields [35,55,56].

The main requirements for the implementation of the VRS system are to have the right technology, the right prescription mapping process and trained specialists. Key VRS technologies include a GPS receiver and a cab display with the ability to load and execute prescription maps, collect data and provide real-time feedback to the tractor operator on drill performance and VRS drill capabilities. VRS options include a hydraulic drive that allows the seeding rate of the seeding machine to be changed throughout the crop rotation [2]. With the development of real-time kinematic (RTK) and other high-grade GPS receivers, accurate topographic measurements can be obtained simultaneously with proximal readings of soil sensors [44]. Variable rate sowing adjustment technologies can operate on prescription maps or real-time (online) sensor readings, which can be adapted to change the sowing rate according to certain algorithms. Site-specific properties in the field (soil properties, OM, landscape differences, etc.) can be detected during the sowing of cereals or before sowing using various sensors. Optical sensors detect the proportion of organic matter and/or moisture in the soil, non-contact EC sensors detect electromagnetic induction, electromagnetic induction sensors detect the electrical conductivity or moisture of the soil (e.g., EM38 TopSoil Mapper GEM-2 and DUALEM), electrical resistance sensors determine the electrical conductivity of the soil from which the soil moisture is determined and gamma ray sensors determine the proportion of organic matter in the soil and texture (e.g., SoilOptix) [45]. Research has shown that the real-time sensor on Veris' multi-sensor platform (MSP) accurately matches the spatial models of organic matter (OM) and cation exchange capacity (CEC). Veris OpticMapper can register the values of soil OM and CEC in real time using a dual wavelength optical sensor mounted on a specially configured row unit [44]. This allows plant residues to be removed and the sapphire window at the bottom of the sensor to adhere to the soil, recording readings every second. Conventional soil samples are obtained from the field after passing with OpticMapper, with soil sample sites monitored by OpticMapper measurements [15,44].

Integration of automatic sensing, modeling and control systems is a complex task. The "VRfertilizer" program and the application of the pesticide Chlorpyriphos can be distinguished as examples of sensor-based site-specific applications [57]. SSSs also have the potential to apply similar principles and thus future research should focus on the development and evaluation of sensor-based SSSs in a variety of soil qualities and crop, local and weather conditions [11].

Measurements of apparent electrical conductivity correlate with soil properties that affect crop productivity, including soil structure, cation exchange capacity, drainage conditions, salinity and subsoil properties [26,58]. The contact method for mapping the soil ECa uses coulters that are in direct contact with the soil to measure its ECa. Another non-contact method uses electromagnetic (EM) induction to measure soil ECa [42]. The latter method can be used to determine the differences in soil properties in the field. To create an accurate soil sampling plan and variable rate seeding map an EM-38 MK-2 scanner can be used (Figure 9). When mounted on plastic sleds, ECa measures from 0 to 150 cm soil depth [22].

Veris Technologies' commercial soil ECa mapping system measures ECa at two depths (0–30 cm, 0–90 cm) when the device is pulled through the field. The ability to measure and obtain results from approximately 25 sample sites per hectare provides the farmer or consultant with a higher resolution dataset to measure the changes in field pH. The usual grid soil sampling practice can provide pH values for each 2.5–4.0 ha area. The sampling resolution of Veris Soil pH Manager is 25 to 40 times higher than that provided by a conventional soil grid. The pH controller can be paired with a soil ECa array to simultaneously capture soil pH, soil ECa and soil color [15]. The sensor operates by lowering the sampling mechanical arm into the soil to take the sample and then lift the sample in front of the electrode set to measure the pH before repeating the process. Studies have shown that one-hectare grids have a wide range of pH values, often ranging from soil that requires lime to soil with a particularly high pH [59].

Figure 9. EM38-MK2 remote sensing device constructed on a plastic sledge [22].

Veris Technologies' Soil pH Manager is equipped with a sensor for real-time determination of soil pH (Figure 10). It automatically collects soil samples and measures soil pH from direct contact with soil material along the way [16].

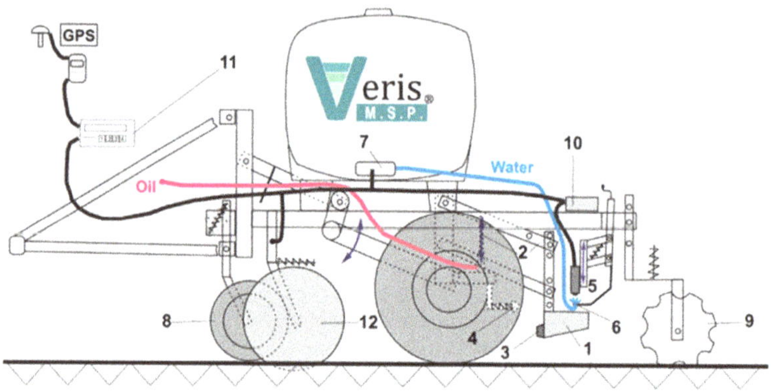

Figure 10. Veris multi sensor platform (MSP) with the Soil pH Manager: 1—soil sampler shoe; 2—parallel linkage; 3—adapter; 4—scraper; 5—pH sensors; 6—water supply with nozzles; 7—water pump; 8—row cleaner; 9—furrow closer; 10—controller; 11—data recorder; 12—sensor of apparent electrical conductivity [16].

As mentioned earlier, two main SSS methods based on maps and touch systems are distinguished. [10]. According to the mapping method, soil and crop identification and sampling, modeling, mapping and the preparation of SSS recommendations are performed in advance of the actual field use and in a sensor-based SSS these different steps are realistically performed by algorithms, hardware and software [11]. The map-based SSS is related to the adjustment of the sowing rate according to the previously created and loaded prescription map in the virtual terminal of the precision seeding machine. GIS and geostatistical analysis allow for the linking of the measured attributes. Based on the yield status of the different field zones, the field is divided into several smaller zones that are assigned a certain seeding rate to create a program application map (AM) or management map. Once the AM is generated, it is converted into a machine-compatible form file and uploaded to the virtual terminal [60]. When working in the field, the variable rate seeding controller delivers the seeding rate according to the optimal quantity and the corresponding location as specified in the AM. A positive aspect of this system is the time it takes to conduct research and then apply VRS. This improves controller responses and smooths out transitional VRS processes. To work under this principle, the equipment uses on-the-go sensors for variable rate seeding (Figure 11). Soil organic matter sensors detect different levels of organic matter and adjust the number of seeds accordingly. There are also soil moisture meters that can be used to adjust the depth and to change the seeding rates [10].

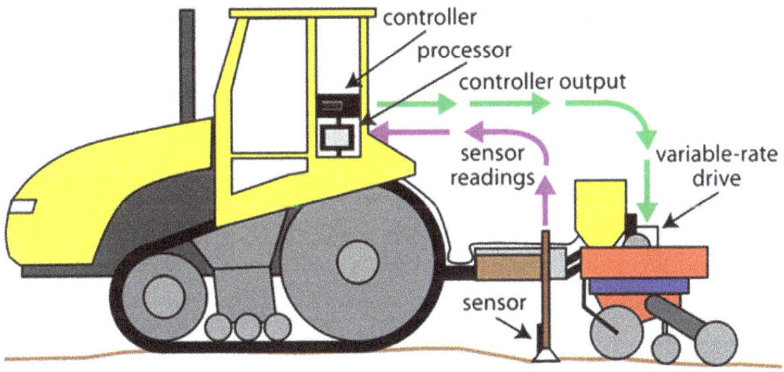

Figure 11. On-the-go soil properties sensor measurement before seeding and adjustment of seeding rate [10].

Another modern soil sensor is the SoilXplorer, which uses innovative methods for real-time soil structure analysis. This is a non-contact sensor that uses electromagnetic signals to measure the soil ECa and is used to justify variable tillage depth, variable seeding depth and variable seeding rate. Soil ECa is highly correlated with the size of soil particles and soil structure. For example, sandy soils have larger particles and low water retention capacity, resulting in lower ECa, while ECa is higher in clay and organic soils. Therefore, the soil sensor can detect compacted areas and readjust the working depth accordingly. Based on this measurement principle, soil type zones, relative water content and compaction zones can be determined. As a result of its four reception coils, ECa can be measured in four different soil layers (0–25 cm, 15–60 cm, 55–95 cm and 85–115 cm) at the same time [61,62].

The SoilXplorer sensor (Figure 12) is installed at a height of 30–40 cm in front of the tractor or other machine so that the drill or tillage machine can be attached at the rear [60]. The sensor autonomously identifies different soil properties, and in real time, the ISOBUS seed drills adjust the seeding rate according to the soil texture and the relative amount of soil water. The highest seeding rate applies to the best soil conditions and the lowest to poor conditions, although the opposite may be the case. With the SoilXplorer sensor, farmers can deepen their knowledge of the soil and increase the efficiency of the tillage and seeding processes. Field tests have shown that reducing the tillage depth from 18 to 10 cm reduces energy expenditure during the tillage process by 45%. Additionally, during the same process, wheel slip can be reduced by about 53% and output can be increased by about 20% [61].

Figure 12. SoilXplorer device for measuring soil structure differences [60].

VRS technology requires the preparation of a prescription map showing how the electronic controller of seed drill should change the gaps between the seeds in a row to make it as convenient as possible to change the seeding rate. Indeed, VRS can be performed by commercial seed drills with electronic seed metering disc controllers that can change the seeding rate by changing the seed distance along the row [63]. There are two main ways to adapt uniform rate seed drills for variable rate seeding, i.e., by changing the active feed-roll length or by changing the seed meter drive shaft speed [62]. With the addition of a controller to a conventional seed drill, it is possible to change the drive speed of the seed metering unit on-the-go. Jafari et al. (2010) presented a system for modifying a conventional seed drill into a variable rate seed drill. It consisted of a DC motor with a fixed speed gearbox, encoders for determining the drive wheel and motor speed of the seed drill, a GPS receiver, a DC electric motor controller and a laptop. The results of the research using different seeding rates from 87.5 (low) to 262.5 (high) kg ha^{-1} showed that the seeding transition rate time from low-to-high and from high-to-low was 7.4 and 5.2 s, respectively [64].

Currently, seed drill manufacturers have adapted to the needs of precision farming and offer farmers a wide range of seed drill models for seeding cereal at a variable rate. Kazlauskas et al. [22] carried out winter wheat seeding at a variable rate with a Horsch Avatar 6.16 SD direct seed drill and the seeding rate varied from 146 to 214 kg ha^{-1}. Cereal seeding at a variable rate can also be conducted with a Vaderstad Spirit 400C, a Horsch Focus 6TD and other seed drills (Figure 13).

(a) (b) (c)

Figure 13. Seed drills for VRS winter wheat seeding: (**a**) Vaderstad Spirit 400C; (**b**) Horsch Focus 6TD; (**c**) Horsch Avatar 6.16 SD.

4. The Influence of VRS on Wheat Growth Characteristics

Differences in seeding rates depend on the field characteristics, capabilities of the seeding equipment, yield objectives, variability of the field, productivity of the soil and understanding of the interaction with the environment. From an agronomic point of view, the seeding rate should vary by at least 4000 seeds per acre (4046.86 m^2 or about 0.4 ha) [9]. The growing number of farmers using VRS and the widespread use of GPS technology on farms are facilitating the implementation of VRS strategies. Farmers need to be well-aware of the variability of their fields in order to apply the right seeding rates [9]. Too low a seeding rate can increase the risk of yield decrease, and too high a seeding rate can increase the total cost of production [65,66].

Precision agricultural technologies can improve the performance of field operations, as errors in the study of field soil characteristics can be very costly depending on the cost of seeds and seeding.

Plant density is a particularly important factor in wheat production systems because it can be controlled. The seeding rate affects the grain yield and is expressed as having a linear and quadratic dependence [67]. The studies of Chen et al. [68] with hard red spring wheat showed that row spacing and seeding rate do not affect each other, i.e., an increase in yield in narrow rows could not be achieved by increasing the seeding rate. Seed density control at a specific field location can be based on soil texture maps. The seeding rate should increase in sandy soils and decrease in clay soils. This principle may allow for an increase in the yield or the preservation of seeds [53].

Ayalew et al. [69] showed that an increasing number of seeds contributes to a lower number of wheat stalks. This could be because a higher seeding rate could increase competition for space, resulting in fewer stems per plant. It was also found that different seeding rates influenced different plant heights, i.e., a higher seeding rate resulted in a higher plant height, and vice versa. Higher plant density increased plant height due to reduced space for horizontal plant development and increased plant competition for light. The decrease in wheat yield with increasing seed yields can be explained by the fact that a denser wheat population forces plants to compete for production resources, which can lead to lower grain yields. Wiesehoff et al. [45] investigated the effect of different seeding rates on the germination of winter wheat in Germany. The results showed that germination decreased with an increase in seeding rate. The main reason for this effect may have been the poor distribution of grains, which led to grains being closely spaced in the soil and competing with each other. Most winter wheat cultivars allow us to compensate for the reduced plant density with higher tillering [45].

Iqbal et al. [70] found that the interaction between seeding rate and nitrogen level does not affect wheat height, number of tillers, spike length or number of spikelets in spike of 1000-grain weight but does affect the number of grains in spike, grain yield and harvest index. Kühling et al. [71] showed that the seeding rate affected two components of the spring wheat harvest—the number of ears per square meter and the number of grains per year in opposite directions. In the areas with a higher seeding rate, the predominant yield was mainly due to the number of ears, and in the areas with a lower seeding rate, there were more grains in each ear.

Optimal VRS recommendations based on a mathematical model are presented according to a predefined mathematical formula of the input seeding rate, taking into account soil quality indicators and/or yield potential. VRS recommendations are generated in real time by identifying and measuring the required soil and crop properties. Real-time optimization of seed input distribution can provide the best yield potential for a given field [12,41,72]. After examining the relationship between maize plant population and yield, a stronger relationship was found between the final number of stems and the yield than between the precision seeding rate and the number of stems. Researchers attempted to incorporate more soil parameters into yield models adjacent to plant populations in order to develop more universal models for seeding rate determination [12,41,72]. In order to obtain higher yields, it may be advisable to use a higher seeding rate in the higher-yielding soil zone, while in lower-yielding soil MZs, better yields may be achieved at lower seeding rates or plant populations [73]. In soils with higher yields and/or higher moisture content, the optimal seeding rate is higher [58]. As the soil moisture content (w) is a deterministic factor for site-specific irrigation, seeding, transplanting and compaction detection, an online measurement system will bring these applications into practice. A fiber-type visible (VIS) and near-infrared (NIR) spectrophotometer with a light reflectance measurement range of 306.5–1710.9 nm was used to measure w during a field operation [22,23]. Seed viability, germination, prevalence of pests, diseases and meteorological conditions all determine the germination and establishment of the crop, i.e., how many plants survive from 100 seeds. Different types of soil MZs based on ECa maps can be used to improve germination and establishment of wheat crops. Field inspection of the soil is also necessary before seeding, as crop germination and establishment are related to the number of stones on the soil surface and soil structure, so predicting seedbed quality and seeding using ECa maps alone will not be reliable [14].

As the plant population in the field increases, the yield per plant decreases, but the increased number of plants may be sufficient to compensate for the decrease in yield per plant [33,74]. However, at higher than optimal seeding rates, the increase in yield due to more plants may not be sufficient to compensate for the losses of one plant [40]. Optimal plant density is a function of the interaction between variety and environment, which varies between different agronomic zones [75]. The influence of plant cultivar on grain yield varies greatly depending on seeding, and the change in the seeding rate has an effect on yield components, especially stem density and the number of ears per m^2 [63,76]. As the seeding rate increases, the number of germinated plants [77] and the crop density increases [63,76], but the grain weight decreases [65,75,78], partly due to the lower efficiency of lighting [67,77]. Varieties have different abilities and adaptations to compensate for low or high seeding rate by modifying crop components such as the number of ears, the number of stems per m^2 [76], the number of grains per ear and the grain weight [65,67]. Geleta et al. [75] and Xue et al. [78] did not find a significant effect of the seeding rate on the grain yield from 65 (245 seeds m^2) to 130 kg ha^{-1} (489 seeds m^2). Stapper and Fischer [79] recorded an increased plant height due to an increase in seeding rate from 50 to 200 kg ha^{-1} (188–753 seeds m^2). It was found that increasing the seeding rate decreases the relative greenery of the leaves [66]. In addition, the highest biomass and N utilization efficiency of common white winter wheat could be achieved by reducing the standard seeding rate by 34–68% [80].

Fang et al. [81] studied the effects of different seeding rates and root pruning on winter wheat grain yield and its components before and after overwintering under field conditions. The aim of the experiment was to determine whether root pruning increased or maintained winter wheat yields at above-optimal seeding densities when growing wheat under semi-dry environmental conditions. The results showed that the yield response of the crop to the seeding rate depends on precipitation, especially on rain in the spring, which coincides with the grain-filling period. In dry years, the decrease in the yields due to higher seeding rate may have been caused by an increased competition for soil water, as more soil water was used before stem elongation due to higher seeding rate, which in turn reduced dry matter production and thus the yield. Many studies have found a positive correlation between the seeding rate, the grain yield and the average kernel weight [75,82], but Fang et al. [81] showed that this correlation depends on the availability of moisture.

In cereal crops, the yield curve is a useful statistical tool for analyzing the effect of seeding rates on plants. The effect of seeding rate on the yield was found to generally correspond to a quadratic response curve. As the seeding rate increases, the yield curve rises abruptly to the optimal seeding rate and then descends slowly [83,84]. Canadian researchers found that the seeding rate had an effect on plant density and the number of ears. Under ecological conditions, the yield increased by 10% when the seeding rate was doubled in wheat and barley cultivars [80]. Other studies have found that when wheat seeding rates were doubled at weed presence/competition locations, yields increased from

23% to 27% [85,86]. In order to optimize the wheat yield, it is important to determine the optimal number of plants per square meter. If the plant population is too large, then the crop will be too dense. This leads to difficulties in crop management and can lead to outbreaks, higher disease prevalence and higher costs. If the number of plants per m^{-2} is too low, the yield will not be optimized due to grain shortages and plants will be vulnerable to pests and will not be able to compete with weeds [14].

Wheat growth and yield depend on environmental conditions and can be regulated to some extent by selecting the right seeding time and seeding rate. Wheat yields increase due to higher seeding rates during the spring seeding [87]. Higher grain yields at higher seeding rates are associated with a higher number of ears per square meter. Therefore, the number of ears per square meter is considered to be the most important component in determining the grain yield. When calculating winter wheat seeding rate in a specific field MZ, Reining et al. [35] took into account the yield obtained from historical records for several years. The GIS-based software module calculated the seeding rate for different yield zones and transformed it directly into the seeding map. The seeding rate can be considered a factor in obtaining higher grain yields and nitrogen efficiency in winter wheat [88]. By optimizing the seeding rate, the yield and N uptake efficiency increased. The increase in N uptake efficiency was mainly due to the optimization of root length density and the synchronous increase in N from the fertilizer and N from the soil. The seeding rate had a significant effect on the number of plants, stems and ears, as well as the leaf area index.

5. The Assessment of Economic Benefits of VRS

VRS has been shown to be profitable for fields with very diverse soil productivity. Quantitative assessments of field variability should be taken into account when deciding whether to use VRS. We concluded that VRT may be of minimal value for fields with maize yields averaging from 9415 to 16,477 kg/ha. If soil types and characteristics do not vary substantially across the field, VRS may not prove to be profitable using currently available technology and agronomic knowledge [2,89]. Researchers in southern Brazil evaluated VRS by creating zones based on data from the producer on field productivity levels, as well as eight-year yield data [73]. The seeding rate of low-yield potential areas decreased and productivity increased by 1197–1900 kg ha^{-1} depending on the yield [73]. The population level of potential MZs with high soil fertility increased and productivity increased by 888–942 kg ha^{-1}, depending on the yield [73].

Economic return is a key issue for farmers interested in VRS. Lowenberg-DeBoer [90] assessed the VRS economic return using zones of yield potential. Potential VRS savings could be around USD 6.25 ha^{-1} in the fields of various seeding populations [6]. These savings from VRS can be effective in the face of rising seed prices. With digital tools and field-based data collection, VRS can enable farmers to explore and fully implement VRS and reap the economic benefits of doing so [2].

The data from the economic analysis of wheat show the effect of different seed quantities and different N levels. It is clear from the data that seeding 120 kg ha^{-1} yields the highest net income while 60 kg ha^{-1} yields the lowest net income. The highest net profit was observed by combining the 120 kg ha^{-1} seeding rate and 120 kg N ha^{-1} quantity of fertilizer, while the minimum profit was obtained when the seeding rate of 120 kg ha^{-1} was applied and nitrogen fertilizer was not used. The highest marginal rate of return was observed by combining 120 kg N ha^{-1} and 120 kg ha^{-1} seeding rate and the lowest marginal rate of return was found in control plots seeded at 60 kg ha^{-1} [91].

Several researchers have analyzed the economics of SSS in terms of economic plant population, degree of field variability, cost-effectiveness of VRS technologies, seed consumption and yield. Bullock et al. [5] found that the optimal density of SSS maize seeds ranged from 44,000 to 104,000 seeds ha^{-1}, and the yields ranged from 5.1 to 18.3 Mg ha^{-1}. A positive Pearson correlation coefficient was found between the soil quality of a particular field area and the optimal seeding rate. Modeling showed that by practicing VRS, farmers could increase their income up to USD 12 ha^{-1} compared to a uniform seeding rate. Robert et al. [92] assessed the economic consequences of maize seeding for a specific area. Their study identified several combinations of different yield options (low, medium and high) and included the costs of seeds and VRS technologies used in the cost–benefit analysis. Two separate strategies were used to provide SSS recommendations, namely agronomic and economic seeding rate recommendations that considered the yield potential of each MZ. According to the agronomic recommendation, the maize seeding rate was 44,460, 69,160 and 74,100 seeds ha^{-1}, and according to the economic recommendation, in low, medium and high yield MZs it was 49,400, 64,220 and 74,100 seeds ha^{-1}, respectively [11].

Taylor et al. [41] assessed the potential of SSS in eastern Kansas, USA for three years and revealed that SSS is not profitable under the cultivation conditions studied. They also suggested looking for a cheaper method to make SSS economically feasible. Elmore and Abendroth [93] critically reviewed several studies and concluded that SSS is an uneconomical technology. The reports show that the total maize sowing rate of 86,450 seeds ha^{-1} is a good field-testing rate but is not necessarily economically ideal. The optimal density of maize plants in a given year can range from 12,350 to 29,640 plants ha^{-1}, depending on the purpose of the crop (i.e., grain, silage) and growing conditions [11].

It is generally accepted that too high or too low a seeding rate is sub-optimal, so there must be some economically justified optimal seeding rate value or range of values [37]. An economic analysis by Holmes [56] revealed that the highest gross yield per hectare is obtained at a lower seeding rate than the one with the highest yield. In the high-yield zone, the optimal seeding rate to maximize gross yields was several thousand lower than the seeding rate at which the highest yields were achieved. The research results also showed that VRS is a valuable tool for reducing losses due to excess nutrients in poor quality crops, i.e., seeding at a variable rate ensures optimal use of other agricultural raw materials, such as fertilizers.

VRS eliminates double sowing in headlands and point rows and redistributes the optimal number of seeds in very heterogeneous fields. In very uniform fields, the return on investment of VRS will be low, while in heterogeneous fields with differentiated soil productivity zones, the return on investment will be much higher [8,36]. Automatic technology for the control of seeding rows (sections) enables automatic control of the seeding sections using the seeding map to reduce the field areas that can be seeded twice due to overlap. Automatic section control can on average save 4.3% of seeds, as well as reduce maize yield losses by 17% compared to seeding methods that inevitably cause double-sown areas [94].

With VRS, the total number of seeds used in the field can be lower, leading to lower GHG emissions from seed production [36]. The positive impact of VRS on GHG emissions may also be due to higher yields [73]. Another GHG reduction measure is related to the lower amount of fuel required to produce the same yield [36].

Summarizing the literature analysis, personal experience and unpublished assessments allows us to conclude that the application of VRS pays off when the field variability is more than 10%. According to German researchers, it is effective to apply VRS when the apparent electrical conductivity changes in the field on average every 25 m and for NPK (nitrogen phosphorus potassium), when the pH changes on average every 50 m. Sowing at a variable rate is worth 10 percent. Variability in apparent electrical conductivity, organic matter and terrain leads to variability in plant density. It is therefore useful to apply a variable rate of nitrogen. The payback of VRS is such that if one earns an average of 100 EUR per ha, then working on 400 ha will pay off in 1 year. Thus, the conclusion is that it is most efficient and most profitable to apply VRS when the farm size is larger than an average of 150 ha. However, this also depends on the crops grown (possibly smaller) as well as the satellite maps used. Due to meteorological conditions, if satellite maps are used, it is often cloudy in Lithuania, which can cause problems. Therefore, it is better to apply remote and proximal sensors close to the ground, which are effective in all meteorological conditions. This increases the use of non-contact apparent electrical conductivity assessment devices due to the unnecessary need for certain conditions. The key conclusion is that the higher the field variability (altitude, terrain, sand, etc.), the higher the VRS efficiency and payback. In Lithuania, the variability of most fields is due to the current cultural, political, historical and climatic situation, so VRS pays off and is effective all over the country.

There are a variety of VRA technologies available that can be used with or without a GPS system. The two basic technologies for VRA are: map-based and sensor-based. Sensor-based VRA requires no map or positioning system. Sensors on the applicator measure soil properties or crop characteristics "on the go." Based on this continuous stream of information, a control system calculates the input needs of the soil or plants and transfers the information to a controller, which delivers the input to the location measured by the sensor. Because map-based and sensor-based VRA have unique benefits and limitations, some SSCM (site-specific crop management) systems have been developed to take advantage of the benefits of both methods. The map-based method uses maps of previously measured items and can be implemented using a number of different strategies. Crop producers and consultants have crafted strategies for varying inputs based on soil type, soil color and texture, topography (high ground, low ground), crop yield, field scouting data, remotely sensed images and numerous other information sources that can be crop- and location-specific. Some strategies are based on a single information source while others involve a combination of sources. Regardless of the actual strategy, the user is ultimately in control of the application rate. These systems must have the ability to determine machine location within the field and relate the position to a desired application rate by "reading" the prescription map. The sensor-based method provides the capability to vary the application rate of inputs with no prior mapping or data collection involved. Real-time sensors measure the desired properties—usually soil properties or crop characteristics—while on the go. Measurements made by such a system are then processed and used immediately to control a variable-rate applicator. The sensor method does not necessarily require the use of a positioning system, nor does it require extensive data analysis prior to making variable-rate applications. However, if the sensor data are recorded and geo-referenced, the information can be used in future site-specific crop management exercises for creating a prescription map for other and future operations, as well as to provide an "as applied" application record for the grower [10].

6. Conclusions and Future Prospects

Precision agricultural technologies allow agricultural producers to improve the management of specific crop resources in crop production. Advances in seeding technology for the implementation of VRS make it possible to make better use of soil variability. VRS allows the population to be adapted to the variability of the field and helps to ensure precision production in the field in order to reduce errors in the seeding process. Another important aspect is that interest in VRS around the world is growing due to the interaction of these technologies with current seed prices. An optimal plant population can improve crop yields while maximizing farm profits. A site-specific seeding (SSS) method, where a variable seeding rate is applied to each field area separately, makes it possible to optimize crop density in anticipation of the best agronomic and economic effect. Various proximal and remote sensor systems, contact and contactless equipment, and mapping and VRS modeling technologies are currently used to determine soil and crop variability. The VRS depends on a good knowledge of the field characteristics, the capabilities of the seeding equipment, the planned crop yield, soil productivity and an understanding of the interaction of machine technology with the environment. Remote and proximal sensors, mounted on tractors or off-road vehicles, help to create field maps of variable properties. The accuracy of these maps and the good assignment of field soil management zones are the successful outcomes of VRS.

The use of VRS equipment pays for itself in one year when it is applied to 400 ha of arable land, and the average benefit is about 100 EUR per ha. By predicting payback over a period of time, it can be concluded that VRS is effective when the farm size is on average at least 150 ha. However, it also depends on the crop and the satellite maps used. The use of satellite maps can be problematic, as it is often cloudy in the Baltic region (where Lithuania is located). It is therefore advisable to make better use of sensory remote and proximal sensors that are effective in all meteorological conditions. One of the least sensitive to meteorological conditions is non-contact devices for apparent electrical conductivity. One of the most important conclusions is that the higher the variability of field and soil properties, the higher the efficiency of the VRS and the faster the payback.

The application of VRS to the seeding of various crops shows positive agro-economic trends, additional yields and higher economic returns. In particular, there is a lack of an optimal VRS index model that would allow for the application of precision seeding depending on various soil and plant productivity factors. It is not clear what the minimum differences between outdoor areas must be for the SSS method to generate a positive return. Therefore, with a view to the near future, scientists, promoters of precision farming, machinery manufacturers and VRS practitioners in agriculture face important challenges in justifying the application of VRS for different crops, considering the variability in soil, crop as well as the environmental. The Baltic States and other countries in the region need to conduct research into the technological operations of VRS of the most common crops, such as winter wheat and winter rapeseed, considering agronomic, technological, energy, environmental and economic indicators.

The prescription map tells the user how much seed to use depending on the location of the seeding equipment in the field. Future research needs to answer the question of which VRS method is most suitable for different regions, as there are two views on seeding with VRS. The first holds that it is better to distribute more seeds in high-productivity soils and less in low-productivity areas. The second, however, posits that it is better to use VRS in reverse, i.e., to distribute more seeds in poor soils and less in high-productivity soils. There is still a lack of knowledge as to which method is more suitable for the most popular plants in the Baltic region.

Our main future aspiration is to present research schemes and methodologies of VRS and seed placement depth control models that provide precise control and allow for the organization of the seeding process. The research methodology should be developed while considering soil heterogeneity using telemetry systems and multifunctional ultraviolet (UV), optical (VIS) and near-infrared (NIR) spectroscopy methods to optimize the number of seeds per unit area and depth of insertion during seeding. The multifunctional model for sustainable precision seed technology control based on UV–VIS–NIR spectrometry will allow us to save seed, to make better use of soil, to increase plant productivity, to protect the environment and to reduce energy consumption and economic costs.

Therefore, in all future studies, it is especially important to test and evaluate the latest methodologies in practice on farms and then to make complex assessments of the soil, plant and environmental parameters, and changes in the proposed methods in order to find the best solution and the most optimal methods for farmers. In the future, we plan to carry out and publish the results of research on various evaluation spectra and to provide accurate instructions and recommendations for farmers and entrepreneurs.

Author Contributions: Conceptualization, E.Š., M.K. and V.N.; methodology, M.K., K.R. and A.J.; formal analysis, M.K., D.S. and K.R.; resources, E.Š. and I.B.; writing—original draft preparation, M.K., V.N., E.Š., A.J. and I.B.; writing—review and editing, E.Š., M.K., K.R., D.S. and V.N.; visualization, M.K. and V.N.; supervision, E.Š.; project administration, I.B. and V.N. All authors have read and agreed to the published version of the manuscript.

Funding: This project received funding from European Regional Development Fund (project No 01.2.2-LMT-K-718-03-0041) under a grant agreement with the Research Council of Lithuania (LMTLT).

Institutional Review Board Statement: Not applicable.

Informed Consent Statement: Not applicable.

Data Availability Statement: The data presented in this study are available from the corresponding author upon request.

Acknowledgments: The authors thank the junior researcher Justinas Anušauskas for his help in conducting experimental research.

Conflicts of Interest: The authors declare no conflict of interest.

References

1. Erickson, B.; Widmar, D.A. *Precision Agricultural Services Dealership Survey Results*; Purdue University Department of Agriculture Economics/Department of Agronomy: West Lafayette, IN, USA, 2015; p. 37. Available online: http://agribusiness.purdue.edu/files/resources/2015-crop-lifepurdue-precision-dealer-survey.pdf (accessed on 7 February 2021).
2. Fulton, J. Variable-rate seeding systems for precision agriculture. In *Precision Agriculture for Sustainability*; Stafford, J., Ed.; Burleigh Dodds Science Publishing Limited, Silsoe Solutions: Cambridge, UK, 2019; pp. 28–297.
3. Hoeft, R.G.; Aldrich, S.R.; Nafziger, E.D.; Johnson, R.R. *Modern Corn and Soybean Production*, 1st ed.; MCSP Publications: Savoy, IL, USA, 2000.
4. Spogis, L.; Steponavičius, D. Methodology for preparing variable seed rate maps. *Agroinžinerija Ir Energ.* **2019**, *24*, 194–200. (In Lithuanian)
5. Virk, S.S.; Fulton, J.P.; Porter, W.M.; Pate, G.L. Row-crop planter performance to support variable rate seeding of maize. *Precis. Agric.* **2020**, *21*, 603–619. [CrossRef]
6. Bullock, D.S.G.; Bullock, D.S.G.; Nafziger, E.D.; Stafford, J.V. Variable rate seeding of maize in the Midwestern USA. In *Precision Agriculture'99, Part 1 and Part 2, Proceedings of the 2nd European Conference on Precision Agriculture, Odense, Denmark, 11–15 July 1999*; Sheffield Academic Press: Sheffield, UK, 1999.
7. Shanahan, J.F.; Doerge, T.A.; Johnson, J.J.; Vigil, M.F. Feasibility of site-specific management of corn hybrids and plant densities in the great plains. *Precis. Agric.* **2004**, *5*, 207–225. [CrossRef]
8. Pedersen, S.M.; Lind, K.M. *Precision Agriculture: Technology and Economic Perspectives*; Springer International Publishing: Cham, Switzerland, 2017; p. 276.
9. Jeschke, M.; Carter, P.; Bax, P.; Schon, R. Putting variable rate seeding to work on your farm. *Crop Insights* **2015**, *25*, 1–4.
10. Grisso, R.B.; Mark, A.; Wade, T.; David, H.; Roverson, G.T. Precision Farming Tools: Variable-Rate Application. *Va. Coop. Ext.* **2011**, 1–7. Available online: https://vtechworks.lib.vt.edu/bitstream/handle/10919/47448/442-505_PDF.pdf (accessed on 17 February 2022).
11. Munnaf, M.A.; Haesaert, G.; Van Meirvenne, M.; Mouazen, A.M. Site-specific seeding using multi-sensor and data fusion techniques: A review. *Adv. Agron.* **2020**, *161*, 241–323.
12. Licht, M.A.; Lenssen, A.W.; Elmore, R.W. Corn (*Zea mays* L.) seeding rate optimization in Iowa, USA. *Precis. Agric.* **2017**, *18*, 452–469. [CrossRef]
13. Kaspar, T.C.; Colvin, T.S.; Jaynes, D.B.; Karlen, D.L.; James, D.E.; Meek, D.W. Relationship between six years of corn yields and terrain attributes. *Precis. Agric.* **2003**, *4*, 87–101. [CrossRef]
14. Griffin, S.; Hollis, J. Using profile soil electrical conductivity survey data to predict wheat establishment rates in the United Kingdom. In *Precision Agriculture'13*; Wageningen Academic Publishers: Wageningen, The Netherlands, 2013; pp. 491–497.
15. Butzen, S.; Gunzenhauser, B.; Shanahan, J. *Putting Variable Rate Seeding to Work on Your Farm*; DuPont Pioneer: Johnston, IA, USA, 2012.
16. Schirrmann, M.; Gebbers, R.; Kramer, E.; Seidel, J. Soil pH mapping with an on-the-go sensor. *Sensors* **2011**, *11*, 573–598. [CrossRef]
17. Hamzehpour, N.; Shafizadeh-Moghadam, H.; Valavi, R. Exploring the driving forces and digital mapping of soil organic carbon using remote sensing and soil texture. *Catena* **2019**, *182*, 104–141. [CrossRef]
18. Kelley, J.; Higgins, C.W.; Pahlow, M.; Noller, J. Mapping soil texture by electromagnetic induction: A case for regional data coordination. *Soil Sci. Soc. Am. J.* **2017**, *81*, 923–931. [CrossRef]
19. Bousbih, S.; Zribi, M.; Pelletier, C.; Gorrab, A.; Lili-Chabaane, Z.; Baghdadi, N.; Ben Aissa, N.; Mougenot, B. Soil texture estimation using radar and optical data from Sentinel-1 and Sentinel-2. *Remote Sens.* **2019**, *11*, 1520. [CrossRef]
20. Lund, E.D.; Christy, C.D.; Drummond, P.E. Practical applications of soil electrical conductivity mapping. *Precis. Agric.* **1999**, *99*, 771–779.
21. Corwin, D.L.; Lesch, S.M. Apparent soil electrical conductivity measurements in agriculture. *Comput. Electron. Agric.* **2005**, *46*, 11–43. [CrossRef]
22. Kazlauskas, M.; Šarauskis, E.; Romaneckas, K.; Steponavičius, D.; Jasinskas, A.; Naujokienė, V.; Bručienė, I.; Žiogas, T.; Vaicekauskas, D.; Anušauskas, J.; et al. Effect of variable rate seeding on winter wheat seedbed and germination parameters using soil apparent electrical conductivity. In Proceedings of the Engineering for Rural Development: 20th International Scientific Conference, Engineering for Rural Development, Jelgava, Latvia, 26–28 May 2021; Volume 20, pp. 1108–1113.
23. Al-Shammary, A.A.; Kouzani, A.Z.; Saeed, T.R.; Lahmod, N.R.; Mouazen, A.M. Evaluation of a novel electromechanical system for measuring soil bulk density. *Biosyst. Eng.* **2019**, *179*, 140–154. [CrossRef]
24. Watson, C.A.; Atkinson, D.; Gosling, P.; Jackson, L.R.; Rayns, F.W. Managing soil fertility in organic farming systems. *Soil Use Manag.* **2006**, *18*, 239–247. [CrossRef]

25. Doran, J.W.; Parkin, T.B. Defining and assessing soil quality. In *Defining Soil Quality for a Sustainable Environment*; Doran, J.W., Coleman, D.C., Bezdicek, D.F., Stewart, B.A., Eds.; Soil Science Society of America and American Society of Agronomy: Madison, WI, USA, 1994; Volume 35, pp. 1–21.
26. Grisso, R.D.; Alley, M.M.; Holshouser, D.L.; Thomason, W.E. Precision farming tools. soil electrical conductivity. *Va. Coop. Ext.* **2009**, *442–508*, 1–6.
27. Gaile, Z.; Ruza, A.; Kreita, D.; Paura, L. Yield components and quality parameters of winter wheat depending on tillering coefficient. *Agron. Res.* **2017**, *15*, 79–93.
28. Wilson, J.P.; Gallant, J.C. *Terrain Analysis: Principles and Applications*; Wilson, J.P., Gallant, J.C., Eds.; John Wiley & Sons, Inc.: New York, NY, USA, 2000.
29. Bunemann, E.K.; Bongiorno, G.; Bai, Z.; Creamer, R.E.; De Deyn, G.; de Goede, R.; Fleskens, L.; Geissen, V.; Kuyper, T.W.; Mader, P.; et al. Soil quality—A critical review. *Soil Biol. Biochem.* **2018**, *120*, 105–125. [CrossRef]
30. Nolin, M.C.; Forand, G.; Simard, R.R.; Cambouris, A.N.; Karam, A. Soil specific relationships between corn/soybean yield, soil quality indicators and climatic data. [CDROM]. In Proceedings of the Fifth International Conference on Precision Agriculture, Bloomington, MN, USA, 16–19 July 2000; Robert, P.C., Rust, R.H., Larson, W.E., Eds.; ASA, CSSA, and SSSA: Madison, WI, USA, 2001.
31. Vasiliniuc, I.; Patriche, C. Selecting parameters for a soil quality index. *Gruntoznabstvo* **2011**, *12*, 46–56.
32. Whetton, R.; Zhao, Y.; Mouazen, A.M. Quantifying individual and collective influences of soil properties on crop yield. *Soil Res.* **2017**, *56*, 19–27. [CrossRef]
33. Van Roekel, R.J.; Coulter, J.A. Agronomic responses of corn to planting date and plant density. *Agron. J.* **2011**, *103*, 1414. [CrossRef]
34. Viscarra Rossel, R.A.; Rizzo, R.; Dematte, J.A.M.; Behrens, T. Spatial modeling of a soil fertility index using visible–near-infrared spectra and terrain attributes. *Soil Sci. Soc. Am. J.* **2010**, *74*, 1293. [CrossRef]
35. Reining, E.; Roth, R.; Kühn, J. Site-specific land use as demonstrated by planning variable seeding rates. In *Precision Agriculture*; Stafford, J.V., Werner, A., Eds.; Wageningen Academic Publishers: Wageningen, The Netherlands, 2003; pp. 567–572.
36. Soto, I.; Barnes, A.; Balafoutis, A.; Beck, B.; Sánchez, B.; Vangeyte, J.; Fountas, S.; Wal, T.V.D.; Eory, V.; Gómez-Barbero, M. *The Contribution of Precision Agriculture Technologies to Farm Productivity and the Mitigation of Greenhouse Gas Emissions in the EU*; EU Science Hub. Publications Office of the European Union: Brussels, Belgium, 2019.
37. Plant, R.E. Site-specific management: The application of information technology to crop production. *Comput. Electron. Agric.* **2001**, *30*, 9–29. [CrossRef]
38. Bunselmeyer, H.A.; Lauer, J.G. Using corn and soybean yield history to predict subfield yield response. *Agron. J.* **2015**, *107*, 558–562. [CrossRef]
39. Smidt, E.R.; Conley, S.P.; Zhu, J.; Arriaga, F.J. Identifying field attributes that predict soybean yield using random forest analysis. *Agron. J.* **2016**, *108*, 637–646. [CrossRef]
40. Lindsey, A.J.; Thomison, P.R.; Nafziger, E.D. Modeling the Effect of Varied and Fixed Seeding Rates at a Small-Plot Scale. *Agron. J.* **2018**, *110*, 2456–2461. [CrossRef]
41. Taylor, R.K.; Staggenborg, S.; Schrock, M.D.; Zhang, N. Using a GIS to evaluate the potential of variable rate corn seeding. In Proceedings of the ASAE Meeting Presentation, Milwaukee, WI, USA, 9–12 July 2000; pp. 9–12.
42. Sudduth, K.A.; Kitchen, N.R.; Drummond, S.T. Soil conductivity sensing on claypan soils: Comparison of electromagnetic induction and direct methods. In Proceedings of the 4th International Conference on Precision Agriculture, St. Paul, MN, USA, 19–22 July 1998; Robert, P.C., Rust, R.H., Larson, W.E., Eds.; ASA, CSSA, and SSSA: Madison, WI, USA, 1999; pp. 979–990.
43. Munnaf, M.A.; Haesaert, G.; Mouazen, A.M. Map-based site-specific seeding of seed potato production by fusion of proximal and remote sensing data. *Soil Tillage Res.* **2021**, *206*, 104801. [CrossRef]
44. Gunzenhauser, B.; Shanahan, J.; Lund, E. Utilizing on-the-go soil sensing devices to improve management zones definition. *Crop Insights* **2012**, *19*, 1–4.
45. Moral, F.J.; Terrón, J.M.; Da Silva, J.M. Delineation of management zones using mobile measurements of soil apparent electrical conductivity and multivariate geostatistical techniques. *Soil Tillage Res.* **2010**, *106*, 335–343. [CrossRef]
46. Heil, K.; Schmidhalter, U. The application of EM38: Determination of soil parameters, selection of soil sampling points and use in agriculture and archaeology. *Sensors* **2017**, *17*, 2540. [CrossRef] [PubMed]
47. Wiesehoff, M.; Müller, J.; Köller, K. Decision support system for map based sowing. In Proceedings of the ASAE Meeting, Chicago, IL, USA, 28–31 July 2002; pp. 1202–1208.
48. Verbeten, B. *Varying Corn & Soybean Populations, Varieties, & Down Force*; Cornell University Cooperative Extension: Ithaca, NY, USA, 2015.
49. Basnyat, P.; McConkey, B.; Meinert, B.; Gatkze, C.; Noble, G. Agriculture field characterization using aerial photograph and satellite imagery. *IEEE Geosci. Remote Sens. Lett.* **2004**, *1*, 7–10. [CrossRef]
50. Fleming, K.L.; Westfall, D.G.; Wiens, D.W.; Brodahl, M.C. Evaluating farmer defined management zone maps for variable rate fertilizer application. *Precis. Agric.* **2000**, *2*, 201–215. [CrossRef]
51. Mzuku, M.; Khosla, R.; Reich, R.; Inman, D.; Smith, F.; MacDonald, L. Spatial variability of measured soil properties across site-specific management zones. *Soil Sci. Soc. Am. J.* **2005**, *69*, 1572–1579. [CrossRef]
52. Schepers, A.R.; Shanahan, J.F.; Liebig, M.A.; Schepers, J.S.; Johnson, S.H.; Luchiari, A. Appropriateness of management zones for characterizing spatial variability of soil properties and irrigated corn yields across years. *Agron. J.* **2004**, *96*, 195–203. [CrossRef]

53. Heege, H.J. *Precision in Crop Farming: Site Specific Concepts and Sensing Methods: Applications and Results*; Heege, H.J., Ed.; Springer Science+BusinessMedia: Dordrecht, The Netherlands, 2013.
54. Ehsani, M.R.; Durairaj, C.D.; Woods, S.; Sullivan, M. Potential application of electrical conductivity (EC) map for variable rate seeding. *CIGR E-J.* **2005**, *7*, 1–17.
55. Dwight, K.; Craig, K.; Grant, H.; Farrell, A. Variable Rate Seeding: Easier Than You Think. Available online: https://www.cropquest.com/variable-rate-seeding/ (accessed on 31 July 2021).
56. Holmes, A. Transforming variability to profitability—Variable seed rates in New Zealand maize. In Proceedings of the 7th Asian-Australasian Conference on Precision Agriculture, Hamilton, New Zealand, 16–18 October 2017; pp. 1–7.
57. Kuang, B.; Mouazen, A.M. Calibration of visible and near infrared spectroscopy for soil analysis at the field scale on three European farms. *Eur. J. Soil Sci.* **2011**, *62*, 629–636. [CrossRef]
58. Bullock, D.S.; Kitchen, N.; Bullock, D.G. Multidisciplinary teams: A necessity for research in precision agriculture systems. *Crop Sci.* **2007**, *47*, 1765–1769. [CrossRef]
59. Brouder, S.M.; Hofmann, B.S.; Morris, D.K. Mapping soil pH: Accuracy of common soil sampling strategies and estimation techniques. *Soil Sci. Soc. Am. J.* **2005**, *69*, 427–441. [CrossRef]
60. Taylor, J.A.; Whelan, B.M.; Mcbratney, A.B. Determining optimum management zone-based seeding rates using on farm experimentation and variable rate seeding technologies. In Proceedings of the 8th International Conference on Precision Agriculture, Minneapolis, MN, USA, 23–26 July 2006.
61. Steyer Soil Explorer Soil Sensor: An Innovative Way to "Deepen" Farmers' Knowledge of Their Soil and Increase Their Tillage Efficiency. Available online: https://www.steyr-traktoren.com/en-distributor/agriculture/News-Site/Pages/2018-11-21-STEYR-SOILXPLORER.aspx (accessed on 29 July 2021).
62. The Agronomic Design 2019 Meeting Passed with Many Farmers and Huge Interest. Available online: https://titanmachinery.bg/en/article/s-mnogo-zemedeltsi-i-pri-silen-interes-premina-agronomic-design-2019 (accessed on 29 July 2021).
63. Silva, E.E.; Baio, F.H.R.; Kolling, D.F.; Júnior, R.S.; Zanin, A.R.A.; Neves, D.C.; Fontoura, J.V.P.F.; Teodoro, P.E. Variable-rate in corn sowing for maximizing grain yield. *Sci. Rep.* **2021**, *11*, 12711. [CrossRef] [PubMed]
64. Jafari, M.; Hemmat, A.; Sadeghi, M. Development and performance assessment of a DC electric variable-rate controller for use on grain drills. *Comput. Electron. Agric.* **2010**, *73*, 56–65. [CrossRef]
65. Spink, J.H.; Semere, T.; Spares, D.L.; Whaley, J.M.; Foulkes, J.M.; Clare, R.W.; Scott, R.K. Effect of sowing date on the optimum plant density of winter wheat. *Ann. Appl. Biol.* **2000**, *137*, 179–188. [CrossRef]
66. Bhatta, M.; Eskridge, K.M.; Rose, D.J.; Santra, D.K.; Baenziger, P.S.; Regassa, T. Seeding rate, genotype, and top-dressed nitrogen effects on yield and agronomic characteristics of winter wheat. *Crop Sci.* **2017**, *57*, 951–963. [CrossRef]
67. Lloveras, J.; Manent, J.; Viudas, J.; López, A.; Santiveri, P. Seeding rate influence on yield and yield components of irrigated winter wheat in a Mediterranean climate. *Agron. J.* **2004**, *96*, 1258–1265. [CrossRef]
68. Chen, C.; Neill, K.; Wichman, D.; Westcott, M. Hard red spring wheat response to row spacing, seeding rate, and nitrogen. *Agron. J.* **2008**, *100*, 1296–1302. [CrossRef]
69. Ayalew, T.; Abebe, B.; Yoseph, T. Response of wheat (*Tritium aestivum* L.) to variable seed rates: The case of Hawassa area, Southern Ethiopia. *Afr. J. Agric. Res.* **2017**, *12*, 1177–1181.
70. Iqbal, J.; Hayat, K.; Hussain, S.; Ali, A.; Bakhsh, M.A.A.H.A. Effect of seeding rates and nitrogen levels on yield and yield components of wheat (*Triticum aestivum* L.). *Pak. J. Nutr.* **2012**, *11*, 531. [CrossRef]
71. Kühling, I.; Redozubov, D.; Broll, G.; Trautz, D. Impact of tillage, seeding rate and seeding depth on soil moisture and dryland spring wheat yield in Western Siberia. *Soil Tillage Res.* **2017**, *170*, 43–52. [CrossRef]
72. Jiang, P.; Thelen, K.D. Effect of soil and topographic properties on crop yield in a North-Central corn–soybean cropping system. *Agron. J.* **2004**, *96*, 252. [CrossRef]
73. Hörbe, T.A.N.N.; Amado, T.J.C.C.; Ferreira, A.O.; Alba, P.J. Optimization of corn plant population according to management zones in Southern Brazil. *Precis. Agric.* **2013**, *14*, 450–465. [CrossRef]
74. Lindsey, A.J.; Thomison, P.R.; Mullen, R.; Geyer, A.B. Corn response to planting date as affected by plant population and hybrid in continuous corn cropping systems. *Crop Forage Turfgrass Manag.* **2015**, *1*, 1–7. [CrossRef]
75. Geleta, B.; Atak, M.; Baenziger, P.S.; Nelson, L.A.; Baltenesperger, D.D.; Eskridge, K.M.; Shipman, M.J.; Shelton, D.R. Seeding rate and genotype effect on agronomic performance and end-use quality of winter wheat. *Crop Sci.* **2002**, *42*, 827–832.
76. Whaley, J.M.; Sparkes, D.L.; Foulkes, M.J.; Spink, J.H.; Semere, T.; Scott, R.K. The physiological response of winter wheat to reductions in plant density. *Ann. Appl. Biol.* **2000**, *137*, 165–177. [CrossRef]
77. Otteson, B.N.; Mergoum, M.; Ransom, J.K. Seeding rate and nitrogen management effects on spring wheat yield and yield components. *Agron. J.* **2007**, *99*, 1615–1621. [CrossRef]
78. Xue, Q.; Weiss, A.; Baenziger, P.S.; Shelton, D.R. Seeding rate and genotype affect yield and end-use quality in winter wheat. *J. Agron. Crop Sci.* **2011**, *2*, 18–25.
79. Stapper, M.; Fischer, R. Genotype, sowing date, and plant spacing influence on high-yielding irrigated wheat in southern New South Wales. I. Phasic development, canopy growth and spike production. *Aust. J. Agric. Res.* **1990**, *41*, 997–1019. [CrossRef]
80. Brown, T.T. Variable Rate Nitrogen and Seeding to Improve Nitrogen Use Efficiency. Ph.D. Thesis, Washington State University, Pullman, WA, USA, December 2015.

81. Fang, Y.; Xu, B.C.; Turner, N.C.; Li, F.M. Grain yield, dry matter accumulation and remobilization, and root respiration in winter wheat as affected by seeding rate and root pruning. *Eur. J. Agron.* **2010**, *33*, 257–266. [CrossRef]
82. Blue, E.N.; Mason, S.C.; Sander, D.H. Influence of planting date, seeding rate, and phosphorus rate on wheat yield. *Agron. J.* **1990**, *82*, 762–768. [CrossRef]
83. Holliday, R. Plant population and crop yield: Part I. *Field Crop Abstr.* **1960**, *13*, 159–167.
84. Mehring, G.H.; Wiersma, J.J.; Stanley, J.D.; Ransom, J.K. Genetic and environmental predictors for determining optimal seeding rates of diverse wheat cultivars. *Agronomy* **2020**, *10*, 332. [CrossRef]
85. Roberts, J.R.; Peeper, T.F.; Solie, J.B. Wheat (Triticum aestivum) row spacing, seeding rate, and cultivar affect interference from rye (Secale cereale). *Weed Technol.* **2001**, *15*, 19–25. [CrossRef]
86. Beavers, R.L.; Hammermeister, A.M.; Frick, B.; Astatkie, T.; Martin, R.C. Spring wheat yield response to variable seeding rates in organic farming systems at different fertility regimes. *Can. J. Plant Sci.* **2008**, *88*, 43–52. [CrossRef]
87. Ozturk, A.; Caglar, O.; Bulut, S. Growth and yield response of facultative wheat to winter sowing, freezing sowing and spring sowing at different seeding rates. *J. Agron. Crop Sci.* **2006**, *192*, 10–16. [CrossRef]
88. Dai, X.; Zhou, X.; Jia, D.; Xiao, L.; Kong, H.; He, M. Managing the seeding rate to improve nitrogen use efficiency of winter wheat. *Field Crops Res.* **2013**, *154*, 100–109. [CrossRef]
89. Nielsen, R.L.B.; Lee, J.; Hetting, J.; Camberato, J. Yield response of corn to plant population in Indiana. In *Applied Crop Production Research Update*; Purdue University Department of Agronomy: West Lafayette, IN, USA, 2017; pp. 1–7. Available online: www.kingcorn.org/news/timeless/CornPopulations.pdf (accessed on 21 March 2021).
90. Lowenberg-DeBoer, J. *Economics of Variable Rate Planting for Corn*; Agricultural Economics Department, Purdue University: West Lafayette, IN, USA, 1998; pp. 1–13.
91. Shah, W.A.; Khan, H.U.; Anwar, S.; Nawab, K. Yield and yield components of wheat as affected by different seed rates and nitrogen levels. *Sarhad J. Agric.* **2011**, *27*, 17–25. [CrossRef]
92. Lowenberg-DeBoer, J. Economics of variable rate planting for corn. In Proceedings of the Fourth International Conference on Precision Agriculture, St. Paul, MN, USA, 19–22 July 1998; Robert, P.C., Rust, R.H., Larson, W.E., Eds.; American Society of Agronomy, Crop Science Society of America, Soil Science Society of America: Madison, WI, USA, 1999; pp. 1643–1651.
93. Elmore, R.; Abendroth, L.J. Variable Rate Seeding Does Not Pay. Available online: https://www.farmprogress.com/variable-rate-seeding-doesn-t-pay (accessed on 29 July 2021).
94. Runge, M.; Fulton, J.P.; Griffin, T.; Virk, S.; Brooke, A. *Automatic Section Control Technology for Row Crop Planters*; Alabama Cooperative Extension System, Auburn University: Auburn, AL, USA, 2014; pp. 1–8. Available online: https://ssl.acesag.auburn.edu/pubs/docs/A/ANR-2217/ANR-2217-low-archive.pdf (accessed on 30 July 2021).

www.ingramcontent.com/pod-product-compliance
Lightning Source LLC
Chambersburg PA
CBHW041937260926
51850CB00018B/215
9783036555836

MDPI
St. Alban-Anlage 66
4052 Basel
Switzerland
Tel. +41 61 683 77 34
Fax +41 61 302 89 18
www.mdpi.com

Agriculture Editorial Office
E-mail: agriculture@mdpi.com
www.mdpi.com/journal/agriculture